Practical Gamma-ray
Spectrometry

Practical Gamma-ray
Spectrometry
2nd Edition

Gordon R. Gilmore
Nuclear Training Services Ltd
Warrington, UK

John Wiley & Sons, Ltd

Other Wiley Editorial Offices

John Wiley & Sons Inc., 111 River Street, Hoboken, NJ 07030, USA

Jossey-Bass, 989 Market Street, San Francisco, CA 94103-1741, USA

Wiley-VCH Verlag GmbH, Boschstr. 12, D-69469 Weinheim, Germany

John Wiley & Sons Australia Ltd, 42 McDougall Street, Milton, Queensland 4064, Australia

John Wiley & Sons (Asia) Pte Ltd, 2 Clementi Loop #02-01, Jin Xing Distripark, Singapore 129809

John Wiley & Sons Ltd, 6045 Freemont Blvd, Mississauga, Ontario, L5R 4J3, Canada

Wiley also publishes its books in a variety of electronic formats. Some content that appears in print may not be available in electronic books.

Library of Congress Cataloging in Publication Data

Gilmore, Gordon.
    Practical gamma-ray spectrometry. — 2nd ed. / Gordon Gilmore.
      p.   cm.
    Includes bibliographical references and index.
    ISBN 978-0-470-86196-7 (cloth : alk. paper)
    1. Gamma ray spectrometry—Handbooks, manuals, etc.   I. Title.
    QC793.5.G327G55 2008
    537.5′352—dc22

                            2007046837

British Library Cataloguing in Publication Data

A catalogue record for this book is available from the British Library

ISBN 978-0-470-86196-7

Dedication

To my friends and family who, I suspect, never really believed I would get this finished,
and to the publishers who patiently tolerated many delays before I did so

Contents

Preface to the Second Edition

During 2005, while this second edition was being prepared, I was totally unprepared to receive a telephone call that my co-author on the first edition, John Hemingway, was seriously ill after suffering a brain-haemorrhage. Only a few days later, on 5th September, he passed away. My original, and obvious, intent was to update the sections allocated to John and myself and publish this second edition as 'Gilmore and Hemingway'. That intent was frustrated by contractual difficulties with John's estate. It became necessary for me to rewrite those sections completely and remove John's name from the second edition. I deeply regret that that was necessary. It has deprived us all of John's often elegant prose and has meant that some topics that John had particular interest in introducing to the new edition have had to be omitted.

Earlier in that year, another reminder of the inexorable passage of time came with the death of someone whose name had been familiar to me throughout my career in gamma spectrometry. On 16th January, Richard Helmer passed away at the age of 70 years. His co-authored work, the justly famous *Gamma and X-Ray Spectrometry with Semiconductor Detectors*, was one of the books that introduced John and myself to the complexities of gamma spectrometry and one which we consistently recommended to others. His influence as an author and in many other roles, such as an evaluator of nuclear data, has left all of us in his debt, whether we all realize it or not.

On a lighter note, during the year 2005 the very title of this book was called into question. The radiochemical mailing list, RADCH-L, agonized, in general terms, over which is the correct term – 'spectrometry' or 'spectroscopy'. Of course, the suffix '-metry' means to measure and '-scopy' means to visualize – and so the discussion went on, to and fro. Eventually, the 1997 IUPAC 'Golden Book', *Compendium of Chemical Terminology*, was quoted: 'SPECTROMETRY is the measurement of such [electromagnetic] radiations as a means of obtaining information about the system and their components'.

That seemed to be the 'clincher'. The prime objective of our activities is to measure gamma radiation, not just to create a spectrum, and so spectrometry' it is, performed by 'gamma spectrometrists'!

Before a second edition is approved, the publishers canvass the opinion of people in the field as to whether a new edition is justified and ask them for suggestions for inclusion. I have taken all of the suggestions offered seriously but, in the event, have had to disappoint some of the reviewers. For example, X-ray spectrometry is such a wide field with a different emphasis to gamma spectrometry and the space available within this new edition so limited, that merely exposing a little more of the 'iceberg' seemed pointless. In other cases, my ignorance of certain specific matters was sufficient to preclude inclusion. I can only offer my apologies to those who may feel let down.

Since the first edition (1995), there have been a number of significant advances in gamma spectrometry. Indeed, some of those advances were taking place while I was writing, meaning re-writes even to the update! In particular, I have included digital pulse processing and I have explained the changes in the way that nuclear data are being kept up to date. On statistics, I have introduced the matter of uncertainty budgets as being of increasing importance now that more laboratories seek accreditation. I have had to re-assess the ideas I espoused in the first edition on peak width and now have a much more comfortable mathematical justification for fitting peak-width calibrations.

Throughout, I have tried to keep to the principles John and I declared in the Preface to the first edition – an emphasis on the practical application of gamma spectrometry at the expense of, if possible, the mathematics. That being the case, I have reproduced most of the Preface to the first edition below. The first edition was very well received. I can only hope that I have done enough to ensure that popular opinion is as supportive of this second edition.

Gordon R. Gilmore

Preface to the First Edition

This book was conceived during one of the Gamma Spectrometry courses then being run at the Universities' Research Reactor at Risley. At that time, we had been 'peddling' our home-spun wisdom for seven or eight years, and transforming the lecture notes into something more substantial for the benefit of course participants seemed an obvious development.

Our intention is to provide more of a workshop manual than an academic treatise. In this spirit, each chapter ends with a 'Practical Points' section. This is not a summary as such but a reminder of the more important practical features discussed within the chapter. We have attempted, not always successfully, it must be admitted, to keep the mathematics to a minimum. In most cases, equations are presented as *faites accomplis* and are not derived.

One practical process that can have a major influence on the reliability of the results obtained by users of gamma-spectrometric equipment is that of *sampling*. It was after much discussion and with some regret that we decided to omit this topic. This is because it is peripheral to our main concern of describing the best use of instrumentation, because we suspect that another book would be necessary to do justice to the subject, and because we do not know much about it. What is clear is that an analyst must be aware that uncertainties introduced by taking disparate samples from an inhomogeneous mass can far outweigh uncertainties in the individual measurements themselves. This is a particular problem when sampling such a diverse and complex mass as the natural environment.

No previous knowledge of nuclear matters or instrumentation is assumed, and we hope the text can be used by complete beginners. There is even a list of names and symbols of the elements; while chemists may smile at this, in our experience not every otherwise scientifically literate person can name Sb and Sn, or distinguish Tb and Yb.

In a practical book, we think it useful to mention particular items of commercial equipment to illustrate particular points. We must make the usual disclaimer that these are not necessarily the best, nor the worst, and in most cases are certainly not the only items available. In general, the manufacturers do a fine job, and choosing one product rather than another is often an invidious task. We can only recommend that the user (1) decides at an early stage what capabilities are required, (2) reads and compares specifications (this text should explain these), (3) is not seduced by the latest 'whizz-bang device', yet (4) bears in mind that more recent products are better than older ones, not just in 'bald' specification but also in manufacturing technology, and should consequently show greater reliability.

Readers may notice the absence of certain terms in common use. The exclusion of some such terms is a deliberate choice. For example, instead of 'photopeak' we prefer 'full-energy peak'; we have avoided the statisticians' use of 'error' to mean uncertainty and reserve that word to indicate bias or error in the sense of 'mistake'. 'Branching ratio' we avoid altogether. This is often used ambiguously and without definition. In other texts, it may mean the relative proportions of different decay modes, or the proportions of different beta-particle transitions, or the ratio of 'de-excitation' routes from a nuclear-energy level. Furthermore, it sometimes appears as a synonym for 'gamma-ray emission probability', where it is not always clear whether or not internal conversion has been taken into account.

We hope sensitive readers are not upset by our use of the word 'program'. This 'Americanized' version is well on its way to being accepted as meaning specifically 'computer program', and enables a nice distinction to be made with the more general (and more elegant-looking) 'programme'.

We have raided unashamedly the manufacturers' literature for information, and our thanks are due particularly to Canberra and Ortec (in alphabetical order) for their co-operation and support in this. The book is not a survey of the latest research nor a historical study, and there are very few specific references in the text. Such that do exist are put at the end of each chapter, where there will also be found a more general short-list of 'Further Reading'.

We also acknowledge our continuing debt to two books: *Radiation Detection and Measurement*, by G.F. Knoll, John Wiley & Sons, Ltd (1979, 1989) and *Gamma- and X-ray Spectrometry with Semiconductor Detectors*, by K. Debertin and R.G. Helmer, North-Holland (1988). These can be thoroughly recommended.

So why write another book? Fine as these works are, we felt that there was a place for a 'plain-man's' guide to gamma spectrometry, a book that would concentrate on day-to-day operations. In short, the sort of book that we wish had been available when we began work with this splendid technique.

Gordon R. Gilmore and John D. Hemingway

Internet Resources within the Book

Throughout this book, I list sources of information of value to gamma spectrometrists. The reality of life in 2007 is that, for very many people, the Internet is the first 'port-of-call' for information. Because of this, I have leaned heavily on Internet sources and quoted links to them as standard URLs – Uniform Resource Locators, i.e. Internet addresses, to suitable websites. URLs are usually not 'case-sensitive'. However, that depends on the type of server used to host the website. It is better to type the URL as given here, i.e. preserving upper/lower-case characters.

A word of caution is necessary. The Internet can be a source of the most up-to-date information and can be far more convenient than waiting for books and articles to be delivered, or a trip to a distant library. However, I feel duty bound to remind readers that, as well as holding the up-to-date information, the Internet is also a vast repository of ancient, irrelevant, inaccurate and out-of date information. It is up to the user to check the pedigree, and date, of all downloaded material. I believe the links that I have quoted to be reliable. Because the Internet is essentially an ephemeral entity, reorganization of a website can result in URLs becoming inactive. Usually, however, the information will still be available on the 'parent site' somewhere, but will need looking for.

As a convenience for readers of this book, I have created a website, *http://www.gammaspectrometry.co.uk*, hosted by Nuclear Training Services Ltd, which holds links to all of the URLs referred to throughout the book, organized by chapter. The site also carries a number of other resources that readers might find useful:

- All the links quoted in Appendix A – Sources of information.
- The data reproduced in Appendices B–E.
- Some of the test spectra referred to in Chapter 15 and a test-spectrum generator.
- Spreadsheet tools to illustrate certain points in the text, including some used to generate figures within the text.
- A number of useful spectra to illustrate points in the text.
- Links to relevant organizations and manufacturers.
- A set of 'taster' modules from the *Online Gamma Spectrometry* course.

This website will also be used to 'post-up' corrections to the text, should any be needed, before they are able to appear in future reprints, which I hope will be useful. In due course, I also intend to create a 'blog' to allow reader feedback and discussion of issues raised.

1

Radioactive Decay and the Origin of Gamma and X-Radiation

1.1 INTRODUCTION

In this chapter I intend to show how a basic understanding of simple decay schemes, and of the role gamma radiation plays in these, can help in identifying radioactive nuclides and in correctly measuring quantities of such nuclides. In doing so, I need to introduce some elementary concepts of nuclear stability and radioactive decay. X-radiation can be detected by using the same or similar equipment and I will also discuss the origin of X-rays in decay processes and the light that this knowledge sheds on characterization procedures.

I will show how the Karlsruhe Chart of the Nuclides can be of help in predicting or confirming the identity of radionuclides, being useful both for the modest amount of nuclear data it contains and for the ease with which generic information as to the type of nuclide expected can be seen.

First, I will briefly look at the nucleus and nuclear stability. I will consider a nucleus simply as an assembly of uncharged neutrons and positively charged protons; both of these are called **nucleons**.

Number of neutrons $= N$

Number of protons $= Z$

Z is the **atomic number**, and defines the element. In the neutral atom, Z will also be the number of extranuclear electrons in their atomic orbitals. An element has a fixed Z, but in general will be a mixture of atoms with different masses, depending on how many neutrons are present in each nucleus. The total number of nucleons is called the **mass number**.

Mass number $= N + Z = A$

A, N and Z are all integers by definition. In practice, a neutron has a very similar mass to a proton and so there is a real physical justification for this usage. In general, an assembly of nucleons, with its associated electrons, should be referred to as a **nuclide**. Conventionally, a nuclide of atomic number Z, and mass number A is specified as ^A_ZSy, where Sy is the chemical symbol of the element. (This format could be said to allow the physics to be defined before the symbol and leave room for chemical information to follow; for example, Co^{2+}.) Thus, $^{58}_{27}$Co is a nuclide with 27 protons and 31 neutrons. Because the chemical symbol uniquely identifies the element, unless there is a particular reason for including it, the atomic number as subscript is usually omitted – as in ^{58}Co. As it happens, this particular nuclide is radioactive and could, in order to impart that extra item of knowledge, be referred to as a **radionuclide**. Unfortunately, in the world outside of physics and radiochemistry, the word **isotope** has become synonymous with radionuclide – something dangerous and unpleasant. In fact, isotopes are simply atoms of the same element (i.e. same Z, different N) – radioactive or not. Thus $^{58}_{27}$Co, $^{59}_{27}$Co and $^{60}_{27}$Co are isotopes of cobalt. Here 27 is the atomic number, and 58, 59 and 60 are **mass numbers**, equal to the total number of nucleons. ^{59}Co is stable; it is, in fact, the only stable isotope of cobalt.

Returning to nomenclature, ^{58}Co and ^{60}Co are **radioisotopes**, as they are unstable and undergo radioactive decay. It would be incorrect to say 'the radioisotopes ^{60}Co and ^{239}Pu . . .' as two different elements are being discussed; the correct expression would be 'the radionuclides ^{60}Co and ^{239}Pu . . .'.

If all stable nuclides are plotted as a function of Z (y-axis) and N (x-axis), then Figure 1.1 will result. This is a *Segrè chart*.

Practical Gamma-ray Spectrometry – 2nd Edition   Gordon R. Gilmore
© 2008 John Wiley & Sons, Ltd

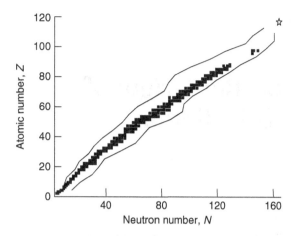

Figure 1.1 A Segrè chart. The symbols mark all known stable nuclides as a function of Z and N. At high Z, the long half-life Th and U nuclides are shown. The outer envelope encloses known radioactive species. The star marks the position of the largest nuclide known to date, $^{277}112$, although its existence is still waiting official acceptance

The **Karlsruhe Chart of the Nuclides** has this same basic structure but with the addition of all known radioactive nuclides. The heaviest stable element is bismuth ($Z = 83$, $N = 126$). The figure also shows the location of some high Z unstable nuclides – the major thorium ($Z = 90$) and uranium ($Z = 92$) nuclides. Theory has predicted that there could be stable nuclides, as yet unknown, called *superheavy nuclides* on an **island of stability** at about $Z = 114$, $N = 184$, well above the current known range.

Radioactive decay is a spontaneous change within the nucleus of an atom which results in the emission of particles or electromagnetic radiation. The modes of radioactive decay are principally alpha and beta decay, with spontaneous fission as one of a small number of rarer processes. Radioactive decay is driven by mass change – the mass of the product or products is smaller than the mass of the original nuclide. Decay is always exoergic; the small mass change appearing as energy in an amount determined by the equation introduced by Einstein:

$$\Delta E = \Delta m \times c^2$$

where the energy difference is in joules, the mass in kilograms and the speed of light in $m\,s^{-1}$. On the website relating to this book, there is a spreadsheet to allow the reader to calculate the mass/energy differences available for different modes of decay.

The units of energy we use in gamma spectrometry are electron-volts (eV), where $1\,eV = 1.602\,177 \times 10^{-19}\,J$.[1] Hence, $1\,eV \equiv 1.782\,663 \times 10^{-36}\,kg$ or $1.073\,533 \times 10^{-9}\,u$ ('u' is the unit of atomic mass, defined as 1/12th of the mass of ^{12}C). Energies in the gamma radiation range are conveniently in keV.

Gamma-ray emission is not, strictly speaking a decay process; it is a de-excitation of the nucleus. I will now explain each of these decay modes and will show, in particular, how gamma emission frequently appears as a by-product of alpha or beta decay, being one way in which residual excitation energy is dissipated

1.2 BETA DECAY

Figure 1.2 shows a three-dimensional version of the low-mass end of the Segrè chart with energy/mass plotted on the third axis, shown vertically here. We can think of the stable nuclides as occupying the bottom of a nuclear-stability valley that runs from hydrogen to bismuth. The stability can be explained in terms of particular relationships between Z and N. Nuclides outside this valley bottom are unstable and can be imagined as sitting on the sides of the valley at heights that reflect their relative nuclear masses or energies.

The dominant form of radioactive decay is movement down the hillside directly to the valley bottom. This is

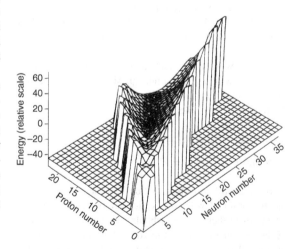

Figure 1.2 The beta stability valley at low Z. Adapted from a figure published by *New Scientist*, and reproduced with permission

[1] Values given are rounded from those recommended by the UK National Physical Laboratory in *Fundamental Physical Constants and Energy Conversion Factors* (1991).

beta decay. It corresponds to transitions along an **isobar** or line of constant A. What is happening is that neutrons are changing to protons (β^- decay), or, on the opposite side of the valley, protons are changing to neutrons (β^+ decay or electron capture). Figure 1.3 is part of the (Karlsruhe) Nuclide Chart.

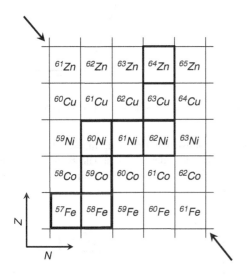

Figure 1.3 Part of the Chart of the Nuclides. Heavy boxes indicate the stable nuclides

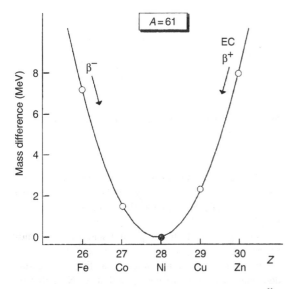

Figure 1.4 The energy parabola for the isobar $A = 61$. ^{61}Ni is stable, while other nuclides are beta-active (EC, electron capture)

If we consider the isobar $A = 61$, ^{61}Ni is stable, and beta decay can take place along a diagonal (in this format) from either side. ^{61}Ni has the smallest mass in this sequence and the driving force is the mass difference; this appears as energy released. These energies are shown in Figure 1.4. There are theoretical grounds, based on the liquid drop model of the nucleus, for thinking that these points fall on a parabola.

1.2.1 β^- or negatron decay

The decay of ^{60}Co is an example of β^- or **negatron** decay (negatron = negatively charged beta particle). All nuclides unstable to β^- decay are on the neutron rich side of stability. (On the Karlsruhe chart, these are coloured blue.) The decay process addresses that instability. An example of β^- decay is:

$$^{60}\text{Co} \longrightarrow {}^{60}\text{Ni} + \beta^- + \bar{\nu}$$

A **beta particle**, β^-, is an electron; in all respects it is identical to any other electron. Following on from Section 1.1, the sum of the masses of the ^{60}Ni plus the mass of the β^-, and $\bar{\nu}$, the anti-neutrino, are less than the mass of ^{60}Co. That mass difference drives the decay and appears as energy of the decay products. What happens during the decay process is that a neutron is converted to a proton within the nucleus. In that way the atomic number increases by one and the nuclide drops down the side of the valley to a more stable condition. A fact not often realized is that the neutron itself is radioactive when it is not bound within a nucleus. A free neutron has a half-life of only 10.2 min and decays by beta emission:

$$\text{n} \longrightarrow \text{p}^+ + \beta^- + \bar{\nu}$$

That process is essentially the conversion process happening within the nucleus.

The decay energy is shared between the particles in inverse ratio to their masses in order to conserve momentum. The mass of ^{60}Ni is very large compared to the mass of the beta particle and neutrino and, from a gamma spectrometry perspective, takes a very small, insignificant portion of the decay energy. The beta particle and the anti-neutrino share almost the whole of the decay energy in variable proportions; each takes from zero to 100 % in a statistically determined fashion. For that reason, beta particles are not mono-energetic, as one might expect from the decay scheme, and their energy is usually specified as $E_{\beta\,\text{max}}$. The term 'beta particle' is reserved for an electron that has been emitted during a nuclear decay process. This distinguishes it from electrons emitted as a result of other processes, which will usually have defined energies. The anti-neutrino need not concern us as it is detectable only in elaborate experiments. Anti-neutrinos (and neutrinos from β^+ decay) are theoretically crucial in maintaining the universality of the conservation laws of energy and angular momentum.

The lowest energy state of each nuclide is called the **ground state**, and it would be unusual for a transition to be made directly from one ground state to the next – unusual, but unfortunately far from unknown. There are a number of technologically important pure beta emitters, which are either widely used as radioactive tracers (^{3}H, ^{14}C, ^{35}S) or have significant yields in fission (^{90}Sr/^{90}Y, ^{99}Tc, ^{147}Pm). Table 1.1 lists the most common.

Table 1.1   Some pure beta emitters[a]

Nuclide	Half-life[b][c]	Maximum beta energy (keV)
^{3}H	12.312 (25) year	19
^{14}C	5700 (30) year	156
^{32}P	14.284 (36) d	1711
^{35}S	87.32 (16) d	167
^{36}Cl	3.01 (2) × 10^5 year	1142
^{45}Ca	162.6 1(9) d[b]	257
^{63}Ni	98.7 (24) year	66
^{90}Sr	28.80 (7) year	546
^{90}Y	2.6684 (13) d	2282
^{99}Tc	2.111 (12) × 10^5 year[b]	294
^{147}Pm	2.6234 (2) year[b]	225
^{204}Tl	3.788 (15) year	763

[a] Data taken from *DDEP* (1986), with the exception of
[b]-latter taken from *Table of Isotopes* (1978, 1998).
[c] Figures in parentheses represent the 1σ uncertainties on the last digit or digits.

The decay scheme of these will be of the form shown in Figure 1.5.

The difficulty for gamma spectrometrists is that no gamma radiation is emitted by these radionuclides and thus they cannot be measured by the techniques described in this text. To determine pure beta emitters in a mixture of radionuclides, a degree of chemical separation is required, followed by measurement of the beta radiation, perhaps by liquid scintillation or by using a gas-filled detector.

However, many beta transitions do not go to the ground state of the daughter nucleus, but to an excited state. This behaviour can be seen superimposed on the isobaric energy parabola in Figure 1.6. Excited states are shown

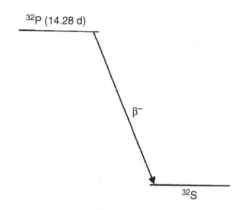

Figure 1.5   The decay scheme of a pure beta emitter, ^{32}P

for both radioactive (Ag, Cd, In, Sb, Te) and stable (Sn) isobaric nuclides, and it should be noted that these states are approached through the preceding or parent nuclide.

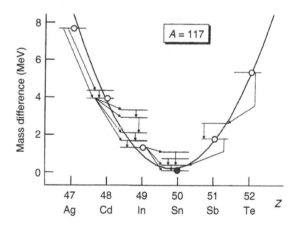

Figure 1.6   The isobar $A = 117$ with individual decay schemes superimposed. ^{117}Sn is stable

The **decay scheme** for a single beta-emitting radionuclide is part of this energy parabola with just the two components of parent and daughter. Figure 1.7 shows the simple case of ^{137}Cs. Here, some beta decays (6.5 % of the total) go directly to the ground state of ^{137}Ba; most (93.5 %) go to an excited nuclear state of ^{137}Ba.

The gamma radiation is released as that excited state de-excites and drops to the ground state. Note that the energy released, 661.7 keV, is actually a property of ^{137}Ba, but is accessed from ^{137}Cs. It is conventionally regarded as 'the ^{137}Cs gamma', and is listed in data tables as such.

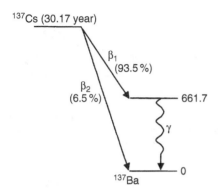

Figure 1.7 The decay scheme of ^{137}Cs

However, when looking for data about **energy levels** in the nucleus, as opposed to gamma-ray energies, it would be necessary to look under the daughter, ^{137}Ba.

In this particular case, 661.7 keV is the *only* gamma in the decay process. More commonly, many gamma transitions are involved. This is seen in Figure 1.6 and also in Figure 1.8, where the great majority of beta decays (those labelled β_1) go to the 2505.7 keV level which falls to the ground state in two steps. Thus, two gamma-rays appear with their energies being the difference between the energies of the upper and lower levels:

$$\gamma_1 = (2505.7 - 1332.5) = 1173.2 \text{ keV}$$

$$\gamma_2 = (1332.5 - 0) = 1332.5 \text{ keV}$$

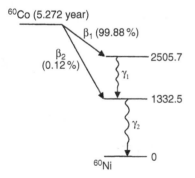

Figure 1.8 The decay scheme of ^{60}Co

The two gammas are said to be **in cascade**, and if they appear at essentially the same time, that is, if the intermediate level (in ^{60}Ni at 1332.5 keV) does not delay emission of the second gamma, then they are also said to be **coincident**. This phenomenon of two gamma-rays appearing

from the same atom at the same instant can have a significant influence on counting efficiency, as will be discussed in Chapter 8.

1.2.2 β^+ or positron decay

Just as β^- active nuclides are neutron rich, nuclides unstable to β^+ decay are neutron deficient. (The red nuclides on the Karlsruhe chart.) The purpose of positron decay, again driven by mass difference, is to convert a proton into a neutron. Again, the effect is to slide down the energy parabola in Figure 1.4, this time on the neutron-deficient side, towards stability, resulting in an atom of a lower atomic number than the parent. An example is:

$$^{64}_{29}\text{Cu} \longrightarrow {}^{64}_{28}\text{Ni} + \beta^+ + \nu \text{ (neutrino)}$$

During this decay a positron, a positively charged electron (anti-electron), is emitted, and conservation issues are met by the appearance of a neutrino. This process is analogous to the reverse of beta decay of the neutron. However, such a reaction would require the presence of an electron to combine with an excess proton. Electrons are not found within the nucleus and one must be created by the process known as **pair production**, in which some of the decay energy is used to create an **electron/positron pair** – imagine decay energy condensing into two particles. The electron combines with the proton and the positron is emitted from the nucleus. Positron emission is only possible if there is a sufficiently large energy difference, that is, mass difference, between the consecutive isobaric nuclides. The critical value is 1022 keV, which is the combined rest mass of an electron plus positron. As with negatrons, there is a continuous energy spectrum ranging up to a maximum value, and emission of complementary neutrinos.

The positron has a short life; it is rapidly slowed in matter until it reaches a very low, close to zero, kinetic energy. Positrons are anti-particles to electrons, and the slowed positron will inevitably find itself near an electron. The couple may exist for a short time as **positronium** – then the process of **annihilation** occurs. Both the positron and electron disappear and two photons are produced, each with energy equal to the electron mass, 511.00 keV (Figure 1.9). These photons are called **annihilation radiation** and the annihilation peak is a common feature in gamma spectra, which is much enhanced when β^+ nuclides are present. To conserve momentum, the two 511 keV photons will be emitted in exactly opposite directions. I will mention here, and treat the implications more fully later, that the annihilation peak in the spectrum will be considerably broader than a peak

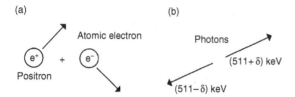

Figure 1.9   The annihilation process, showing how the resultant 511 keV photons could have a small energy shift: (a) possible momenta before interaction giving (b) differing photon energies after interaction

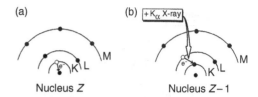

Figure 1.10   (a) Electron capture from the K shell, followed by (b) electron movement (X-ray emission) from L to K, and then M to L, resulting in X-radiations

produced by a direct nuclear-generated gamma-ray of the same energy. This can help in distinguishing between the two. The reason for such broadening is due to a Doppler effect. At the point where the positron–electron interaction takes place, neither positron nor electron is likely to be at complete rest; the positron may have a small fraction of its initial kinetic energy, the electron – if we regard it as a particle circling the nucleus – because of its orbital momentum. Thus, there may well be a resultant net momentum of the particles at the moment of interaction, so that the conservation laws mean that one 511 keV photon will be slightly larger in energy and the other slightly smaller. This increases the statistical uncertainty and widens the peak. Note that the sum of the two will still be (in a centre of mass system) precisely 1022.00 keV.

1.2.3   Electron capture (EC)

As described above, β^+ can only occur if more than 1022 keV of decay energy is available. For neutron deficient nuclides close to stability where that energy is not available, an alternative means of decay is available. In this, the electron needed to convert the proton is captured by the nucleus from one of the extranuclear electron shells. The process is known as **electron capture** decay. As the K shell is closest to the nucleus (the wave functions of the nucleus and K shell have a greater degree of overlap than with more distant shells), then the capture of a K electron is most likely and indeed sometimes the process is called **K-capture**. The probability of capture from the less strongly bound higher shells (L, M, etc.) increases as the decay energy decreases.

Loss of an electron from the K shell leaves a vacancy there (Figure 1.10). This is filled by an electron dropping in from a higher, less tightly bound, shell. The energy released in this process often appears as an X-ray, in what is referred to as **fluorescence**. One X-ray may well be followed by others (of lower energy) as electrons cascade down from shell to shell towards greater stability.

Sometimes, the energy released in rearranging the electron structure does not appear as an X-ray. Instead, it is used to free an electron from the atom as a whole. This is the **Auger effect**, emitting **Auger electrons**. The probability of this alternative varies with Z: at higher Z there will be more X-rays and fewer Auger electrons; it is said that the **fluorescence yield** is greater. Auger electrons are mono-energetic, and are usually of low energy, being emitted from an atomic orbital (L or M) where the electron binding energies are smaller. There is a small probability of both Auger electrons and X-rays being emitted together in one decay; this is the radiative Auger effect. Note that whenever X-rays are emitted, they will be characteristic of the daughter, rather than the parent, as the rearrangement of the electron shells is occurring after the electron capture.

For neutron deficient nuclides with a potential decay energy somewhat above the 1022 keV threshold, both positron decay and electron capture decay will occur, in a proportion statistically determined by the different decay energies of the two processes. Figure 1.11 shows the major components of the decay scheme of ^{22}Na, where both

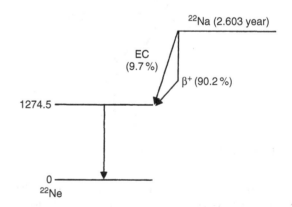

Figure 1.11   The decay scheme of ^{22}Na. Note the representation of positron emission, where 1022 keV is lost before emission of the β^+

positron decay and electron capture are involved. We can deduce from this that the spectrum will show a gamma-ray at 1274.5 keV, an annihilation peak at 511.0 keV (from the β^+), and probably X-rays due to electron rearrangement after the EC.

1.2.4 Multiple stable isotopes

In Figures 1.4 and 1.6, I suggested that the ground states of the nuclides of isobaric chains lay on a parabola, and the decay involved moving down the sides of the parabola to the stable point at the bottom. The implication must be that there is only one stable nuclide per isobaric chain. Examination of the Karlsruhe chart shows quite clearly that this is not true – there are many instances of two, or even three, stable nuclides on some isobars. More careful examination reveals that *what* is true is that every *odd*-isobar only has one stable nuclide. It is the even numbered isobars that are the problem. If a parabola can only have one bottom, the implication is that for even-isobars there must be more than one stability parabola. Indeed that is so. In fact, there are two parabolas; one corresponding to even-Z/even-N (even–even) and the other to odd-Z/odd-N (odd–odd). Figure 1.12 shows this. The difference arises because pairing of nucleons give a small increase in stability – a lowering of energy. In even–even nuclides there are more paired nucleons than in odd–odd nuclides and so the even–even parabola is lower in energy. As shown in Figure 1.12 for the $A = 128$ isobaric chain, successive decays make the nucleus jump from odd–odd to even–even and back. There will be occasions, as here,

where a nucleus finds itself above the ultimate lowest point of the even–even parabola, but below the neighbouring odd–odd points. It will, therefore be stable. (It is the theoretical possibility that a nuclide such as ^{128}Te could decay to ^{128}Xe, which fuels the search for double beta decay, which I will refer to from time to time.) In all, depending upon the particular energy levels of neighbouring isobaric nuclides, there could be up to three stable nuclides per even-A isobaric chain.

In the case of $A = 128$, there are two stable nuclides, ^{128}Te and ^{128}Xe. ^{128}I has a choice of destination, and 93.1 % decays by β^- to ^{128}Xe and 6.98 % decays by EC to ^{128}Te. The dominance of the ^{128}Xe transition reflects the greater energy release, as indicated in Figure 1.12. This behaviour is quite common for even mass parabolas and this choice of decay mode is available for such well-known nuclides as ^{40}K and ^{152}Eu. Occasionally, if the decay energy for β^+ is sufficient, a nuclide will decay sometimes by β^- and sometimes by EC and β^+.

1.3 ALPHA DECAY

An alpha particle is an He-4 nucleus, $^4_2$He$^+$, and the emission of this particle is commonly the preferred mode of decay at high atomic numbers, $Z > 83$. In losing an alpha particle, the nucleus loses four units of mass and two units of charge:

$$Z \longrightarrow Z - 2$$

$$A \longrightarrow A - 4$$

Typical is the decay of the most common isotope of radium:

$$^{226}_{88}\text{Ra} \longrightarrow {}^{222}_{86}\text{Rn} + {}^4_2\text{He} + Q$$

The product in this case is the most common isotope of radon, ^{222}Rn (usually just called 'radon' and which incidentally is responsible for the largest radiation dose from a single nuclide to the general population). A fixed quantity of energy, Q, equal to the difference in mass between the initial nuclide and final products, is released. This energy must be shared between the Rn and the He in a definite ratio because of the conservation of momentum. Thus, the alpha-particle is mono-energetic and alpha spectrometry becomes possible. In contrast to beta decay, there are no neutrinos to take away a variable fraction of the energy.

In many cases, especially in the lower Z range of α decay, the emission of an alpha particle takes the nucleus directly to the ground state of the daughter, analogous to the 'pure-β' emission described above. However, with

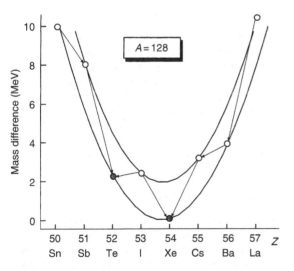

Figure 1.12 The two energy parabolas for the isobar $A = 128$. ^{128}Te and ^{128}Xe are stable

heavier nuclei, α decay can lead to excited states of the daughter. Figure 1.13, the decay scheme of ^{228}Th, shows gamma emission following alpha decay, but even here it will be seen that most alpha transitions go directly to the ^{224}Ra ground state.

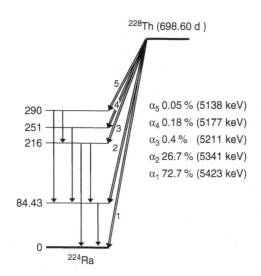

^{228}Th (698.60 d)

290
251
216

α_5 0.05 % (5138 keV)
α_4 0.18 % (5177 keV)
α_3 0.4 % (5211 keV)
α_2 26.7 % (5341 keV)
α_1 72.7 % (5423 keV)

84.43

0

^{224}Ra

Figure 1.13   The decay scheme of ^{228}Th

Calculation of the alpha decay energy reveals that even nuclides, such as ^{152}Eu and the stable ^{151}Eu, are unstable towards alpha decay. Alpha decay of ^{151}Eu would release 1.96 MeV of energy. The reason that this, and most other nuclides, do not decay by alpha emission is the presence of an energy barrier – it takes energy to prise an alpha particle out of the nucleus. Unless the nucleus is excited enough or is large enough so that the decay energy is greater than the energy barrier, it will be stable to alpha emission. That does not preclude it from being unstable to beta decay; ^{151}Eu is stable, ^{152}Eu is radioactive.

1.4 SPONTANEOUS FISSION (SF)

Spontaneous fission is a natural decay process in which a heavy nucleus spontaneously splits into two large fragments. An example is:

$$^{252}_{98}\text{Cf} \longrightarrow {}^{140}_{54}\text{Xe} + {}^{108}_{44}\text{Ru} + {}^{1}_{0}\text{n} + Q$$

The two product nuclides are only examples of what is produced; these are fission fragments or (when in their ground states) fission products. The range of products, the energies involved (Q) and the number and energies of

neutrons emitted are all similar to those produced in more familiar neutron-induced fission of fissile or fissionable nuclides. ^{252}Cf is mentioned here as it is a commercially available nuclide, which is bought either as a source of fission fragments or as a source of neutrons.

Once more, the driving force for the process is the release of energy. Q is of the order of 200 MeV, a large quantity, indicating that the fission products have a substantially smaller joint mass than the fissioning nucleus. This is because the binding energy per nucleon is significantly greater for nuclides in the middle of the Periodic Table than at the extremes. ^{108}Ru, for example, has a binding energy of about 8.55 MeV per nucleon, while the corresponding figure for ^{252}Cf is about 7.45 MeV per nucleon. Despite the emission of neutrons in this process, fission products are overwhelmingly likely to find themselves on the neutron rich, β$^-$ active side of the nuclear stability line. They will then undergo β$^-$ decay along an isobar, as, for example, along the left-hand side of Figure 1.12, until a stable nucleus is reached. During this sequence, gamma emission is almost always involved, as described earlier. The distribution of fission product masses will be discussed in Section 1.9.

As with alpha decay, calculation of mass differences for notional fission outcomes suggest that even mid-range nuclides, in terms of mass, would be unstable to fission. Fission is prevented in all but very large nuclei by the **fission barrier** – the energy needed to deform the nucleus from a sphere to a situation where two nearly spherical fission product nuclei can split off.

1.5 MINOR DECAY MODES

A number of uncommon decay modes exist which are of little direct relevance to gamma spectrometrists and I will content myself with just listing them: delayed neutron emission, delayed proton emission, double beta decay (the simultaneous emission of two β$^-$ particles), two proton decay and the emission of 'heavy ions' or 'clusters', such as ^{14}C and ^{24}Ne. Some detail can be found in the more recent general texts in the Further Reading section, such as the one by Ehmann and Vance (1991).

1.6 GAMMA EMISSION

This is not a form of decay like alpha, beta or spontaneous fission, in that there is no change in the number or type of nucleons in the nucleus; there is no change in Z, N or A. The process is solely that of losing surplus excitation energy, and as I have shown is usually a by-product of alpha or beta decay. First – what is a gamma-ray?

1.6.1 The electromagnetic spectrum

Gamma radiation is electromagnetic radiation, basically just like radio waves, microwaves and visible light. In the enormous range of energies in the electromagnetic spectrum, gammas sit at the high-energy, short-wavelength, end, as shown in Figure 1.14.

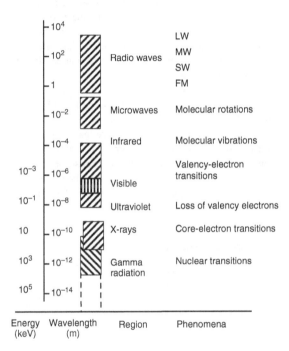

Figure 1.14 The electromagnetic spectrum

Wavelength, λ, or frequency, ν, are, in principle, equally valid as energy units for characterizing these radiations, and indeed are the preferred units in other parts of the electromagnetic spectrum. Relationships between these quantities for all electromagnetic radiation are:

$$E = h \times \nu \tag{1.1}$$

and

$$\lambda \times \nu = c \tag{1.2}$$

where h (the Planck constant) $= 4.135 \times 10^{-15}$ eV Hz^{-1} and c (the velocity of light, or any electromagnetic radiation, in a vacuum) $= 2.997926 \times 10^{8}$ m s^{-1}. Thus, 1000 keV $\equiv 1.2398 \times 10^{-12}$ m, or 2.4180×10^{20} Hz. There is some overlap between higher-energy X-rays (the X-rays

range is from just under 1 to just over 100 keV) and lower-energy gammas (whose range we will assume here to be from 10 to 10 000 keV). The different names used merely indicate different origins.

The 10^8 eV in the figure is by no means the upper limit to energy. Astronomers detect so-called 'cosmic gamma-rays' (more strictly photons) at much higher energies. Our common energies of around 10^6 eV would be their 'soft' gammas. Above that is 'medium energy' to 3×10^7 eV, 'high energy' to 10^{10} eV, 'very high energy' to 10^{13} eV and 'ultra high energy' to $> 10^{14}$ eV. Measurement of the higher energies is via the interaction of secondary electrons which are produced in the atmosphere; large scale arrays of electron detectors are used.

We have already seen that gamma emissions are the result of transitions between the excited states of nuclei. As the whole technique of gamma spectrometry rests on (a) the uniqueness of gamma energies in the characterization of radioactive species, and (b) the high precision with which such energies can be measured, it is of interest to consider briefly some relevant properties of the excited states.

1.6.2 Some properties of nuclear transitions

It is sometimes useful to think of nucleons in a nucleus as occupying different shells in much the same way as electrons are arranged in shells outside the nucleus. Then, exactly as quantum theory predicts that only particular electron energies are available to extranuclear electrons giving K, L, M shells, etc., so calculations for the nucleus only allow the occupation of certain energy shells or energy levels for neutrons and, independently, for protons. An excited nuclear state is when one or more nucleons have jumped up to a higher-energy shell or shells. Our interest here is in movement between shells and in what controls the probability of this occurring.

Nuclear energy states vary as charge and current distributions in the nucleus change. Charge distributions result in electric moments; current distributions give rise to magnetic moments (the neutron may be uncharged but it still has a magnetic moment). Consider first the electric moment. Oscillating charges can be described in terms of spherical harmonic vibrations, which may be expressed in a multipole expansion. Successive terms in such an expansion correspond to angular momenta in definite quantized units. If one unit of angular momentum is involved, this is called electric dipole radiation and is indicated by E1; if two units are involved, we have electric quadrupole radiation, E2, and so on. Likewise, there is a parallel system of magnetic multipoles corresponding to changes in magnetic moments, which give rise to M1 for the magnetic dipole, M2 for the magnetic quadru-pole, etc.

As well as changes in angular momentum, there is also the possibility of a change in parity, π. This concept is a property of wave functions and is said to be either + or − (even or odd), depending on the behaviour of the wave function as it is mathematically reflected in the origin. So, there are three properties of a nuclear transition:

- Is it an electric or magnetic transition, E or M?
- Which multiplicities are involved, or, what is the change in angular momentum, e.g. E1, E2, E3, etc.?
- Is there a change of parity?

These ideas are used in formulating selection rules for gamma transitions. This gives a sound theoretical basis to the apparently arbitrary probability of the appearance of particular gamma emissions. Sometimes, decay schemes have energy levels labelled with spin and parity, as well as energy above the ground state. Figure 1.15 shows examples of this, with the type of multipole transitions expected according to the selection rules.

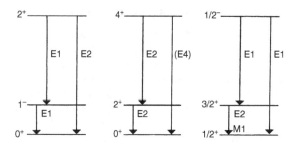

Figure 1.15 Representation of some gamma decay schemes, showing spins, parities and expected multipole transitions

1.6.3 Lifetimes of nuclear energy levels

Nuclear states also have definite lifetimes, and where transitions would involve a large degree of 'forbiddenness' according to the selection rules, the levels can be appreciably long-lived. If the lifetime is long enough to be easily measurable, then we have an **isomeric state**. The half-life of the transition depends on whether it is E or M, on the multiplicity, on the energy of the transition and on the mass number. Long half-lives are strongly favoured where there is high multipolarity (e.g. E4 or M4) and low transition energy. Most gamma transitions occur in less than 10^{-12} s. As to what is readily measurable in practice is something of a moot point, but certainly milliseconds and even microseconds give no real problems. Some would take 1 ns as the cut-off point.

These nuclear isomers, sometimes said to be in **metastable states**, are indicated by a small 'm' as superscript. An example is the 661.7 keV level in ^{137}Ba (see Figure 1.7); this has a half-life of 2.552 min and would be written as ^{137m}Ba (sometimes seen as ^{137}Ba^{m}, deprecated by this author). Note that in the measurement of ^{137}Cs there is no indication that this hold-up in the emission process exists. Only a rapid chemical separation of barium from caesium, followed by a count of the barium fraction, would show the presence of the isomer. Normally, the 661.7 keV gamma-ray appears with the half-life of the ^{137}Cs because ^{137}Cs and ^{137m}Ba are in secular equilibrium (See Section 1.8.3 below).

Some half-lives of isomeric states can be very long, for example, ^{210m}Bi decays by alpha emission with a half-life of 3.0×10^6 year. Alpha decay is, however, a rare mode of decay from a metastable state; gamma-ray emission is much more likely. A gamma transition from an isomeric state is called an **isomeric transition** (IT). On the Karlsruhe Nuclide Chart, these are shown as white sections within a square that is coloured (if the ground state is radioactive) or black (if the ground state is stable).

1.6.4 Width of nuclear energy levels

A nuclear energy level is not at an infinitely precise energy, but has a certain finite width. This is inversely related to the lifetime of the energy level through the Heisenberg Uncertainty Principle, which may be expressed as:

$$\delta E \times \delta t \geq h/2\pi (= 6.582\,122 \times 10^{-16}\,\text{eV s}) \quad (1.3)$$

where:

- δE is the uncertainty in the energy, which we will assume to be equivalent to an energy resolution (FWHM).
- δt is the uncertainty in time, taken as the mean life of the level; mean life is $1/\lambda$ or $1.4427 \times t_{1/2}$.
- h is the Planck constant.

Thus, δE for the 661.7 keV level of ^{137m}Ba whose half-life is 2.552 min, will be about 3×10^{-18} eV – exceedingly small. The level involved in the decay of ^{60}Co at 1332.5 keV (see Figure 1.8) has a lifetime of 7×10^{-13} s; this implies an energy width of about 9×10^{-4} eV. This is still very small compared to the precision with which gamma energies can be measured and to the FWHM of spectrum peaks, typically 1.9 keV at 1332.5 keV. In general, the widths of the nuclear energy levels involved in gamma emission are not a significant factor in the practical determination of gamma energies from radioactive-decay processes. This is considered further in Chapter 6.

1.6.5 Internal conversion

The emission of gamma radiation is not the only possible process for de-excitation of a nuclear level. There are two other processes: internal conversion (IC) and pair production.

Pair production as a form of gamma decay is uncommon, and I will only touch upon it here. There are close similarities with the process described in detail in the section of this text on interactions of gamma radiation with matter, where pair production is, by contrast, of major importance. It is only possible if the energy difference between levels is greater than 1022 keV, when that part of the total energy is used to create an electron–positron pair. These two particles are ejected from the nucleus and will share the remainder of the decay energy as kinetic energy. An example of decay by pair production is the isomeric transition of ^{16m}O, which has a half-life of 7×10^{-11} s, and a decay energy of 6050 keV.

Internal conversion, on the other hand, is very common. In this process, the energy available is transferred to an extranuclear electron, which is ejected from the atom. This is called an *internal conversion electron*. It is mono-energetic, having an energy equal to the transition energy less the electron binding energy and a small nuclear recoil energy. Measurement of the distribution of electron energy (i.e. an electron spectrum) would reveal peaks corresponding to particular electron shells, such as K, L and M. Loss of an electron from a shell leaves a vacancy and this vacancy will be filled by an electron dropping into it from a higher shell. Thus, as with electron capture, an array of X-rays and Auger electrons will also be emitted.

However, note that because IC is a mode of de-excitation and there is no change in Z, N or A, the X-radiation that is produced is characteristic of the *parent* isomeric state. Both 'parent level' and 'daughter level' are the same element. This is in contrast to electron capture, where the X-rays are characteristic of the daughter. If X-ray energies are to be used as a diagnostic tool, the user must know which decay process is occurring.

Internal conversion operates in competition with gamma-ray emission, and the ratio of the two is the **internal conversion coefficient**, α:

$$\alpha = \frac{\text{number of IC electrons emitted}}{\text{number of gamma-rays emitted}} \qquad (1.4)$$

This may be subdivided into α_K, α_L, etc., where electrons from the individual K and L shells are considered. Values of α depend on the multipolarity, transition energy and atomic number. In broad terms, α increases as the half-life and Z increase, and as ΔE decreases. At high Z, isomeric transitions with small transition energies may be 100% converted.

A practical point arises regarding the use of information taken from decay scheme diagrams. It cannot be assumed that because $x\%$ of disintegrations are feeding a certain energy level, then the same $x\%$ of disintegrations will produce a gamma-ray of that energy. An example of this is the decay of ^{137}Cs. The decay scheme (see Figure 1.7) shows that 93.5% of decays populate the 661.7 keV level in the daughter. However, tables of decay data state that the emission probability of the 661.7 keV gamma is only 85.1%. Thus, 8.4% of the gamma decays are internally converted; a number which could be calculated from the coefficient α. Another example is the decay of ^{228}Th. It may be deduced from Figure 1.13 that some 27% of the α decays feed the ^{224}Ra level at 84.4 keV, yet the emission probability of the 84.4 keV gamma-ray is only 1.2%, and instead there are many Ra X-rays.

1.6.6 Abundance, yield and emission probability

It is common for the number of gamma-rays emitted by a nuclide to be referred to as 'the abundance', sometimes as the 'yield'. Both of these terms lack precision. Historically, confusion was often caused because an author or data source would quote abundances that were effectively beta transition data – the 93.5% figure quoted above. In fact, the proportion of decays that give rise to 661.7 keV gamma-rays in the example above is 85.1% when internal conversion is taken into account. In this text, I will use the term **gamma emission probability** on the basis that it says exactly what it means – the probability that a gamma-ray will be emitted, all other factors being taken into account.

1.6.7 Ambiguity in assignment of nuclide identity

We have seen how gamma-rays are emitted with very precisely defined energies; these energies being characteristic of particular radionuclides. The majority of ambiguities that arise in allotting nuclides to energies would probably be overcome if the energy resolution of detector systems were improved. (In passing, it must be said that a significant improvement in the resolution of germanium detectors is unlikely.) However, there is a not uncommon situation where discrimination between radionuclides by gamma energy alone is in principle not possible. This is where isobaric nuclides are decaying to the same stable product from either side. The gamma radiation is the result of transitions within energy levels in the stable nuclide; there can only be one set of energy levels and thus the gamma energies must be the same, regardless of how the energy levels are fed.

Figure 1.16 shows the region of the (Karlsruhe) Chart of the Nuclides showing how both ^{51}Cr and ^{51}Ti decay to ^{51}V. The relevant decay schemes are summarised in Figure 1.17. Data compilations give gamma energies and emission probabilities, as shown in Table 1.2. There is no way of distinguishing between the major gamma energies at 320 keV as they are identical. The situation in

Table 1.2 Gamma energies shown by two isobaric nuclides

Nuclide	Gamma energy (keV)	Emission probability, P_γ
^{51}Ti	320.084	0.931
	608.55	0.0118
	928.63	0.069
^{51}Cr	320.084	0.0987

this case is of practical interest as both ^{51}Ti and ^{51}Cr are the only gamma-emitting thermal neutron activation products of these particular elements. If there were sufficient activity present, then the other lower emission probability gammas from ^{51}Ti should be visible, so allowing discrimination from ^{51}Cr, but if there is only a small peak at 320 keV, these other gamma peaks may not be apparent. ^{51}Cr, as an EC nuclide, emits X-rays; these are of vanadium, but with energies at 4.95 and 5.43 keV may well be below the energy range of the detector. The other nuclear parameter is half-life, and in this case the half-lives are very different, i.e. ^{51}Ti, 5.76 min and ^{51}Cr, 27.71 d. However, a single count does not give any half-life information.

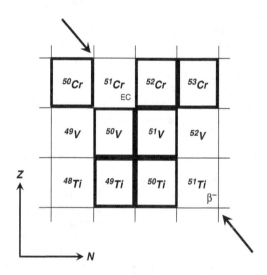

Figure 1.16   Chart of the Nuclides, showing part of the isobar $A = 51$. Heavy boxes indicate the stable nuclides

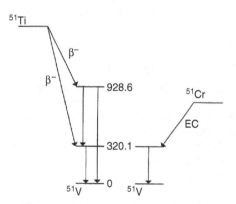

Figure 1.17   Decay schemes of ^{51}Ti and ^{51}Cr

1.7 OTHER SOURCES OF PHOTONS

1.7.1 Annihilation radiation

In Section 1.2.2, I explained how positron decay gives rise to an annihilation peak at 511.00 keV, and how this is distinguishable by being much broader than would be expected. The Doppler effect could add 2 keV to other uncertainties contributing to the width of spectrum peaks. There are reports of very small variations in energy (about 1 in 10^5) depending on the atomic mass of the material; this will be of no practical significance. Some positron emitters, such as ^{22}Na, have been used as energy calibration standards and ^{22}Na, in particular, has the advantage of a very simple spectrum with only two widely spaced peaks. However, care must be taken if the calibration program also takes the opportunity of measuring peak width or peak shape, for which purpose the 511 keV peak is quite unsuited. In general, the annihilation peak must be regarded as a special case needing some thought in its interpretation. Further information is given later in Chapter 2, Sections 2.2.3 and 2.5.3, and Chapter 6, Section 6.5.4.

1.7.2 Bremsstrahlung

Bremsstrahlung is a German word meaning 'slowing-down radiation'. It is electromagnetic radiation produced by the interaction of fast electrons with the Coulombic field of the nucleus. The electron energy loss appears as a continuum of photons, largely apparent in the X-ray region, although in principle the maximum energy is that of the beta-particle. Other energetic particles lose energy in a similar way, but bremsstrahlung is only significant with light particles, the effect being inversely proportional to the square of the charged particle mass. The effect on a gamma spectrum is to raise the general background continuum, hence making the detection of a super-imposed gamma-ray more difficult. There is an inverse relationship between the number of quanta emitted and the photon energy, so that the bremsstrahlung background level decreases with increasing energy.

There is a larger bremsstrahlung interaction with higher atomic number absorbers and higher electron (beta) energies. In practice, for 1000 keV betas in lead ($Z = 82$), there is an appreciable effect; for 1000 keV betas in aluminium ($Z = 13$), the effect is unimportant. It follows that any structure near the detector, a sample jig, for example, should be constructed of a low Z material such as a rigid plastic. The use of a graded shield (see Chapter 2, Section 2.5.1) will minimize bremsstrahlung in the same way. In complete contrast, the effect is put to positive use when a source of high-intensity 'X-radiation' is required for photon activation analysis or medical purposes. There are occasional reports of attempts to reduce the bremsstrahlung effect in gamma spectra by using an electromagnetic field near the source to divert betas away from the detector. This is clearly a cumbersome procedure and has been found to be of limited value in practice.

1.7.3 Prompt gammas

These are gamma-rays emitted during a nuclear reaction. If we consider the thermal neutron activation of cobalt, which in the short notation is expressed as:

$$^{59}Co(n, \gamma)^{60}Co$$

then the gamma-ray shown is a prompt gamma released as the excited ^{60}Co nucleus falls to the ground state. This happens quickly, in less than 10^{-14} s, is a product of a nuclear excitation level of ^{60}Co itself and is unrelated to the gamma emissions of the subsequent radioactive decay of the ^{60}Co, which as we have seen are a property of ^{60}Ni. Any measurements of prompt gammas must necessarily be made on-line and special equipment is needed, for

example, to extract a neutron beam from a reactor. Energies are often greater than those of beta decay gammas, going up to over 10 MeV. Analytically, the method is useful for some low Z elements which do not give activation products with good decay gammas. Elements, such as H, B, C, N, Si, P and Ca, have been determined by such means. I will discuss later how prompt gammas can appear from detector components if systems are operated in neutron fields (Chapter 13), even very low naturally occurring neutron fields.

1.7.4 X-rays

I have described how X-radiation appears as a result of rearrangement of the extranuclear atomic electrons after electron capture and internal conversion. X-rays are mono-energetic, the energy being equal to the difference between electron energy levels (or very close to this, as conservation of energy and momentum mean that a very small recoil energy must be given to the whole atom, thus reducing the X-ray energy slightly).

X-ray nomenclature

X-radiation has been known from the very earliest days of nuclear science; Röntgen named the rays in 1895, before Becquerel's discovery of radioactivity, and the classification of X-rays inevitably proceeded in a piecemeal fashion. Hence while Figure 1.18(a) is the sort of logical nomenclature one might pursue with today's hindsight, in practice this is only an approximation to the complicated real situation of Figure 1.18(c). Broadly, K_{α}, K_{β}, L_{α} and L_{β} fit the format more or less correctly, but the logical system fails even with some of these, and does so most decidedly with more distant transitions.

The K-shell has only one energy level. Higher shells have sub-shell levels; they are said to be degenerate. This means that the electrons in each shell, apart from the K shell, do not all have exactly the same energy. There is *fine structure* present, giving s, p, d, etc., sub-shells, and as exemplified in Figure 1.18(b), this gives rise to fine structure in the X-rays; the sub-divisions are labelled $K_{\alpha 1}$, $K_{\alpha 2}$, $L_{\gamma 6}$ and so on. Not all possible transitions occur in reality as selection rules operate based on angular momentum changes. Thus, the transition from L_1 to K is very unlikely to occur, and is said to be 'hindered'.

X-ray energies

Present-day germanium detectors have sufficient resolution to separate the energies involved in the fine structure of elements with high values of Z, where the differences are marked, but would not resolve fine structure

(a)

(b)

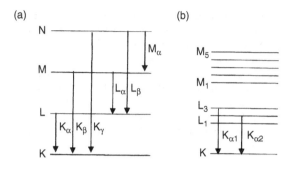

(c)

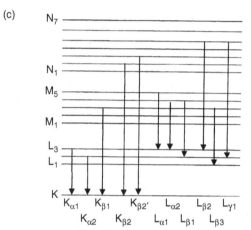

Figure 1.18 X-ray nomenclature: (a) how in an ideal world, the broad groups of X-rays should be named (see text); (b) an example of fine structure; (c) how in the real world, some X-rays are named

at low Z. Because of the widespread use of lead as a shielding material, that particular set of X-rays occurs very commonly in gamma spectra. L-series and M-series X-radiation is much lower in energy than the K-series shown. A comprehensive listing of energies is given by Browne and Firestone (1986); a few representative values are shown in Table 1.3.

X-rays and identification

It should be noted that X-ray measurements can give information as to the element involved, but will not identify the isotope of the element. This is because the arrangement of the atomic electrons is determined only by the number of protons in the nucleus (Z) and not by its mass (A). I repeat the point made earlier, that in order to identify a radioactive element from X-ray energies, we need to know the type of decay involved. Only in internally converted isomeric transitions are the X-rays characteristic of the radionuclide itself. In electron capture they identify the daughter; if the daughter has atomic number Z, then that of the decaying nuclide is $Z + 1$.

The energy widths of X-rays

We have seen how nuclear energy levels have very small widths that have negligible impact on the width or shape of gamma peaks in spectra. This is not always the case with X-rays. The width of atomic electron levels is described by a Lorentzian function (as are nuclear levels) and this contains a width parameter Γ similar to an FWHM (see Chapter 6, Section 6.1). The Lorentzian function is:

$$L(x) = \frac{\Gamma/2\pi}{(x - x_0)^2 + (\Gamma/2)^2} \qquad (1.5)$$

while the Gaussian is:

$$G(x) = \frac{1}{\sigma \times \sqrt{2\pi}} \exp \frac{-(x - x_0)^2}{2\sigma^2} \qquad (1.6)$$

and FWHM $= 2.355 \times \sigma$.

Figure 1.19 shows both functions, drawn so as to make $\Gamma = $ FWHM. Peak broadening in the detection process is described by Gaussian functions (Chapter 6). The pulse size distribution going through the detection system reflects both the initial Lorentzian energy distribution and the imposed Gaussian broadening. A convolution of these two, in practice usually with wildly differing 'widths', gives a **Voight function**.

Table 1.3 Some K X-rays

Element	Z	X-ray energy (keV)			
		$K_{\alpha 1}$	$K_{\alpha 2}$	$K_{\beta 1}$	$K_{\beta 2}$
Copper	29	8.047	8.027	8.904	8.976
Germanium	32	9.885	9.854	10.981	11.10
Cadmium	48	23.174	22.984	26.084	26.644
Lead	82	74.969	72.805	84.784	97.306
Uranium	92	98.434	94.654	111.018	114.456

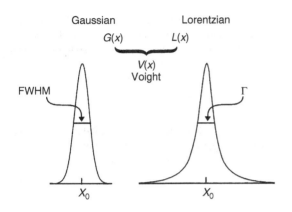

Figure 1.19 The Lorentzian and Gaussian distributions, drawn with the same *x*-axes and with Γ = FWHM. A convolution of these gives a Voight function

As an X-ray energy is the difference between an initial and a final state, the intrinsic width of both levels must be taken into account. With Lorentzian functions, these are simply added directly. Table 1.4 shows the intrinsic width of some atomic electron levels and the corresponding X-ray, and compares these to the actual X-ray energy and the resolution of a good detector at that energy. It can be seen that:

- The intrinsic width of electron energy levels (column (b)) is many orders of magnitude greater than the width of nuclear levels (typical examples in Section 1.6.4 are 3×10^{-18} eV and 9×10^{-4} eV).
- The width of the X-ray is the sum of the widths of the two relevant energy levels (columns (b) and (c)).
- The widths are a strong function of *Z*.
- Intrinsic X-ray widths are small ($\leq 0.1\%$) compared to X-ray energies, even at high *Z* (column (d)).
- Intrinsic X-ray widths have a considerable influence on the widths of their peaks in spectra at higher *Z* (column

(f)). The ratios in this column can be read as some measure of the proportions of Lorentzian and Gaussian in the mixed Voight function. More pertinently, the Lorentzian function has long tails (Figure 1.19) and a significant contribution from this intrinsic line shape will lead to peak shapes that also have long tails. This is known as **Lorentzian broadening** and may well affect computer analysis of peak areas. If counts in the tails are not included, then areas of high *Z* X-rays will be underestimated by up to a few percent.

1.8 THE MATHEMATICS OF DECAY AND GROWTH OF RADIOACTIVITY

Radionuclides are, by definition, unstable and decay by one, or more, of the decay modes; alpha, beta-minus, beta-plus, electron capture or spontaneous fission. Although strictly speaking a de-excitation rather than a nuclear decay process, we can include isomeric transition in that list from the mathematical point of view. The amount of a radionuclide in a sample is expressed in **Becquerels** – numerically equal to the rate of disintegration – the number of disintegrations per second. We refer to this amount as the **activity** of the sample. Because this amount will change with time we must always specify at what time the activity was measured.

From time to time we will have to take into account the fact that one radionuclide will decay into another, but for the moment we can consider only the simple decay process.

1.8.1 The Decay equation

Radioactive decay is a first order process. The rate of decay is directly proportional to the number of atoms of radionuclide present in the source, i.e. the activity, *A*, is directly proportional to the number of atoms, *N*, of nuclide present:

$$A = -dN/dt = \lambda N \tag{1.7}$$

Table 1.4 The effect of the intrinsic width of atomic energy levels[a,b]

Element	Z	(a) $K_{\alpha 2}$ X-ray energy (keV)	(b) Width of level (eV)		(c) Width of $K_{\alpha 2}$ X-ray (eV)	(d) $\dfrac{\text{width } K_{\alpha 2}}{\text{energy } K_{\alpha 2}}$	(e) FWHM of good detector (eV)	(f) $\dfrac{\text{width } K_{\alpha 2}}{\text{FWHM}}$
			K	L_2				
Nickel	28	7.45	1.44	0.52	1.96	0.00026	155	0.013
Cadmium	48	22.98	7.28	2.62	9.9	0.00042	200	0.049
Lead	82	72.80	60.4	6.5	66.8	0.00092	350	0.19
Uranium	92	94.65	96.1	9.3	105.4	0.0011	415	0.25

[a] Level width data from Krause and Oliver (1979), *J. Phys. Chem. Ref. Data*, **8**, 329.
[b] Detector resolution assumed to vary linearly from 150 eV at 5.9 keV to 500 eV at 122 keV.

The proportionality constant, λ, is called the **decay constant** and has the units of reciprocal time (e.g. s^{-1}, h^{-1}, etc.). The reciprocal of the decay constant is the **mean lifetime**, τ, of the radionuclide, the average time which an atom can be expected to exist before its nucleus decays:

$$\tau = 1/\lambda$$

This time represents a decay of the source by a factor of e (i.e. 2.718). It is more convenient and meaningful to refer to the **half-life**, $t_{1/2}$, of the radionuclide – the time during which the activity decreases to half its original value:

$$\lambda = \ln 2/t_{1/2} = 0.693/t_{1/2}$$

For example, for ^{60}Co:

Decay constant: $3.60 \times 10^{-4}\,d^{-1}$
Half-life: $1925.5 \pm 0.5\,d$ (5.27 years)
Mean lifetime: $2777.9\,d$

The 'year' has alternative definitions, and there is a move towards standardizing on the 'day' as the unit for quoting long half-lives. Equation (1.7) leads to the more commonly used decay equation relating number of atoms (N_t) at time (t) and half-life ($t_{1/2}$):

$$N_t = N_0 \exp([-\ln 2]\,[t]/t_{1/2}) \text{ or } N_t = N_0 \exp(-\lambda t) \quad (1.8)$$

where N_0 is the number of atoms at time $t = 0$. In practice, it is more useful to replace number of atoms by activity, bearing in mind that activity is proportional to the number of atoms:

$$A_t = A_0 \exp(-\lambda t) \quad (1.9)$$

Figure 1.20(a) illustrates the shape of the decay curve. If we take logarithms of our activity (Equation (1.9)) we transform it into a linear relationship:

$$\log A_t = \log A_0 - \lambda t$$

So, when plotted on a logarithmic scale, the activity of a source over a period of time would be a straight line (Figure 1.20(b)). Indeed, if a straight line is not obtained then one can be sure that more than one nuclide is being measured. In favourable cases, it is possible to resolve composite decay curves to estimate the relative amounts of the component radionuclides and their half-lives. (Resolution is easier, of course, if the half-lives are known.)

1.8.2 Growth of activity in reactors

Most radionuclides are created by nuclear reactions. A typical example would be neutron activation within a nuclear reactor. The activity at a point in time after the start of irradiation represents a balance between the rate of creation of radioactive atoms and the rate of decay,

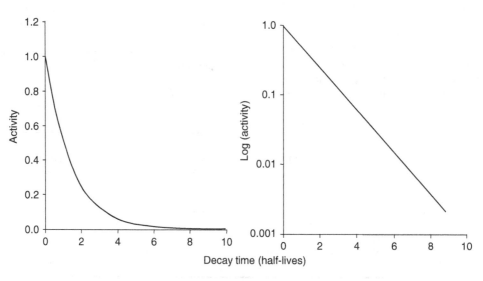

Figure 1.20   The decay of a radioactive nuclide: (a) linear scale; (b) logarithmic scale (activity is plotted relative to that at the start of the decay)

which increases as the number of atoms increases. The rate of growth of activity can be expressed as:

$$dN/dt = N_T \sigma \phi$$

where N_T is the number of target atoms, ϕ is the neutron (or, in general, the particle) flux and σ is the cross-section for the reaction, which we might imagine in this case to be that for thermal neutron capture, (n,γ), reaction. Within a particular system all these terms are constant (or at least they are until a significant fraction of target atoms are 'burned up'). This means that, in the short term, the rate of growth of activity is constant.

The rate of decay is governed by Equation (1.7) and will increase as the number of atoms of radionuclide increases. Common sense suggests that the rate of decay can never be greater than the rate of growth and that at some point in time the rate of decay must become equal to the rate of growth. Solving the resulting differential equation leads to the following:

$$A_t = A_S \left[1 - \exp\left(-\lambda t\right) \right] \qquad (1.10)$$

where t is here the time of irradiation and A_S is the **saturation activity** – the maximum activity that can be achieved during an irradiation. This is illustrated in Figure 1.21. The activity rises to within 0.1 % of saturation after 10 half-lives of the product. A simple practical consequence of the growth equation is that short-lived nuclides reach saturation very quickly, long-lived nuclides very slowly.

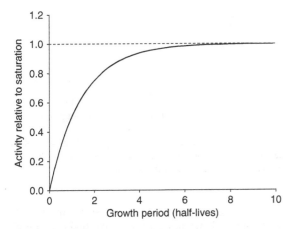

Figure 1.21   The growth of radioactivity in a nuclear reactor

1.8.3 Growth of activity from decay of a parent

When one radionuclide (the parent) decays into another radionuclide (the daughter), the rate of change of the number of daughter atoms must be the difference between the rate of growth from the parent and the rate of decay of the daughter:

$$dN_D/dt = \lambda_P N_P - \lambda_D N_D$$
$$= \lambda_P N_{P0} \exp\left(-\lambda_P t\right) - \lambda_D N_D$$

where the subscripts D and P refer to daughter and parent, respectively, and the subscript 0 indicated the number of atoms at $t = 0$. Solving this linear differential equation gives:

$$N_D = N_{P0}[\exp\left(-\lambda_P t\right) - \exp\left(-\lambda_D t\right)]\lambda_P/(\lambda_D - \lambda_P)$$
$$+ N_{D0}\exp\left(-\lambda_D t\right) \qquad (1.11)$$

Bearing in mind that $A = \lambda N$, we can rewrite this in terms of activity:

$$A_D = A_{P0}[\exp\left(-\lambda_P t\right) - \exp\left(-\lambda_D t\right)]\lambda_D/(\lambda_D - \lambda_P)$$
$$+ A_{D0}\exp\left(-\lambda_D t\right) \qquad (1.12)$$

To gain an understanding of this relationship as the ratio of parent to daughter half-lives varies, it is useful to imagine that we chemically separate the parent and daughter, in which case the second term of Equation (1.12) is zero. We can then follow the change in their activities in the initially pure parent and calculate the total activity. There are three particular cases depending upon whether the parent half-life is greater or less than the daughter half-life.

Transient equilibrium – $t_{1/2}$ parent > $t_{1/2}$ daughter

In a **transient equilibrium**, the activity of the daughter nuclide is in constant ratio to that of the parent nuclide and apparently decays with the half-life of the parent. In Figure 1.22, we see the decay of the parent unaffected by the absence or presence of daughter and the growth of the daughter activity. (The time scale is in units of half-life of the daughter nuclide.) The total activity in the system is the sum of the parent and daughter activities. Transient equilibrium is established after about 10 half-lives of the daughter nuclide after which the daughter apparently decays with the half-life of the parent.

This should not be forgotten when measuring nuclides that are sustained by decay of a parent. For example, if ^{95}Nb is measured in a fission product mixture it is likely,

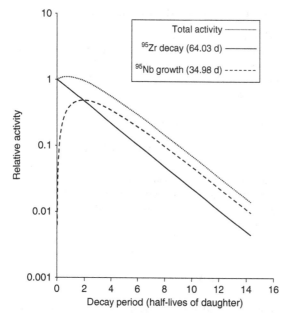

Figure 1.22 Transient equilibrium – relative activities of parent and daughter nuclides after separation

using the commonly available commercial gamma spectrum analysis packages, that the normal half-life of ^{95}Nb, 34.98 d, will be extracted from the package's nuclide library and used to correct ^{95}Nb activities. In fact, if the age of the fission product mixture is greater than a year or so then the appropriate half-life is that of the parent ^{95}Zr, i.e. 64 d. At shorter decay times, the accurate calculation of ^{95}Nb activity will be beyond the capabilities of a standard analysis program. Fission product mixtures contain many isobaric decay chains and many such parent/daughter pairs of nuclides; this problem is not at all uncommon.

If we take Equation (1.12) and set t to a value much greater than the half-life of the daughter, we can calculate the relative numbers of parent and daughter atoms at equilibrium:

$$N_D/N_P = \lambda_P/(\lambda_D - \lambda_P) \tag{1.13}$$

and the equilibrium activity of the daughter relative to that of the parent:

$$A_D = A_P\lambda_D/(\lambda_D - \lambda_P) \text{ or, in terms of half-lives,}$$

$$A_D = A_P t_{1/2P}/(t_{1/2P} - t_{1/2D}) \tag{1.14}$$

where $t_{1/2P}$ and $t_{1/2D}$ are the half-lives of parent and daughter, respectively.

Figure 1.22 also shows that, as equilibrium is approached, the activity of the daughter nuclide becomes greater than that of the parent. This may, at first sight, seem odd. Take the case of ^{140}La in equilibrium with its parent ^{140}Ba (half-lives of 1.68 and 12.74 d, respectively) and apply the equations above. At equilibrium, the number of ^{140}La atoms in the source will be only 0.15 times the number of ^{140}Ba atoms. However, because ^{140}La has a shorter half-life its activity will be 15 % greater than that of the parent ^{140}Ba. The difference can be substantial; in an equilibrium mixture of ^{95}Zr/^{95}Nb the activity of the ^{95}Nb is $64/(64-35)$, i.e. 2.2 times that of the ^{95}Zr.

Secular equilibrium – $t_{1/2}$ parent $>>$ $t_{1/2}$ daughter

If the half-life of the parent nuclide is very long compared to that of the daughter, the equilibrium state is referred to as **secular equilibrium**. In such situations, where $t_{1/2D}$ becomes negligible, Equation (1.13) becomes $A_D = A_P$, i.e. the daughter activity equals the parent activity (Figure 1.23).

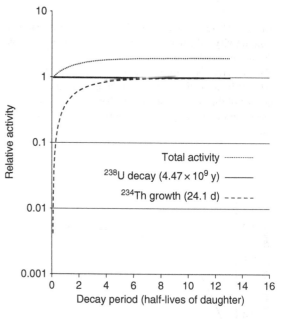

Figure 1.23 Secular equilibrium – relative activities of parent and daughter nuclides after separation where the parent half-life is much greater than that of the daughter

Take, for example, the first three stages of the ^{238}U decay series shown in Figure 1.24. We find that in each parent/daughter case, the half-life of the parent is very

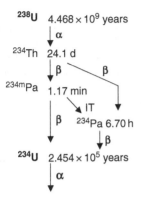

Figure 1.24   The first few stages of the ^{238}U decay series (IT, isomeric transition)

much greater than the daughter and therefore there will be a secular equilibrium established between each pair. The activity of each daughter will be equal to that of its parent and the total activity, for this portion of the decay chain, will be three times that of the ^{238}U. Note that as far as the ^{234}Pa and the ^{234m}Pa are concerned, at equilibrium, the half-life of their parent is effectively that of ^{238}U. The branching of the ^{234}Th decay means that the total ^{234}Pa + ^{234m}Pa will be, at equilibrium, equal to that

of the ^{238}U. The activity will be shared between the two nuclides according to the branching ratio.

If we look at the complete ^{238}U decay scheme, we find 14 daughter nuclides, all of whom have much shorter half-lives than their ultimate parent. The total activity, assuming radioactive equilibrium is established, will be 14 times the ^{238}U activity. (A more complete discussion of gamma spectrometry of the uranium and thorium decay series nuclides can be found in Chapter 16, Section 16.1.2)

No equilibrium – $t_{1/2}$ parent < $t_{1/2}$ daughter

If the daughter half-life is greater than that of the parent then, obviously, the parent will decay, leaving behind the daughter alone. Figure 1.25 shows the growth of daughter activity within an initially pure parent. No equilibrium is established; ultimately the decay curve will be that of the grown-in daughter.

1.9 THE CHART OF THE NUCLIDES

The general layout of the *Karlsruhe Chart of the Nuclides* has been described; it has been used to explain beta decay and parts have been illustrated in Figures 1.3 and 1.16. In this section, I discuss its use in diagnosis, firstly as a data source, and then as an indicator of the location of probable nuclides. There are other versions of this chart, but none have made it onto the walls of counting rooms as often as the Karlsruhe Chart.

1.9.1 A source of nuclear data

I would certainly not recommend printed versions of the chart as the best source of nuclear data; the numbers presented are too lacking in detail and are (probably) not up to date. There is an interactive on-line version (See Further Reading and the book's website) created by the US National Nuclear Data Center – a very useful resource containing much more up-to-date information than the printed versions, but without nuclear cross-section data. The website does, however, have easy links to energy level data and gamma-ray emission data. A software version has been developed for PCs, which should contain good data based on the OECD Nuclear Energy Agency's Data Bank. However, access to that CD-ROM seems not to be straightforward.

The printed chart is nevertheless very useful for rapid assessment. For each element, there is the element symbol, atomic weight and thermal neutron absorption cross-section. Within the horizontal strip of isotopes, stable species are shown with a black background and contain the mass number, natural isotopic abundance of the

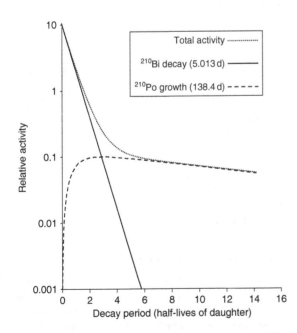

Figure 1.25   Relative activities of parent and daughter nuclides after separation where there is no equilibrium

isotope in atom%, and thermal neutron absorption cross-section, σ, in barns (b). Where appropriate, a stable square will have a white section to indicate the presence of a metastable state of the stable nuclide with its half-life and the energy of the IT gamma decay in keV. Radioactive species are colour-coded: blue, β⁻; red, EC or β⁺; yellow, α; green, spontaneous fission. If two modes of decay occur, two colours are shown.

The ^{60}Co square would be blue for β⁻ and contains the half-life, the major maximum beta energies in MeV (useful for bremsstrahlung estimation), major gamma energy (in keV) in order of emission probability and the thermal neutron cross-section. The isomer is shown as a white section with decay mode(s) and energies. On this particular chart, electron capture is shown as ε, isomeric transition as I and conversion electron emission as e⁻, along with standard symbols.

In Figure 1.26(b), 'σ 20 + 17' indicates 20 b for the (n, γ) cross-section to form ^{60m}Co ($t_{1/2} = 10.47$ min) and 17 b for the (n, γ) reaction to give ^{60}Co directly in its ground state. For most purposes, these numbers would need to be summed if ^{60}Co activity was sought, as essentially all of the metastable states will end up as the ground state within a couple of hours.

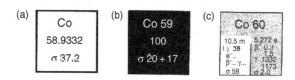

Figure 1.26   Typical data from a chart of the nuclides: (a) the element; (b) a stable isotope; (c) a radioisotope with a metastable state

The chart contains sufficient information for simple calculations as to whether the quantities of radionuclides found are those that might be expected. The overall activation equation is:

$$A = N_T \times \sigma \times \phi \times [1 - \exp(-\lambda \times t_{irr})] \times \exp(-\lambda \times t_{dk})$$

(1.15)

where:

- A = induced activity (in Bq);
- N_T = number of target atoms;
- σ = cross-section (probability of reaction taking place; an area, in units of barns (b) where $1\,b = 10^{-28}\,m^2$);

- ϕ = flux of activating particle, in most cases, neutrons (units: particles per unit area per second, e.g. n cm^{-2}s^{-1}; note that units of area used must be the same as those used for the cross-section);
- t_{irr} = irradiation time (in same units as half-life used to calculate λ);
- t_{dk} = cooling time (decay time) from the end of the irradiation to the time of measurement (again same units as half-life).

The shape of this expression is shown in Figure 1.27.

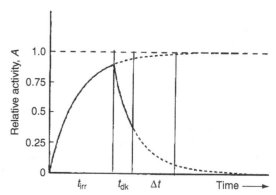

Figure 1.27   The quantity of radionuclide increases during irradiation and then decreases by radioactive decay to the time of measurement: t_{irr}, irradiation period; t_{dk}, decay time; Δt, measurement period

The three parts of the equation correspond as follows:

- $[N_0 \times \sigma \times \phi]$ = the saturation activity, indicated by the dashed line at $A = 1.0$; the maximum activity obtainable, asymptotically approached when $t_{irr} > t_{1/2}$.
- $[1 - \exp(-\lambda \times t_{irr})]$ = the approach to saturation; a useful property is that when $t_{irr} = t_{1/2}$, this factor = 0.5, i.e. the activity is half the saturation value; at $t_{irr} = 2 \times t_{1/2}$, the factor is 0.75.
- $[\exp(-\lambda \times t_{dk})]$ = normal radioactive decay from the end of the irradiation to the time of measurement.

1.9.2 A source of generic information

The chart is helpful when tracking down unknown radionuclides detected in a measurement to know what type of nuclear reaction may have been responsible for the production of the activity. In most cases, this will be a thermal neutron reaction resulting in activation by (n, γ), or if the target material is fissile, a fission reaction (n, f).

Thermal neutron capture (n, γ)

If this is the reaction, then the search for nuclides is narrowed dramatically. From more than 2000 radionuclides displayed on the chart, we need scan only 180 or so. It is clear that as the (n, γ) reaction merely adds one neutron to a stable nuclide we must look at isotopes just one square to the right of a (black) stable one. For example, instead of considering all 21 radioisotopes of arsenic, we need look at only one; the stable arsenic is ^{75}As and therefore we look at ^{76}As. There are just two qualifications to this simple picture:

- There is one element where there is a significant chance of finding that two neutrons have been added – gold.

$$^{197}\mathrm{Au} + \mathrm{n} \longrightarrow {}^{198}\mathrm{Au} + \mathrm{n} \longrightarrow {}^{199}\mathrm{Au}$$

$$\text{stable} \qquad \beta^- \qquad\qquad \beta^-$$

$$(\sigma = 98.8\,\mathrm{b}) \quad (\sigma = 25\,100\,\mathrm{b})$$

This is due to the very large neutron absorption cross-section of the first product, ^{198}Au. In most cases, while there may be a considerable amount of activity formed by an (n, γ) reaction, this will correspond to relatively few atoms. Thus, the amount of target material available for a second reaction is small, and a large reaction probability (as we have here with ^{198}Au) is needed to give significant amounts of the second reaction product.
- The second complication is when the (n, γ) product does not decay to a stable nuclide but to a radioactive one. Then nuclides that are an extra transformation away (β$^-$ normally) need to be considered. An example is:

$$^{130}\mathrm{Te}\,(\mathrm{n}, \gamma) \quad {}^{131}\mathrm{Te} \longrightarrow \quad {}^{131}\mathrm{I} \longrightarrow \quad {}^{131}\mathrm{Xe}$$

$$\text{stable} \qquad \beta^-\,(25\,\text{min}) \quad \beta^-\,(8.0\,\text{d}) \quad \text{stable}$$

or the highly significant sequence:

$$^{238}\mathrm{U}\,(\mathrm{n}, \gamma)\ {}^{239}\mathrm{U} \longrightarrow \quad {}^{239}\mathrm{Np} \longrightarrow {}^{239}\mathrm{Pu}$$

$$\text{'stable'} \quad \beta^-\,(23\,\text{min}) \quad \beta^-\,(2.3\,\text{d}) \quad \alpha\,(2.4 \times 10^4\,\text{y})$$

These transformations can be readily traced on the chart.

A further point: most elements have more than one stable isotope. So, if one activation product is found, check with the chart for others that could be formed by the same mechanism. For example, if 35.3 h ^{82}Br is seen, formed by (n, γ) from stable ^{81}Br, look at the chart which will tell you that there is plenty of stable ^{79}Br with an adequate cross-section, so that, time scales permitting, there should be 4.4 h ^{80m}Br present and its daughter ^{80}Br as well.

Fast neutron reactions, (n, p) etc.

Other neutron-induced reactions will normally involve energetic or fast neutrons, where the extra kinetic energy is needed to knock out extra particles. Common reactions are (n, p), (n, α) and (n, 2n), and Figure 1.28 shows these transformations on the Z against N nuclide chart format. The quantity of radioactivity formed by these reactions is often small because of relatively low fluxes of fast neutrons and small cross-sections. However, reactor operators and persons involved in reactor decommissioning will be aware of the significant amounts of activity that can be formed by certain reactions, such as, ^{54}Fe(n, p)^{54}Mn, ^{58}Ni(n, p)^{58}Co and ^{27}Al (n, α)^{24}Na. The likelihood of the production of all these radionuclides can again be followed on the nuclide chart.

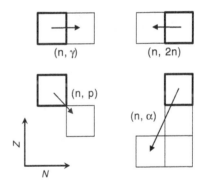

Figure 1.28 Location of products of neutron reactions. Heavy boxes indicate the stable target nuclides

Fission reactions (n, f)

Figure 1.29 shows that fission products range over masses from about 75 to 165, which correspond to elements Ge to Dy. Most probable nuclides are grouped into two distinct mass regions (the asymmetric mass distribution), as also shown in Figure 1.30.

With ^{235}U(n, f) the distribution peaks at $A = 90$ to 100, and $A = 134$ to 144. The cumulative fission yield data for each mass number are given on the Karlsruhe Chart as percentage yields on the right-hand edge of the chart.

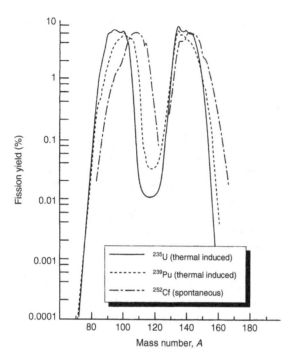

Figure 1.29 Mass yield curves for the thermal neutron fission of ^{235}U and ^{239}Pu, and spontaneous fission of ^{252}Cf

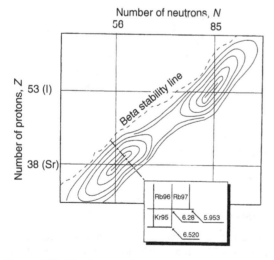

Figure 1.30 The location of fission products on the Nuclide Chart, indicating regions of high independent yield. The inset shows how data are presented for the cumulative yield of each isobar

The identification problem appears difficult, but bear in mind that:

- Most fission products fall into narrow bands around the two peaks described above, comprising only some 20 isobars.
- All are neutron rich, so that all EC/β^+ nuclides can be ignored.
- A large fraction of possible contenders have short half-lives, and these you may well not need to consider. On the other hand, all fission products are members of isobaric chains and decay will usually result in other activities.

Fission product yields from fast neutron fission show a very similar distribution to Figure 1.29, the main difference being that yields in the minimum between the two peaks increase by a factor of about three.

Information on the history of the 'unknown', in particular, its age or cooling time from the end of irradiation, is most useful. Figure 1.31 shows how the total activity of some elements formed in fission varies with time. These are not all gamma emitters.

PRACTICAL POINTS

- Gamma spectrometry using germanium detectors is the best technique for identifying and quantifying radionuclides. This is due to the very sharply defined and characteristic energies of gamma-rays which are produced by the great majority of radionuclides.
- *However*, there are a small number of 'pure beta emitters', which do not emit gamma radiation. These cannot be identified by gamma spectrometry. Some are technologically important (^{3}H, ^{14}C, ^{90}Sr).
- Most gamma-rays are a consequence of beta decay. Gamma-emission probabilities are not necessarily the same as beta decay probabilities because of internal conversion. This latter can result in X-radiation.
- Gammas and X-rays are usually properties of the daughter nucleus (but not with isomeric transitions). This can lead to identical gamma energies being shown by two isobaric nuclides.
- X-ray energies overlap the low-energy range of gamma-rays. X-ray peak shapes can be different from gamma-ray peak shapes.
- X-ray energies will tell you the element present, but not which isotope. This identification presupposes knowledge of the decay mode: IT $\longrightarrow$ nuclide directly; EC, β^-/IC $\longrightarrow$ daughter.
- Decay schemes give vital information on whether gammas are in 'cascade'. This has great significance in true coincidence summing.

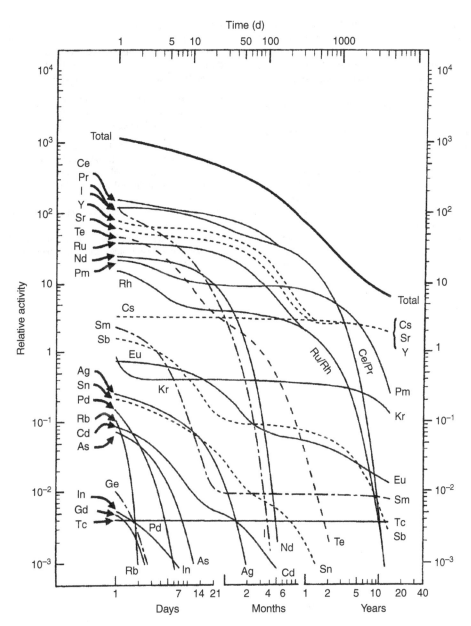

Figure 1.31 Relative radioactivity from fission products as a function of decay time. Data are for thermal neutron fission of ^{235}U, flux of 10^{13} n cm^{-2} s^{-1} and irradiation time of 2 years. Several different nuclides may contribute to the curve for each element. Adapted with permission from Choppin and Rydberg (1980)

- The Karlsruhe Chart of the Nuclides is a useful tool in helping to identify nuclides, both with regard to classes of nuclide present and for the nuclear data it shows. The data should not necessarily be relied upon for accurate work.

FURTHER READING

- There are a number of general books that cover many of the topics of this chapter:

Keller, C. (1988). *Radiochemistry* (English edition), Ellis Horwood, Chichester, UK.

Ehmann, W.D. and Vance, D.E. (1991). *Radiochemistry and Nuclear Methods of Analysis*, John Wiley & Sons, Inc., New York, NY, USA.

Choppin, G.R. and Rydberg, J. (1980). *Nuclear Chemistry, Theory and Applications*, Pergamon Press, Oxford, UK.

• A classic authoritative text on the physics of the atom that has run to several later editions:

Evans, R.D. (1955). *The Atomic Nucleus*, Mcgraw-Hill, New York, NY, USA.

• The following is currently the best single-volume complete compilation of nuclear decay data. (More information on sources of data, printed and Internet, are given in Appendix A):

Browne, E., Firestone R.B., Baglin, C.M. and Chu, S.Y.F. (1998). *Table of Isotopes*, John Wiley & Sons, Inc., New York, NY, USA (This book is now accompanied by a CD containing the nuclear data tables. See *http://www.wiley.com/toi*).

• Decay schemes, and hence gammas in cascade, are shown in the forerunner to the above. Numerical data are obviously older and less reliable, and the format is not user-friendly:

Shirley, V.S. and Lederer, C.M. (1978). *Table of Isotopes*, 7th Edn, Wiley Interscience, New York, NY, USA.

• Charts of the nuclides can be found online at:

An interactive online version from the US NNDC (*http://www.nndc.bnl.gov/chart/*).

• Printed versions can be found on Amazon as:

Magill, J. and Galey, J. (2004). *Radioactivity, Radionuclides, Radiation* (with the Fold-out Karlsruhe Chart of the Nuclides) (Hardcover), Springer-Verlag, Berlin, Germany (a CD-ROM accompanies the book).

The *General Electric* printed version (USA) can also be found on Amazon.

• The Karlsruhe Chart of the Nuclides can be purchased on-line at: http://www.nucleonica.net/nuclidechart.aspx

2

Interactions of Gamma Radiation with Matter

2.1 INTRODUCTION

In this chapter, I will discuss the mechanisms of interaction of gamma radiation with matter. That will lead
directly to an interpretation of the features within a gamma
spectrum due to interactions within the detector itself
and within the detector surroundings. Finally, the design
of detector shielding will be considered. Although the
discussion will centre on gamma radiation, it should not
be forgotten that gamma radiation is electromagnetic in
nature, as is X-radiation, and that to a detector they are
indistinguishable.

The instrumental detection of any particle or radiation depends upon the production of charged secondary
particles which can be collected together to produce an
electrical signal. Charged particles, for example, alpha-
and beta particles, produce a signal within a detector by
ionization and excitation of the detector material directly.
Gamma photons are uncharged and consequently cannot
do this. Gamma-ray detection depends upon other types of
interaction which transfer the gamma-ray energy to electrons within the detector material. These excited electrons
have charge and lose their energy by ionization and excitation of the atoms of the detector medium, giving rise to
many electron–hole pairs. The absorption coefficient for
gamma radiation in gases is low and all practical gamma-
ray detectors depend upon interaction with a solid. As
we shall see, the charged pairs produced by the primary
electron are electron–hole pairs. The number produced is
proportional to the energy of the electrons produced by
the primary interaction. The detector must be constructed
of suitable material and in such a way that the electron–
hole pairs can be collected and presented as an electrical
signal.

2.2 MECHANISMS OF INTERACTION

It would not be unexpected that the degree of interaction of gamma radiation with matter would depend upon
the energy of the radiation. What might not be expected,
however, is the detailed shape of that energy dependence.
Figure 2.1 shows the attenuation coefficient of a number
of materials relevant to gamma-ray spectrometry as a
function of gamma-ray energy.

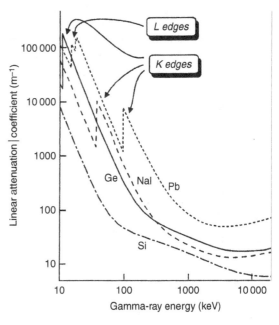

Figure 2.1 Attenuation coefficient of materials as a function
of gamma-ray energy

Practical Gamma-ray Spectrometry – 2nd Edition   Gordon R. Gilmore
© 2008 John Wiley & Sons, Ltd

The features in these curves which are most striking are the sharp jumps in attenuation coefficient at low energy and the rise at high energy after a fall over most of the energy range. These features can be explained by a detailed examination of the interaction processes involved and more importantly, from a gamma spectrometry point of view, this examination will allow the shape of the gamma spectrum itself to be explained. It is also apparent in the diagram that the probability of an interaction, as expressed by the attenuation coefficient, depends upon the size of the interacting atom. The attenuation coefficient is greater for materials with a higher atomic number. Hence, germanium is a more satisfactory detector material for gamma-rays than silicon and lead is a more satisfactory shielding material than materials of a lower atomic number.

At the outset, I should, perhaps, make plain the difference between attenuation and absorption. An attenuation coefficient is a measure of the reduction in the gamma-ray intensity at a particular energy caused by an absorber. The absorption coefficient is related to the amount of energy retained by the absorber as the gamma radiation passes through it. As we shall see, not all interactions will effect a complete absorption of the gamma-ray. The result of this is that absorption curves lie somewhat below attenuation curves in the mid-energy range. Figure 2.2 compares the mass absorption and mass attenuation curves for germanium. Mass absorption and attenuation will be considered in more detail at the end of this chapter.

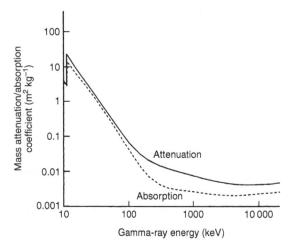

Figure 2.2 Comparison of absorption and attenuation coefficients in germanium

Each of the curves in Figure 2.1 is the sum of curves due to interactions by **photoelectric absorption**, **Compton scattering** and **pair production**. The relative magnitude of each of these components for the case of germanium is shown in Figure 2.3.

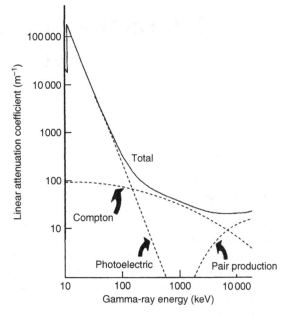

Figure 2.3 The linear attenuation coefficient of germanium and its component parts

Photoelectric interactions are dominant at low energy and pair production at high energy, with Compton scattering being most important in the mid-energy range. Gamma radiation can also interact by coherent scattering (also known as Bragg or Rayleigh scattering) and by photonuclear reactions. Coherent scattering involves a re-emission of the gamma-ray after absorption with unchanged energy but different direction. Such an interaction might contribute to attenuation of a gamma-ray beam, but because no energy is transferred to the detector, it can play no part in the generation of a detector signal and need not be considered further. The cross-sections for photonuclear reactions are not significant for gamma-rays of energy less than 5 MeV and this mode of interaction can be discounted in most gamma-ray measurement situations.

It is important to be aware that each of the significant interaction processes results in the transfer of gamma-ray energy to electrons in the absorbing medium, i.e. the gamma-ray detector. In all that follows, therefore, the energy transferred to the electrons represents the energy absorbed by the detector and is, in turn, related to the output from the detector.

2.2.1 Photoelectric absorption

Photoelectric absorption arises by interaction of the gamma-ray photon with one of the bound electrons in an atom. The electron is ejected from its shell (Figure 2.4(a)) with a kinetic energy, E_e, given by:

$$E_e = E_\gamma - E_b \qquad (2.1)$$

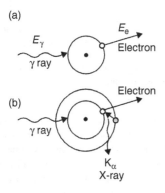

(a)

E_γ
γ ray
E_e
Electron

(b)

γ ray
Electron
K_α
X-ray

Figure 2.4   (a) The mechanism of photoelectric absorption, and (b) the emission of fluorescent X-rays

where E_γ is the gamma-ray energy and E_b the energy binding the electron in its shell. The atom is left in an excited state with an excess energy of E_b and recovers its equilibrium in one of two ways. The atom may de-excite by redistribution of the excitation energy between the remaining electrons in the atom. This can result in the release of further electrons from the atom (an Auger cascade) which transfers a further fraction of the total gamma-ray energy to the detector. Alternatively, the vacancy left by the ejection of the photoelectron may be filled by a higher-energy electron falling into it with the emission of a characteristic X-ray which is called **X-ray fluorescence** (see Figure 2.4(b)). This X-ray may then in turn undergo photoelectric absorption, perhaps emitting further X-rays which are absorbed, in turn, until ultimately all of the energy of the gamma-ray is absorbed. (In order to conserve momentum when an electron is ejected, a very small amount of energy must be retained by the recoiling atom. This is very small and can be ignored for all practical purposes.)

The energy level from which the electron is ejected depends upon the energy of the gamma-ray. The most likely to be ejected is a K electron. If sufficient energy is not available to eject a K electron, then L or M electrons will be ejected instead. This gives rise to the discontinuities in the photoelectric absorption curves. These absorption edges occur at the binding energies corresponding to the electron shells. For example, in the curve for germanium (Figure 2.1) the K absorption edge occurs at 11.1 keV. For caesium iodide, there are two K edges, one corresponding to the iodine K electron at 33.16 keV and the other to the caesium K electron at 35.96 keV. Below these energies, only L and higher order electrons can be photoelectrically ejected. Since there is then one less way in which energy can be transferred to the interacting atom, the attenuation coefficient falls in a stepwise manner at the precise energy of the K electron. Similar edges corresponding to L and other less tightly bound electrons can be seen at lower energies in the curve for lead. The L electron shell has three sub-levels and this is reflected in the shape of the L edge.

The probability that a photon will undergo photoelectric absorption can be expressed as a cross section, τ. This measure of the degree of absorption and attenuation varies with the atomic number, Z, of the absorber and the gamma-ray energy, E_γ, in a complicated manner:

$$\tau \propto Z^n / E_\gamma^m \qquad (2.2)$$

where n and m are within the range 3 to 5, depending upon energy. For example, functions such as $Z^5/E_\gamma^{3.5}$ and $Z^{4.5}/E_\gamma^3$ have been quoted. The significance of this equation is that heavier atoms absorb gamma radiation, at least as far as the photoelectric effect is concerned, more effectively than lighter atoms. It follows that ideal detector materials would be of high Z, given that their charge collection characteristics were satisfactory.

The photoelectric attenuation coefficient, μ_{PE}, can be derived from the related cross-section in the following manner:

$$\mu_{PE} = \tau \times \rho \times N_A / A \qquad (2.3)$$

where ρ is the density of the absorbing material, A its average atomic mass and N_A the Avogadro constant. In the literature, there is some confusion over the use of 'coefficient' and 'cross-section'. In some texts, the two are taken to be identical. Here, I shall consistently maintain the distinction implied above in Equations (2.2) and (2.3).

It is normally assumed that photoelectric absorption results in the complete absorption of the gamma-ray. However, for those events near to the surface of the detector there is a reasonable probability that some fluorescent X-rays, most likely the K X-rays, might escape from the detector. The net energy absorbed in the detector would then be:

$$E_e = E_\gamma - E_{K\alpha} \qquad (2.4)$$

where $E_{K\alpha}$ is the energy of the Kα X-ray of the detector material. This process is known as **X-ray escape**. Since a precise amount of energy is lost, this gives rise to a definite peak at the low-energy side of the full energy peak. In a germanium detector, it would be called a **germanium escape peak** and in a sodium iodide detector an **iodine escape peak**. (Because of the relative sizes of sodium and iodine, most absorption by sodium iodide is by interaction with iodine atoms.) Such peaks are usually only significant for small detectors and low-energy photons but can be found associated with higher-energy gamma-ray peaks when these are very well defined. Spectra measured on detectors designed for low-energy gamma- and X-rays may well also show evidence of L escape X-rays.

2.2.2 Compton scattering

Compton scattering (Figure 2.5) is a direct interaction of the gamma-ray with an electron, transferring part of the gamma-ray energy. The energy imparted to the recoil electron is given by the following equation:

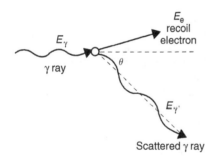

Figure 2.5   The mechanism of Compton scattering

$$E_e = E_\gamma - E'_\gamma \tag{2.5}$$

or:

$$E_e = E_\gamma \left\{ 1 - \frac{1}{[1 + E_\gamma(1 - \cos\theta)/m_0 c^2]} \right\} \tag{2.6}$$

Putting different values of θ into this equation shows how the energy absorbed varies with the scattering angle. Thus, with $\theta = 0$, i.e. scattering directly forward from the interaction point, E_e is found to be 0 and no energy is transferred to the detector. At the other extreme when the gamma-ray is backscattered and $\theta = 180°$, the term within brackets in the equation above is still less than 1 and so only a proportion of the gamma-ray energy will

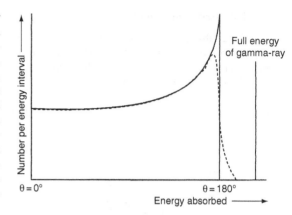

Figure 2.6   Energy transferred to absorber by Compton scattering related to scattering angle

be transferred to the recoil electron. At intermediate scattering angles, the amount of energy transferred to the electron must be between those two extremes. (Figure 2.6 is a schematic diagram showing this relationship.) The inescapable conclusion is that, at all scattering angles, less than 100 % of the gamma-ray energy is absorbed within the detector.

Simplistically, I have assumed that the gamma-ray interacts with a free electron. In fact, it is much more likely that the electron will be bound to an atom and the binding energy of the electron ought to be taken into account. Most interactions will involve outer, less tightly bound, electrons and in many cases the binding energy will be insignificant compared to the energy of the gamma-ray (a few eV compared to hundreds of keV). Taking binding energy into account alters the shape of the Compton response function to some extent, making the sharp point at the maximum recoil energy become more rounded and the edge corresponding to 180° backscatter acquires a slope. This is indicated by the dotted curve in Figure 2.6.

The Compton scattering absorption cross-section, often given the symbol σ, is related to the atomic number of the material and the energy of the gamma-ray:

$$\sigma \propto f(E_\gamma) \tag{2.7}$$

An energy function of $1/E_\gamma$ has been suggested as appropriate. Using an analogous relationship to that of Equation (2.3), we can calculate a Compton scattering coefficient, μ_{CS}. If we also take into account the fact that over a large part of the Periodic Table the ratio A/Z is reasonably constant with a value near to 2 we can show that:

$$\mu_{CS} = \text{constant} \times \sigma \times f(E_\gamma) \tag{2.8}$$

the implication being that the probability of Compton scattering at a given gamma-ray energy is almost independent of atomic number but depends strongly on the density of the material. Moreover, there is little variation of the mass attenuation coefficient, μ_{CS}/ρ, with atomic number, again at a particular energy – a fact which ameliorates the difficulties of making a correction for self-absorption of gamma-rays within samples of unknown composition.

2.2.3 Pair production

Unlike photoelectric absorption and Compton scattering, pair production results from the interaction of the gamma-ray with the atom as a whole. The process takes place within the Coulomb field of the nucleus, resulting in the conversion of a gamma-ray into an electron–positron pair. In a puff of quantum mechanical smoke, the gamma-ray disappears and an electron–positron pair appears. For this miracle to take place at all, the gamma-ray must carry an energy at least equivalent to the combined rest mass of the two particles – 511 keV each, making 1022 keV in all. In practice, evidence of pair production is only seen within a gamma-ray spectrum when the energy is rather more than 1022 keV.

In principle, pair production can also occur under the influence of the field of an electron but the probability is much lower and the energy threshold is 4 electron rest masses, making it negligible as a consideration in normal 0 to 3 MeV gamma spectrometry. The electron and positron created share the excess gamma-ray energy (i.e. the energy in excess of the combined electron–positron rest mass) equally, losing it to the detector medium as they are slowed down. As I explained in Chapter 1, when the energy of the positron is reduced to near thermal energies, it must inevitably meet an electron and the two will annihilate, releasing two 511 keV annihilation photons. This is likely to happen within 1 ns of creation of the pair and, taking into consideration the fact that the charge collection time of typical detectors is 100 to 700 ns, the annihilation can be regarded as instantaneous with the

pair production event. The complete sequence of events is described in Figure 2.7. The net energy absorbed within the detector by the immediate consequences of the pair production event is (with energies expressed in keV):

$$E_e = E_\gamma - 1022 \tag{2.9}$$

The cross-section for the interaction, κ, depends upon E_γ and Z in a complicated manner which can be expressed as:

$$\kappa \propto Z^2 f(E_\gamma, Z) \tag{2.10}$$

The attenuation coefficient, μ_{PP}, is calculated in a similar manner to the photoelectric attenuation coefficient (Equation (2.2)). The variation of κ with atomic size is dominated by the Z^2 term, the function in parentheses changing only slightly with Z. The energy dependence of κ is determined by the function $f(E_\gamma, Z)$ which increases continuously with energy from the threshold at 1022 keV so that at energies greater than 10 MeV pair production is the dominant mechanism of interaction (see Figure 2.3).

It is more than likely that the electron with which the positron annihilates will be bound to an atom. It is necessary, therefore, for some energy to be shared with the atom in order to remove the electron. This means that the energy available to be shared between the annihilation quanta will be lower than expected. For example, in aluminium the annihilation radiation has been estimated to be 510.9957 keV instead of the theoretical 511.0034 keV. In everyday gamma-ray spectrometry, the difference is unlikely to be noticed. What is certainly noticeable is the extra width of annihilation gamma-ray peaks due to **Doppler broadening**, the reason for which I explained in Chapter 1 (Section 1.2.2).

2.3 TOTAL ATTENUATION COEFFICIENTS

The curves plotted in Figure 2.1 are the sum of the coefficients for each of the significant interaction processes:

$$\mu_T = \mu_{PE} + \mu_{CS} + \mu_{PP} + \mu_{RS} \tag{2.11}$$

where the final term represents the loss of gamma radiation by elastic (Rayleigh) scattering. In terms of cross-sections, Equation (2.11) can be rewritten as follows:

$$\mu_T = (\rho \times N_A/A)(\tau + \sigma + \kappa + \sigma_{RS}) \tag{2.12}$$

A more useful coefficient in practical terms is the mass-attenuation coefficient, the ratio of attenuation coefficient to the density of the material:

$$\mu_T/\rho = (N_A/A)(\tau + \sigma + \kappa + \sigma_{RS}) \tag{2.13}$$

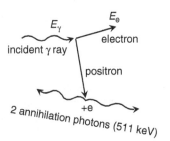

Figure 2.7   The mechanism of pair production

This is the parameter plotted in Figure 2.2, comparing attenuation and absorption. The attenuation coefficient only expresses the probability that a gamma-ray of a particular energy will interact with the material in question. It takes no account of the fact that as a result of the interaction a photon at a different energy may emerge as a consequence of that interaction. The total absorption coefficient, μ_A, must, of course, take into account those incomplete interactions:

$$\mu_A = (\rho \times N_A/A)(\tau \times f_{PE} + \sigma \times f_{CS} + \kappa \times f_{PP}) \quad (2.14)$$

In this expression, each 'f' factor is the ratio of the energy imparted to electrons by the interaction to the initial energy of the gamma-ray. Rayleigh scattering does not contribute to absorption of energy and does not appear in Equation (2.14). The detailed calculation of these factors need not concern us here but they will include such considerations as energy lost to bremsstrahlung and fluorescence. More detail can be found in a useful compilation of mass absorption and attenuation coefficients by Hubell (1982), referred to at the end of this chapter.

2.4 INTERACTIONS WITHIN THE DETECTOR

We have seen how all significant interactions between gamma-rays and detector materials result in the transfer of energy from the gamma-ray to electrons, or, in the case of pair production, to an electron and a positron. The energy of these individual particles can range from near zero energy to near to the full energy of the gamma-ray. In gamma-ray spectrometry terms, energies may be from a few keV to several MeV. If we compare these energies with the energy needed to create an ion pair in germanium – 2.96 eV – it is obvious that the energetic primary electrons must create electron–hole pairs as they scatter around within the bulk of the detector. We can calculate the expected number of ion pairs created by one such energetic electron as follows:

$$N = E_e/\varepsilon \quad (2.15)$$

where E_e is the electron energy and ε the energy needed to create the ion pair. Multiplying this number by the charge on an electron would give an estimate of the charge created within the detector. It is these secondary electrons and their associated positively charged holes which must be collected in order to produce the electrical signal from the detector. This aspect of the detection process will be followed up in Chapter 3.

The detail of the manner in which gamma-rays interact with matter determines the size of the detector signal for each particular gamma-ray. This will depend upon the energy of the photon, the atomic number of the absorber atom and, for Compton events, the angle between the incident gamma radiation and the scattered gamma-ray. Bearing in mind that, in most cases, a single interaction will not completely absorb the gamma-ray, we might expect that the location of the interaction within the detector might be important (X-ray escapes) and that the size of the detector might be a consideration.

2.4.1 The very large detector

I shall define the 'very large detector' as one so large that we can ignore the fact that the detector has a surface. Consider bombarding this detector with a large number of gamma-rays of exactly the same energy (greater than 1022 keV so that we can take into account pair production events). Because the detector is large, then we can expect that every gamma-ray will have an opportunity to interact by one or other of the three processes we have already discussed. Figure 2.8 shows representative interaction histories for gamma-rays interacting by each of the processes described above.

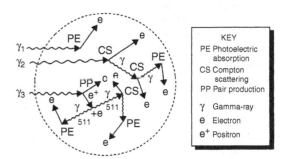

Figure 2.8 Examples of interaction histories within a very large detector

If the interaction happens to be by the photoelectric effect, the result will be complete absorption with the release of photoelectrons and Auger electrons sharing between them the total gamma-ray energy. Each and every gamma-ray interacting in this manner will deliver up its whole energy and, since the gamma-rays are identical, will produce an identical detector response.

The Compton scattering history in Figure 2.8 shows an initial interaction, releasing a recoil electron, followed by further Compton interactions of the scattered gamma-ray releasing more recoil electrons. After each successive scattering, the scattered gamma-ray carries less energy. Eventually, that energy will be so low that photoelectric absorption will be inevitable and the remaining gamma-ray energy will be transferred to photoelectrons. Thus, the total energy of the gamma-ray is shared between a number of recoil electrons and photoelectrons. The time scale for these interactions is much shorter than the charge collection time of any practical detector and to all intents and purposes all of the primary electrons are released at one instant. From event to event, the actual number of Compton events taking place before the final photoelectric event will vary but in every case the total gamma-ray energy will be transferred to primary energetic electrons within the detector. Again we can expect a constant detector response to all gamma-rays of the same energy.

Similarly, the pair production interaction history demonstrates that all of the energy of the gamma-ray can be transferred to the detector. In this case, the total gamma-ray energy is, in the first place, shared equally between the electron and the positron created by the interaction. Both the electron and the positron will lose energy, creating electron–hole pairs in the process. When the positron reaches thermal, or near thermal energies, it will be annihilated by combination with an electron, releasing the two annihilation photons of 511 keV. Figure 2.8 shows these being absorbed by a combination of Compton scattering and photoelectric absorption in the normal way. Ultimately, by a combination of the initial pair production, eventual annihilation of the positron and absorption of the annihilation photons, the complete gamma-ray energy is absorbed. Again, although individual interaction histories will differ from gamma-ray to gamma-ray, the detector response to identical gamma-rays will be the same.

Since for each identical gamma-ray we now expect the same detector response, irrespective of the initial mode of interaction, we would expect that the gamma-ray spectrum from such a detector would consist of single peaks, each corresponding to an individual gamma-ray energy emitted by the source. In some quarters, the peaks in gamma-ray spectra are referred to as 'photopeaks', with the implication that such peaks arise only as a consequence of photoelectric events. As we have seen, the events resulting in total absorption can also involve Compton scattering and pair production and the term **full energy peak** to describe the resulting peaks in a spectrum is to be preferred and will be used here.

2.4.2 The very small detector

If we go to the opposite size extreme and consider the same interactions in a very small detector – defined as one so small that only one interaction can take place within it – a different picture emerges (Figure 2.9). While the very large detector referred to above is entirely hypothetical, the very small detector now being discussed is not too different from the small planar detectors manufactured for the measurement of low-energy gamma and X-radiation and the necessarily small room-temperature semiconductor detectors that will be discussed in Chapter 3 Section 3.2.5. Again, we can consider various interaction histories for the three modes of interaction.

Now, only photoelectric interactions will produce full energy absorption and contribute to the full energy peak. Because of the small size of the detector, all Compton scattering events will produce only a single recoil electron carrying a portion of the gamma-ray energy. The scattered gamma-ray will inevitably escape from the detector, taking with it the remaining gamma-ray energy. The detector response to Compton interactions will, therefore, mirror the

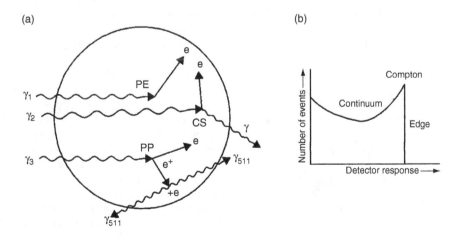

Figure 2.9   (a) Interaction histories within a very small detector, and (b) the detector response from Compton interactions

curve shown in Figure 2.6 and the corresponding gamma-ray spectrum would exhibit the characteristic **Compton continuum** extending from zero energy up to the **Compton edge**, illustrated in Figure 2.9(b). There would be no Compton scattering contribution to the full energy peak.

The energy absorbed by pair production events would be limited to the energy in excess of the electron–positron rest masses. It can be assumed that both electron and positron will pass on their kinetic energy to the detector but in this notional small detector the loss of energy absorbed caused by the escape of the annihilation gamma-rays will give rise to the so-called **double escape peak**. ('Double' because both 511 keV annihilation photons escape from the detector.) This peak, 1022 keV below the position of full energy absorption, would be the only feature in the spectrum attributable to pair production (see Figure 2.11(a) below). Well-defined double escape peaks tend to be slightly asymmetric towards high energy.

With a small detector, the higher surface-to-volume ratio means that the probability of a photoelectric absorption near to the detector surface is much greater than in a larger detector with a corresponding increase in the probability of X-ray escape. In small germanium detectors, we can expect, therefore, to find germanium escape peaks 9.88 keV (the energy of the K_α X-ray) below each full energy peak in the spectrum.

2.4.3 The 'real' detector

Of course, any 'real' detector represents a case somewhere between the above two extremes. We can expect that some Compton scattering events and even some pair production events might be followed by complete absorption of the residual gamma-ray energy or, more likely, by a greater partial absorption. There are other specific possibilities (Figure 2.10) which give rise to identifiable features in

a gamma-ray spectrum. Compton scattering events may be followed by one or more further Compton interactions, each absorbing a little more of the gamma-ray energy, before the scattered gamma-ray escapes from the detector.

If we imagine that this multiple scattering follows an initial event that would have produced a response near to the Compton edge, then we can appreciate that the extra energy absorbed could, in some cases, result in events that would appear in the spectrum between the Compton edge and the full energy peak. These are referred to as **multiple Compton events**.

If the gamma-ray energy is greater than the 1022 keV threshold, a further feature in the gamma-ray spectra may be seen due to pair production. If, after annihilation of the positron, only one of the annihilation photons escapes while the other is completely absorbed, precisely 511 keV will be lost from the detector. This will result in a separate peak in the spectrum representing $E_\gamma - 511$ keV, called the **single escape peak**. Of course, both photons may be partially absorbed, giving rise to counts elsewhere in the spectrum with no particular spectral signature. Single escape peaks have their own Compton edge, 170 keV below the single escape energy.

2.4.4 Summary

Figure 2.11 shows the gamma-ray spectra expected from the three detectors discussed. It is obvious that the bigger the detector, the more 'room' there is for the gamma-rays to scatter around in and transfer a bigger proportion of their energy to the detector and the hence the larger the full energy peaks. These conceptual spectra may be compared to the actual gamma-ray spectra of ^{137}Cs and ^{28}Al measured using an 18 % Ge(Li) detector in Figure 2.12. All of the features mentioned above can be clearly seen.

To summarize, an ideal 'very large detector' response would contain only full energy peaks corresponding to the energies of the gamma-rays emitted by the source. In a 'real' detector, other features appear in the spectrum as a consequence of incomplete absorption of the gamma-ray energy. In some circumstances, the loss of precise amounts of energy results in peaks (single and double escape peaks and X-ray escape peaks) or, when random losses occur, in a *continuum*. The degree of incomplete absorption depends upon the physical size of the detector and the energy of the gamma-ray. The larger the detector, the more 'room' there is to accommodate multiple scattering, and the lower the gamma-ray energy the greater the probability of complete absorption by the photoelectric

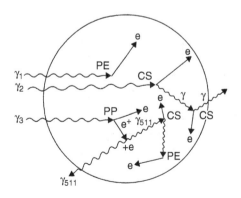

Figure 2.10 Additional possibilities for interaction within a 'real' detector

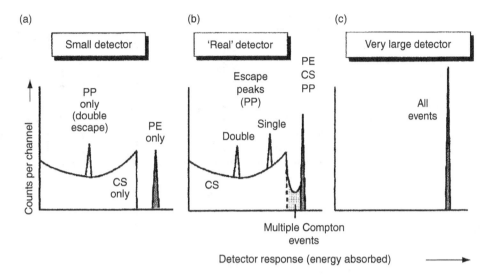

Figure 2.11 Spectra expected from detectors of different sizes. The larger the detector, the higher the proportion of events resulting in complete absorption: PE, photoelectric effect; CS, Compton scattering; PP, pair production

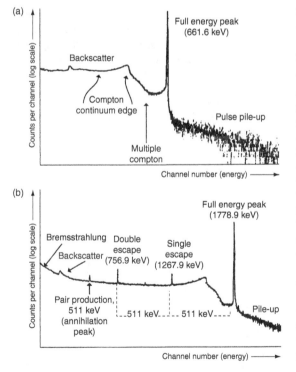

Figure 2.12 Example spectra illustrating the various spectral features expected: (a) ^{137}Cs; (b) ^{28}Al

effect. Gamma-ray detector manufacturers use the **peak-to-Compton** ratio of a detector as a figure of merit. This

is discussed, along with other parameters, in a detector specification in Chapter 11.

2.5 INTERACTIONS WITHIN THE SHIELDING

Within the spectra shown above, certain other features are referred to which are not a consequence of gamma-ray interactions within the detector itself. These are artifacts. In the first place, we would not expect counts to appear in the spectrum above the full energy peak (apart from the natural background). These arise from summing of the energy of more than one gamma-ray arriving at the detector simultaneously. The continuum above the full energy peaks in Figure 2.12 is due to **random summing** (sometimes referred to as **pile-up**), determined by the statistical probability of two gamma-rays being detected at the same time and therefore on the sample count rate.

Another type of summing, referred to as **true coincidence summing**, is a function of the nuclide decay scheme and the source/detector geometry and will be dealt with in some detail in Chapter 8. All of the other features in the spectrum can be attributed to unavoidable interactions of gamma-rays from the source with the surroundings of the detector – the shielding, cryostat, detector cap, source mount, etc.

2.5.1 Photoelectric interactions

The most troublesome photoelectric interactions will be those with the shielding, usually lead. As shown in

Figure 2.4, a photoelectric absorption can be followed by the emission of a characteristic X-ray of the absorbing medium. There is a significant possibility that this fluorescent X-ray may escape the shielding and that it will be detected by the detector, as indicated in Figure 2.13. The result will be a number of X-ray peaks in the gamma spectrum in the region 70–85 keV. This may or may not be a problem in practice, depending upon the type of gamma-ray spectrum measured. However, if low-energy gamma-ray measurements are contemplated, fluorescent X-rays are an unnecessary complication.

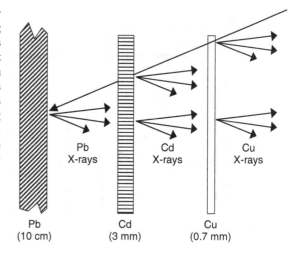

Figure 2.14　The composition of a graded shield

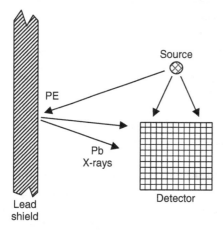

Figure 2.13　Photoelectric interactions with the shielding producing lead X-rays

2.5.2 Compton scattering

The normal geometric arrangement of source–detector-shielding (Figure 2.15(a)) means that most gamma-rays are scattered through a large angle by the shielding. They are, in fact, backscattered. Examination of the relationship between the energy of the scattered gamma-rays and scattering angle (Figure 2.15(b)) reveals that, whatever the initial energy the energies of backscattered gamma-rays (say, all those scattered through more than 120°) are within the broad range 200–300 keV. The result is that backscattered radiation appears as a broad, ill-shaped peak in the spectrum. There is little that can be done about the backscatter peak, although a larger shield may help.

Fortunately, there is a ready solution to the problem in the form of a **graded shield** (Figure 2.14). The lead shielding is covered by a layer of cadmium to absorb the lead X-rays. This will result instead in the production of cadmium fluorescent X-rays that can in turn be absorbed by a layer of copper. In most circumstances, copper fluorescent X-rays of 8–9 keV are too low in energy to be a problem, but a layer of plastic laminate would absorb these and provide a convenient 'wipe-clean' surface. An alternative partial solution would be to construct a larger volume shield to move the lead further away from the source–detector arrangement and thus reduce both the intensity of gamma radiation reaching the lead and the X-radiation reaching the detector. This, however, would entail considerable extra cost in lead. The graded shield is a more cost effective solution. Suitable thicknesses of material are discussed later in Section 2.8.

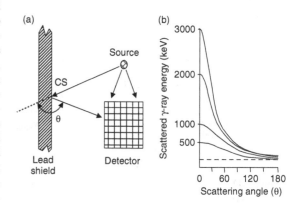

Figure 2.15　(a) Backscatter of radiation from the shielding, and (b) energy of backscattered gamma-rays as a function of scattering angle

2.5.3 Pair production

The consequences of pair production in the surroundings to the detector give rise to what is often referred to as the annihilation peak at 511 keV in the spectrum. This is caused by the escape of one of the 511 keV photons from the shielding, following annihilation of the pair production positron (Figure 2.16). This is analogous to the single and double escape mechanisms within the detector but, of course, only one of the 511 keV photons can ever be detected because they are emitted in opposite directions. The annihilation peak is clearly visible in the spectrum of ^{28}Al (Figure 2.12(b)) but not in that of ^{137}Cs (Figure 2.12(a)) because the latter does not emit gamma-rays greater than the 1022 keV pair production threshold.

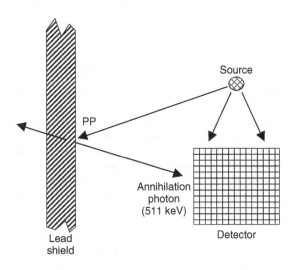

Figure 2.16 Annihilation radiation reaching the detector as a consequence of pair production in the shielding

When interpreting spectra, it is worth remembering that a 511 keV photon can also be expected whenever a radionuclide emits positrons as part of its decay process. Common examples of such nuclides are ^{22}Na, ^{65}Zn and ^{64}Cu. The interpretation of the presence of a 511 keV peak is not, therefore, as obvious as it might appear. There are three possible explanations, which are not mutually exclusive:

- positron decay of a radionuclide;
- pair production in the shielding by high-energy gamma-rays from the source;
- pair production in the shielding by high-energy cosmic rays.

It is not wise, therefore, automatically to dismiss the presence of an annihilation peak without considering its source.

2.6 BREMSSTRAHLUNG

The one remaining unexplained feature of Figure 2.12 is the bremsstrahlung continuum. As described in Chapter 1 Section 1.7.2, any source emitting β particles will have a bremsstrahlung spectrum superimposed on the gamma-ray spectrum. In practice, this is only significant if the β-particle energy is much greater than 1 MeV. (In the case of ^{28}Al, this energy is 2.8 MeV.) The presence of this radiation causes a considerable increase in peak background at low energy and reduces the precision of measurement (Figure 2.17.)

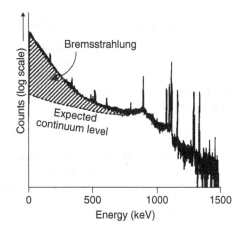

Figure 2.17 Bremsstrahlung due to ^{32}P beta particles (1.711 MeV) in an irradiated biological sample

Bremsstrahlung cannot be avoided completely. The β particles emitted by the source must be absorbed somewhere; all one can do is to arrange matters so that they are absorbed close to the source rather than close to the detector. A simple measure, as long as there is a reasonable distance between source and detector, is then to use an absorber near to the source and rely upon the inverse square law to reduce the bremsstrahlung intensity at the detector. There will, of course, be some absorption of lower-energy gamma-rays but under most circumstances the benefits should outweigh the losses. This benefit will be all the greater if a low Z material is used to absorb the beta particles, minimizing the absorption of gamma- and X-rays. The use of a 6.5 mm thick $(1.2\,\mathrm{g\,cm^{-2}})$ beryllium absorber for this purpose has been demonstrated by Gehrke and Davidson (2005). (Note that the beryllium window of an n-type HPGe detector is only $0.09\,\mathrm{g\,cm^{-2}}$ and has little effect in limiting bremsstrahlung.)

If, for reasons of sensitivity, it is essential to count sources close to the detector and absorbers offer little relief, then there is little to be done except, perhaps, the under-used last resort – radiochemical separation of the nuclide of interest. Systems using magnets to divert the β particle away from the detector have been demonstrated but appear to offer only small improvements in precision and again demand a substantial source–detector distance to be effective.

(The reader may recall that the gamma-ray interactions produce fast electrons which scatter within the detector. As they decelerate, a proportion of their energy will be emitted as bremsstrahlung rather than used to create electron–hole pairs. We need not worry about this as it is already taken into account when the various absorption and attenuation coefficients are calculated.)

2.7 ATTENUATION OF GAMMA RADIATION

Equations (2.11) and (2.12) defined the total attenuation coefficient for gamma radiation passing through matter. Using this coefficient, we can calculate the degree of attenuation of a narrow beam of gamma radiation by using the following simple equation:

$$I = I_0 \, e^{-\mu t} \tag{2.16}$$

where t is the thickness of the absorber in units consistent with the units of μ (i.e. cm if the attenuation coefficient, μ, is in $cm^2 \, g^{-1}$ and m if in $m^2 \, kg^{-1}$). This equation relates the intensity of gamma-rays at a specified energy after attenuation, I, to that without attenuation at the same energy, I_0. This relationship is only valid under 'good geometry' conditions (see Figure 2.18(a)) with a thin absorber and a collimated gamma-ray source. Under the open conditions indicated in Figure 2.18(b) the equation fails because of scattering from the absorber. Gamma-rays which, on the

basis of the geometrical arrangement of source, absorber and detector, might be expected to miss and may be Compton scattered back into the detector, hence increasing the true gamma-ray intensity.

When gamma-rays above the pair production threshold energy are considered, there may also be an annihilation gamma radiation contribution to the dose rate beyond the absorber.

This phenomenon is referred to as *build-up* and can be accounted for by a correction to Equation (2.16):

$$I = I_0 \, e^{-\mu t} \times B \tag{2.17}$$

The build-up factor, B, is the ratio of the total photons at a point to the number arriving there without being scattered. There are a number of empirical equations in use for estimation of the build-up factor, references to which are given later.

2.8 THE DESIGN OF DETECTOR SHIELDING

The purpose of detector shielding is to reduce the amount of radiation from background sources reaching the detector. This background derives from radioactive nuclides within the environment, [40]K in natural potassium and the uranium decay chain nuclides, for example, and to a certain extent, cosmic radiation, which will be discussd in Chapter 13, Section 13.4.6. Let us take as a principal aim reduction of this external radiation by a factor of 1000. If we assume the energy of the gamma radiation is 1 MeV for convenience, then, ignoring build-up and using Equation (2.16), we can estimate the thickness of shielding needed to produce this degree of attenuation. Table 2.1 lists the calculated thicknesses of materials needed to attenuate gamma radiation of various energies.

We can see that a greater thickness of iron or copper would be needed to provide the same degree of shielding as lead. Even so, on the grounds of cost, iron might be regarded as a better choice. Unfortunately, modern iron is often contaminated with ^{60}Co and, unless aged iron is available, is not normally the first choice. Adequate reduction in the external gamma radiation intensity is not the only criterion that must be considered. As the atomic number of an absorber increases, the importance of Compton scattering as the primary interaction decreases relative to photoelectric absorption and pair production. If a shield is made of lead rather than iron, fewer gamma-rays will be Compton scattered as opposed to absorbed. That, in turn, means that there will be fewer scattered gamma-rays to penetrate the shielding from outside and, perhaps more importantly, fewer backscattered gamma-rays from within the shield. Conventionally, detector shielding is constructed of

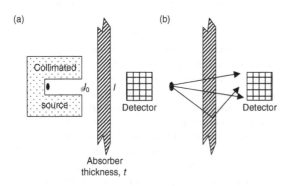

Figure 2.18 (a) Attenuation of a beam of gamma-rays under 'good geometry' conditions, and (b) Build-up under open geometry conditions

Table 2.1 Attenuation of gamma radiation by shielding materials

Photon energy (keV)	Absorber element	Mass attenuation coefficient ($m^2 kg^{-1}$)	Density ($kg\,m^{-3}$)	Thickness for 1000-fold attenuation (mm)
1000	Fe	0.005994	7860	147
	Cu	0.005900	8920	131
	Pb	0.007103	11350	86
80 (Pb X-rays)	Cu	0.07587	8920	10.2
	Cd	0.2736	8650	2.9
	Sn	0.3013	7280	3.1
30 (Cd X-rays)	Cu	1.083	8920	0.7

100-mm thicknesses of lead. Although a greater thickness of lead would provide a greater reduction in background peak heights, the greater mass of lead available for interaction with cosmic rays would lead to an increase in the overall background continuum level – 100 to 150 mm is regarded as optimum.

As I noted above, photoelectric absorption of gamma-rays from the source by lead shielding can result in significant and potentially troublesome lead fluorescent X-ray peaks in the gamma-ray spectrum. The general advice would be to make sure that the shielding is at least 10 cm away from the detector in order to limit fluorescence. Fluorescent X-rays can easily be absorbed by a layer of a lighter element mounted on the inside of the shield. As Table 2.1 indicates, 10 mm of copper would be needed to reduce the intensity by a factor of 1000 but only 3 mm of cadmium or tin. If cadmium is used in preference to copper, as is usual, then it would be desirable to remove the fluorescent cadmium X-rays generated in the lining itself. An outer layer of only 0.7 mm of copper will achieve this. In practice, bearing in mind the high cost of cadmium and tin relative to that of copper and the fact that the copper layer will itself contribute to absorption of the lead X-rays, a compromise is usually adopted. For example, commercial systems offer graded shields comprising only 0.5 mm or 1 mm of cadmium but 1–2 mm of copper. (In the former case, there is also a not unreasonable compromise on the degree of attenuation of the lead X-rays.) From a practical point of view, the mechanical properties of cadmium make it preferable to tin in that a cylinder of 1 mm cadmium is self supporting, whereas the same thickness of tin is soft and tends to collapse. The cost is, however, considerably higher. ANSI N42 (1991) suggests 2 mm of tin and 0.5 mm of copper or, if tin is not to be used, 1 cm of copper. For low-background spectrometry, the latter would not be appropriate because low Z materials near the detector cause an increase in scattered radiation that raises the continuum at low energy. The matter is discussed further in Chapter 13, Section 13.4.4

You should be aware that cadmium has a high cross-section for the absorption of thermal neutrons. During this absorption, or thermal neutron capture reaction, gamma radiation is emitted, the most noticeable of which is at an energy of 558 keV. If a detector system is to be used in a neutron field, then cadmium in the graded shield should be avoided. Fortunately, tin has a much lower thermal neutron cross-section and can be used instead. Difficulties due to neutron capture might be expected, and indeed have been observed, when operating a detector close to a nuclear reactor but the problem can also occur in environmental measurements (see Chapter 13, Section 13.3.4.2.)

Its high cost makes the re-use of cadmium an attractive proposition. However, care should be taken when contemplating the re-use of cadmium that has been in use on nuclear sites. Cadmium is usually present on such sites to provide neutron shielding. The neutron capture reactions referred to above result in the activation of the various cadmium isotopes. Although the cadmium may be regarded as inactive from a health physics and disposal point of view it may be that in a low background system, [109]Cd, with a half-life of 453 days emitting a gamma-ray at 88.03 keV, is noticeable.

Cadmium is a toxic metal and should not be handled without actively considering the hazards involved. Under no circumstances should cadmium be flame-cut or soldered unless proper ventilation is provided. Once the cadmium has been formed to the correct shape for the shield, coating it with a varnish will reduce the handling hazard.

Apart from commenting on the possible presence of ^{60}Co in steel we have said nothing about the contribution to detector background from impurities in the shielding and construction materials. These are of particular importance in systems designed for low-activity counting and will be discussed in detail in Chapter 13.

PRACTICAL POINTS

Gamma-rays interact with matter by elastic (Rayleigh) scattering and inelastic processes. Only the latter contribute to absorption of energy. Both contribute to attenuation of gamma intensity. Energy is transferred to the material via energetic electrons or positrons. The kinetic energy of these particles is dissipated by creating secondary ion pairs which provide a basis for the detector signal.

The inelastic interactions are as follows:

- Photoelectric absorption – total absorption of the gamma-ray energy, possibly followed by escape of fluorescent radiation (X-ray escape peaks).
- Compton scattering – partial absorption giving rise to the Compton edge and Compton continuum.
- Pair production – total absorption, followed by possible partial or complete loss of annihilation quanta. Complete loss produces the single and double escape peak. Counts appearing between the Compton edge and the full energy peak are due to multiple interactions of whatever type.

Gamma-ray interactions with the detector surroundings produce features which can be assigned as follows:

- Fluorescent X-rays (usually lead) – photoelectric absorption and emission of fluorescent radiation.
- Backscatter peak – Compton scattering through a large angle, giving rise to a broad distribution at about 200 keV.
- Annihilation peak (511 keV) – pair production within the detector surroundings, followed by escape of one of the annihilation gamma-rays in the direction of the detector. Be aware that many neutron-deficient nuclides may emit positrons, the annihilation of which will also give rise to counts in the annihilation peak.

The larger the detector, the greater the probability of complete absorption of the gamma-ray and hence a larger full energy peak and lower Compton continuum (i.e. higher peak-to-Compton ratio).

Sources emitting high-energy beta-particles are likely to give rise to a bremsstrahlung continuum at low energy.

Attenuation of collimated beams of gamma radiation follows a simple exponential relationship involving the attenuation coefficient, μ:

$$I = I_0 \, e^{-\mu t}$$

Under open geometry conditions, a build-up factor must be included, as in Equation (2.17).

Optimum shielding for typical gamma spectrometry applications needs no more than 100 mm of lead, 3 mm of cadmium or tin and 0.7 mm of copper.

FURTHER READING

- Very good general discussions of interactions of gamma radiation are given in the following textbooks on measurement of radiation in general:

Knoll, G. F. (2000). *Radiation Detector and Measurements*, 3rd Edn, John Wiley & Sons, Inc., New York, NY, USA.

Tsoulfanidis, N. (1995). *Measurement and Detection of Radiation*, McGraw-Hill, New York, NY, USA.

- The following book covers gamma spectrometry specifically:

Debertin, K. and Helmer, R. G. (1988). *Gamma and X-Ray Spectrometry with Semiconductor Detectors*, North-Holland, Amsterdam, The Netherlands.

- An underestimated source of background information is the manufacturers' literature. The introductory sections of the Canberra, PGT and ORTEC catalogues are good, and their various *Applications Notes* are worth acquiring. These are available for downloading from the Internet.

- A good compilation of mass attenuation and absorption coefficients is:

Hubell, J. H. (1982). Photon mass attenuation and energy-absorption coefficients from 1keV to 20 MeV, *Int. J. Appl. Radiat. Isotopes*, **33**, 1269–1290.

An excellent Internet source of attenuation and absorption data is: *http://physics.nist.gov/PhysRefData/XrayMassCoef/cover.ihtml* (Table 3 is particularly useful).

- Useful summaries of the factors to be taken into account when setting up a detector system to achieve the best quality spectra are:

ANSI (1991). *Calibration and Use of Germanium Spectrometers for the Measurement of Gamma-ray Emission Rates of Radionuclides*, ANSI/N42.14-1991, IEEE, New York, NY, USA.

Gehrke, R.J. and Davidson, J.R., (2005). Acquisition of quality γ-ray spectra with HPGe spectrometers, *Appl. Radiat. Isotopes*, **62**, 479–499.

3

Semiconductor Detectors for Gamma-Ray Spectrometry

3.1 INTRODUCTION

In Chapter 2, the manner in which gamma radiation interacts with matter was explained in terms of various mechanisms, each of which transfers energy from the gamma-ray to electrons and, in the case of pair production, positrons. These particles lose their kinetic energy by scattering around within the detector, creating ionized atoms and ion pairs. This population of secondary entities forms the basis of the detector signal. In this chapter, I shall discuss the properties of materials suitable for constructing detectors for use in gamma-ray spectrometry, the principles behind their operation and the manufacture of detectors. In this chapter, I will also concentrate on the most common types of detector; other, more specialized, detectors will be mentioned from time to time elsewhere.

Perhaps, before continuing, I should consider the properties of an ideal detector for gamma spectrometry. The requirements, which would be less demanding if mere detection rather than spectrometry were the aim, can be summarized as follows:

- output proportional to gamma-ray energy;
- good efficiency, i.e. high absorption coefficient, high Z;
- easy mechanism for collecting the detector signal;
- good energy resolution;
- good stability over time, temperature and operating parameters;
- reasonable cost;
- reasonable size.

First of all, it is very desirable that the detector should have a high enough absorption coefficient for gamma radiation so that there is a reasonable probability of complete absorption. As I showed in the last chapter, in principle this can always be ensured if the detector is made large enough. In practice, the material chosen must provide complete absorption within a detector of achievable size. This consideration alone rules out gas detectors. (Health physics instrumentation often uses Geiger–Müller detectors to measure both beta- and gamma-radiation. However, in the case of gamma radiation, detection depends upon interactions with the body of the detector rather than the gas.) Bearing in mind that the absorption coefficients for all the significant interaction processes increase with atomic number, we would seek a high atomic number material.

Having absorbed the gamma-ray and created many charged species (electron–hole pairs), the detector material must allow the charge to be collected in some manner and presented as an electrical signal. The most obvious way to do this is to supply an electric field across the detector material to 'sweep' the charge carriers out of the detector. This, of course, can only be done if the detector has suitable electrical characteristics. As we will see shortly, this is the basis of the semiconductor type of detector. The other widely used type of detector, the scintillation detector, depends upon transfer of the energy of the charged secondary species to atoms that de-excite by emitting light which is then collected by a photomultiplier. Scintillation detectors will be discussed in detail in Chapter 10. If the charged species are to be collected by an electric field, then a mechanism must be available to allow the charged species to migrate in the field in a controlled manner.

3.2 SEMICONDUCTORS AND GAMMA-RAY DETECTION

3.2.1 The band structure of solids

In a free atom, the electrons are disposed in precisely determined energy levels. Combining a collection of atoms together into a solid structure broadens those energy levels into energy bands, each of which can contain a fixed number of electrons. Between these bands are energy regions that are forbidden to electrons. The uppermost occupied energy band, the inhabitants of which are responsible for chemical reactions, is known as the **valence band**. In order for an electron to migrate within the material, it must be able to move out of its current energy state into another in order to move from atom to atom. (This is illustrated schematically in Figure 3.1.) If electrons can jump into suitable energy levels, then an external electric field applied to the material would cause a current to flow. There are three types of material: **insulators**, **conductors** and **semiconductors**. These differ in their electronic structures.

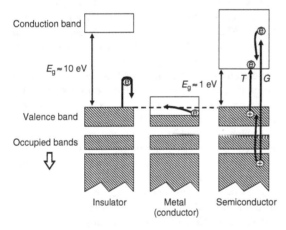

Figure 3.1 Schematics of the electronic band structures in insulators, metals and semiconductors

In an insulator, the valence band is full and the next available energy states are in a higher band, called the **conduction band**, separated by a forbidden region. For an electron to migrate through the material, it must gain sufficient energy to jump from the valence band across the band gap, E_g, into the conduction band. In an insulator, this gap is of the order of 10 eV, much greater than can be surmounted by thermal excitation. The electrons are immobile and the material is unable to pass an electrical current, however great the electric field (short of electrical breakdown). In a metal, the valence bands are not full and in effect the conduction band is continuous with the valence

band. Thermal excitation ensures that the conduction band is always populated to some extent and the imposition of an electric field, however small, will cause a current to flow. From the point of view of constructing a practical gamma-ray detector, this is of little use because the extra current caused by the gamma-ray interaction would be insignificant compared to the normal background current.

The band structure of semiconductors is not dissimilar to that of insulators. The valence bands are full but the band gap is much smaller, of the order of 1 eV, similar to the energies achievable by thermal excitation. Under normal conditions there will always be a small population of electrons in the conduction band and the material will exhibit a limited degree of conductivity. The probability that an electron will be promoted to the conduction band is strongly influenced by temperature (T in Equation (3.1); k is the Boltzmann constant):

$$p(T) \propto T^{3/2} \exp\left(-E_g/2kT\right) \tag{3.1}$$

Cooling the material will reduce the number of electrons in the conduction band, thereby reducing the background current (in detector terms, the leakage current) and make it much easier to detect the extra excitation due to the gamma-ray interactions. This is the basis of the semiconductor gamma-ray detector.

3.2.2 Mobility of holes

When an electron is promoted from the valence, or any other, band to the conduction band, a vacancy is left behind in the otherwise full band. This vacancy is effectively positively charged and is referred to as a **hole**. Holes are also mobile. An electron within the valence band may replace that lost from the vacancy, thus filling the hole. That will leave, in turn, another vacancy. In the presence of an external electric field, the hole can appear to move towards the cathode (see Figure 3.2). Since both electrons and holes carry charge, both will contribute to the conductivity of the material.

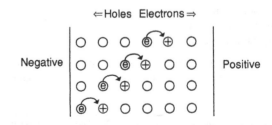

Figure 3.2 A model for hole mobility in solids

3.2.3 Creation of charge carriers by gamma radiation

The interaction of a gamma-ray with the semiconductor material will produce primary electrons with energies considerably greater than thermal energies. Interaction of these can raise electrons from deep occupied bands well below the valence band into energy levels well above the base of the conduction band. These deeply embedded holes and the excited electrons will tend to redistribute themselves within the available energy bands until the holes lie at the top of the valence band and the electrons at the base of the conduction band indicated on the right-hand side of Figure 3.1). In this process, further excitation can occur, giving a cascade of electron–hole pairs for each primary electron interaction. Under normal circumstances, the extra excited electrons in the conduction band might be expected to eventually de-excite and return to the valence band, restoring the conduction band population to that expected from thermal excitation alone. In the presence of an electric field, they will instead migrate up (electrons) or down (holes) the field gradient. The number of electron–hole pairs produced, n, will be related directly to the gamma-ray energy absorbed, E_{abs}, i.e. if ε is the average energy needed to create an electron–hole pair:

$$n = E_{abs}/\varepsilon \qquad (3.2)$$

One important component of the detector resolution is a function of n (this is discussed in Chapter 6) and, other things being equal, one would choose the detector material with a low ε so as to maximize n.

Although semiconductor materials provide a ready means by which the electron–hole pairs can be collected, in a practical detector this must be accomplished within a reasonably short time. The electrons and holes must have good mobility within the material and there must be no traps that might prevent them reaching the collecting contacts. Trapping centres can be of different types and are a consequence of:

- impurities within the semiconductor lattice;
- interstitial atoms and vacancies within the lattice caused by structural defects within the crystal;
- interstitial atoms caused by radiation damage.

In practical terms, this means that the detector material must be available, at reasonable cost, with a high purity and as near perfect as possible crystalline state. Mobility and trapping will be discussed in more detail later.

3.2.4 Suitable semiconductors for gamma-ray detectors

To summarize the previous sections we can say that the ideal semiconductor detector material will:

- have as large an absorption coefficient as possible (i.e. high atomic number);
- provide as many electron–hole pairs as possible per unit energy (i.e. low ε);
- allow good electron and hole mobility;
- be available in high purity as near perfect single crystals;
- be available in reasonable amounts at reasonable cost.

Taking all these items into account leaves only a few possible candidates, selected data for which are shown in Table 3.1. The most obvious candidate is silicon that,

Table 3.1 Parameters for some materials suitable for gamma-ray detectors

Material	Atomic number	Operating temperature	Band gap (eV)[a]	ε (eV)[a,b]	Density (g cm^{-3})	Mobility(cm^2 V^{-1} s^{-1})[a]	
						Electrons	Holes
Si	14	RT	1.106	3.62	2.33	1350	480
Ge	32	Liquid N_2 (77 K)	0.67	2.96	5.32	3.6×10^4	4.2×10^4
CdTe	48, 52	RT	1.47	4.43	6.06	1000	80
CdZnTe	48, 30, 52	RT	1.57	4.64	5.78	1000	50–80
HgI$_2$	80, 53	RT	2.13	4.22	6.30	100	4
GaAs	31, 33	RT	1.45	4.51	5.35	8000	400
TlBr	81, 35	$-20\,^{\circ}$C	2.68	?	7.56	—	—
PbI$_2$	82, 53	—	2.6	7.68	6.16	8	2
GaSe	31, 34	—	2.03	6.3	4.55	—	—
AlSb	13, 51	—	1.62	5.05	4.26	—	—
CdSe	48, 34	—	1.75	?	5.74	—	—

[a] Values are given at 77 K for Ge and 300 K otherwise.
[b] Electron–hole creation energy.

because of the efforts put into the preparation of high purity material for the electronics industry, is readily available at reasonable cost. Its only disadvantage is its low atomic number, which means that in practice it is only used for the measurement of low-energy photons. Detectors based upon silicon are in routine use in X-ray spectrometry. This is a field in its own right, with its own specific problems, and I will not consider silicon detectors further in this work.

Germanium is by far the most common gamma-ray detector material. Its higher atomic number than silicon makes it practicable to use it for the detection of higher-energy gamma radiation. Over recent years, in response to the demand for this type of detector, the technology for the manufacture of high-purity germanium with a suitable degree of crystal perfection has improved considerably. This is emphasized by comparing the first germanium detectors commercially available (a few cubic centimetres in size, using lithium drifting to compensate for inadequate purity with a resolution of 4 to 5 keV at 1332 keV) with modern state-of-the-art detectors (hundreds of cubic centimetres of hyper-pure germanium offering better than 1.8 keV resolution). Germanium is unique in Table 3.1 in that detectors made from it must be operated at low temperature in order to reduce the leakage current sufficiently, as explained earlier, by virtue of Equation (3.1).

3.2.5 Newer semiconductor materials

The other potential semiconductor detector materials have a larger band gap than germanium and consequently would have the advantage of room temperature operation assuming that their other properties were satisfactory. Of these, only cadmium telluride, cadmium zinc telluride (CZT) and mercuric iodide have found their way into commercial production but only then for limited applications. The higher atomic numbers of these materials, and hence larger absorption coefficients, make them attractive detector materials. For example, 2 mm of cadmium telluride is equivalent to 10 mm of germanium in terms of gamma-ray absorption. However, in practice, a number of factors limit their use. In the first place is the availability of material with a satisfactory crystalline perfection, but restrictions that are more fundamental arise from the mobility of the charge carriers.

As Table 3.1 shows, the charge carrier mobilities in these materials are considerably lower than those for germanium. This, compounded by the fact that the mobility of the holes, which are susceptible to trapping, is much lower than that for electrons, imposes severe charge collection problems on the detectors. In fact, trapping of the holes is enhanced to such an extent that complete charge collection is very difficult to achieve over distances of more than 1 mm. This means that only small detectors can be made from these materials. Because of their small size they are best used for measuring low-energy gamma-rays. The performance of cadmium telluride and cadmium zinc telluride detectors are enhanced by novel electronic means, which compensate for the poor charge collection. These will be discussed in Chapter 4, Section 4.3.5, after I have explained the electronics of charge collection.

The energy needed to create each charge carrier in these newer materials is also somewhat higher than in germanium and, given all other things being equal, cadmium telluride, CZT and mercuric iodide can never achieve the same resolution. Table 3.2 compares the resolution and energy range for highest efficiency for a number of detectors that might be used for low-energy measurements. This demonstrates that the resolution of these

Table 3.2   Comparison of resolution and energy range of various low-energy gamma-ray detectors

Material	Size of detector	Resolution (keV) at:			Optimum energy range (keV)[a]
		5.9 keV	122 keV	661.66 keV	
Ge	50 mm^2 × 5 mm planar	0.145	0.5	—	2–100
CdTe	25 mm^2 × 2 mm (RT)	—	10–12	—	6–350
CdTe	10 mm × 10 mm × 10 mm (−30 °C)	—	1.93	2.89	10–1000
CdZnTe	10 mm × 10 mm × 10 mm	—	9	23	30–1400
CdZnTe	15 mm × 15 mm × 15 mm	—	8.5	12	20–1400
HgI$_2$	100 mm^2 × 1 mm	1.5	—	—	2–60
TlBr	2 mm × 2 mm × 2 mm	—	6	—	10–1000
NaI(Tl)	51 mm (diam.) × 2.5 mm	2.9	—	—	2–70
NaI(Tl)	51 mm (diam.) × 25 mm	—	31	—	20–200

[a] The approximate energy range over which usable detection efficiency can be expected. The actual range will depend upon the thickness and material of the detector window.

detectors, although much worse than germanium semiconductor detectors, is better than that of scintillation detectors. The data in the table are incomplete in that, in some cases, there are significant factors not listed. For example, CdTe detectors can give resolution much better than CdZnTe ones, especially when cooled. However, the peaks in CdTe spectra have energy-dependent tails on the low-energy side. In some circumstances, that could cause difficulties when trying to deconvolute multiplet peaks. Using co-planar grid technology, the peaks in a CdZnTe spectrum will be Gaussian in shape, although of poorer resolution. (This will be discussed further in Chapter 4, Section 4.3.5.)

Although the size of the detectors available is limited, this has been turned to market advantage by aiming such detectors at such applications as medical tracing systems where small size is a necessity but resolution not of prime importance. They are, of course, generally useful where limitations in space, or availability of liquid nitrogen, would preclude the use of a germanium detector and are also finding applications as portable probes in nuclear safeguard programmes.

Thallium bromide is a material showing some promise. Its high density and high band gap make it an attractive proposition. Working devices have been constructed demonstrating the feasibility of spectrometry at least up to the 661.6 keV gamma-ray of ^{137}Cs, even with detectors as small as a few cubic millimetres in size.

Gallium arsenide is a material with some theoretical promise as a semiconductor detector which although abandoned for the present could ultimately have a place in gamma spectrometry due to improvements in the technology of manufacture of the material for the electronics industry. The other materials listed in Table 3.1 have all been considered as potential materials for gamma spectrometry, but discarded for one reason or another. In practice, the material almost universally used for gamma-ray spectrometry is germanium and we shall not deal in detail with other types of detector, with the exception of scintillation detectors, which will be discussed separately in Chapter 10.

3.3 THE NATURE OF SEMICONDUCTORS

Before discussing the preparation of gamma-ray detectors, it is necessary, without delving into solid state physics in any great detail, to understand the basic nature of semiconductor materials. In an absolutely pure semiconductor material, thermal excitation would promote a certain number of electrons from the valence band to the conduction band, leaving behind an equal number of positively charged holes. A material of this kind containing equal numbers of electrons and holes is described as an **intrinsic semiconductor**.

It is, of course, not possible to prepare any material completely free of impurities. In semiconductors, these can have a significant effect upon the conductivity. Consider germanium: it is four valent and in a crystal lattice will be surrounded by four other germanium atoms, each equally contributing electrons to the bonding between them. If one of these germanium atoms is replaced by an impurity atom of a different valency this will disturb the electronic balance of the lattice. For example, if the impurity is three valent gallium or boron, then at the impurity lattice site there will be one electron too few to maintain the overall electronic configuration. In effect, we have a *hole*. Such impurities are referred to as *acceptor impurities* and when distributed throughout the semiconductor material give rise to extra energy states just above the valence band, called **acceptor states**. Germanium with this type of impurity would be called **p-type** germanium ('p' for positive acceptor impurities).

On the other hand, five valent impurities, such as arsenic or phosphorus, will have one electron in excess of that required for electronic uniformity. The impurity atom will be a **donor atom** sitting in a donor site and will introduce **donor states** just below the conduction band. Germanium with such impurities is **n-type** germanium ('n' for negative donor impurities).

It is, of course, possible that any particular piece of germanium will contain both types of impurity. Each of these impurities will effectively negate one of the opposite type and the net semiconductor character of the material will depend upon the type of impurity in excess. In the unlikely event of an exact cancellation, the material would be called **compensated** germanium. There is, of course, scope for adjusting the nature of the semiconductor by adding small amounts of impurity of an appropriate type; a process known as *doping*.

As I noted above, the introduction of impurity atoms introduces extra states, either just above the valence band or just below the conduction band. The effect of this is to narrow the band gap and since the conductivity depends upon the number of electrons in the conduction band, the conductivity of a doped material will be higher than the intrinsic conductivity. At very high dopant concentrations, and high conductivity, the semiconductor would be designated p+ or n+ as appropriate. Such a material is sometimes produced *in situ* by evaporation or ion bombardment of appropriate impurities to produce electrical contacts for detectors. (On the other hand, extremely high purity p-type materials are sometimes known as π-type and high purity n-type as ν-type – 'π' and 'ν' being the Greek alphabetic equivalents of 'p' and 'n', respectively.)

The semiconductor gamma-ray detector depends upon the curious electronic redistribution which takes place when masses of dissimilar semiconductor types are placed in contact with each other. (This contact must be intimate and we will discuss how this is achieved in practice in Section 3.4.1.) Figure 3.3 shows this diagrammatically. The p-type material has an excess of holes and the n-type an excess of electrons. As these diffuse under thermal influence, holes may 'stray' from the p side to the n side of the junction and electrons from the n side to the p side. Excess holes meeting excess electrons will combine together, mutually annihilating. The result will be a region around the physical junction of the two types of material where the excess charge carriers have cancelled each other out. This is called a **depletion region** (Figure 3.3(b)). The migration of the charge carriers gives rise to a space charge in this region and the generation of a voltage across the junction called the **contact** or **diffusion voltage**, about 0.4 V high in germanium (Figure 3.3(c)). The depletion region is the active element of the detector. This region is very thin, but if a positive voltage is connected to the n side of the junction the width of the depletion layer increases as the electrons are withdrawn from the material. The negative voltage applied to the p side of the junction will withdraw the holes. Because the positive voltage is connected to the negative type semiconductor, this is called a **reverse biased junction**. (It is interesting to examine the output from a germanium spectrometer with the bias voltage turned off and a radioactive source on the detector. An output signal, albeit noisy and far from satisfactory, can usually be clearly seen originating from interactions within the depletion region. Applying only a few volts of bias to actively collect the charges immediately improves the signal and reduces the noise.) The width of the depletion layer, d, can be estimated by using the following equation:

$$d \approx [2\kappa\mu\rho\,(V_0 + V_b)]^{1/2} \qquad (3.3)$$

where V_0 and V_b are the contact and bias voltages, respectively, κ is the dielectric constant, ρ ($\Omega\,\mathrm{m}$) is the resistivity of the material and μ ($\mathrm{m^2\,V^{-1}\,s^{-1}}$) is the mobility of the majority charge carrier in the material. In p-type semiconductors, because the impurities cause an excess of holes these are the majority charge carriers; in n-type materials, electrons are the majority charge carriers. Now V_b is very much greater than V_0 and Equation (3.3) leads us to the conclusion that $d \propto V_b$.

In charged particle detection systems, the ability to alter the thickness of the depletion layer is a useful way in which the detector can be matched to the range of the particles to be measured. In gamma spectrometry, the active size of the detector is always maximized by increasing the bias voltage so as to extend the depletion region across the whole available detector volume. (In fact, the bias will be raised somewhat above this depletion voltage in order to improve the charge collection process.)

The resistivity, ρ, is a function of the concentration of dopant atoms in the material, N, and can be calculated as follows (where e is the charge on the electron):

$$\rho = 1/(eN\mu) \qquad (3.4)$$

Ideally, we would wish to achieve as great a depletion depth as possible at as low a voltage as possible. Equation (3.4) suggests that it would be advantageous to utilize germanium with as low a concentration of impurity as possible. Indeed, as the size of the detector increases the impurity concentration must be reduced, otherwise the necessary bias voltages would be too great. (For example, increasing the diameter from 3 to 7 cm requires more than a tenfold improvement in purity.) The great improvement in the efficiency of detectors in recent times has been as a consequence of the increased availability of extremely high-purity germanium. Germanium for the manufacture of gamma-ray detectors is, worldwide, the purest material produced in bulk.

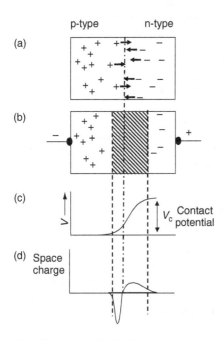

Figure 3.3 (a) p–n Junction before charge carrier redistribution. (b) Depletion region created by carrier redistribution. (c) Variation in potential across junction. (d) Variation in space charge across junction

3.4 THE MANUFACTURE OF GERMANIUM DETECTORS

3.4.1 Introduction

In practice, gamma-ray detectors are not constructed by placing differing types of semiconductor materials in contact but by conversion of one face of a block of germanium to the opposite semiconductor type by evaporation and diffusion or by ion implantation. Figure 3.3 demonstrated the mechanism of depletion. If the concentration of impurities on either side of the junction is different (in Figure 3.3, there are more p-type impurities on the left-hand side of the junction than n-type on the right), then the space charge distribution will not be symmetrical about the junction. As indicated in Figure 3.3(d), the width of the depletion region is greater on the side of lower impurity concentration. (In principle, the product of the impurity concentration and depletion width must be the same on both sides of the junction.)

It follows that if we take a block of suitably high-purity p-type germanium and create on one face an n+ layer (as in Figure 3.4), then applying a reverse bias to the detector will create a depletion layer throughout the p-type material. This is the basis of all germanium detector manufacture. At one time, such detectors were often called, incorrectly, **intrinsic detectors** (implying that the germanium used was of intrinsic quality). This term has been superseded by the more accurate **hyperpure,** or simply high-purity, germanium (**HPGe**) detector.

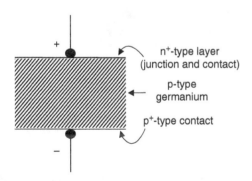

n+-type layer
(junction and contact)

p-type
germanium

p+-type contact

Figure 3.4   The basic construction of a germanium detector

It is said that the relative efficiency of an HPGe detector can be estimated by dividing the active volume of the detector by 4.33. Taking the specific gravity of germanium to be $5.33\,g\,cm^{-3}$, this amounts to about 23 g of germanium per 1% relative efficiency. In 2003, the largest HPGe detector produced was a p-type coaxial detector made from 4.4 kg of germanium with a relative efficiency of 207.6 % and a resolution, FWHM, of 2.4 keV at 1332.54 keV (Sangsingkeow *et al.*, 2003). Figure 3.5(f) visually compares the size of a 15 % detector and a 150 % detector with a standard golf ball.

3.4.2 The manufacturing process

While there are a number of manufacturers of detector systems, there are only three manufacturers of detector-grade germanium: ORTEC, PGT and Canberra. Other detector 'constructors' would use material from one or other of these. For the germanium manufacturer, the starting point is electronic-grade polycrystalline germanium. Although already of very high purity, this is further purified by *zone refining*. The germanium is melted in a pyrolytic graphite coated quartz crucible using radio-frequency (RF) heating coils. It is a well-established fact that as a liquid freezes and solid appears, impurities will concentrate in the liquid phase hence leaving the solid purer than the original melt. (A principle that has been applied illicitly to the concentration of alcoholic beverages for a very long time!)

Each zone refiner coil (see Figure 3.5(a)) melts a small portion of the germanium in the crucible. As the coil is slowly moved along the length of the crucible, the molten zone moves with it. The germanium melts as the coil approaches and freezes as the coil moves away, leaving a higher concentration of impurities in the liquid than the solid. In this way, the impurities are 'swept' along in the molten zone to the end of the bar. Many sweeps are needed, helped by using three coils, after which the impurity concentration is reduced by a factor of 100 or more. The tapered end of the ingot (Figure 3.5(b)) now contains most of the impurities and can be cut off. The germanium at this stage is still polycrystalline and not yet suitable for detector use. The zone refined ingot is checked for purity using Hall effect measurements and sliced up ready for conversion to single crystal material.

Large single crystals are grown by the Czochralski method. The germanium is melted in a quartz crucible by RF coils and maintained at a temperature just above the melting point (937 °C). A small seed crystal, precisely cut with respect to its crystal planes, is dipped into the molten germanium and slowly withdrawn (Figure 3.5(c)). The size of the crystal can be controlled by the rate of withdrawal and the temperature of the melt. The whole growing process takes place in an atmosphere of hydrogen. Germanium wets quartz and if excess were left in the crucible and allowed to freeze, expansion of the germanium would break the crucible. Therefore, the melt must be used completely and

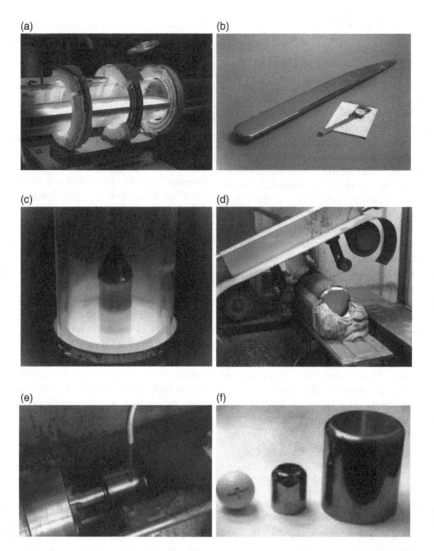

Figure 3.5   Stages in the manufacture of germanium detectors: (a) a three coil zone refiner; (b) a zone refined billet of germanium; (c) growing ('pulling') a germanium crystal by the Czochralski technique; (d) a mounted crystal being sliced by a string saw; (e) grinding the germanium crystal; (f) left to right – standard golf ball, 15 % detector and 150 % detector. Reproduced by permission of ORTEC

as the drawing finishes, the crystal must be tapered to minimize thermal strain. Crystal growing demands considerable expertise and, to some extent, this aspect of germanium production limits the size of detectors. The preparation of large diameter crystals demands very close control of the crystal pulling parameters to achieve a product that has a uniform impurity concentration across the diameter of the crystal. The difficulties are particularly acute with n-type germanium. Redistribution of the impurities can result in a change to p-type some way down the crystal.

Not surprisingly, the details of such processes are a closely guarded commercial secret.

Having prepared the single crystal, it must be cut to suitable dimensions. This is by no means as straightforward as it sounds. Simple sawing operations will damage the crystal structure at the cut faces, causing imperfections that could act as charge carrier traps. Special saws have been developed to allow shaping without causing such damage. (Figure 3.5(d) shows a string saw that uses a silicon carbide and water slurry to abrade the germanium block.)

Hall effect measurements are again used to determine the impurity concentration and semiconductor type (n or p) along the length of the germanium block. Detector-grade germanium is then cut out and the reject (but still extremely valuable) material returned to the zone refining stage of preparation. The selected germanium is ground to a perfectly cylindrical shape. If the detector is to be a coaxial detector, the edge at one end of the crystal is then rounded to improve the charge collection; this process is refined to as **bulletization**. (This will be explained fully in due course.) A coaxial detector crystal will then have a hole machined into the unmodified end to provide a location for the central contact. The crystal is then lapped to remove much of the surface mechanical damage caused by machining.

The n+ contact is formed by diffusing lithium onto the appropriate parts of the detector surface. This depends upon the type of detector intended. A p-type detector will have an n+ layer over the whole of the outside except for the flat base. The detector will then be lapped once more, chemically polished and a protective surface coating of germanium hydride applied by sputtering. The p+ contact is created by ion-implantation of boron atoms onto the surface. Figure 3.5(f) shows two completed detectors, a very small one and a very large one – 15% and 150% relative efficiency, respectively.

The final detector crystal must be kept extremely clean. Typical operating voltages may be 1000 to 3000 V applied across only a few centimetres of germanium. Even slight traces of surface impurity could give rise to undesirable surface leakage currents. For this reason, and to provide thermal insulation for the cooled detector, the final step in manufacture is to mount the detector in an evacuated housing that also carries electrical connections to the detector. More often than not, the preamplifier electronics will be mounted inside this housing, allowing the critical electronic components to be cooled. The whole will then be mounted on a cryostat to provide cooling. (Cryostats will be discussed separately in Section 3.7.)

3.4.3 Lithium-drifted detectors

Lithium-drifted detectors, usually referred to as 'jellies' (i.e. Ge(Li)) are no longer manufactured but might still be found in gamma spectrometry laboratories. Lithium drifting was a response to the unavailability of large crystals of high-purity germanium in the early days of semiconductor detector manufacture. Lithium atoms are small and can easily sit interstitially within the germanium lattice where they act as n-type (i.e. donor) impurity atoms. Normally germanium of indifferent purity is p-type and if lithium is distributed through the germanium lattice

then the lithium atoms cancel out the p-type acceptor impurities of whatever type. As it happens, lithium has an exceedingly high mobility in germanium and can easily be drifted through the crystal.

Lithium was coated onto the surface of the p-type crystal by any suitable means – vacuum deposition or even by painting on a lithium/oil suspension. By applying a reverse bias at a temperature of 50 °C, the lithium ions migrate through the crystal. At the lithium doped face there is a large excess of n-type sites (an n+ region) and in the centre is a large volume where there is almost exact cancellation of the existing acceptor sites by the lithium. Contacts for the detector were made by electrodeposited gold or by using indium alloys. Whereas the Ge(Li) has been completely superseded by the HPGe detector, silicon photon detectors must still be prepared by lithium-drifting silicon (Si(Li) or 'silly' detectors). The manufacturing process is basically the same as for the germanium equivalent, although the details of the drifting process differ.

The mobility of lithium in germanium at room temperature is so high that under normal circumstances the lithium would continue to migrate within the crystal and destroy the carefully created balance of impurities. Therefore, unlike Si(Li) detectors, Ge(Li) detectors must be kept at liquid nitrogen temperature for the whole of their lives, whether in use or not.

3.4.4 The detector configurations available

Germanium detectors are available in a number of different configurations to suit particular applications. The simpler standard ones are shown diagrammatically in Figure 3.6, together with an indication of the energy range over which they might be used. (The efficiency curves in this figure are intended to provide a general impression. Curves for actual detectors would differ depending upon their size and type.) The high-energy capability of a detector is limited by the fall off in absorption coefficient at high energy and by size of the detector crystal. (Remember that the probability of total absorption of a gamma-ray depends upon the space available for multiple interactions.) The limit of 3 MeV implied in Figure 3.6 is a purely arbitrary cut-off which could justifiably be extended to 10 MeV or more. At low energies, the gamma-ray absorption coefficient is very high and high efficiency would be expected. This is limited by absorption of the photons before they reach the detector itself.

3.4.5 Absorption in detector caps and dead layers

In the first place, we can expect absorption in the detector cap and secondly in the contact layer on the face of

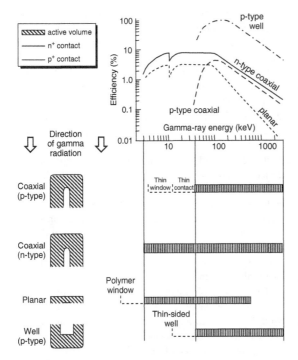

Figure 3.6 Configurations of detector generally available, together with schematic efficiency curves and an indication of the energy range over which they might be used

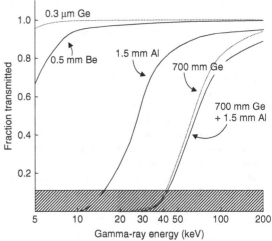

Figure 3.7 Transmission curves demonstrating the absorption of low-energy photons by detector dead layers and cap materials

the detector facing the incoming radiation. Figure 3.7 shows the calculated proportion of gamma-rays transmitted by a number of relevant materials. Detector caps are constructed of aluminium (sometimes magnesium) about 1.5 mm thick, or sometimes with a beryllium window about 0.5 mm thick. Manufacture of a typical lithium n+ contact produces a dead layer of impure germanium about 700 μm thick. In contrast, the dead layer caused by the ion-implanted p+ contact is only 0.3 μm thick. If we take 10 % transmission as a (completely arbitrary) limit, then we can see that although an aluminium detector cap limits use of the detector to greater than about 15 keV, much greater absorption in an n+ type of detector dead layer will raise this limit to 40 keV. Thus for low-energy work, detectors with beryllium windows and an ion-implanted p+ contact layer facing the gamma radiation is required.

If any detector could be described as having a 'standard configuration', it would be the closed-end p-type coaxial detector mounted in an aluminium outer cap. This type of detector has an outer 700 μm n+ contact and cannot be used much below 40 keV regardless of the type of detector cap used. It is possible to buy modern p-type

HPGe detectors with the n+ layer made particularly thin on the top face. Together with a beryllium window, such a detector might have an energy response extending to below 10 keV. (An example would be the Canberra XtRa extended range detector.)

The limitation caused by the thick n+ layer on the outside of the detector can be eliminated by turning the detector 'inside out', so to speak. If n-type germanium is selected for manufacturing the detector, the n+ contact will be on the inside of the detector and a very thin p+ layer, which can easily be manufactured by ion-implantation of boron, will be on the outside. Such detectors are often known as **reverse electrode detectors**. With a beryllium window, reverse electrode detectors can be used down to a few keV, as indicated by the efficiency curve in Figure 3.6. In these curves there is a very obvious 'nick' in the otherwise horizontal part. This sudden change in efficiency can be related to the K absorption edge in the photoelectric response of germanium at about 11 keV, discussed in Chapter 2. As we will see later, the n-type detector, in addition to its extended low-energy measurement capability, has advantages in that it is more resistant to neutron damage. Offset against this is its higher cost (as a consequence of the greater difficulties of manufacture) and the greater potential for true coincidence summing problems. Those will be discussed in Chapter 8.

Traditionally, end caps have been made of aluminium or maganesium. Recent developments have seen the emergence of the end caps made completely of **carbon fibre**, or **carbon composite**. These caps are of inherently low activity, compared to the alternatives – they are aimed at the low background market. Their low density means greater transmission of the gamma-rays and, therefore, higher efficiency at low energy.

3.4.6 Detectors for low-energy measurements

There is little point in buying more germanium than is needed to absorb fully the gamma-rays you want to measure. If a detector is only to be used for the measurement of lower-energy gamma-rays, then a small detector will be adequate. Planar detectors are thin detectors with typical diameters up to 50 mm and typical depths up to 20 mm. They are in principle parallel plate detectors and may be of p- or n-type germanium. Their construction is carefully tailored to produce a detector with the best possible resolution. For example, the area of the n+ contact may be reduced to a small spot in order to reduce the capacitance and hence improve the resolution of the detector. In some detectors (e.g. Canberra LEGe and the ORTEC LO-AX™ – see below), the p+ contact extends around the side of the detector. For extended capability at photon energies below 3 keV, it is possible to buy planar detectors with a reinforced polymer film window. This will take the range of operation well within that normally served by Si(Li) detectors. The manufacturers claim that improvements in technology mean that germanium detectors with superior resolution to Si(Li) detectors can now be made.

Another variant for low-energy measurements is the ORTEC Instruments 'LO-AX' detector, which in essence is a short n-type coaxial detector (see Figure 13.7). The reduction in capacitance over a conventional 'long' detector gives the detector a significant resolution advantage at the expense of an impaired high-energy response due to the smaller size.

3.4.7 Well detectors

All of the detectors discussed so far have a maximum possible efficiency of 50 % due to geometry considerations. (Slightly less, in fact, because the crystal itself is mounted a few millimetres below the detector cap.) The germanium well detector is basically a p-type coaxial detector with the negative contact hole drilled out large enough to fit small samples within the detector itself (see also Chapter 13, Section 13.2.2). This type of detector provides nearly 4π geometry and, over part of the energy range, nearly 100 % detection efficiency. The actual efficiency curve will depend very much on the dimensions of the well, its diameter and annular thickness, but also on the wall thickness of the well in the outer cap. With a 0.5 mm-thick wall, reasonable efficiency might be achieved down

to 10 keV or so. When purchasing such a detector it is important to be sure that the well is just that.

Although it might seem perverse to make a well detector with a hole all the way through, PGT do just that and claim a number of 'through-hole advantages'. Although germanium is lost beneath the sample in the well, this is compensated for by the fact that a larger sample can be placed within the region of highest efficiency – in effect, the whole sample can be centred in the detector crystal. Blind holes are said to be microphonic (see below), because they need a long lead (the gate lead) to connect to the contact within the detector well. With the through-hole design, the lead can be short, leading to improved performance and better resolution. Through-holes are easier to manufacture and larger through-wells can be constructed than blind-wells.

3.5 DETECTOR CAPACITANCE

The construction of a junction detector, p+ and n+ conductors separated by what is effectively an insulating layer, is not dissimilar to that of a capacitor. The capacitance of a detector is of importance because of the effect it has on the resolution of the detector. (This will be discussed further in the next chapter.) The capacitance will depend upon the shape and size of the detector. For example, if we compare a planar detector to a simple parallel plate condenser, then we can estimate the capacitance, C, from

$$C = \kappa A/(4\pi d) \tag{3.5}$$

where A is the area of the detector and again d is the thickness of the depletion layer and κ the dielectric constant. Equation (3.3) showed that the depletion depth depends upon the bias voltage applied to the detector. Because in germanium spectrometry we would always use sufficient bias to deplete the whole thickness of the detector, we can take d as the detector thickness. If we replace the area in Equation (3.5) by $\pi D^2/4$, with D being the detector diameter, then for germanium ($\kappa = 16$) this equation reduces to:

$$C\,(\text{pF}) = 0.111 \times [D\,(\text{mm})]^2/d\,(\text{mm}) \tag{3.6}$$

For example, a planar detector of diameter 36 mm and thickness 13 mm would be expected to have a capacitance of about 11 pF.

Comparing a detector to a cylindrical capacitor (but bearing in mind the fact that the central hole of the detector does not pass all the way through) gives the following equation relating capacitance to the height of the detector, h, and the detector and core radii (r_2 and r_1, respectively):

$$C = \kappa h / 2 \ln(r_2/r_1) \tag{3.7a}$$

or:

$$C\,(\mathrm{pf}) = 0.888 \times h\,(\mathrm{mm}) / \ln(r_2/r_1) \tag{3.7b}$$

Since the detector capacitance reduces as the ratio r_2/r_1 increases, there is every reason to keep the central contact hole as small as possible. Again, as an example to illustrate the capacitance expected of a detector we can consider a coaxial detector, 64 mm long and 50 mm diameter, with a core diameter of 8 mm (dimensions typical of a detector of 38 % relative efficiency). Equation (3.7b) provides an estimate of 31 pF, not accounting for the closed end of the detector.

As the thickness of a planar detector increases, a point is reached at which the detector would be thick enough to take a core. Would there be any advantage in making a coaxial detector instead of a planar detector? If the capacitance is calculated for the two configurations (assuming true coaxial geometry for simplicity instead of a closed end) then it becomes apparent that for the same diameter and thickness a coaxial detector has a lower capacitance than a planar detector. We could expect this lower capacitance to then feed through into better resolution. Indeed the ORTEC Instruments LO-AX™ short coaxial detector referred to above is claimed to provide better resolution than the equivalent planar detector (370 eV at 5.9 keV for a 51 mm diameter detector compared to 550 eV) as a consequence of halving the capacitance. Canberra fabricate detectors of similar size and resolution with a p+ contact extending around the side using a spot contact on the back face rather than the well contact of a coaxial geometry (see also Figure 13.8 below).

3.5.1 Microphonic noise

Microphonic noise refers to mechanically generated noise that degrades spectrum resolution. Small movements of the components of the detector in its mounting caused by vibration can alter the capacitance of the arrangement. Changes in capacitance mean changes in the noise characteristics of the detector and increased peak width. A change of capacitance between the field effect transistor (FET) gate and the detector bias supply of only $5 \times 10^{-7}\,\mathrm{pF}$ can cause an electrical signal equivalent to a 10 keV X-ray. In discussing well detectors in Section 3.4.7, the suggestion made was that the necessarily long lead connecting the contact inside the well is

prone to movement. Vibration sources regularly experienced are as follows:

- Bubbling of liquid nitrogen under normal quiescent conditions.
- Turbulence and boiling of liquid nitrogen while filling the Dewar flask. It is not a good idea to measure spectra while topping up the Dewar flask.
- Vibration of equipment in contact with the cryostat. This was the reason for the loss of resolution of detectors fitted with first generation electrically cooled cryostats. (See Section 3.7.5).
- Environmental noise transmitted through floors and, in extreme cases, the air.

Detectors vary in their sensitivity to vibration. PGT market a detector with an ultra-low vibration mounting, which they call 'The Quiet One™'.

3.6 CHARGE COLLECTION IN DETECTORS

3.6.1 Charge collection time

In an earlier section, I briefly mentioned the collection of the charge carriers produced by interaction of the gamma-ray with the detector crystal. If we are to understand some of the problems in gamma spectrometry, it is necessary to discuss charge collection and the mobility of the charge carriers in more detail.

To start with, I will consider a parallel plate (i.e. planar) type of detector (Figure 3.8). I will consider the movement of an electron–hole pair, created at a distance x from the positive (i.e. n+) electrode. Without any external persuasion, the pair will move within the lattice under the influence of thermal excitation. If an electric field, E, created by a bias voltage, V_b, is applied across the detector, an additional drift motion will be added parallel

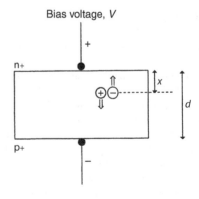

Figure 3.8   Charge carrier motion in a semiconductor detector

to the direction of the field. (We will assume that V_b is sufficient to deplete the detector fully and so the depletion depth, d, is its full thickness.) The electrons will migrate to the positive collector and the holes to the negative.

The time taken for each charge carrier to reach its destination, the **charge collection time**, will depend upon the distance it must travel and the speed at which it travels. For each entity we can define a **mobility**, μ, equal to the ratio v/E, where v is the **drift velocity** of the electron or hole and E the electric field strength. The mobilities of electrons and holes are different, although in germanium this difference is only about 15 %. (This is fortunate. In gas detectors, the mobility of electrons and their accompanying positive ions are different by orders of magnitude, a factor which causes considerable charge collection problems. The same problem besets room temperature detectors made of semiconductor materials with very different electron and hole mobilities, listed in Table 3.1.)

At low field intensity the mobility is almost constant, but at high field intensity the drift velocity does not increase proportionately with field and eventually reaches a **saturation velocity**. It is then independent of field strength (Figure 3.9). If we provide a detector with a great enough bias voltage to ensure that the field strength is sufficient to give all charge carriers their saturation velocity throughout, we can easily calculate the charge collection time. Taking the simple planar detector geometry depicted in Figure 3.8:

$$\text{electron collection time, } t_e = x/v_e \qquad (3.8a)$$

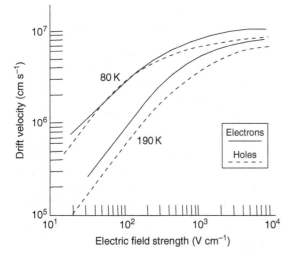

Figure 3.9 Mobility of electrons and holes in germanium as a function of electric field strength and temperature

$$\text{hole collection time, } t_h = (d-x)/v_h \qquad (3.8b)$$

where v_e and v_h are the saturation velocities of electrons and holes, respectively, and x and d are defined in Figure 3.8. The fact that the mobility decreases with increasing temperature is yet another justification, in addition to reducing thermal excitation across the band gap, for operating germanium detectors at low temperatures.

If we look now at the position at which the electron–hole pair is created relative to the positive and negative collectors of the detector, then we can appreciate that the electrons and holes will arrive at different times. The electron from an event close to the positive electrode will be collected well before the hole has time to travel to the negative electrode. This, in turn, means that the way in which the output pulse from the detector rises will depend upon the position at which the charge carriers are produced.

3.6.2 Shape of the detector pulse

The contribution that a charge carrier makes to the external electrical signal depends upon the charge it carries and the fraction of the electric field it travels across. In the situation we have just considered where the carrier pair are created close to the positive collector, the electron, although arriving first, will make little contribution to the electrical signal because it only travels across a small fraction of the total field. On the other hand, the hole will traverse almost the complete field and will contribute most to the signal. It follows then that the rise in the external electrical signal with time, that is, the shape of the leading edge of the output pulse, depends upon where in the detector the interaction takes place.

Figure 3.10 shows the calculated shapes of the rising edges of the output pulse for interactions within an idealized n-type coaxial detector. These diagrams show separately the contributions made to the electrical signal by electrons and holes and the different arrival times. The actual shape of the rising edge will depend upon the shape of the electric field within the detector. The total charge collected per electron–hole pair is, ignoring for the moment trapping and recombination, constant and is equivalent to the charge carried by an electron.

In calculating Figure 3.10, it was assumed that all the charge carriers from a gamma-ray interaction are produced at the same point. This is patently not so because, even ignoring the fact that most gamma-rays are absorbed by multiple interactions, the primary electrons formed in the interaction will not transfer their energy to the detector at a single point. Nevertheless, the figure does give some idea of the way in which the

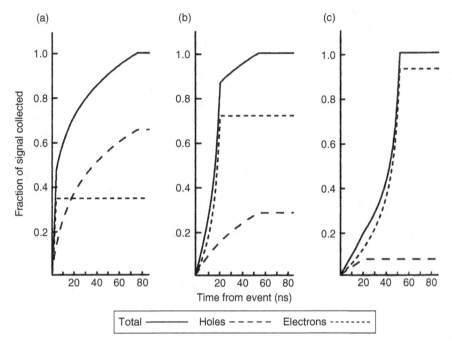

Figure 3.10 Calculated shapes of the rising edges of an n-type germanium detector for interactions at different points within a coaxial detector: (a) near to core; (b) mid-way; (c) near to outside

leading edge of the detector pulse can vary and is gratifyingly consistent with what can be seen on an oscilloscope when the actual output pulses from the preamplifier of a detector system are monitored (see Chapter 4, Figure 4.7 for examples).

The shape of the detector pulse and the charge collection time have an influence on the way in which the pulse processing electronics is set up. In particular, long charge collection times, especially if extended by trapping (see Section 3.6.6), have important implications for counting rates. These matters will be discussed more fully when the electronic parts of the system (Chapter 4) and high count rate systems (Chapter 14) are considered.

3.6.3 Timing signals from germanium detectors

Event timing is not a great concern in everyday gamma-ray spectrometry. The emphasis is on how much energy is absorbed by the gamma-ray event and how many events occur within a particular counting period, not precisely when within that period they occurred. However, if the gamma-ray detector pulse is to be used in a coincidence system, then the time relationship between pulses from different detectors becomes important.

As we saw in Section 3.6.2, the shape of the leading edge of the detector pulse varies with the location of the event within the detector. Since the electronic methods for generating a timing signal depend upon the rise time of the pulse, there will be an uncertainty in the time between the gamma-ray event and the appearance of the timing signal. (Rise time is defined as the time for a pulse to rise from 10 to 90 % of its full height – see Figure 4.15.)

There will, in fact, be a distribution of time differences, which, depending upon the method used for deriving the signal, might be expected to have a width at half maximum of 3–10 ns, depending upon the gamma-ray energy and detector size. For example, 3.7 ns is quoted in a manufacturer's catalogue for a 61% relative efficiency detector at 1332 keV gamma-ray energy.

3.6.4 Electric field variations across the detector

The variation in electric-field strength across a detector depends upon the shape of the detector and the type of intrinsic region. For example, in a p-type high purity planar detector at a point a distance x from the p+ (negative) contact, the electric field, $E(x)$, is given by the following relationship:

$$|E(x)| = \frac{V}{d} + \frac{eN}{\kappa}\left(x - \frac{d}{2}\right) \qquad (3.9)$$

where N is the concentration of acceptor atoms (for an n-type detector the donor concentration would be used), e is the charge on the electron and κ again the dielectric constant. This is shown diagrammatically in Figure 3.11(a). For true coaxial detectors (Figure 3.11(b)), the field strength profile across the detector is much more complicated and depends upon the size of the core relative to the overall diameter of the detector:

$$|E(x)| = \frac{eNr}{2\kappa} + \frac{V - (eN/4\kappa)(r_2^2 - r_1^2)}{r \ln (r_2/r_1)} \qquad (3.10)$$

Here, r is the distance of the interaction from the centre line of the detector. This equation, in which the electric field varies as the reciprocal of r, was used to calculate the curve in Figure 3.11. Examination of Equation (3.10) suggests that greater field strength will be obtained for a given size of detector if the size of the central hole is made as small as possible. In all configurations of detector, the electric field strength decreases with increasing detector size for a fixed bias voltage, V_b. If the field falls sufficiently, then the speed of the charge carriers may fall below their saturation velocity. The combination of long migration distances and possible slower speed means that charge collection times in long detectors are much greater than in small ones.

complicated because of the influence of the active volume at the end of the detector. Originally, such detectors were manufactured with a complete cylindrical shape but it was realized, particularly as detector size increased, that in such detectors there were regions of especially low field. Interactions taking place in these regions gave rise to pulses with rise times much greater than the average – so much longer, in fact, that such pulses could *never* be collected completely within reasonable integration times. Particular problems were recognized in the corner regions on the detector face. In principle, such pulses might be filtered out by electronic selection based on rise time but the practical solution is delightfully simple – remove those regions altogether. Hence, modern detectors have a rounded edge on the front face, the process being known as **bulletization** (Figure 3.12).

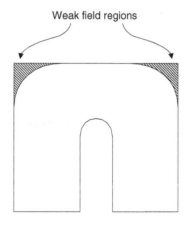

Figure 3.12 Bulletization of detectors

PGT do not agree that the weak field regions of a non-bulletized detector are a problem. In Section 1.15.1 of their Nuclear Product Catalog, they state that '*The variation in the electric field in the crystal, which such a shape supposedly avoids, is minor*'. They go on to explain that, what they call, the 'straight-across' design of their detectors makes for more secure clamping of the detector inside the end cap, which minimizes microphonics resulting in lower resolution.

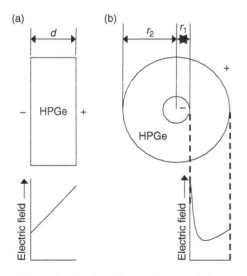

Figure 3.11 The electric field strength across different types of detector: (a) planar detector; (b) true coaxial detector

3.6.5 Removing weak field regions from detectors

The most common type of detector in use is the closed-end coaxial in which the field variation is somewhat more

3.6.6 Trapping of charge carriers

We have assumed so far that the electrons and holes are free to migrate to the collector contacts unhindered. In any detector crystal there is likely to be a small population of traps. These might be a consequence of crystal imperfections, of interstitial impurities or of radiation damage. If a charge carrier migrates to one of these traps it may be held until thermal excitation releases it again. The average time that a carrier might spend in such a trap depends upon the depth of the trap in terms of energy. A shallow trap might have little discernible effect on the overall charge collection but a deep trap could well hold up the carrier until the electronic system has measured the charge. This would represent a loss of charge. The situation is illustrated in Figure 3.13, where one of the charge collection curves from Figure 3.10 is used to indicate the possible consequences of shallow trapping and de-trapping and complete loss by permanent trapping.

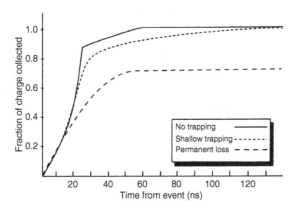

Figure 3.13 The effect of shallow and permanent trapping on charge collection

The lowest curve in this figure represents a permanent loss of charge caused by permanent trapping or recombination of charge carriers. The middle curve is the less serious, but still inconvenient, case where the charge eventually succeeds in migrating to the collecting electrode, albeit somewhat later than expected. Whether it arrives too late to be taken into account by the external electronic circuits depends upon how their integration times relate to the charge collection time. (This will be covered in Chapter 4.)

3.6.7 Radiation damage

Irradiation of germanium detectors by neutrons, particularly fast neutrons, can given rise to the displacement of atoms from their lattice positions forming an interstitial–vacancy pair, called a **Frenkel defect**. These are particularly effective traps and if germanium detectors are used in a neutron field for any length of time the spectrum will

Table 3.3 Fast neutron radiation damage thresholds

Detector type	Relative efficiency (%)	Threshold fast neutron dose (cm^{-2})
p-type	20	2×10^8
	70	1×10^7
n-type	30	4×10^9
	70	1×10^8
Planar	—	1×10^9

deteriorate because of the increasing proportion of these defects. The effect is a pronounced tail on the low-energy side of every gamma-ray peak caused by incomplete charge collection. Table 3.3 lists the threshold fast neutron dose above which resolution degradation can be expected for a number of different types of detector.

The reverse electrode detector (n-type) is substantially more tolerant of radiation damage and, apart from its enhanced low-energy response (in a suitable windowed cap), is recommended for use where that is likely. Also immediately obvious is the fact that large detectors are much more susceptible to damage. When choosing a detector for use in this situation it may be better to trade-off efficiency against a better damage tolerance and longer detector life.

The reasons for the better tolerance of n-type detectors lie in the particular way in which the charge carriers migrate. Figure 3.14(a) shows a normal p-type detector crystal. Of every electron–hole pair produced, the electron will move outwards to the positive electrode and the hole will move inwards to the negative core. It is likely that most of the gamma-ray interactions will occur near the surface of the crystal and therefore, on average, the hole must travel further than the electron. In contrast, the neutron damage to the detector will be uniformly distributed throughout the volume of the detector. Because of this there is a reasonable probability that the holes will be trapped while traversing the damaged sensitive region.

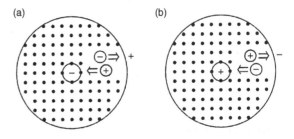

Figure 3.14 Carrier migration in (a) p-type (normal) and (b) n-type (reverse electrode) coaxial detectors

The situation with the n-type detector is quite the reverse (Figure 3.14(b)). In this type of detector, the electrons must travel the furthest. Since electron trapping is not as important as hole trapping, there will be fewer of those charge carriers that contribute most to the electrical signal lost to trapping. Another slight bonus is that since the saturation drift velocity of electrons is greater than that for holes, the overall charge collection time should be slightly shorter with the prospect of improved timing performance. Obviously the bigger the detector, regardless of the semiconductor type, the further all charge carriers must travel, the greater the probability of trapping and the greater the susceptibility to radiation damage.

The interstitial atom responsible for the Frenkel defect is in an unstable position and is only prevented from moving back to its lattice position by an energy barrier. Raising the temperature of the detector by quite a modest amount will anneal out the defect. For a p-type detector, the temperature is 120 °C but this must be held for one week. Annealing of n-type detectors is somewhat easier, needing only 100 °C for 24 h. P-type detectors are likely to suffer some loss of efficiency because the outer lithium n+ layer will tend to diffuse into the crystal, increasing the thickness of the dead layer, with an associated increase in absorption of low-energy photons, and an overall decrease in active volume. We can also expect an outward diffusion from the inner lithium contact of an n-type detector but this has little effect on the effective active volume of the detector, because the area of the dead layer is much less and, in any case, most gamma-ray interactions take place in the outer parts of the detector. Gamma-rays reaching the detector never see the n+ layer and so there is no increased absorption at low energy.

Radiation-damaged detectors can be repaired in-house, but some of the manufacturers and independent companies do provide a repair service. Such services are certainly worth taking advantage of. Apart from the loss of efficiency in p-type detectors, the performance of repaired detectors is likely to be almost as good as new. Perhaps surprisingly, there appears to be no limit to the number of times a detector can be annealed in this way.

If an n-type detector is only slightly impaired, warming to room temperature and then re-cooling may improve the resolution to some extent. However, there is a risk involved. Depending upon the degree of damage there may be instead a gross loss of performance, thought to be due to the clustering of defects and consequent increased trapping. The general advice would be to maintain the detector at liquid nitrogen temperature until proper repair can be undertaken (see Chapter 12).

3.7 PACKAGING OF DETECTORS

We have established in previous sections that germanium detectors are operated at low temperature in order to reduce electronic noise and thereby achieve as high a resolution as possible. The detector must, then, be mounted in a cryostat. The construction of the cryostat must take into account a number of factors, as follows:

- The detector must be maintained at a temperature close to 77 K.
- The detector must be kept under a clean vacuum to prevent condensation on the detector. There must be electrical feed-throughs to take the signal from the detector.
- The detector cap must be thin enough to allow the gamma radiation to penetrate but still withstand the vacuum and provide a reasonable degree of protection to the detector.
- As far as possible, the cryostat construction should isolate the detector from mechanical vibration (anti-microphonic mountings). It has been suggested that even the slight vibrations caused by bubbles from the boiling nitrogen can cause a certain amount of electronic noise.
- The materials from which the cryostat is constructed may have to be specially selected if the detector system is intended for low background measurements (see Chapter 13).

The most common means of providing a suitably low temperature is cooling with liquid nitrogen (boiling point 77 K). Liquid nitrogen is readily available to most gamma-spectrometry laboratories but for use in locations remote from a liquid nitrogen supply other arrangements may have to be used. The general arrangement of the detector within its cryostat is shown in Figure 3.15. As we shall see in Chapter 4, it is important that the total input capacitance to the preamplifier, which includes the detector and cabling, is as small as possible. In addition, there are benefits in reduced noise if certain components of the preamplifier are cooled. Because of this, it is routine to mount the preamplifier close to the detector. In modern systems, it is usual to mount the preamplifier within an extension to the detector housing, as shown in Figure 3.16. This figure also shows more details of one manufacturer's detector housing.

3.7.1 Construction of the detector mounting

The detector is mounted within a thin aluminium retaining sleeve, which also forms an outer contact with

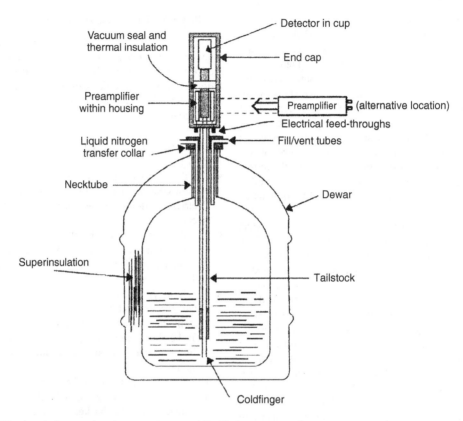

Figure 3.15   A typical germanium detector, cryostat and liquid nitrogen reservoir

the detector. (You should be aware of the extra absorption caused by this extra aluminium if the detector is to be used sideways on to the incoming radiation.) The core contact is made with either a conical pin or a spring-loaded pin extending within the hollow core. The whole of this arrangement is fixed to a pedestal that is, in turn, fixed to the copper cold finger, which extends through the whole cryostat to the liquid nitrogen reservoir. The complete assembly is then covered by the end cap to form a sealed chamber. The upper part of the detector housing is evacuated and thermally insulated from the rest of the housing. A pack of charcoal or molecular sieve absorbent will be mounted in the detector chamber to absorb traces of gases left after evacuation when the detector is cooled. Beneath the detector pedestal are secured the preamplifier field effect transistors (FETs) that need to be cooled.

At one time, preamplifiers projected from the side of the detector housing but nowadays, as electronic components and integrated circuits have reduced in size, the normal arrangement is to house the preamplifier around the cold finger below the detector chamber. When this is covered by its cylindrical shroud, the whole forms a compact cylindrical arrangement. (Even so, where there are particular reasons, perhaps lack of space or a desire to distance the preamplifier from a low background detector, a side-mounted preamplifier can still be specified.) In systems designed for low background measurements, it is likely that a high-purity lead shield would be put between the detector housing and the preamplifier to shield the detector from the small amount of radioactivity in the materials of which the preamplifier is constructed. The complete detector and preamplifier assembly mounted on its cold finger can then be fixed onto a suitable liquid-nitrogen reservoir by an arrangement, such as that shown in Figures 3.15 and 3.16.

In some cases, it may be necessary to position shielding or other detectors behind the detector itself (for example, in an anti-Compton spectrometer system - see Chapter 13). For this purpose, detector systems are available with the preamplifier mounted close to the dewar leaving a clear length of cold finger between it and the detector.

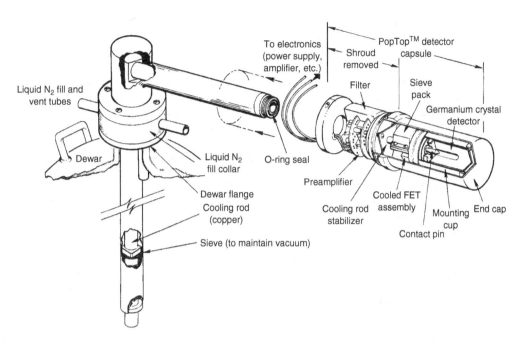

Figure 3.16 A modern arrangement of detector and preamplifier within the cryostat housing – an exploded view of the ORTEC Pop-Top™ detector capsule with a horizontal dipstick cryostat and a 20 l Dewar. Reproduced by permission of ORTEC

The most common type of detector system is likely to have an upward facing detector with a vertical dipstick Dewar, as shown in Figure 3.15. There are, however, a range of possible orientations facing horizontally, downwards or at an angle and a range of possible Dewar types to suit the particular needs of the user. Dewars with offset access and with access from the side or bottom are off-the-shelf options. According to the manufacturers, typical liquid nitrogen evaporation rates for a good Dewar might be 0.5 to 1 L per day. Practical experience suggests that that may be optimistic but a holding time of two weeks might be expected for the standard 30 L Dewar. For portable systems, it is possible to buy detectors with small Dewars of 1.5 to 5 L capacity.

It is important to have an assured supply of liquid nitrogen and to organize a routine for filling the Dewar. This routine should take into account the usage of the detector. It is not a good idea to re-fill during a count. Even with an anti-microphonic construction, the electronic noise generated during the filling may ruin a measurement in progress. In laboratories with many detectors, there is a case for an automatic system for filling each Dewar on demand when level monitors detect a low level

of nitrogen in the Dewar. In such cases, it would be worthwhile providing gating signals to prevent, temporarily, the acquisition of counts during filling.

3.7.2 Exotic detectors

The demands of the nuclear physics community over the years has led to the production of many highly specialized arrangements of detector, preamplifier and Dewar; for example, systems with several detectors within one housing which allow angular correlations between events in the different detectors to be examined. The *clover detector* has four separate large coaxial HPGe detectors mounted in the same enclosure, each with its own preamplifier. Segmented detectors have multiple contacts, each feeding a separate preamplifier; that, in effect, splits the detector into a number of separate regions. Interactions within different regions from the same detector event can provide information about the nature of the interaction. Taking the principle one stage further, the *TIGRE detector* has contacts splitting it into six segments radially and four down its length, giving 24 segments, called *voxels*. Such an arrangement allows three dimensional tracking of gamma-ray interactions. The review by Sangsingkeow *et al.* (2003) discusses these detectors briefly, along with such joys as RHESSI, PT6X2 and C-TRAIN, all of which have been created in connection with sophisticated physics experiments – one of them (RHESSI) has been launched into space.

These systems are undoubtedly technically very advanced and fascinating. I venture to suggest that it will be some time before their advantages in terms of better day-to-day spectrometry will outweigh their cost.

3.7.3 Loss of coolant

It is not unknown for Dewars and cryostats to spring a leak. If this happens, the loss of vacuum results in a loss of thermal insulation. The increase in evaporation rate will then lead to complete loss of liquid nitrogen over a short time and warm-up of the detector. If a detector is allowed to warm up while the bias supply is connected, damage will be caused to the preamplifier. To prevent this, every detector system should have a temperature sensor and some means of switching off the bias if the detector warms up, whatever the failure – mechanical or human. Most bias supplies suitable for germanium spectrometry now have the appropriate cut-out built in and the detector manufacturers now routinely provide a temperature sensor. However, these systems do not work unless the user bothers to connect them together!

A word of caution is perhaps relevant here. There is a documented record of an incident in which a detector cap exploded on warming up after a loss of coolant. It would appear that the vacuum chamber developed a pinhole leak which let in air. The oxygen in the air, having a higher boiling point than nitrogen, condensed on the cold detector. As the thermal insulation was lost, the liquid nitrogen boiled off rapidly and, when was all gone, the detector began to warm up. The rate of warming appears to have been so rapid that, as the condensed liquid oxygen boiled, the pressure within the detector chamber became high enough to burst the chamber open. I would recommend that if a detector is seen to have lost liquid nitrogen and the cap is frosted (rather than just cold), that it be left in its closed shielding until completely warmed to room temperature before any attempt is made to investigate the cause of nitrogen loss. If the detector is not shielded, it should be moved to an isolated position in the laboratory and personnel warned to keep clear!

3.7.4 Demountable detectors

In some respects, the permanent attachment of a detector and cryostat to the liquid nitrogen Dewar is a disadvantage. For example, if a laboratory also has a need to make measurements in the field (perhaps literally!), then two detector systems would be needed – e.g. a permanently installed system with, say, a 301 Dewar and a system with a portable Dewar for the field measurements. Not only does this mean twice the cost but twice the effort in calibration. In response to this, ORTEC introduced the PopTop™ detector mounting (see Figure 3.16). The detector capsule, including the preamplifier, can be unscrewed from the cold finger and transferred to a different Dewar without breaking the vacuum of the detector chamber. This is not an instant changeover. Contraction of the cold finger thread on cooling means that the detector must be allowed to warm up to room temperature before the capsule can be unscrewed. It should be pointed out that impatience leading to forcible removal of the capsule will inevitably lead to serious, and often costly, cryostat damage.

In spite of early claims and counterclaims, there are no disadvantages to the demountable capsule and other manufacturers have followed suit and now also supply demountable detectors. For most users, this offers no advantage other than a flexibility that they may never need. Certainly, for the manufacturers there are significant advantages in that a detector need not be committed to a particular Dewar configuration until just before shipping. Because of this, it is a possibility that the demountable detector will become the norm for all manufacturers.

3.7.5 Customer repairable detectors

In Section 3.6.7, I discussed the damage caused to detectors by fast neutrons. Normally, a detector would be returned to the manufacturer, often to their central manufacturing facilities, for annealing. Of course, this necessarily involves waiting some weeks for its return and the inconvenience of having to dismount and despatch the detector. If neutron damage is a predictable certainty, rather than an unfortunate accident, there is now a self-help option available. Taking advantage of the demountable detector format, detectors can now be bought in a capsule with a built-in heating element to allow *in situ* annealing (ORTEC NDR detector mounting). As part of the 'kit', a vacuum port is provided so that a good vacuum can be maintained during the heating and a controller for the heating element. Repair can be accomplished within a matter of hours.

From time to time, a detector may start to exhibit poor resolution with no apparent cause. This could be due to de-gassing of the charcoal pack or molecular sieve within the detector housing, resulting in leakage currents across the detector. In such cases, resolution can be restored by thermal cycling the detector, as described later in Chapter 12, Section 12.3.

3.7.6 Electrical cooling of detectors

Although liquid nitrogen is a delightfully simple means of cooling, there are situations where it may not be appropriate. The most obvious, of course, is where a liquid nitrogen supply cannot be assured. In other instances, there may be insufficient space for a Dewar large enough to provide a long enough holding time or a need for unattended remote operation. In such circumstances, electrical cooling is the alternative. At one time, this was a costly alternative – electrical cooling might double the cost of a detector system. Recent models are much more affordable and, in some circumstances, might actually be more economical in overall running costs than liquid nitrogen cooling. However, it should not be forgotten when deciding to invest in electrical cooling that, if continuous cooling is to be assured, an uninterruptible mains power supply must be available and making sure that the HT cut-off is connected is essential.

In early electrical systems, cooling was provided by a helium-filled refrigerator mounted underneath the detector. These were notoriously microphonic and significantly degraded the resolution of the detector. In more modern cooling systems, the compressor is located some distance from the expansion head on which the detector is fixed, and connected only by a flexible metal hose. The manufacturers now claim less than 10 % degradation of the resolution of a standard HPGe detector below 500 keV and none at all above that energy. Another disadvantage of the early systems was the high maintenance cost and relatively short maintenance interval. Once again, that is no longer the case and systems such as the ORTEC X-Cooler II have a designed life of five years, which appears to be supported in practice by the performance of coolers of similar design.

Recently, various new designs of electrical cooler based on the Solvay-cycle, Joule–Thompson cooling and the Stirling cycle, and on refrigerants other than helium have been introduced. A design based on the Stirling cycle has led to the ORTEC trans-SPEC, the first handheld portable detector system complete with cooler. Although suffering from microphonic degradation of the detector resolution, this has been compensated for by the introduction of a specially designed digital filter into the Digital Signal Analysis system (see Chapter 4, Section 4.11).

It might be noted that, although electrical cooling is certainly a viable option where ready supplies of liquid nitrogen are not available, it is now no longer necessary to depend upon conventional suppliers of bulk liquid nitrogen. Laboratory scale liquid nitrogen generators are available and, depending upon local circumstances, might even be competitive with bulk suppliers in terms of running costs (see Further Reading for information).

Miniature electrically powered Peltier cooling systems are now available to refrigerate items such as small detectors and preamplifiers. At $-30\,°C$, the temperature is not low enough for HPGe detectors, but adequate for CdTe detectors, giving an impressive improvement in resolution.

PRACTICAL POINTS

- The fortuitous combination of absorption coefficient, semiconductor properties and availability in a suitably pure state provides us with germanium as the predominant material for high-resolution gamma-ray detectors.
- The normal choice for high-resolution spectroscopy of X-rays would normally be the Si(Li) detector but recent special HPGe designs may be a better option.
- Detectors for use at low energy should be reverse electrode (n)-type and be provided with a thin beryllium window.
- Ge detectors are operated at liquid nitrogen temperature to reduce the leakage current and to increase the mobility of the charge carriers. If liquid nitrogen supply is a problem, electrical cooling systems are available.
- The only detectors currently available which are operable at room temperature are the CdTe, CdZnTe and HgI_2 detectors. However, these are limited in size and are best suited to low-energy photon measurements.
- If neutron damage is likely, an n-type detector should be selected and consideration given to having this supplied in a DIY repair mounting (e.g. ORTEC NDR).

FURTHER READING

The most easily available and relevant information on the construction and configuration of gamma-ray detectors is the manufacturers' current equipment catalogues. The detector market is so competitive that new detector developments are rapidly communicated to potential purchasers. The technical details of implementing the new technology will be regarded as commercially confidential.

For general background reading on principles, the first three works referred to in the reading list for Chapter 2 are as good a source as any.

- There is an excellent description of detector construction at: *http://www.ortec-online.com/detectors/photon/a1_1.htm.*

- Information on new electrical cooling systems:
Broerman, E., Upp, D., Twomey, T. and Little W. (2001). Performance of a new type of electrical cooler for HPGe

detector system, presented at the *Institute of Nuclear Materials Management Conference*, Indian Head, CA, USA.

Upp, D., Keyser, R.M. and Twomey, T. (2005). New cooling methods for HPGe detectors and associated electronics, *J. Radioanal. Nucl. Chem.*, **264**, 121–126.

● Information on CdZnTe detectors:

eV Products: *http://www.evproducts.com* (This site has links to a number of literature sources dealing with the theoretical and technical aspects of the material at *http://www.evproducts.com/white_papers_news.html*).

XRF Corporation: *http://www.xrfcorp.com*.

Cardoso, M.J., Simoes, B.J., Menezes, T. and Correia, M.B.A. (2003). CdZnTe spectra improvement through digital pulse amplitude correction using the linear sliding method, *Nucl. Instrum. Meth. Phys. Res., A*, **505**, 334–337.

Owens, A., Buslaps, T., Gostilo, V., Graafsma, H., Hijmering, R., Kozorezov, A., Loupilov, A., Lumb, D. and Welter, E. (2006). Hard X- and γ-ray measurements with a large volume coplanar grid CdZnTe detector, *Nucl. Instrum. Meth. Phys. Res., A*, **563**, 242–248.

● Comparison of CdTe and HPGe detectors:

Perez-Andujar, A. and Pibida, L., (2004). Performance of CdTe, HPGe and NaI(Tl) detectors for radioactivity measurements, *Appl. Radiat. Isotopes*, **60**, 41–47.

There is a considerable amount of information, albeit of a very technical nature, about other new semiconductor materials accessible via search engines on the Internet.

● Information on a liquid nitrogen generator:

Rigaku/MSC: *http://www.rigakumsc.com/cryo/nitrogen.html*.

● For a brief review on advances in detector technology:

Sangsingkeow, P., Berry, K.D., Dumas, J., Raudorf, T.W. and Underwood, T.A. (2003). Advances in germanium detector technology, *Nucl. Instrum. Meth. Phys. Res., A*, **505**, 183–186.

4

Electronics for Gamma-Ray Spectrometry

4.1 THE GENERAL ELECTRONIC SYSTEM

4.1.1 Introduction

The output from a gamma-ray detector is, in essence, an amount of electrical charge proportional to the amount of gamma-ray energy absorbed by the detector. The function of the electronic system is to collect that charge, measure the amount and store the information. In this chapter, I will discuss the problems associated with the pulse measurement and then go on to examine the function and mode of operation of each of the components of the system in detail. While some of the concepts are somewhat mathematical, I will attempt to explain them in a simple, non-mathematical, manner.

A typical simple electronic system for gamma-ray spectrometry might be as shown in Figure 4.1. A more comprehensive arrangement might include a *pulser*. The bias supply provides the electric field to sweep the electron–hole pairs out of the detector, which are then collected by the preamplifier. In spectrometry systems, the collected charge is then converted to a voltage pulse. The linear amplifier changes the pulse shape and increases its size. The **multichannel analyser** (MCA) sorts the pulses by pulse height and counts the number of pulses within individual pulse height intervals. Each of these units will be explained fully in due course.

As we saw in Chapter 3, in modern systems the detector and preamplifier are manufactured as a single unit, the first stage of the preamplifier being intimately attached to the detector capsule. This arrangement has advantages in that critical components of the preamplifier can be cooled to reduce thermal noise contributions. Traditionally, the other items in the system, sometimes including the MCA, are purchased in the US defined standard *Nuclear Instrumentation Module* (NIM) modular format. (Other modular standards have been defined, for example, the CAMAC standard, but these are not generally used for gamma spectrometry systems.) In recent times, there has been a trend to providing non-NIM systems containing the whole of the electronic system within a single 'black box'. The flexibility of the modular concept is traded for a greater simplicity of setting up and operation. In particular, the digital signal processing systems described in Section 4.11 are single, multifunction units.

NIM units, or NIM modules, are manufactured to standard physical dimensions and fit into 'slots' in a NIM-bin that supplies standard electrical power supplies. The NIM standard also defines the pulse and logic specifications for the signals passing between modules. In this way it is possible to install a variety of modules from different manufacturers within the same bin and for them to work together as part of a complete system. As with all standards, there are situations where this ideal state cannot be achieved. A typical example is where gating signals to and from an MCA are involved. Due to differences within different MCAs, such signals, although standard in size and polarity, might have different time relationships with other pulses within the system. The main advantage of the modular system is that the configuration can be changed at will and if one unit becomes faulty it takes only a few minutes to replace it with a spare working unit. (In a multi-user laboratory this very flexibility can be a disadvantage as units from temporarily inactive systems are 'borrowed' to repair faulty systems.)

It is now usual for a computer, often a desktop 'personal' computer (PC), to be used to control data acquisition, store the data and, eventually, to process it into a useful form. Portable gamma-ray detector systems interfaced to laptop and notepad computers are commonplace.

Practical Gamma-ray Spectrometry – 2nd Edition   Gordon R. Gilmore
© 2008 John Wiley & Sons, Ltd

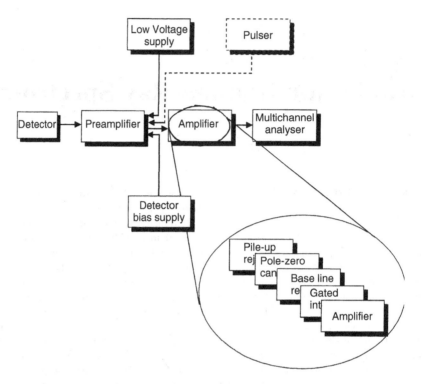

Figure 4.1   A simple schematic electronic system for gamma spectrometry

4.1.2 Electronic noise and its implications for spectrum resolution

The objective of the electronic system is to transfer the signal from detector to MCA with as little alteration as possible. Ultimately, the information carried by the pulses, the amount of gamma-ray energy absorbed, will be collected together in the gamma-ray spectrum. We would like the peaks in our spectrum to be as narrow as possible. I explained in Chapter 1 that the natural energy spread, or width, of a gamma-ray is based upon the lifetime of the energy level from which the gamma-ray is emitted and is exceedingly small, much less than 1 eV. The actual width of a gamma-ray peak in the final spectrum is caused by various sources of 'noise' or uncertainty in the creation, collection, transmission and measurement of the charge created by the gamma-ray event. Every extra item of uncertainty introduced, by whatever means, will inevitably lead to a broadening of the gamma-ray peak.

In Chapter 6, I will discuss how the total width of the peak, w_T, can be split up into a number of components as in the following manner:

$$w_T^2 = w_P^2 + w_C^2 + w_E^2 \text{ (+ other terms?)} \tag{4.1}$$

Of these terms w_E, the electronic noise over which we have some control, will be our main concern in this chapter although we shall have to consider uncertainty of charge collection, w_C, later. The remaining term, w_P, the charge production uncertainty, will be covered in Chapter 6.

Noise in an electronic circuit is generated in many ways. Even the humble, apparently passive, resistor is an important source. Consider a detector/amplifier system without any external radiation signal (source or background). At the output from the amplifier, there will be a randomly fluctuating voltage level due to these inherent noise sources. Now, any true signal from the detector must be measured, ultimately as a voltage, in the presence of this underlying uncertain noise level. The smaller the true signal is, relative to the noise level, the greater the effect of the noise on the final measurement.

Figure 4.2 illustrates the effect of electronic noise on the measurement process. Figure 4.2(a) shows the ideal situation where the pulse height is measured relative to a constant baseline. The only uncertainty in the pulse height recorded would be that inherent in the measurement procedure. Figure 4.2(b) shows the same pulse measured against a noisy background. There is now an additional,

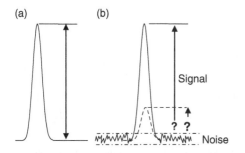

Figure 4.2 (a) Pulse height measured relative to invariant baseline. (b) pulse heights measured in the presence of electronic noise

and potentially large, uncertainty due to the variability of the baseline level to contribute to the overall peak width. It

follows, then, that we have a vested interest in maintaining a good **signal-to-noise ratio** (S/N) throughout the signal chain from detector to MCA. This will be a preoccupation throughout this chapter. Notice that noise does not alter the size of the signal, only the precision with which it can be measured.

4.1.3 Pulse shapes in gamma spectrometry systems

Before considering the parts of the system in detail, it would be sensible to examine the pulses which are transmitted within the system. There are three broad classes of pulse (Figure 4.3):

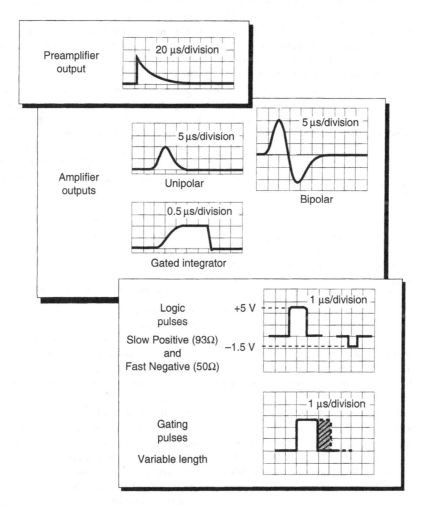

Figure 4.3 Shapes of electronic pulses likely to be found in gamma spectrometer systems

- Linear pulses carry information in their size, that is, the pulse height or the pulse area. The output pulses from the preamplifier and amplifier are of this type. The linearity of the linear pulses is clearly crucial and much of the expense of the system comes down to maintaining this linearity. As an example, if we wish to measure an energy of say 2000 keV to a precision of 0.2 keV or better we must be sure that the gain of the system will be constant to better than 0.01 % over the period of a count. (Compare this with the integral linearity specification of typical spectroscopy amplifiers of 'better than 0.025 %'.)
- Logic pulses are control pulses that might start or stop a process, or indicate the presence of a valid pulse, or reset a circuit; the information is in the presence or absence of the pulse, the pulse itself being a fixed standardized height and width. There are two standards for logic pulses, slow positive and fast negative, which are used in different types of system. ('Slow' and 'fast' here refer to the rise time of the pulse.) In normal gamma spectrometry systems it is most likely that logic pulses will be slow positive, 93 Ω matched.
- Gating pulses are a special form of logic pulse of fixed height but variable width, whose function is to hold open or hold closed an electronic gate for a certain time. Such signals may be used to indicate periods when an electronic module is busy performing a task.

4.1.4 Impedance – inputs and outputs

When putting together a spectroscopy system, the term **impedance** will be encountered from time to time. For example, an amplifier output may be specified as 93 Ω impedance and it is important to have some appreciation of the significance of such statements. The impedance of a circuit or component, usually given the symbol Z, is a combination of resistance, capacitance and inductance. The units of impedance are ohms (Ω) and, for most practical gamma spectrometry purposes, impedance can be thought of as a resistance.

In some circumstances, it is important to take account of the impedance of the various parts of a system. Figure 4.4(a) shows a schematic diagram of an output circuit. A source voltage, V_S, is transmitted through an output, or source, impedance, Z_S, to a load with impedance Z_L. If V_S is attenuated in any way in passing from output to input, it is probable that the signal-to-noise ratio will be degraded. The larger noise component of the signal would then worsen the final spectrum resolution. The load voltage, V_L, can be calculated as:

$$V_L = V_S \times Z_L/(Z_S + Z_L) \tag{4.2}$$

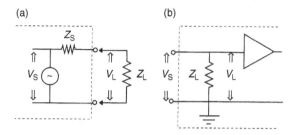

Figure 4.4 (a) Schematic diagram of an output circuit. (b) Schematic diagram of an input circuit (the dotted lines represent the physical limits of the circuits)

If we are to avoid 'loading' the signal (so that V_L is significantly less than V_S), it follows that Z_S must be very much less than Z_L. For this reason, output impedances are generally arranged to be low. Typically, values of less than 0.1 Ω might be specified. An exception to this general rule is when pulses must be transmitted through long cables when larger output impedances may be desirable, as we will see in due course.

Figure 4.4(b) shows a schematic input circuit. The source voltage V_S is supplied to an input with impedance, Z_L, which is effectively in parallel with the source. In this case, because the impedance of the main part of the circuit is likely to be high, if Z_L is low in comparison, then once again the signal will be 'loaded'. A significant current will flow through the input impedance and V_L will be very much lower than V_S, resulting in a poorer signal-to-noise ratio. It follows therefore that input impedances should be high. Typical values for amplifiers would be more than 1000 Ω. (Oscilloscope input impedances are usually even higher, of the order of megohms, for the same reason.)

4.1.5 The impedance of cabling

Pulses are transmitted between separate parts of a gamma spectrometry system along a screened coaxial cable. This cabling has resistance and capacitance, and, therefore, impedance. Cables are specified by their characteristic impedance Z_0. This is a notional parameter defined as the ratio of voltage to current when the voltage is applied to an infinitely long cable.

Cabling, which might be thought as the least likely component of a system to cause problems, is a potential source of serious signal distortion and attenuation. The behaviour of a pulse in a cable may vary depending upon the shape of the pulse, and in particular on its rise time. (Rise time is defined as the time it takes a pulse to rise from 10 % of its full height to 90 %; see Figure 4.7 below.) A pulse whose rise time is short compared to the time the pulse takes to travel along the cable is referred to as a 'fast' pulse. If the rise time were much greater than the transit time, then the pulse would be 'slow'. The speed of a pulse depends on the materials from which the cable is made and, in particular, on the dielectric constant of the insulator between core and screen. In the coaxial cable usually used in gamma spectrometry systems, RG62, a pulse will travel at a speed of about four nanoseconds per metre. A convenient rule is that if:

$$\text{Length (m)} \times 4/\text{Rise time(ns)} << 1 \text{ the pulse is slow,}$$

otherwise it is fast   (4.3)

The reason for considering whether a pulse is fast or slow is that problems arise when fast pulses are transmitted through unmatched cables. Consider a positive step pulse being transmitted down a long cable to an input that terminates the cable with impedance Z_T. Figure 4.5 illustrates the situation.

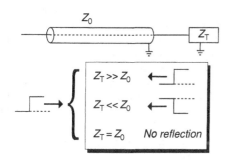

Figure 4.5   Reflection of pulses transmitted down a long cable (fast pulses)

When the pulse reaches the termination, a proportion will be reflected back down the cable to a degree depending upon the relationship between the terminating impedance and the characteristic impedance of the cable. At the extreme, when the terminating impedance is infinite (i.e. the cable feeds an open circuit) the pulse will be reflected back unchanged. On the other hand, if the cable feeds a short circuit (Z_T very low) the pulse will be reflected back in the opposite phase, that is, as a negative pulse. In other circumstances, the pulse will be reflected attenuated to a degree, and with a polarity, which depends on how near Z_T is to Z_0 and whether it is greater or less than Z_0. One can appreciate that pulses being reflected back down the cable will have an effect on the pulses being transmitted up the cable and must lead to undesirable effects in the gamma-ray spectrum. At the least,

the spectrum resolution is likely to be adversely affected and, in more extreme personal experience, peak doubling within the spectrum is possible.

The only condition under which no reflection takes place and the pulse is absorbed completely is when the terminating impedance is equal to the characteristic impedance of the cable (i.e. $Z_T = Z_0$). In effect, the pulse then passes from cable to input without noticing the interface. The correctly matched input appears to the pulse source as part of an infinite cable. As the reader may have guessed by now, typical cable used in gamma spectrometry systems has a characteristic impedance of 93 Ω (such cable would have 'RG62' stamped on it) and amplifiers often have optional outputs and, in some cases, inputs to suit. Obviously, if matching is of concern then a 93 Ω output would be used as well, as matching at the receiving end of the cable. You should be aware, though, that 93 Ω is not a universal standard and some manufacturers, to satisfy their target market, make units matching to 50 Ω or 75 Ω to suit cable of that characteristic impedance.

4.1.6 Impedance matching

If the electronic units in the system do not have appropriately matched sockets, it is a simple matter to arrange this. The first step would be to match the input to the cable by using a T-connector to connect a plug with an appropriate resistor (e.g. 93 Ω) connected across it to the input socket. Suitable ready made terminating plugs can be obtained from electronics suppliers. This places the resistor in parallel with the input cable. If the matching problems persist, it may then be necessary to fix a terminating resistor in series with the output socket. This is most neatly done internally in the unit and internal jumpers are often provided for this purpose.

There is a penalty to be accepted if impedance matching is necessary. We have already seen that if pulses are not to be loaded at the input, the input impedance must be high. A terminated input does not satisfy this condition and the signal will be attenuated by a factor of two with an associated worsening of the signal-to-noise ratio. However, if impedance matching is necessary the possible loss of resolution caused by matching will be more than offset by the gains. Even so, there is little point in using matched impedance inputs unless there is a demonstrable need.

To give this a practical perspective, if we consider a preamplifier pulse with a rise time of, say, 200 ns, and were to transmit it along a 2 m cable the pulse would be 'slow' (2 m × 4/200 ns is much less than 1). If the cable were 200 m long, the pulse would be fast and careful attention to matching would have to be considered. Amplifier output pulses have much longer rise times of a few microseconds and there should be little problem transmitting pulses over some hundreds of metres. It is worth pointing out that matching is more of an art than a science and experience has shown that sometimes pulses which might be expected, according to theory, to be slow might in practice behave as if they were fast. It is not wise to rely too heavily on 'rules-of-thumb' in this particular case. If good resolution cannot be achieved, and the rest of the system is set up correctly, then checking the cable matching might be the solution. Other problems caused by cabling will be discussed in the chapter on troubleshooting (see Chapter 12, Section 12.4).

4.2 DETECTOR BIAS SUPPLIES

The detector bias supply for a semiconductor gamma-ray detector is the least critical unit in the electronic system. Units would normally be able to supply up to 5000 V with about 3000 V being required by a typical high-purity germanium detector. As long as the bias is well above the depletion voltage, charge collection is not greatly affected by changes in bias, and the stability of the bias supply is not critical. (As we shall see in a later chapter, the specification for the bias supply for a scintillation system is much more demanding.) It is advisable for the bias to be adjusted slowly and, for this reason, bias supplies are normally controlled by a single multi-turn potentiometer. The front panel would also have an ON/OFF switch (in addition to the ON/OFF switch of the NIM-bin) and polarity indicating lights which operate as soon as the mains is switched on to the system. The polarity of the output is usually altered by making an internal plug adjustment and these lights allow the operator to check that the polarity is correct for the detector before switching on the bias (as opposed to the mains).

Although modern front-end electronics are protected against voltage surges it is as well to adhere to the conventional wisdom and turn the bias voltage up to the operating level slowly. When switching on a system from cold, it is as well to follow a few simple good practice rules before switching on the mains. First check that the bias ON/OFF switch is OFF, ensure that the potentiometer is turned down to zero volts and only then switch on the mains. If the detector is newly connected, make a positive check that the polarity indication is correct for the detector. Only then switch the bias on and turn slowly up to the operating level.

At one time, when only lithium-drifted detectors were available, detectors were never allowed to warm up to room temperature. Nowadays, it is common for high-purity detectors to be allowed to warm up, especially portable systems with a small liquid nitrogen reservoir. Unfortunately, if the detector bias is not switched off before the detector is allowed to warm, high currents flowing within the preamplifier components will cause damage. All modern bias supply modules include automatic shutdown circuitry that can be connected to a temperature sensor mounted close to the detector crystal via the preamplifier. I do not recommend operating an HPGe gamma spectrometer without the shutdown circuit connected. In some cases, an internal bias disabling circuit may be mounted in the cryostat itself that prevents the bias being applied unless the detector is cold. The bias supply might be provided with a current-limiting trip circuit that prevents the bias being supplied unless the detector is cold and the leakage current low. After a trip, such modules must be manually reset by a pushbutton.

The bias supply is delivered to the detector via the preamplifier and it is convenient to supply it to the preamplifier through coaxial cable. The 93 Ω cable referred to above, RG62, is not rated for high voltage. Instead, cable RG59, designed to withstand high voltages, is the type that should be used to connect the bias supply to the preamplifier. The high-voltage connectors are also different from the standard BNC bayonet plug and socket used for pulse cables. Modern systems use SHV plugs and sockets. Older modules might still be found with a slightly different type of socket – designated MHV. These are similar to the BNC socket in design and it is possible (albeit with some persuasion) to connect BNC cables onto MHV sockets. My advice would be to standardize and replace MHV sockets on older units with SHV connectors that are completely incompatible with BNC plugs.

4.3 PREAMPLIFIERS

The charge created within the detector by interaction with the gamma radiation is collected by the preamplifier. In spite of its name the function of the preamplifier is not to amplify the pulse – it merely goes before (i.e. pre-) the amplifier – but to interface the detector to the amplifier and collect the charge generated by absorption of the gamma-ray. It provides a high impedance load for the detector and a low impedance source for the amplifier. Preamplifiers, in general, can have various modes of operation: current-sensitive, voltage-sensitive and charge-sensitive. Only the latter type is used in high-resolution gamma spectrometry using semiconductor detectors. It has advantages in terms of noise performance and because the gain is independent of detector capacitance.

4.3.1 Resistive feedback preamplifiers

The schematic arrangement of a conventional charge-sensitive preamplifier is shown in Figure 4.6. Charge from the detector is collected on the capacitor C_f over a period of time, effectively integrating the detector charge pulse. As the charge is collected, the voltage on the capacitor (and ultimately at the preamplifier output) rises, producing a step change in voltage. Without further action, the voltage at the input would remain at that level. To allow the charge to leak away and prepare the input for the next pulse, a large resistor, R_f, called the **feedback resistor**, is connected in parallel with the capacitor. This type of preamplifier is referred to as a **resistive feedback preamplifier**.

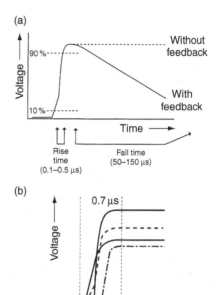

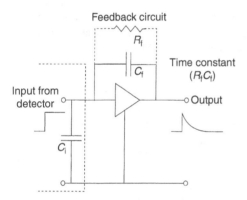

Figure 4.6 Schematic diagram of a resistive feedback charge coupled preamplifier

Figure 4.7 Shape of the output pulse from a resistive feedback preamplifier: (a) definition of rise time and fall time; (b) actual rising edge shapes derived from a 45 % detector

The output pulse shape is characterized by a fast rise time, 100 to 700 nanoseconds, determined by the charge collection characteristics of the detector, and long decay time (several tens of microseconds) determined by the time constant of the feedback circuit, $R_f C_f$. The shape of the output pulse is shown in Figure 4.7.

As I explained at length in Chapter 3, Section 3.6.2, the detailed shape of the leading edge of the pulse will depend on the position of the interaction within the detector. The overall charge collection time depends upon detector size, 300–400 ns being typical of detectors of, say, 50 % relative efficiency. The information carried by the preamplifier pulse is in its rising edge. Ideally, its height is proportional to the gamma-ray energy absorbed by the detector.

The resistive feedback preamplifier has two major limitations:

- Because of the long decay time of the pulse, at other than very low count rates, successive pulses pile up one on top of another (Figure 4.8). This is inevitable and in itself is not of great consequence since, as we shall see later, the amplifier is capable of extracting the pulse height information from the rising edge of each pulse. More importantly, at a high count rate the average DC voltage level at the input will rise above that at which linearity between charge and pulse height

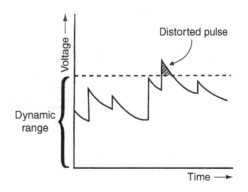

Figure 4.8 Pile-up at the output of a resistive feedback preamplifier

can be assured, referred to as its **dynamic range**. Ultimately, at higher count rates, the DC voltage will rise to near the supply voltage for the preamplifier (24 V), at which point the transistors within the preamplifier will cease to operate; it will 'lock-up' and no pulses will be output at all. This condition occurs when the product of detector current and feedback resistance exceeds the dynamic range. Most modern spectrometry preamplifiers provide a visible indication of count rate overload in the form of an LED indicator mounted on the preamplifier casing (Figure 4.9). This will illuminate when the input count rate is greater than 75 % of the maximum. This is an excellent idea but unfortunately the normal physical arrangement of detector and shielding means that this is invisible to all but contortionists!

Lock-up, or preamplifier saturation, is a matter of some concern when the instrumentation has the function of watching for emergency situations. If the emergency were severe enough, a system, perfectly satisfactory at low count rate, might lock-up, giving an outward appearance of a safe condition. The maximum rate at which the preamplifier can satisfactorily handle pulses depends upon the average height of those pulses. Fewer larger pulses will exceed the linearity threshold rather than smaller pulses. Preamplifier count rate performance is quoted by the manufacturers in terms of an **energy rate** – $\mathrm{MeV\,s^{-1}}$. More than $10^5\,\mathrm{MeV\,s^{-1}}$ would be typical (for example, a performance of 2×10^5 counts per second of $^{60}\mathrm{Co}$ is claimed for the Canberra 2002 preamplifier). When assessing count rate capability, two things should be borne in mind. First, the average energy absorbed in a detector is considerably lower than the gamma-ray energy because the majority of interactions will transfer less than full energy to the detector. The average energy depends upon the spectrum 'shape' but might be as little as one third of the highest major gamma-ray energy in the spectrum. Secondly, a high preamplifier throughput will not necessarily be maintained through the rest of the electronic system. The limitations in high count rate systems will be discussed in detail in Chapter 14.

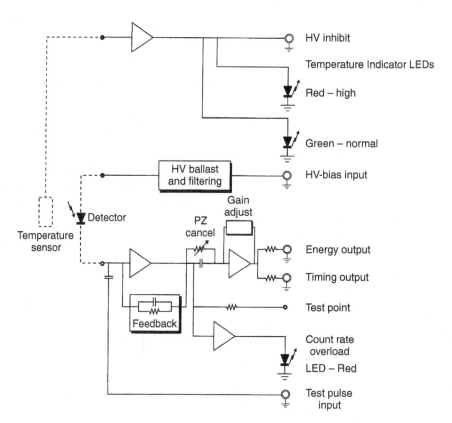

Figure 4.9 Block diagram of Canberra 2002 resistive feedback preamplifier indicating the function of the various panel sockets and indicators

• The second limitation of the resistive feedback preamplifier is that the feedback resistor R_f has an intrinsic noise associated with it (*Johnson noise*) and this can be a significant problem with pulses of very small size. In order to minimize this source of noise the value of R_f is chosen to be high. This, in turn, means that the decay time of the output pulse is long, which exacerbates the pile-up problem. In principle, it would be possible to reduce the time constant by reducing C_f but doing that would affect the linearity of the preamplifier. There is, however, scope for reducing the value of R_f, in order to trade off resolution for count rate performance and this will be referred to in Chapter 14, Section 14.3.2.

4.3.2 Reset preamplifiers

One solution to both these problems is to dispense with the feedback resistor altogether. Remember that the information carried by the pulse is in the height of the step change in voltage and that the feedback resistor is only there to return the DC voltage level at the input to normal in readiness for the next pulse. If we ignore the feedback and allow the DC level to build up stepwise, as in Figure 4.10, we can rely on the filtering in the amplifier to recover the height information at a later stage. Eventually, though, the voltage level must approach the limit of

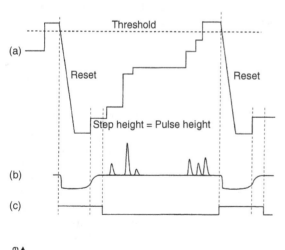

Figure 4.10 Schematic waveforms involved in a reset preamplifier system: (a) output from the preamplifier; (b) output from the linear amplifier; (c) inhibit signal from preamplifier to amplifier

linearity and at that point, the DC level must be returned rapidly to zero (i.e. reset) after which the stepwise integration process can be continued. This is the basis of the automatic reset preamplifier. At a predetermined voltage limit, the pulse being integrated is allowed to continue to full height and then the reset process takes over to return the output level to zero. The effect of the sudden fall in DC level as the reset takes place can cause untoward effects within the amplifier. The preamplifier must provide a gating signal to inform the rest of the electronic system so that any pulses output by the amplifier during the reset, likely to be spurious, can be ignored. There are two ways of performing the reset:

• In low-energy X-ray systems, resetting will most commonly be by a pulsed optical device with an LED triggering a light sensitive solid-state switch.
• For higher energies at high count rates, a transistor reset circuit is used. For gamma-ray spectrometry, one would use a **transistor reset preamplifier** or TRP.

A significant major advantage of the TRP is that it cannot lock up even at extremely high energy rates. Against this is a more complicated gating arrangement to prevent the amplifier attempting to analyse spurious pulses that might be induced during the reset interval. The TRP can also impose a significant extra dead time on the system. The time during which the rest of the system is inhibited is perhaps two or three times the pulse width and might be tens of microseconds per pulse. This represents a significant extra dead time over and above the MCA dead time and ultimately limits the throughput of the preamplifier. If the gamma-ray energy is high, then on average, fewer pulses will be acquired before the output voltage level exceeds the reset threshold and this will increase the number of resets per second, making the extra dead time penalty greater.

4.3.3 The noise contribution of preamplifiers

The specification for a spectrometry preamplifier will often include a statement of its noise characteristics. The sources of noise in the preamplifier and its effect on the overall resolution of the system will be discussed in Chapter 6. A major source is the external capacitance, C_i in Figure 4.6. So for example, Canberra specifies the performance of its 2002 preamplifier as 0.57 keV for 0 pF input capacitance and 2.2 keV for 100 pF. Two figures are provided because there is a continuous variation of noise contribution with capacitance, as shown in Figure 4.11(a).

This external capacitance arises from the detector itself and the interconnections between it and the preamplifier

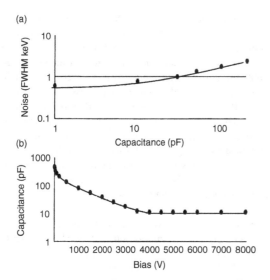

(a)

(b)

Figure 4.11 (a) The variation of preamplifier noise with input capacitance. (b) The measured capacitance of a detector at different bias settings

input. The detector capacitance is a function of the size of the depletion layer, as I explained in Chapter 3, Section 3.5. The example of a coaxial detector used in that section was estimated to have a capacitance of 31 pF and, if it were used with the 2002 preamplifier, a noise contribution of about 1 keV would be expected. You should be aware that unless the detector is fully depleted by providing the appropriate bias, the capacitance might be much greater than expected. Figure 4.11(b) shows the capacitance of a typical 30 % detector as a function of bias voltage. Clearly, a low bias would have a profound detrimental effect on the system resolution. Perhaps it would be appropriate to explain how one may measure the noise performance of a preamplifier in practice. It is first necessary to determine the system gain in terms of voltage output per electron-volt input. The procedure is as follows:

- Set up the system, perform a normal energy calibration and then remove all sources from the vicinity of the detector.
- Connect a step pulse generator to the test pulse input on the preamplifier and adjust the pulse height so that the pulses appear at a convenient energy in the spectrum. Measure the equivalent energy of the pulses, E_p, using the spectrometer calibration.
- Using an oscilloscope, measure the height of the pulses, V_p, at the output from the linear amplifier. The system gain is then V_p/E_p.

- Disconnect the pulse generator and, again using the oscilloscope, measure the rms (root-mean-square) noise voltage, V_{rms} at the linear amplifier output.
- The preamplifier noise is then given by the following equation:

$$FWHM_{noise} = 2.35 \times V_{rms} \times E_p/V_p \qquad (4.4)$$

The measurement can be made using a mono-energetic gamma-ray source instead of a pulse generator but the greater spread of pulse heights makes measurement of the pulse height more difficult.

An alternative, and simpler, estimate of the electronic noise can be made from the FWHM vs. energy calibration (Chapter 6, Section 6.5). Equation (4.1) relates the FWHM of a peak to the various uncertainty contributions. Common sense tells us that if the gamma-ray energy is zero, the uncertainty on the charge production and on the charge collection must also be zero. Hence, if we extrapolate the width calibration to zero energy we will have an estimate of the electronic noise.

4.3.4 The rise time of preamplifiers

Ideally, we wish to collect the charge as quickly as possible and it would not be satisfactory if the preamplifier itself limited that process. The specification of a preamplifier will include a statement of the rise time of its output, again related to the input capacitance (e.g. < 20 ns at 30 pF input capacitance for the instrument referred to above). It is sufficient if this is small compared to the rise time of the detector pulses so that the effective rise time is determined by the detector, not by the preamplifier.

4.4 AMPLIFIERS AND PULSE PROCESSORS

In this section and those that follow, I will describe the functions of analogue pulse processing systems. Until recently, they have been the norm. Since the first edition of this book, digital systems have been introduced with many advantages. In time, they could very well supplant analogue systems for routine 'out-of-the-box' gamma spectrometry. I will persist with a detailed explanation of analogue processing, not least because there are still many analogue systems around, but also because an understanding of analogue processing helps one to appreciate how the digital systems accomplish the same functions.

4.4.1 The functions of the amplifier

The very sharply peaked pulses emanating from the traditional resistive feedback preamplifier are not suitable for direct measurement of peak height. (From a measurement point of view alone, the ideal would be a pulse that gradually approached a relatively flat top and then fell away as rapidly as possible to the baseline – quite the opposite from a preamplifier pulse.) In any case pileup, which is inevitable at anything other than low count rates, prevents a simple measurement of pulse height relative to the amplifier baseline voltage. This is demonstrated in Figure 4.12 where a number of pulses with identical step height, when piled-up, produce peak voltages at different heights. We would prefer to have narrow peaks of an ideal shape to work with. I will talk later about what this ideal shape might be and concentrate for now on extracting the pulse height information, which we should remember is proportional to gamma-ray energy absorbed, from the rising edge of the pulse. This is accomplished by electronic filtering, also referred to as **shaping**. Shaping is the primary function of the amplifier but in order to correct various undesirable consequences of the shaping, which would impair resolution, an amplifier for high-resolution spectrometry must also provide pole-zero cancellation and baseline restoration. Spectrometry amplifiers may also provide pile-up rejection. The sophistication of the modem amplifier, in which amplification is more or less an incidental function, is such that the very term 'amplifier' is an understatement. Perhaps 'pulse processor' would be a more expressive sobriquet. I shall explain each of the amplifier functions separately.

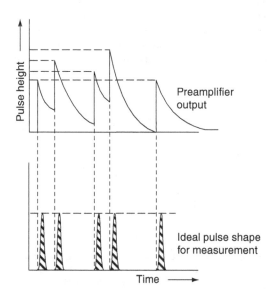

Figure 4.12 Piled-up resistive feedback preamplifier output and the desired converted pulses

4.4.2 Pulse shaping

Figure 4.13 shows, schematically, two basic filters and the effect they have on a step pulse applied to their inputs. For now, we can take the step pulse as approximating the sharp rising edge of the preamplifier pulse. The **differentiator**, also referred to as a **high-pass filter**, allows only the high-frequency components of the pulse to pass, blocking off the DC component of the step and resulting in a sharply peaked output. The fall time of the output pulse is determined by the product of R and C and in the context of gamma spectrometry, pulse processing is likely to be in the range 1 to 12 μs.

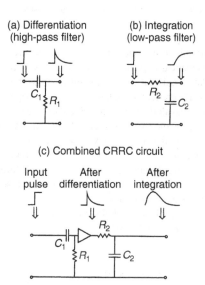

Figure 4.13 Effect of filters on step pulses: (a) differentiator; (b) integrator; (c) combined differentiator and integrator

Figure 4.13(b) shows an **integrator**, a series resistor followed by a capacitor in parallel with the pulse. This is a **low-pass filter** – it passes only the low-frequency components of the pulse. It would modify a step pulse by slowing down the rise of the leading edge. The rate of rise would again be determined by the value of RC for the circuit. If we present the output from a differentiator to the input of an integrator, the step pulse is converted to a short, somewhat asymmetric pulse, as shown in Figure 4.13(c). A preamplifier pulse passing through this combined circuit would be converted to a pulse that is much shorter (a few microseconds long rather than hundreds of microseconds) and of a shape much more easily handled by the pulse height measuring circuits in the analogue-to-digital converters (ADCs). This simple type of shaping would be referred to as **RC shaping**. (RC shaping is very easily demonstrated experimentally by using the simple circuits shown in Figure 4.13. Capacitors of 0.01 μF and resistors of 100 Ω will give time constants of 1 μs. Apply a square wave from a pulse generator, a tail pulse from a pulser unit, or possibly a preamplifier output, to the input and examine the output with an oscilloscope.)

4.4.3 The optimum pulse shape

I have emphasized the importance of noise in spectrometry systems and perhaps I should consider at this point what would be the ideal shape for our pulses. Figure 4.14 shows a number of different pulse shapes, together with a theoretical parameter, 'relative noise', which the shaping network might introduce. There is no need here to present a mathematical explanation; suffice it to say, the smaller the relative noise, the better the ultimate resolution of the system. Theoretically, the optimum pulse shape would be a cusp and so all the figures are given relative to this shape. It is not possible to produce exactly the cusp shape using practical circuits and, in any case, such a pulse would not be much more satisfactory for ADC height measurement than the original preamplifier pulse. Although not achievable in analogue systems, the cusp filter, somewhat modified, can be used in digital signal processing systems where the filter is defined digitally rather in terms of electronic components (Section 4.11).

Early amplifiers for gamma spectrometry (using scintillation spectrometers at the time) used RC shaping and allowed some control over the output pulse shape by providing independent control of the differentiation and integration time constants. However, theory suggests that the lowest noise contribution is found when the integration and differentiation times are made equal. On all modern amplifiers, the shaping time constants are made equal and are controlled by a single selector knob.

The simple passive RC shaping circuit with a single integration used in those units is much inferior in terms of noise (1.36 relative to the cusp). If a second integrator is added to the circuit, the relative noise drops to 1.22. Theoretically, if an infinite number of integration stages were added the best noise performance of this type of shaper, 1.12, would be achieved. This is equal to the performance of a Gaussian pulse-shaping network and only slightly worse than that of triangular shaping. A true Gaussian pulse shape is not realizable in practice and all amplifiers for gamma spectrometry provide instead a 'semi-Gaussian' shaped output. This is normally equivalent to a single differentiation followed by two integrators. The integration will be done by an active circuit (involving operational amplifiers), rather than by the simple RC filter described above, providing a network with a relative noise contribution close to that of true Gaussian shaping. On the amplifier itself, the output will usually be labelled 'UNI' or 'UNIPOLAR'.

The width of the unipolar output pulse obviously depends upon the shaping time constants but there does seem to be some disagreement in the published data on the dimensions of semi-Gaussian unipolar pulses. To clarify the matter, Figure 4.15 and Table 4.1 summarize a number of measurements, made by John Hemingway,

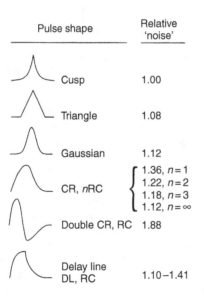

Pulse shape	Relative 'noise'
Cusp	1.00
Triangle	1.08
Gaussian	1.12
CR, *n*RC	1.36, *n* = 1 1.22, *n* = 2 1.18, *n* = 3 1.12, *n* = ∞
Double CR, RC	1.88
Delay line DL, RC	1.10–1.41

Figure 4.14 Relative noise contribution of different pulse shapes

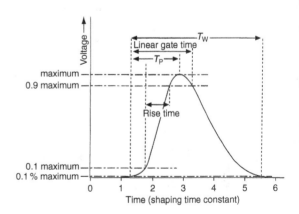

Figure 4.15 Definition of the various time factors relating to pulse shape

Table 4.1 Measured timing factors for semi-Gaussian output pulses

Factor	Time interval	Symbol	Time[a]
Rise time	0.1 to 0.9 of pulse maximum	—	$1.26 + 0.05$
Peaking time	threshold[b] to maximum	T_P	$2.1 + 0.1$
Linear gate time	threshold to 0.9 of max. beyond max.	T_{LG}	$2.6 + 0.2$
Width	threshold to threshold	T_W	$5.6 + 0.5$

[a] Time is specified in units of time constant.
[b] The threshold used was, as near as possible, 0.1 % of peak maximum.

using three different amplifiers (ORTEC 673, Canberra 2020 and Tennelec 244) with time constants varying from 0.5 to 12 μs and pulse heights from 0.5 to 7 V. The pulses were derived from ^{60}Co counted on a standard detector system, an ORTEC n-type coaxial detector of 23 % relative efficiency and about 2 keV resolution at 1332.54 keV. The various time factors, when scaled to the appropriated time constant, T_C, appeared to be independent of time constant, pulse height and amplifier. The only exception to this was that the Tennelec amplifier gave smaller values of total width T_W than the other two: $(4.9 \pm 0.3) \times T_C$, as compared to $(5.9 \pm 0.3) \times T_C$ for the other two. Some figures within the manufacturers' reports suggest a factor of 8.

If an extra differentiator is added to the shaping circuit, a pulse that crosses the baseline with a negative portion is produced. This is the BIPOLAR pulse output found on almost all spectrometry amplifiers. As we can see, this has a higher noise contribution than the semi-Gaussian pulse and is not normally used for spectrometry. It can, however, be used as a source of pulses when an accurate measure of height is not needed and where the amplifier is AC coupled to the following circuits. In such a situation, the fact that the pulse possesses positive and negative excursions minimizes baseline shift. The very clear transition from positive to negative also makes bipolar pulses appropriate for crossover timing purposes.

Triangular pulse shaping (perhaps more accurately referred to as 'quasi-triangular' shaping because the output pulse is only an approximation to a triangle) is available on some modern high specification amplifiers. It is particularly useful when operating the system with a shaping time constant shorter than the optimum (see below). For the same peaking time, the triangular pulse width is somewhat narrower than the semi-Gaussian pulse

(for their 2025 research amplifier, Canberra quote 2.7 × the peaking time as opposed to 2.9 for the semi-Gaussian, although these figures are not consistent with the measurements in Table 4.1). For the same overall dead time per pulse (related to the pulse width), triangular shaping can be expected to contribute 8 % or so less noise.

4.4.4 The optimum pulse shaping time constant

In Section 4.3.3, I pointed out that the electronic noise at the preamplifier input makes a significant contribution to the energy resolution of a semiconductor detector system. This noise contribution can be minimized by choosing an appropriate amplifier shaping time constant. Figure 4.16 shows, schematically, the variation in noise (expressed as a contribution to the resolution FWHM) as a function of shaping time constant. The reason why this curve has a minimum will be discussed in Chapter 6. For now, we need only be aware that this minimum, the so-called **noise corner**, exists and when setting up our amplifier we must check the resolution over a range of time constants to seek this minimum.

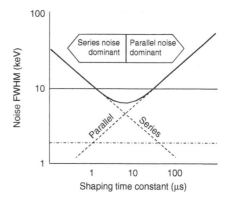

Figure 4.16 Variation of noise, expressed as the contribution to overall resolution, with shaping time constant (schematic)

We cannot ignore the role of the collection processes in the detector. As we saw in Chapter 3, it takes rather longer to collect all of the charge in a larger detector than in a small one. Although the rise time of the input to the shaping network is less than 1 μs, the charge integration time must be somewhat longer than this to ensure complete charge collection. Incomplete charge collection gives rise to a proportion of pulses rather smaller than they should be and would lead to tailing at the low-energy side of the spectrum peak. This tailing may be so slight as to have little visual effect on the peak but would be detectable by a measurement of peak width. In

Figure 4.16, the effect of poor charge collection would be added to the series noise effect at low time constant.

Regardless of the electronic noise considerations there is, therefore, a minimum shaping time demanded by charge collection considerations. It is recommended that the shaping time should be several times the longest pulse rise time. As a general 'rule-of-thumb', the optimum shaping time for small HPGe detectors is 2–4 μs and for larger high-purity Ge detectors 4–10 μs. Now, the longer the shaping time, the longer the amplifier output pulse (see Figure 4.15). It is self-evident that the longer the pulse, the fewer pulses per second can be transmitted through the system. There is, then, a conflict if we wish to use a large detector at high count rate. The maximum of the resolution versus shaping time curve is quite broad and it would be common sense to choose as small a shaping time within that minimum as possible. If the detector is to be used at high count rate, there is scope for selecting an even shorter time constant and accepting a slightly impaired resolution. For example, a reduction in time constant from 3 μs to 2 μs would improve throughput by 33 % at the expense of perhaps only a few percent lower resolution.

4.4.5 The gated integrator amplifier

It is worth examining the problems of poor charge collection further. Figure 4.17 shows two curves, the higher is the output from a shaping circuit with a time constant very much longer than that needed to allow full charge collection. The curve below is the pulse shape that might be obtained with a more practical time constant. The difference in the pulse height is referred to as **ballistic deficit**. If this deficit were constant and proportional to pulse height, there would be no problem. Unfortunately, it varies with rise time and, as we know, there can be a considerable variation in the rise time of pulses emanating from a semiconductor detector. These variations in ballistic deficit will inevitably lead to a loss of peak resolution in the final spectrum.

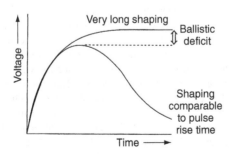

Figure 4.17   Definition of ballistic deficit

Although the charge may not be collected within the effective integration time of the shaping circuit, thus contributing to peak height, charge will continue to be collected during the remainder of the pulse length. Thus, we may have a pulse with a height which does not represent the total charge collected but whose *area* does. This is exploited in the **gated integrator (GI) amplifier**. The shaped input pulse is integrated on a capacitor for the full duration of the shaped pulse. At the end of the integration time, switches isolate the input and return the output to baseline as rapidly as possible. This results in the peculiar shaped output shown at the top of Figure 4.18. In spite of this odd shape, the pulse is still a linear pulse and retains the proportionality between energy absorbed in the detector and pulse height.

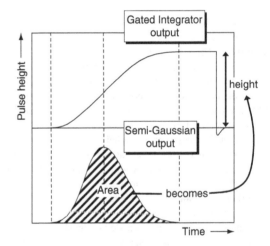

Figure 4.18   Comparison of semi-Gaussian and gated integrator output pulses

The shaping circuit preceding the gated integrator, here referred to as a **pre-filter**, would be an active RC shaping circuit but would not necessarily present a semi-Gaussian pulse to the integrator. The typical active shaping used provides an approximately *trapezoidal* pulse. The noise performance of the gated integrator can approach that of normal triangular shaping but various losses of signal-to-noise ratio occur because of limitations in the practical realization of the integrator. This means that, although the gated integrator provides much superior performance at high count rate when short shaping times are used, its performance under more normal conditions is likely to be slightly inferior to semi-Gaussian shaping. For that reason, commercial gated integrator amplifiers usually provide an alternative semi-Gaussian output as well. There are many particular problems associated with high count rate measurements and Chapter 14 has been devoted to an in-depth discussion.

4.4.6 Pole-zero cancellation

Figure 4.13 demonstrated the effect of simple shaping circuits on a step function that was taken, for the time being, to be a reasonable representation of the leading edge of the preamplifier pulse. However, normal resistive feedback preamplifiers do not provide a step function but a pulse with a long falling tail. If the relevant pulses are examined with an oscilloscope, there may appear to be little difference from the step pulse response. It is only when the part of the pulses just before they return to the baseline is examined on an expanded voltage scale that it becomes apparent that the effect of the differentiator on a tail pulse is somewhat different (Figure 4.19).

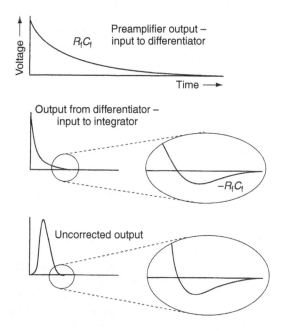

Figure 4.19 Schematic illustration of the pole-zero problem

The tail of the pulse passing through the differentiator forces the output to fall below and then rise towards the baseline with a time constant equal to that of the feedback circuit in the preamplifier. This is then transmitted through the integrator to the output pulse. A second pulse following close behind the first may find itself in the trough of the depressed baseline and be measured incorrectly by the MCA system. This problem can be corrected by a **pole-zero cancellation** (PZ) circuit. The term 'pole-zero' relates to the mathematical representation of the effect and has no spectrometric significance. In effect, this correction matches the differentiator circuit to the fall time of the preamplifier pulse.

In principle, the circuit introduces a variable resistor across the capacitor of the differentiator that can be adjusted by means of a multi-turn potentiometer on the amplifier front panel (Figure 4.20). Some amplifiers are provided with two potentiometers – coarse and fine adjustment.

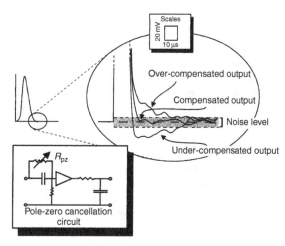

Figure 4.20 Pole-zero cancellation in practice, showing the schematic circuit and the effect of cancellation on the tail of the pulse

The adjustment is normally made using an oscilloscope to examine the tail of the pulse on an expanded voltage scale (say, 20 mV/division). If the pulse undershoots the baseline, compensation is needed by a (usually) clockwise adjustment. If the control is turned too far, then the output pulse will be overcompensated – corrected by an anticlockwise turn. With a little trial and error, the control can be adjusted so that the pulse returns to the baseline as directly as possible. The pole-zero cancellation should be checked whenever gross changes are made to the amplifier settings. In particular, it is essential that it be checked every time the shaping time constant is changed. Some would argue that checking is necessary after any gain change is made. For small gain adjustments, it is not essential. Certainly, a poorly adjusted PZ cancellation is the most immediate and effective route to poor resolution and should be the first check whenever there are problems of deteriorating resolution. It is worth emphasizing the point that unless the appropriate part of the pulse is examined in detail poor pole-zero cancellation may not be apparent.

Some high specification amplifiers have built-in facilities to help with pole-zero cancellation. For example, the ORTEC 672 Spectroscopy Amplifier provides automatic cancellation at the press of a button. Some amplifiers provide LED over/under indicators to assist with the adjustment. Nevertheless, because of the importance of this setting I would advocate that every gamma spectrometry laboratory should have ready access to an oscilloscope. With experience, a quick look at the pulses coming from the preamplifier or the amplifier can quickly reassure one that everything is normal, or lead one to a solution if it isn't.

Note that the transistor reset preamplifier with its step pulse output has no need for PZ adjustment. If such a preamplifier is used, the pole-zero potentiometer should be adjusted to its fully anticlockwise position which corresponds to infinite time constant. Certain amplifiers provide a switch to disable the PZ cancellation.

4.4.7 Baseline shift

The differentiator in the shaping circuit contains a capacitor in series with the pulse flow (see Figure 4.13). A capacitor will not allow the passage of a DC current and so this is called an *AC coupling*. When a pulse passes through a capacitor, unless there are other correcting factors, the baseline must be suppressed slightly in order to make the net area of the pulse above and below the baseline equal. This is baseline shift and for a single pulse will be negligible but for a succession of pulses may be significant (Figure 4.21). Now the ADC in the MCA will measure the height of the pulse relative to a fixed reference voltage and so variations in the baseline

to the pulse will cause errors in the measurement of pulse height.

For a regular periodic series of identical pulses, the baseline shift would be constant but the random manner in which pulses appear means that this baseline shift will be variable and will depend upon the average number and size of pulses passing through the filter on a millisecond-by-millisecond basis. (This effect can be seen quite clearly by observing the output pulses from the amplifier on an oscilloscope. Normal DC coupling will give a steady display. Switching in AC coupling on the oscilloscope will produce a display which 'dances' up and down randomly.) The effect of this random uncertainty on the spectrum will be a degradation of the resolution rather than a shift in the energy calibration. Note that the problem can be avoided by using the bipolar amplifier output. The two lobes of the bipolar pulse have a self-cancelling effect. However, using bipolar pulses carries a resolution penalty that is likely to be at least as large as the baseline shift penalty.

High-quality spectrometry amplifiers will have a **baseline restorer** (BLR) built into the final stage of the amplifier. The most effective type at high count rate (when the problem is particularly severe) is the gated restorer in which the baseline is maintained at ground potential during the period between pulses. Usually the BLR will have at least some of the following operating options:

- AUTO/MANUAL threshold (may also be labelled AUTO/VAR). The BLR must distinguish between the gross baseline shift and deviations caused by noise on the input signal. This is achieved by setting a threshold below which the BLR is inactive. The AUTO switch position allows the BLR restorer to set its noise discriminator threshold automatically. MANUAL provides manual control of the threshold by means of a multi-turn potentiometer.
- HIGH/LOW/AUTO(/PZ) rate. Since the BLR effectively operates between pulses, the speed of correction must be increased as the count rate increases. This switch allows the user to optimize the BLR for different count rate situations. If the AUTO position is selected, the instrument will adjust this automatically according to the input count rate. Otherwise, select HIGH for a high count rate and LOW for a low count rate. In most cases, the AUTO position will be found to be most useful, but if low-frequency noise is a problem, then selecting the HIGH position may help. Sometimes, there will be a PZ position on this switch. This selects the lowest correction rate and should be used when setting up the pole-zero cancellation. It may also provide slightly better resolution when counting at low count rate with long shaping time constants.

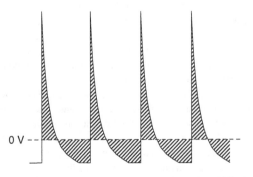

Figure 4.21 Baseline shift due to a regular series of identical pulses. The total areas above and below the baseline are equal

In this context, a high count rate would be when the duty cycle of the amplifier is more than 20 % and a low count rate when below 5 %. **Duty cycle** is the proportion of the time when the amplifier is busy and, assuming that the amplifier dead time is 6 times the shaping time constant, it can be estimated (as a percentage) as:

Pulses per second × shaping time constant (μs)

$$\times 6 \times 10^{-4} \qquad (4.5)$$

For example, an output count rate of 2800 pps with 3 μs shaping would give a duty cycle of about 5 % while 17 000 pps with 2 μs shaping would give 20 %.

- SYM/ASYM correction mode. The symmetrical (SYM) mode provides identical correction rates, no matter whether the baseline has shifted above or below the baseline. The asymmetrical position (ASYM) supplies a higher correction rate if the shift is positive than if negative. In general, the asymmetric mode would be preferred but the symmetric correction is said to be better if there are baseline problems due to microphonic or external noise pickup. (It would be better, of course, to eliminate these external problems rather than rely on the BLR restorer to cope with the problem.)

A normal setup would be AUTO threshold, AUTO rate and ASYM mode. If the required resolution cannot be achieved, then other settings may be beneficial. These can only be determined by experiment.

4.4.8 Pile-up rejection

Pile-up, also referred to as **random coincidence** or **random summing**, is the consequence of two, or more, gamma-rays being detected almost simultaneously. If they all arrive within the width of the amplifier output pulse, they will not be recognized as separate events. The resulting output pulse will be equivalent to the height of the first pulse received plus a proportion of the height of the second pulse depending upon how close the pulses are. The situation is demonstrated in Figure 4.22, which plots the calculated preamplifier and amplifier output pulse shapes when two pulses of equal size arrive at the preamplifier input. (A simple inactive pulse shaping is assumed and the pulses are about twice as wide as the equivalent normal semi-Gaussian pulses.) The shapes are calculated for different degrees of overlap. The diagram shows that for very close coincidence the height of the combined output pulse, which the MCA will attempt to measure, is almost the sum of the two pulse heights.

Such random coincidence is undesirable because it causes counts to be lost from the full energy peaks in the

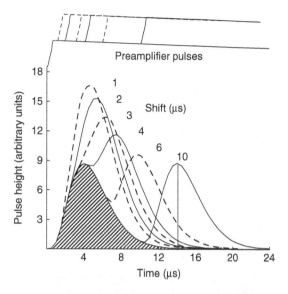

Figure 4.22 Calculated pulse shapes resulting from pile-up. Amplifier shaping is semi-Gaussian (differentiation followed by four passive integrations) with time constants of 1 μs

spectrum. If we imagine that both gamma-rays arriving at the detector within the resolving time of the amplifier were fully absorbed, and therefore destined to contribute to their respective full energy peaks, the coincidence will result in the loss of one count from each peak and the appearance of a count somewhere else in the spectrum. In the example spectra in Figure 2.12 all the counts above the full energy peak are the result of pile-up. The probability of random summing increases with the square of the total count rate and will be discussed in more detail in Chapter 7, Section 7.6.8.

It is evident from Figure 4.22 that the pile-up amplifier output pulse is misshapen, the peak of the pulse occurs later and the pulse is wider than expected for a single pulse. Therein lays the basis for a hardware solution to the problem. Without delving into the detailed electronic circuitry to any depth, Figure 4.23 explains the procedure. In parallel with the normal pulse shaping circuits, which produce the normal amplifier output, is put a fast-shaping amplifier (the fast differentiator shown below in Figure 4.25). Using very short time constants, this produces a pulse with a high series noise component but which is merely used to indicate, as rapidly as possible, that a pulse has been detected. The effect of the noise is then eliminated by using a discriminator to derive a short logic pulse. This then triggers a time period, called the **inspection interval**, which is equivalent to the expected

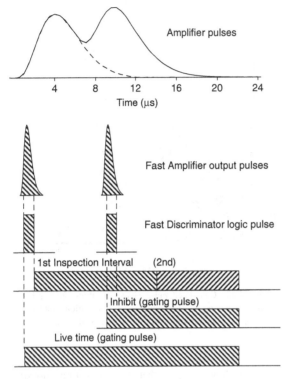

Figure 4.23 The principle of the spectroscopy amplifier pile-up rejection system

pulse length. If another pulse arrives within this inspection period (which will be detected by its fast amplifier pulse), a gating pulse will be generated which can be used to inhibit the storing of the original pulse. The loss of these pulses from the spectrum must be accounted for by a slight increase to the MCA dead time and this is accomplished by a dead time gating signal from the amplifier.

The actual manner in which this gating pulse is derived depends upon whether the pile-up is on the leading or trailing edge of the pulse. For statistical reasons, a dead time gate pulse from the trailing edge pile-up must be terminated by the next pulse to be detected. At low count rates, this could be some time, resulting in an exceedingly long, and unrealistic, dead time pulse. Because of this, it is advisable to disable the pile-up rejection circuitry at low count rate when there is no need for it anyway. Some pile-up rejectors take a sensible view of matters and open the dead time gate after a default delay time.

As described, it is not possible to resolve pulses that are closer than the expected width of a single amplifier output pulse, 2.5 to 3 times the peaking time depending upon pulse shape. Any pairs of pulses closer than this will be rejected. Clearly, there are situations where a pulse

may arrive before the previous pulse has returned to its baseline but will peak afterwards. This pulse would have a valid height but be rejected anyway (see the 10 µs delayed pulse in Figure 4.22). Reducing the resolution time by taking account of such matters can yield a substantial reduction in unnecessary rejections.

Amplifiers are available which also provides an alternative means of pile-up rejection by detecting the delayed peaking time of a pile-up pulse. Since pulses affected by ballistic deficit or charge trapping will also be delayed, these pulses will also be rejected.

Setting up the standard pile-up reject system involves adjusting the threshold of the fast discriminator to eliminate spurious rejections due to the noise associated with the fast-amplifier output. The procedure is fairly simple. With no source on the detector, turn the PUR threshold control counter-clockwise until the PUR LED glows continuously. Turn the control slowly clockwise until the LED flashes only in response to each input pulse. (The transition from glowing to flashing may take only a few degrees of adjustment.) At this point, the PUR threshold is set just above the relevant noise level. In practice, I find it better to turn the control a further third of a turn. Having set up the threshold, it would be wise to check that the correction is satisfactory, up to the maximum count rate the system is to be used for, by using the moving source method described in Chapter 7, Section 7.6.8.

The effectiveness of pile-up rejection is impressive. Figure 4.24 shows the pile-up in a spectrum of ^{137}Cs at a moderate count rate of 3300 pps. It is obvious that even though pile-up rejection removes a considerable proportion of pile-up pulses it can never be 100 % effective (note that Figure 4.24 is plotted on a logarithmic scale). The sum peak at 1323.3 keV is due to a random coincidence of two completely absorbed 661.66 keV gamma-rays arriving so closely in time that the total pulse height is equivalent to the total gamma-ray energy. Because these arrive within the resolution time of the pile-up rejector (which might be as little as 250 ns), they can never be removed by an electronic method based upon peak shape. A numerical method of correction for this residual summing, which would only be needed at very high count rate, will be discussed in Chapter 7, Section 7.6.8.

4.4.9 Amplifier gain and overview

Perhaps surprisingly, there is little comment to be made on the only remaining function of the amplifier – amplification. Almost all spectroscopy amplifiers provide two amplification (gain) controls: coarse, providing switched gain factors, and fine, providing continuous gain control on a multi-turn potentiometer. The gain is adjusted to

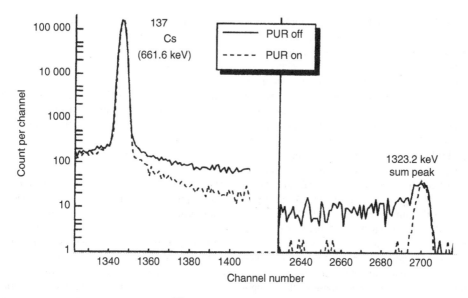

Figure 4.24 The effect of pile-up rejection on a ^{137}Cs spectrum

match the maximum voltage output of the amplifier to the voltage range of the MCA. Amplification is performed early in the pulse processing chain to avoid amplification of the extra noise introduced by the various operations. Figure 4.25 shows the block diagram of a typical amplifier, putting these into the context of a complete instrument.

Points to note are the pole-zero cancellation circuit at the start of the chain, so that the amplifier is matched to the preamplifier as early as possible, and the base line restorer

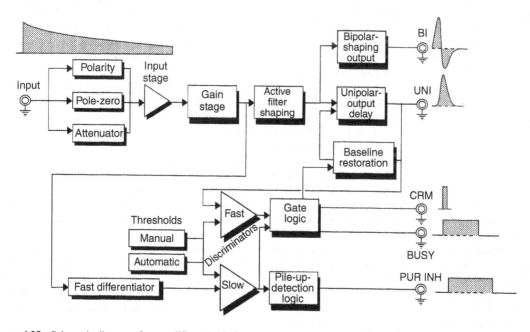

Figure 4.25 Schematic diagram of an amplifier suitable for gamma-ray spectrometry. Adapted from *Detectors and Instruments for Nuclear Spectrometry*, by ORTEC

at the end of the chain, leading to the unipolar output so that DC coupling to the MCA can be assured. In the lower part of the diagram are gating and pile-up rejection logic circuits. The latter is dependent upon a signal indicating the start of the pulse. This is derived by applying a very fast differentiator to the normal pulse. This very short pulse can also be used to provide a logic pulse output whenever a pulse is detected by the amplifier and whether or not this is subsequently gated out by the pile-up rejector. In this particular case, this pulse, indicating total pulse rate and suitable for supplying to a count rate meter, is labelled CRM output. Additional outputs are provided for the pile-up rejection gate and a BUSY signal to permit proper live time correction by the MCA.

Within complex multi-detector systems it may sometimes be necessary to combine the outputs from different detectors before submitting the result to the MCA. This demands very careful matching of the energy calibrations of the different sub-systems. In response to this, amplifiers may offer an SFG (super fine gain) control to provide much more precise gain adjustment.

4.5 RESOLUTION ENHANCEMENT

There are two situations where there may be unavoidable losses in resolution due to poor charge collection. In Section 4.4.5, I discussed ballistic deficit caused by the long rise time (i.e. collection time) of the detector pulses compared to the peaking time of the amplifier output pulse. At low count rates, ballistic deficit can be eliminated by using a long shaping time but this would be unacceptable at moderate to high count rates and loss of resolution would be inevitable. The second situation is that of the damaged detector. In Chapter 3, Section 3.6.6, we saw how charge carrier trapping could delay, or prevent, the collection of charge carriers and, in Chapter 3, Section 3.6.7 how this is exacerbated by neutron damage. (Adjusting the shaping time would be ineffective for countering permanent charge trapping which is independent of the pulse processing.) Both of these effects lead to a variable reduction in pulse height for a certain proportion of detector events and will cause a degree of broadening of the spectrum peak.

Resolution enhancement works by making an empirical correction to the height of all pulses that are identified as being subject to these effects. It depends upon the fact that, apart from reducing the pulse height, delayed charge collection also delays the point in time at which the pulses reaches their maximum height. That is, it lengthens the peaking time by a certain time, Δt. To correct for ballistic deficit, the height of the pulse is increased by an amount proportional to the original pulse height, V_0, and the square of the delay:

$$V_C = V_0(1 + k\Delta t^2) \tag{4.6}$$

where V_C is the 'true' pulse height and k is an empirically determined factor. A similar equation has been derived for the correction of charge trapping losses:

$$V_C = V_0(1 + k\Delta t^n) \tag{4.7}$$

The power, n, depends upon the detector. A value of 3 is appropriate when there is a high concentration of traps near to the circumference, while a value of 2 is optimal for the more usual situation where the traps are uniformly distributed throughout the detector. This principle was implemented in analogue correction circuits in modules called **resolution enhancers,** which could correct for most simple types of majority carrier trapping in both n- and p-types of detector.

In Chapter 3, Section 3.6.7, I explained that fast neutron irradiation creates a uniform 'field' of traps throughout the detector. This being so, the resolution enhancer can correct for this additional trapping in the same way. However, neutron damage in n-type detectors creates hole traps and it turns out that because these are the minority carriers the method will not work. Even so, this would have been a useful device to have available for p-type damaged detectors which are less successfully repaired by annealing. The manufacturer's literature quoted improvements of more than a factor of two in resolution, for example, from 4.26 to 2.02 keV in a particular case. Unfortunately, these modules seem to have been withdrawn from the market. This may be because digital pulse processors are able to perform the same sort of function as part of their normal operation.

4.5.1 New semiconductor materials

All of the new semiconductor materials, CdTe, CdZnTe, HgI_2, etc., suffer from serious charge collection problems. The mobility of electrons in these materials is much lower than in germanium and the mobility of the holes is orders of magnitude lower than that (see Chapter 3, Table 3.1). This limits the size of these detectors and the energy range over which they can be used. Even worse, the poor charge collection gives rise to very prominent tails on the low-energy side of the peaks. This can be minimized in various ways.

Amptek cool their CdZnTe detector to $-30\,°C$ so that they can increase the bias voltage. This helps charge collection, and they then use rise time discrimination, only recording pulses corresponding to complete charge collection. This works well for pulses corresponding to photons below 50 keV. Above that, peaks become increasingly tailed until the detector becomes unusable above 600 keV. This method of charge loss correction means a reduction in efficiency because of the lost counts.

ORTEC's Radiant 2000™ CdTe detector, also cooled, uses a method of charge correction (possibly similar to that described above) which corrects pulse heights, rather than rejects them. This particular instrument has an operational range of 10 to 1000 keV and has a resolution of 2.89 keV at 661.6 keV – not much more than twice the width of peaks we might expect from a typical HPGe detector.

eV Products use a completely different way of generating good spectra. Their CdZnTe detectors are manufactured as square blocks of material with a full-area cathode on one side and a co-planar grid anode on the opposite. The grid, shown schematically in Figure 4.26, has two anode grids, described as 'interdigitated'. One of those is the collection anode and is at 1700 V bias relative to the cathode. The other anode is held at a voltage 60 V below the collection grid. The two grids are connected via independent preamplifiers to a subtraction circuit and the output from that to a shaping amplifier. Charge motion in the bulk of the detector is sensed equally by the two grids, giving a null output, but as the charge approaches the anodes, because of their different potentials, the signals on the anodes differs. In a well-designed detector, in which there is no electron trapping, the magnitude of the difference signal is the same no matter where the charge was generated within the detector. That means that peaks in the spectrum are symmetrical and Gaussian. At 23 keV (661.66 keV) the resolution of such a detector cannot match that of the CdTe detector described above. Nevertheless, the co-planar grid technique does seem to be worth keeping an eye on.

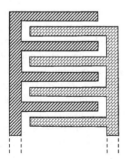

Figure 4.26 Schematic diagram of the double anode co-planar grid used by eV Products for their CZT detectors

4.6 MULTICHANNEL ANALYSERS AND THEIR ANALOGUE-TO-DIGITAL CONVERTERS

4.6.1 Introduction

The output from the amplifier is a stream of shaped and conditioned pulses, random in height and random in spacing. The task for the MCA – the **Multichannel Analyser** – is to measure the height of each of these pulses and count the numbers occurring within small voltage ranges. Because the height of each pulse is proportional to the amount of energy absorbed in the detector, the resulting list of numbers of counts is our gamma-ray spectrum.

The simplest way of measuring a gamma spectrum would be to use a **single channel analyser**(SCA), as suggested in Figure 4.27. The SCA has two electronic thresholds: the **lower level discriminator** (LLD) at H_1, below which pulses are not allowed to pass, and the **upper level discriminator** (ULD) at H_2, above which pulses are blocked. Pulses between these two limits are allowed to pass and ultimately each one generates a logic output pulse. In the example in Figure 4.26 above, only pulses 2 and 5 would give an output pulse – all the others being rejected. One can imagine that if the window – the voltage slot between H_1 and H_2 – were made small enough, we could progressively move it across the energy range, stopping at each point to measure the number of pulses. We would certainly achieve a spectrum but at considerable cost in terms of time. At any point in time, all pulses not within the window are lost – wasted one might say. In fact, in the early days of gamma spectrometry this was the only way of creating a spectrum and the technique may still be used in setting up simple systems where only the single channel output is required. It is also worth noting that the first stage of all analogue MCA systems consists of, in effect, a single channel analyser to reject unwanted low-energy pulses below an LLD and those high-energy pulses above the ULD.

In practice, we need a system which would monitor a large number of 'windows' simultaneously – a parallel, rather than a series system. Early spectrometers did, in fact, consist of a series of SCAs with their outputs feeding individual counters, from which we get the term multichannel analyser. At the time, detectors were of poor resolution and relatively few channels could be tolerated, but even so, such stacking of SCAs became impracticable as ambition expanded. The effort in precise setting up of each channel and the expense of such a large number of individual instruments limited such an approach. Salvation came with the introduction of the **analogue-to-digital converter** (ADC). As it happens, the stacked SCA principle persists in the form of the **flash ADC**. Until recently, these devices were not suitable for spectrometry purposes

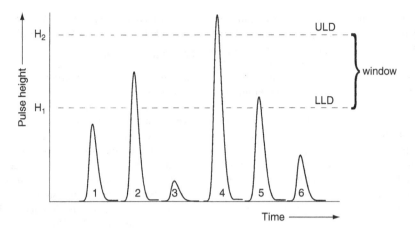

Figure 4.27   A single channel analyser with upper and lower discriminator levels defining a 'window'

but we will meet them later in Section 4.11 when we discuss digital signal processing.

Viewing a gamma spectrum as the outcome of a series of stacked SCAs reminds us that a gamma spectrum is not a smooth mathematical function but a histogram – a series of individual counts collected within small consecutive pulse height intervals (Figure 4.28). Strictly speaking, we should refer our spectra as **differential pulse height spectra**: a plot of dN/dH versus H, dN being the number of counts and dH the pulse height interval. The idea that pulses are sorted by energy gave rise to the name kicksorter early on in the history of gamma spectrometry.

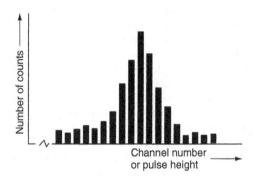

Figure 4.28   Multichannel analysis produces a histogram of counts against channel number – a differential pulse height spectrum

The multichannel analyser has a number of functions:

• from the output from the amplifier, it rejects out-of-range pulses;

• it measures the height of each of those accepted and adds a count into the memory location corresponding to the channel representing the voltage range;

• it displays the data as a spectrum and allows the data to be printed or saved to a data storage device.

A general idea of the component parts of an MCA is shown in Figure 4.29.

While the MCA is technically the hardware device which collects the spectrum data, modern systems will be controlled by software and may be intimately associated with the spectrum analysis program. So much so, that the term may often refer to the whole system, hardware plus software. We can now consider the MCA operations in detail, using Figure 4.30 as a basis.

4.6.2 Pulse range selection

It is likely that the output from the amplifier will carry pulses that are of no value within the spectrum. Very small pulses may simply be a consequence of electronic noise or be X-rays too low in energy to be of interest, and very large pulses may represent gamma-rays beyond the range of interest or be very large cosmic-ray generated gamma-rays. None of these is wanted. The first task of the MCA system is to prevent them being analysed.

To do this, the pulses are passed through an SCA with lower and upper discriminator levels that can be adjusted by the user, often by means of screwdriver-adjustable potentiometers. Both will provide adjustment over the full range (0 to 10 V, typically). Pulses falling within the window will be allowed to pass through the **linear gate** (so-called because it passes linear pulses, rather than logic pulses) to the ADC. However, in order to allow time for

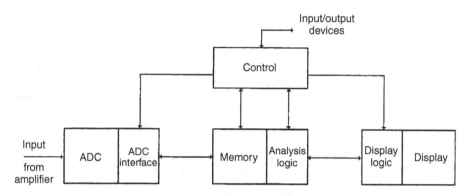

Figure 4.29 Functional block diagram of a traditional multichannel analyser

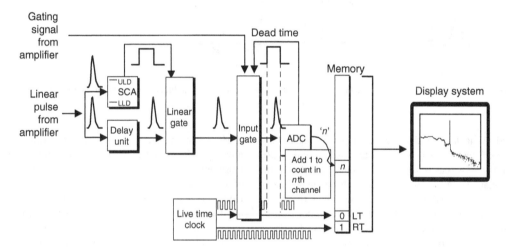

Figure 4.30 Schematic operating sequence of an MCA

the SCA to make its decision as to whether a pulse is valid or not and open the gate, the input pulses are delayed slightly.

Note that the lower level discriminator (LLD) has a different function from that of the user-adjustable **ADC zero** control, although under some circumstances the effect may appear to be the same in removing small pulses from the spectrum. The latter, which only has a small adjustment range (perhaps ± 0.3 V) is provided in order to make the energy calibration of the spectrometer pass through the origin (see Figure 4.31 below). Since our calibrations and the programs that use them will take into account the zero offset, this is not an adjustment that is critical and is often ignored.

It may not always be appropriate to use the lower level discriminator to remove gross numbers of counts from the low-energy end of the spectrum. In measurements of plutonium isotopes, the spectra will be dominated below

70 keV by L Xrays and by ^{241}Am at 59.54 keV. These activities are usually removed by a heavy metal filter to absorb them before they reach the detector, easing the count rate load on the detector. Raising the LLD would remove the peaks from the spectrum but do nothing to lower the number of counts handled by the detector.

4.6.3 The ADC input gate

Having successfully negotiated the linear gate, our spectrum pulse meets the **input gate** to the ADC. The problem for any ADC is that it can only handle one pulse at a time. During the time it is measuring a pulse, many microseconds in many cases, other pulses must be prevented from entering. That is the function of the input gate.

Let us assume that when our pulse arrives at the input gate the ADC is inactive and the gate is, therefore, open. The pulse will pass to the ADC and the conversion process

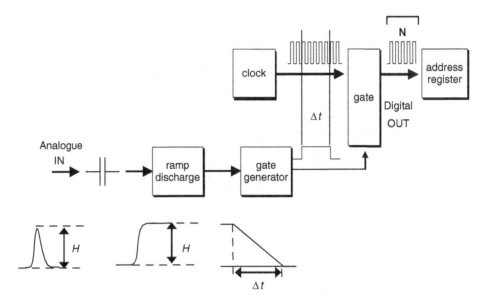

Figure 4.31   The Wilkinson ADC, showing how the digital number generated is a direct function of an intermediate time-dependent step

started. As it does so, the ADC creates a gating pulse – the 'busy' signal – which is used to close the input gate. When the ADC has finished and a count recorded in the appropriate MCA channel, the gate will be opened again to await the next pulse. The time during which the gate is closed is referred to as **dead time** (DT) and, not surprisingly, the time it is open is called **live time** (LT). (Dead time in other radiometric contexts may be referred to as resolving time or paralysis time. Neither of those terms is apt for gamma spectrometry). The normal physical time is usually called **real time** (RT), but sometimes **clock time** (CT) or **true time** (TT).

$$DT = RT - LT$$

In a particular measurement, the time from start of the count to the end of the count may be $1000\,s$ (real time), but of that $235\,s$ might have been needed to convert pulses (the dead time) resulting in the input gate only being open for $765\,s$ live time. To calculate the count rate of the whole spectrum or of peaks within the spectrum, we must divide the counts recorded by the live time. In that way, we account for the pulses during dead time periods. MCA systems will often present the dead time as the count progresses as a percentage – $100 \times (RT - LT)/RT$. In our example, the dead time would have been 23.5%. How do we measure the live time?

MCA systems always incorporate a **live time clock**; this might provide a pulse every few milliseconds. Feeding those pulses straight into a register would give a measure of the duration of the count – the real time, in fact. However, if those pulses are fed to the register via the input gate, the number recorded will represent the time during which the gate is open – the live time of the count. This is the conventional way in which live time is measured. In old hardwired MCA systems, it was usual to use channel zero to count the gated clock (live time) pulses and channel one to count the real time pulses. Nowadays, separate registers are used, but the counts may still be inserted into channels 0 and 1 for backwards compatibility.

Notice that any gating signal applied to the input gate will cause a dead time interval. This is, therefore, the logical place to send gating pulses from an amplifier to indicate a pile-up rejection period or a transistor reset period. The ADC gating pulse and the amplifier gating pulse will be ORed together to close the input gate, whatever the particular reason.

4.6.4 The ADC

Conventional analogue MCA systems use one of two types of ADC: the *Wilkinson ADC* and the *Successive Approximation ADC*. Digital signal processing systems use the *Flash ADC*. The latter will be discussed in Section 4.11; here I will discuss the other types separately.

The Wilkinson ADC

Figure 4.31 shows the basis of the Wilkinson ADC. The pulse height measurement proceeds as follows:

- As the analogue pulse rises above a threshold, it begins to charge a capacitor. This continues until the pulse passes through its maximum height. That maximum voltage, H, the pulse height, is retained on the capacitor.
- Once the voltage on the capacitor has stabilized, a linear discharge is triggered and at the same time, a timing gate is opened. When the voltage on the capacitor reaches zero, that gate is closed. Because the ramp discharge is linear, the time taken to fall to zero will be proportional to the voltage and then to the gamma-ray energy. In effect, the height of the input pulse has been converted to time.
- This time is measured by a high frequency pulse stream, generated by a crystal-controlled clock, which passes through the timing gate. The pulse stream is blocked until the gate opens, at which point a register starts to count the pulses, stopping when the gate closes again. The number of pulses passing through the gate is proportional to the height of the input pulse – the analogue pulse height has been converted to a digital number.

It is obvious that the time to effect this measurement – the **conversion time** – is proportional to the pulse height. This variable dead time must be taken into account within the live time measurement system. The resolution of the ADC depends upon the relationship between the rate of discharge of the capacitor and the clock rate. If the discharge rate is decreased to shorten the conversion time, the resolution will be decreased because fewer clock pulses will be recorded. A faster clock would be needed to maintain the resolution. Clock rates of 100 MHz and 450 MHz are typical. (Note that this clock is separate from and independent of the live time correction clock.) Wilkinson ADCs are considered to have excellent linearity, and at one time were regarded as significantly better than other types of ADC.

The successive approximation ADC

The components of this device are shown in Figure 4.32, and the operating mechanism is illustrated in Figure 4.33. The pulse height being measured is compared against a multistage reference voltage, and at each successive stage, an increasingly precise estimate of the pulse maximum is obtained:

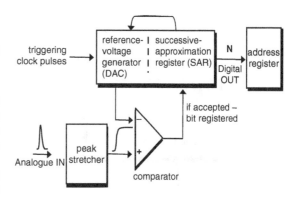

Figure 4.32 The components of a successive approximation ADC

- The analogue input pulse is first stretched or 'held', so that the pulse maximum is available for comparison over the time needed for the conversion.
- A comparator compares the pulse voltage to a reference voltage generated by a DAC (digital-to-analogue converter) triggered by a clock pulse. Initially, this voltage will be $0.5 \times$ the full voltage range of the ADC. In Figure 4.33, this is stage 1.
- The comparator asks the simple question *is the analogue pulse maximum greater than this voltage?* If the answer is 'Yes', the comparator sends a logic pulse to the successive approximation register (SAR) to set a bit in the address register to 1. If no pulse is received, the SAR sets 0. This is the first digital approximation to the pulse height. In the example in Figure 4.33, it would be '1'.
- Having set that bit to '1', the DAC raises the reference voltage to half way between $0.5 \times$ and $1 \times$ full range. The comparison is repeated for stage 2. In the example, the pulse is lower than the new reference voltage and the SAR sets the next address bit to '0'. The digital approximation is now '10'.
- The DAC now alters the reference voltage again, but this time, because the last bit set was zero, lowers it to half way between $0.75 \times$ and $0.5 \times$ the full range. The comparison is again made and, in the example, the third bit becomes '0'.

At each stage, the binary number being generated becomes a more accurate representation of the actual pulse height. For 4096 (2^{12}) channel resolution, 12 comparison stages would be required to generate the 12 bits, 8192 channels would require 13 stages, and so on. The whole process is conducted by a series of shift registers triggered by a clock. Obviously, the faster the clock, the faster the conversion, but at a higher cost.

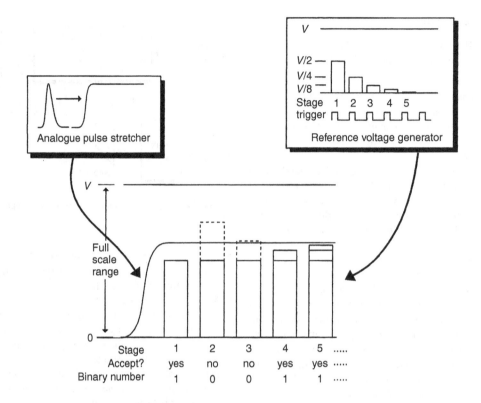

Figure 4.33   The mechanism used in the successive approximation ADC

Note that, whatever the pulse height from zero to the full range of the ADC, *all* conversions will involve *all* comparison stages. Therefore, the conversion time will be the same for all pulses. These ADCs are sometimes referred to as **fixed dead time** or **fixed conversion time** ADCs. Conversion times of 0.5, 1 and 10 μs per pulse are readily available. Successive approximation ADCs can be faster than Wilkinsons but, as we will see below (Section 4.6.6), not necessarily so unless the spectrum has predominantly high-energy gamma-rays. Historically, these ADCs were regarded as having worse linearity than Wilkinsons but there seems to be little to choose between modern versions.

4.6.5 MCA conversion time and dead time

The time that an MCA takes to measure and store a pulse height, i.e. the dead time, is the sum of the following three times, which, with others, are indicated in the profile of a pulse in Figure 4.34:

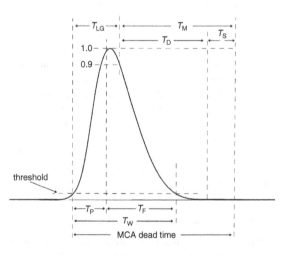

Figure 4.34   Timing specification for a semi-Gaussian pulse and its digitization. The pulse threshold may be the same as the LLD: T_P, peaking time; T_{LG}, linear gate time; T_F, fall time; T_W, pulse width, threshold to threshold; T_D, ADC conversion time (digitization time); T_S, memory storage time; $T_M = T_D + T_S$

(1) The **linear gate time** (T_{LG}) – the time the input stage (the SCA) takes to recognize a pulse and open the linear gate.
(2) The **ADC conversion time** (T_D) – the time taken to convert the analogue pulse height to a digital number.
(3) The **memory storage time** (T_S) – also called the memory cycle time.

The SCA at the input to the MCA must wait until the pulse height has started to fall from its maximum before it can determine whether the pulse has exceeded the LLD and fallen below the ULD before it can open the linear gate. Figure 4.34 assumes that the pulse falls by 10 % from its peak before triggering the gate, a figure quoted by one of the ADC manufacturers. The time for that to happen depends upon the rise time of the pulse and, therefore, on the shaping time constant set on the amplifier. The linear gate time, for a number of amplifiers, has been shown to be about 2.5 × the time constant. It means that for a system with 3 µs shaping time, about 8 µs must elapse before the linear gate opens to accept the pulse.

The conversion time (T_D) of the ADC depends on the type. For a Wilkinson, it depends on conversion clock rate. To convert a pulse corresponding to channel N the time would be:

$$T_D = (N + X)/\nu + R \tag{4.8}$$

where ν is the frequency of the clock (s^{-1}), X is any digital offset imposed (usually zero), and R is a fixed overhead associated with generating the linear ramp. For a Canberra 8701 ADC, where ν is 10^8 per second (100 MHz) and R is 1.5 µs, conversion times over an 8k spectrum would range from 1.5 to 83.4 µs.

For a successive approximation ADC, the conversion time is, of course, independent of pulse height and channel number. Instruments are available with conversion times from about 1 to 25 µs, with 10 µs for an 8k spectrum being typical. At the time of writing, the fastest readily available ADCs are the 8715 from Canberra with 0.8 µs (8k spectrum) and the ORTEC ASPEC 927 with 1.25 µs (16k) conversion times, including the memory transfer times.

Memory storage time (T_S), the time needed to add one count to the content of the channel corresponding to the pulse height, is said to range from 0.5 to 2 µs. It can taken into account by extending the dead time period beyond that defined by the ADC busy signal. However, in some systems incrementing the channel content can be done in parallel with the start of conversion of the next pulse, making its impact negligible.

4.6.6 Choosing an ADC

There is a range of ADCs available commercially – 100 and 450 MHz Wilkinsons, 0.8, 1.5 and 8 µs successive approximations, and others. Inevitably, higher specification will be reflected in cost. Which is best? Which is right for a particular purpose? Figure 4.35 plots the conversion time for four ADCs, two Wilkinsons (W) and two successive approximations (SA), against channel number. In this figure, the conversion time includes an allowance for T_{LG}, assuming a shaping time constant of 3 µs. The numbers may not be representative of any particular system, but they do illustrate the principle. Let us consider measuring the 1332 keV peak of ^{60}Co in a 4k spectrum at around 0.5 keV per channel. The peak would appear at about channel 2600. Which of the various ADCs would give the fastest conversion?

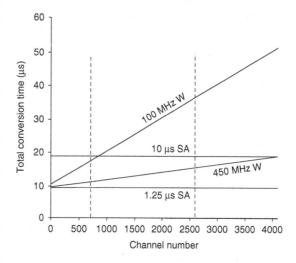

Figure 4.35 Comparison of total conversion times for four ADCs. An amplifier shaping time of 3 µs was assumed, giving $T_{LG} = 7.8 \, \mu s$. $T_S = 0.75 \, \mu s$

At first sight, it seems the choice is clear; the high specification 1.25 µs SA ADC is fastest. Why bother with the rest? Although, of course, that high specification is only achieved at a significant financial cost. If one's expected count rates are low, there is no need to buy a high specification ADC. If the count rate were 50 cps, the average time between pulses would be about 8600 µs, during which time the ADC would be twiddling its electronic thumbs with nothing to do. A conversion time of 36 µs for the 100 MHz Wilkinson ADC would not be significant.

Even at higher count rates, the conclusion might not be so obvious. We saw in Chapter 2 (Figure 2.12) that only a small proportion of the detected gamma-rays are fully absorbed and appear in the full energy peak. Most of the gamma-rays are only partially absorbed and appear elsewhere in the spectrum, mainly on the Compton continuum. It is this majority of the events that determine the dead time of a count. As a rule-of-thumb, we can say that the average energy of single gamma-rays absorbed by a detector is a third of the peak energy. On that basis, when assessing ADC performance with respect to effective conversion time of our ^{60}Co spectrum, we should be looking at channels nearer to 700. Figure 4.35 shows that even though at higher channels the 1.25 μs SA ADC is much faster than the 450 MHz Wilkinson, at channel 700 their speed is comparable. On the same basis, the 100 MHz Wilkinson and the 10 μs SA are also comparable and at low pulse height the Wilkinson is faster.

The conclusion is that when choosing an ADC, perhaps one should not be blinded by the specification, but its cost and expected use should be considered a little more deeply. Since the first edition of this book, a browse through the manufacturer's catalogues suggests a shift towards successive approximation ADCs rather than Wilkinson ADCs.

4.6.7 Linearity in MCAs

In principle, the relationship between pulse height (and therefore energy) and channel number would be exactly linear, passing through zero. In practice, although we can readily represent that relationship by a straight line, it is very likely that it will not pass through the origin. If that were necessary, the ADC zero offset control would be used to make that so. In general, the line would be characterized by a slope and an intercept – this is, in effect, the energy calibration of the spectrometer. Altering the amplifier gain or time constant would alter the pulse heights, resulting in a change in the slope of the pulse height/channel number relationship (Figure 4.36).

Two measures of linearity are specified by the manufacturers of ADCs – integral linearity and differential linearity.

Integral linearity is a measure of the deviation from an ideal response, as shown in Figure 4.37. A typical specification might quote '< ± 0.05 % over 99 % of the range' or '< ± 0.025 % of full scale over the top 99.5 % of the range'. Deviations are likely to be greatest at the extremes of the range. Such specifications are acceptable, but note that 0.05 % of a spectrum energy range of 2000 keV is 1 keV – similar in magnitude to the FWHM of peaks in the spectrum.

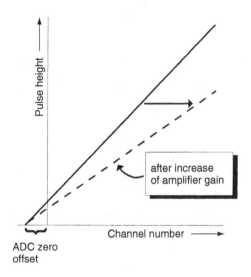

Figure 4.36 The ideal response of an MCA, showing ADC zero offset and the effect of increasing the gain of the amplifier

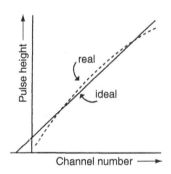

Figure 4.37 Integral linearity as a measure of the difference from the ideal response of pulse height versus channel number

Differential linearity is a measure of the constancy of channel width. It is measured by attempting to place equal numbers of counts into each channel by using a sliding-pulse generator. In an ideal system, that would be achieved. In a real system, there will be slight differences in the numbers of counts, as suggested in Figure 4.38. Figures such as '< ± 1 %' and '< ± 0.7 % over the top 99.5 % of range' are typical. In practice, it is very unlikely that 1 % of counts being redistributed to other channels will alter the energy calibration significantly and in terms of peak area measurement the effect is unlikely to be noticed.

On top of these linearity limitations, all of the electronic components of the pulse handling and measurement chain will be subject to time and temperature effects. Gamma spectrometry measurements of environmental samples can

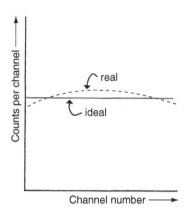

Figure 4.38 Differential linearity. A measure of the constancy of channel width. The result of attempting to put an equal number of counts into every channel

involve count times of several days, during which time the laboratory temperature is quite likely to change as day turns to night and the weather changes from rain to sun. That being so, it makes sense to have a temperature stabilized laboratory. Temperature stability of amplifiers and ADCs might be quoted as '< 0.009 % of full scale per °C' for gain drift and '< 0.0025 % of full scale per °C' for zero drift, suggesting that a 10° C temperature change could shift a peak by one or two channels. It is wise to allow the temperature of gamma spectrometer electronics to stabilize after switching on before meaningful measurements are made.

Experience shows that long-term drift at constant temperature is negligible in modern equipment, the MCA being responsible for only a small fraction of a channel in 24 h.

4.6.8 Optimum spectrum size

MCAs allow the user to select a spectrum size, which might be anything from 1k up to 16k in steps of a factor of two. Small spectrum size means a larger number of counts per channel, but with the disadvantage of tending to merge peaks together. Large spectrum size will, in principle, allow peaks to be defined better but with fewer counts in each channel to offset that advantage. Is there an optimum spectrum size? The dilemma is illustrated in Figure 4.39 where a doublet (the crosshatched shape in the background) is digitized in spectra of different sizes. The peaks of the doublet are separated by 1 FWHM.

In spectrum (a), the 2048 channel spectrum, corresponding to 1 keV per channel, the peaks are barely

resolved. Clearly, we need more channels of information to define the shape of the doublet. At 0.5 keV per channel, the doublet is resolved reasonably but the number of counts in each channel is halved. Increasing the spectrum size to 8192 maintains the shape of the doublet, but again the number of counts per channel is halved. In the 16k channel spectrum (d), 0.125 keV per channel, the peaks are resolved but there are now so few counts in each channel that there is noticeable scatter even at the tops of the peaks and the scatter on the background continuum is much greater. At such a spectrum size, we would be in greater danger of loosing small peaks in that scatter. In addition, because of the smaller number of counts our spectrum analysis software would not be able to estimate the background to the peaks as precisely, resulting in greater uncertainty on the peak area estimations. These statistical effects are discussed further in Chapter 5, Section 5.5.2.

A frequently used guideline is to set the spectrum size so that 4 channels are equivalent to 1 FWHM. If we take the resolution of a germanium detector as 2 keV (as it will be in somewhere in the region 1000 to 1500 keV), our optimum energy scale should be 2 keV/4, i.e. 0.5 keV channel. If we require a spectrum range of 0–2000 keV we will need 4000 channels – a 4k spectrum. If we are more concerned about adequate spectrometry at low energies, where the resolution is likely to be near to 1 keV, we would need an energy scale of 0.25 keV/channel and a spectrum size of 8k. The same rule applied to a scintillation system would suggest a spectrum size of only 200 channels if the resolution at 661.6 keV were 7 %.

4.6.9 MCA terms and definitions

There are a number of terms used when discussing MCAs that can cause confusion. These are listed here together in order to avoid this:

- **Lower level discriminator** (LLD) – pulses below this level will not be analysed. Use this to reject electronic noise and low-energy X-rays.
- **Upper level discriminator** (ULD) – pulses above this level will not be analysed. Use this to reject very high-energy pulses. This will often be left at its maximum, but still performs a useful function in rejecting high-energy cosmic gamma-rays.
- **ADC zero level** – use this to adjust the energy calibration so that it passes through 0 keV. Not ideal for eliminating the effect of noise.
- **Digital offset** – this is a means of shifting the spectrum to lower channel numbers by subtracting a fixed number (the offset) from every channel number output by the

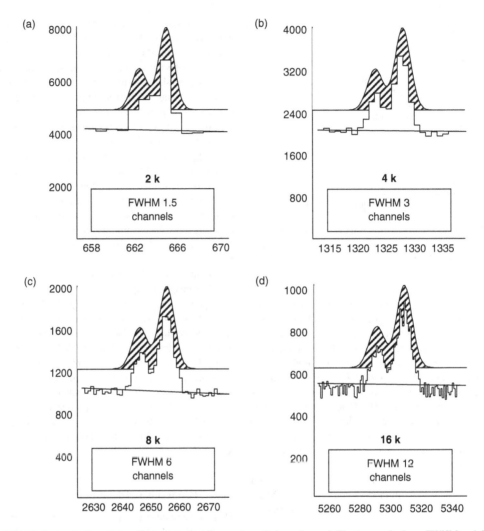

Figure 4.39 A demonstration of the effect of varying the number of channels used. The two peaks have FWHM = 1.5 keV, are 1 × FWHM apart, and have areas of 2500 and 5000 counts on a background of 4000 counts per keV (*y*-axis, counts; *x*-axis, channel number)

ADC. It should *not* be used simply to eliminate trouble-some pulses at low energy. Digital offset takes place *after* the ADC has measured the pulse and any pulses lost from the spectrum will still contribute to dead time.

Digital offset might be used to expand the upper part of a spectrum by offsetting it by 50 %, say, and then increasing the gain to expand the remaining spectrum to the full spectrum size.

● **Conversion range** – the maximum pulse height the MCA can accept, typically 10 V. An MCA should be able to accept unipolar (semi-Gaussian), triangular, bipolar and gated integrator pulses.

● **ADC resolution** is the total number of channels available within the ADC. It varies from model to model, but MCAs for germanium systems might incorporate a 16k (16 384), 8k (8192), or 4k (4096) channels ADC. It would be normal for an MCA to have available as many memory channels as there are ADC resolution channels.

● **ADC conversion gain** is simply the number of channels actually used in a particular application – in everyday parlance, the spectrum size. It is possible, of course, for an MCA to have a 16k ADC but the user opts to use only 8k or 4k. With a conversion gain of 4k, pulses are accumulated in channels from 0 to 4095. Conversion gain may be switch or software-selectable, and may go down to 256 channels for scintillation spectrometry purposes. When using a conversion gain of less than the ADC resolution, it is sometimes possible, particularly on older systems, to acquire several separate spectra stored within the whole of the available memory. For example, in a system with 16k ADC resolution, using only 4k conversion gain, four spectra could be stored within the 16k memory. This little used facility seems to have been abandoned on modern day-to-day systems.

4.6.10 Arrangement of the MCA function

The MCA has seen many incarnations over its lifetime. 'Old timers' who used the first experimental instruments would talk of improbably small spectrum sizes and spectrum data recorded by photographing the CRT display. Fortunately, at that time high-resolution detectors were still an unfulfilled dream. Even when the first lithium drifted germanium detectors became available, the MCA might be a (large) box on wheels that contained only the ADC and the memory functions. Later, as integrated circuits replaced individual transistors, portability was accomplished without the wheels. An MCA was then likely to be a complete instrument: high voltage, amplifier, ADC and memory, although the amplifier would not have been suitable for high-resolution spectrometry.

There then followed a period when NIM modular instruments became the norm with each individual function in a separate 'box'. Connection to a computer for MCA control and spectrum display started to become widespread. Initially, the computer would be a minicomputer, perhaps running a spectrum analysis program written in FORTRAN. As the PC (personal computer) began to take the world by storm it became, and still is, the most common user interface for MCA systems. Some instruments were based on standard PC cards, carrying the ADC, which could be plugged into the computer interface bus. Others, with external memory, relied upon a dual port interface to the computer. Recent advances have reverted to the 'all-in-one-box' model with every part of the pulse processing chain tightly integrated into a package the size of a large book and connection to the computer by a USB (Universal Serial Bus) cable. Modular instrumentation is still available but increasingly the manufacturers are promoting complete instruments, often with portability and ruggedness in mind. MCAs which interface directly to laptop computers are commonplace.

Nowadays, choosing an appropriate MCA probably comes down to choosing the right ADC for your type of work and finding a system with a user interface that suits you, possibly taking into account other features, such as portability.

4.6.11 Simple MCA analysis functions

It is difficult nowadays to separate the MCA as an electronic device and the software used to control it and handle the spectra it generates. Hardwired MCA instruments, such as the Canberra S100 and the ORTEC MCB/Maestro MCA Emulator, did little more than acquire the spectra and store them, with facilities for two point energy calibration and ROI readout. One can expect all MCA systems to do the following:

- Provide energy calibration – at its most basic using two points to define a straight line; other devices allow a multi-point straight line, or, multi-point fitting to a second order expression. Simultaneous peak width calibration would be a useful bonus.
- All systems use the idea of a *region of interest* (ROI), which the user sets up around a peak in the spectrum by visual inspection. The more elaborate systems have peak search routines that find the peaks and set the ROIs automatically.
- Once a ROI is defined, all systems will calculate gross area and net area. The gross area is the sum of all counts in the ROI; the net area is the sum after subtraction of a background continuum. It is useful if the calculation provides the counting uncertainty on the net area.
- The centroid of the peak – a fractional number – is usually calculated in both channel and energy units.
- PC-based systems will often contain a relatively simple library of gamma energies and can attempt to allocate a nuclide identity to the peak energy. Note: these are not particularly reliable.
- Most systems can give information on peak width in the form of FWHM and often more detail may be deduced on peak shape from FWTM (full width at one-tenth maximum) and FWFM (full width at one-fiftieth maximum) – see Chapter 11, Section 11.4.2.

More sophisticated software systems, such as the Genie 2000™ (Canberra) and GammaVision™ (ORTEC), allow complete calibration of energy, peak width and detector efficiency, are able to search through a spectrum seeking out statistically significant peaks, assigning them to nuclides, and calculating sample activity. Many optional features, such as decay correction and allowance for random summing, will be available. Computer analysis of spectra is covered in full in Chapter 9.

4.7 LIVE TIME CORRECTION AND LOSS-FREE COUNTING

4.7.1 Live time clock correction

In Section 4.6.3, I explained that an allowance must be made for detector pulses that reach the MCA system but are not analysed because the ADC is already busy measuring a previous pulse. Because of the lost pulses, simply dividing the number of pulses within a channel, a region-of-interest, a peak, or indeed the whole spectrum, by the real count period will underestimate count rates. The lost pulses can be accounted for by dividing numbers of counts by the live time, as measured by the **live time clock** (LTC – see Section 4.6.3 and Figure 4.30).

This works well at low to moderate count rates but is limited at high count rate. Dead time is discussed further in Chapter 14 in this connection. For now, it is sufficient to say that, in general, high dead times are to be avoided. Each laboratory has its own arbitrary limit. My own was about 30 %, but many laboratories have much lower limits. At very high dead times, measurement of live time may be inaccurate because of differences in shape between the detector pulses and the live time clock pulses, which are not blocked in the same way by the input gate. Inaccurate live time measurement does not prevent nuclide identification but does affect the quality of quantitative nuclide measurements.

The normal LTC system also has limitations when the count rate alters rapidly during the count period. Such situations might arise when measuring rapidly decaying sources or when the gamma spectrometer is used to monitor flow of material through a pipe, for example. A sudden 'slug' of high activity material might cause the count rate to rise over a matter of fractions of a second and then fall again equally rapidly. Under such circumstances, depending upon the clock rate, the count rate might be altering significantly within the period of a clock pulse, causing inaccurate compensation for the lost pulses.

The LTC method is often referred to as the **extended live time method** because the count is extended in real time to take account of the dead time, and it is quite usual to count to a preset live time. When counting a decaying source, the count rate at the start of a count will be, obviously, greater than that at the end. If the count period is extended in such a situation, it will be into a lower count period. Not a particularly satisfactory state of affairs. In such cases, it is better to count for a preset real time so that every measurement has the same decay factor. Whether counting for preset real or live time, it would, in any case be necessary to correct for decay during counting as will be explained in Chapter 7, Section 7.6.10.

4.7.2 The Gedcke–Hale method

This method of live time correction, favoured by ORTEC, also compensates for losses due to leading edge pile-up in the amplifier. The method is based upon the following logic, which refers to the time periods illustrated in Figure 4.34. The probability of a pulse being processed by the ADC has two components:

(1) The probability that, when the pulse arrives, the ADC is not already processing a pulse. This means that there are no pulses within a time equal to $(T_{LG} + T_M)$ or T_W, whichever is greater. (T_M is the time it takes the ADC to measure and store the pulse; T_W is the pulse width.)

(2) The probability that no other pulses arrive during the time the pulse is being read into the ADC, T_{LG}. Another pulse arriving within that period would trigger a pile-up rejection event and the pulse would not be measured.

Both those probabilities can be quantified using Poisson statistics, from which the expected number of pulses accepted by the ADC during a count period can be estimated and the relationship between live time (LT) and real time (RT):

$$LT = RT \times \exp\left[-r \times (2 \times T_{LG} + T_M)\right] \qquad (4.9)$$

where r is the count rate. The simple live time clock correction could be represented by a similar equation but without the linear gate term. The principle advantage of the Gedcke–Hale method is that it does take losses at the linear gate into account. It is implemented by circuitry similar, at first sight, to the normal live time clock, in that it accumulates time ticks when the system is not busy and stops when it is. However, it differs in that, as soon as the pulse is detected at the input to the MCA, the Gedcke–Hale live time clock start to count backwards either until a pile-up event is registered or until the linear gate is closed. This, in effect, gives a double weighting to the dead time interval associated with the linear gate. It can be shown that, statistically, such a procedure emulates Equation (4.9).

The ORTEC ultra-high count rate 'Mercury' system uses this method and the accuracy of the live time clock at maximum throughput is said to be better than 3 %. There is further discussion of high pulse rates and dead time in Chapter 14.

4.7.3 Use of a pulser

The principle is simple. Pulses from a pulse generator, shaped to simulate detector pulses, are injected into the test pulse input of the preamplifier. Figure 4.9 shows that those pulses would mix with the detector pulses at the earliest possible stage of the pulse processing chain and appear in the spectrum as a 'pulser peak'. Any processes leading to loss of pulses within that chain should affect the artificial pulses in the same way as the detector pulse. For example, if the count rate is such that the dead time is 10 %, then 10 % of detector pulses will not be recorded by the MCA and we would expect 10 % of pulser pulses to be lost as well. Comparing the actual peak area of the pulser peak with the known number of pulses injected provides a correction for the dead time losses within the system. In fact, any other processes leading to pulse loss, for example, random summing, would also be corrected for at the same time.

There are, however, difficulties. Accurate correction for pulse loss depends upon the pulser pulses accurately mimicking the detector pulses. The rise time and the fall times of the pulser pulses should be identical to those of detector pulses. Leaving aside the fact that preamplifier output pulses have a variable rise time, none of the readily available pulsers allow detailed control of the fall time. Bear in mind that the fall time of the preamplifier pulses depends upon the time constant of the feedback circuit in the preamplifier, and that pole-zero cancellation within the amplifier matches the shaping circuits to the input pulse fall time. The consequence of this is that it may not be possible to pole-zero correct the pulser pulses and the detector pulses together. At anything more than a low count rate, many detector pulses may be incorrectly measured by the ADC if they occur close in time to a pulser pulse.

Another way in which pulser pulses may not resemble detector pulses is that, unless special measures are taken, they do not occur randomly. This has statistical ramifications that cast doubt on the accuracy of the correction. For example, pulser pulses cannot be in coincidence with themselves. So, although they may prevent a proportion of detector pulses from being measured this will not be reflected in the same fraction of pulser pulses being lost. The correction due to random summing will be inaccurate. To some extent, the significance of this depends upon the pulser rate compared to the detector pulse rate. Conventional advice is to keep the pulser rate below 10 % of the detector pulse rate. It is possible to achieve randomness in the pulser output by using a subsidiary detector with a small radioactive source to provide the trigger. It would also be necessary to provide a scaler to count the number of trigger pulses; the whole system then starts to become cumbersome.

Although, at first sight, a pulser does seem to be a very direct way of correcting for all pulse losses in the pulse-processing system, at high count rate, where correction for losses is most important, there are the most difficulties. I am aware of laboratories where the pulser method is used routinely, but only at low count rate. A recommended specification and procedure for setting up a pulser is given in Chapter 11, Section 11.3.5.

4.7.4 Loss-free counting (LFC)

The term **loss-free counting** refers to systems where there are, in effect, no dead time losses. All the various loss-free counting systems achieve this apparently utopian state by determining the instantaneous count rate through the ADC and, as each pulse is measured, adding additional counts into the spectrum (instead of a single count as in Figure 4.29) to account for dead time losses.

The **Harms procedure** was a pioneering effort. In this system, the pulses that are rejected during dead time periods were counted. This was then used to derive an integer weighting factor 'n' so that, when the next real event was processed, n counts would be added instead of one. The problem with this procedure is that the ADC processing time is not the only reason for loosing pulses. At higher count rates, pulse pile-up (random summing) can dominate and the fact that this is not taken into account by the Harms procedure is a serious limitation.

The **virtual pulse generator** (VPG) was devised by Westphal in 1982. This takes the idea of the pulser correction but dispenses with the pulse generator itself. In principle, the electronics takes stock of the situation at particular times and asks the question. *'If a pulse were injected at this moment, would it be processed or not?'* The system then takes appropriate steps to take account of any losses. In Figure 4.40, line A is the pulse stream from the amplifier and line D represents the moments in time when the question is asked – the virtual pulse stream. Clearly, looking at the times when the ADC is busy in line B, we can expect the virtual pulse only to be accepted during the time interval from t_3 to t_5. However, if a real pulse were to arrive at time t_3 it would not begin to be converted until t_4. During that time, called the **pulse evolution time**, no other pulses would be accepted. So the time period during which pulser pulses would be accepted is effectively t_4 to t_5. Unlike the real pulser method, the virtual pulser method can use a very high pulse rate with no disturbance to the real pulse stream. In Canberra's Loss-free Counting Module 599, the virtual pulse generator has a frequency of 5 MHz and can handle n from 1 to 255. Westphal's 1982 paper claims successful correction of counting losses as high as 98 % up to 800 000 cps.

While a loss-free counting spectrum is fine for determining peak positions and accurate peak areas, the counting statistics of each channel of the spectrum will have been disturbed by the addition of the extra counts. We can no longer say, following Poisson statistics, that the standard

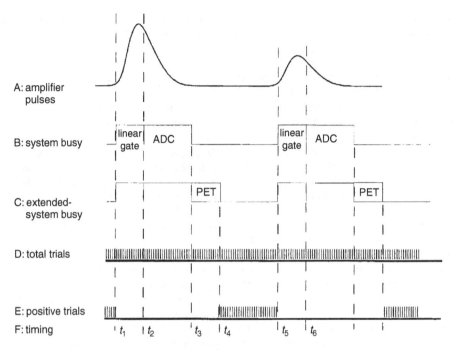

Figure 4.40   The function of the virtual pulse generator. Adapted by permission from a Canberra Nuclear *Application Note*

uncertainty of a channel count is the square root of that count. To allow the correct statistics to be determined, it is essential that an uncorrected spectrum be also acquired, from which correct uncertainties can be calculated. Loss-free counting systems usually provide the means to acquire the corrected and uncorrected spectra simultaneously.

4.7.5 MCA throughput

As the count rate applied to an MCA increases, the dead time will increase. Common sense suggests that there must be some point at which the MCA cannot handle the pulse stream effectively. **Throughput** relates the pulse rate entering the MCA to the number of conversions per second actually achieved. For the purposes of illustration, if we consider a system with an amplifier shaping time of $3\,\mu s$ and an ADC conversion time of $10\,\mu s$, Equation (4.9) can be used to calculate the throughput at a range of input count rates, as shown in Figure 4.41.

Taking that particular data, the maximum throughput would be about $14\,400\,s^{-1}$ for an input count rate of $40\,000$ pps. Beyond that modest maximum, the pulse measurement rate actually decreases. In some situations, there may be advantage to be gained by arranging one's measurements so as to reduce the count rate. Note that Figure 4.41 represents the throughput of the MCA alone.

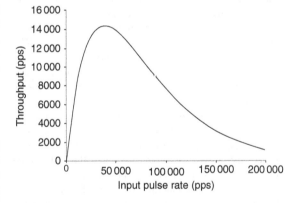

Figure 4.41   Calculated throughput of an MCA system: shaping time constant, $3\,\mu s$; fixed conversion time, $10\,\mu s$

The reader should remember that throughput will also be constrained by the amplifier and that, at very high count rates, a resistive feedback preamplifier may lock up. Throughput of complete systems is discussed in Chapter 14.

4.8 SPECTRUM STABILIZATION

Over a period of time, it is possible that a spectrum will 'drift' so causing the peaks to broaden and, in extreme cases, render the energy calibration invalid. The reason lies in the instability of the pulse processing system all the way from detector to MCA. The most likely cause would be temperature change. Both the gain of the system, affecting the energy scale, and the zero level, affecting the energy calibration intercept, can drift. One way of avoiding this is spectrum stabilization. Lack of drift is particularly important when measuring very low activities, which might need count periods of several days when, as it happens, stabilization is more problematic. Even at high count rate, some means of removing the peak shifts caused by limitations within the electronics might be appreciated.

My own experience has been that, in a temperature-controlled laboratory for routine measurements, spectrum stabilization is not necessary. Unexpected spectrum shifts are occasionally experienced but these have usually been attributable to malfunction of the temperature control. However, in environments less comfortable than a clean temperature-controlled counting room and at high count rates the ability to stabilize the spectrometer may be welcome.

4.8.1 Analogue stabilization

An **analogue stabilizer** uses the shape of monitor peaks in the spectrum to control the gain of the system. Figure 4.42 shows the principle of analogue gain stabilization. A peak, ideally a singlet, is selected high in energy. Pulses are taken from the amplifier, in parallel to those going to the MCA, and two SCAs are set up to cover the small energy windows, on either side of the peak centroid, so that the count rates in the windows are the same. Counters keep track of these count rates and, if there is an imbalance, the gain of the amplifier is altered in such a way as to restore the balance.

Analogue stabilization is only suitable for low-resolution spectrometry. Scintillation detectors are particularly prone to drift because of the temperature sensitivity of the electronics and to instability of the high voltage. This form of

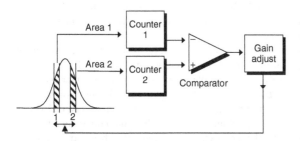

Figure 4.42 The operation of gain stabilization in an analogue stabilizer

stabilization was notorious for 'hunting'. Statistically, it is very unlikely for the counts in the two monitoring counters to be equal and for high-energy peaks and low count rates the uncertainty on these counts is high. This could lead to frequent and unjustified gain alterations even when no actual drift had occurred.

In order for the analogue stabilizer to operate successfully it is, of course, necessary for there to be a prominent high-energy peak in the spectrum. If there were no such peak, one would have to be introduced by an external source; not a particularly good solution because of the Compton contribution to the rest of the spectrum. Because it takes place external to the MCA, analogue stabilization does not take into account MCA drift. It is difficult to set up and operate and, even on low-resolution systems, the job could be better done by using digital stabilization.

For NaI(Tl) scintillation systems, it is possible to buy a detector that contains a small amount of ^{241}Am. This is said to be 'seeded'. The 59.54 keV gamma-ray from the ^{241}Am provides a well-defined peak at low energy for the spectrum stabilizer to work on, and the alpha particles provide a peak above 3 MeV as a high-energy marker.

4.8.2 Digital stabilization

Once one has a spectrum, any decisions about the position of peaks can be determined by making calculations using data within the spectrum, ensuring that all drift, whatever the source, is compensated for. Both gain and zero drift can be compensated and, operating after the ADC, will take into account drift throughout the pulse processing chain. Note that the spectrum stabilizer is a function of the MCA system, even if controlled from within the spectrum analysis software. Not all software will be able to control the MCA stabilizer and not all MCA systems will have stabilization available.

In operation, two prominent peaks are chosen – one at low energy and one at high energy. If a series of samples are to be measured automatically, these peaks must be present in every spectrum. Before the stabilizers are enabled, they would be reset to the centre point of their adjustment range, the pole-zero cancellation checked and an energy calibration performed. Then for each stabilization peak, an ROI is defined by specifying a central channel and a width. A full ROI width of $2 \times$ FWHM would be appropriate. Once these ROIs are defined, when counting starts the spectrum stabilizer will alter the gain and zero offset of an amplification stage within the MCA so as to make the calculated peak centroids, based on current channel contents, equal to the defined centroid channels. Before calibrating the system and setting up the stabilizer ROIs , it would make sense to adjust the external amplifier gain of the system so that both monitored peaks were centred on a channel. Otherwise, the calibration might alter as soon as the stabilizer begins to operate. The specification for the ORTEC 919 Multichannel Buffer suggests that its stabilizer would be capable of compensating for spectrum shifts within its stated temperature stability.

4.9 COINCIDENCE AND ANTICOINCIDENCE GATING

The MCA or ADC will have a signal input labelled GATE or GATE IN. This allows other parts of the pulse processing chain to select or prevent pulses from being processed by providing a standard gating pulse. A switch or internal jumper will allow this signal to operate in *coincidence* or *anticoincidence* modes:

- In the anticoincidence mode, the pulse applied to the MCA linear pulse input will only be processed if there is NO signal on the gate input at the time the linear gate is about to be opened. This is the more common use. It might be used by the pile-up rejector to prevent piled-up pulses being processed or by a Compton or background suppression system to indicate that the pulse is, in some way, invalid.
- In the coincidence mode, the linear pulse would only be accepted if there WAS a signal on the gate input. This might be used, for example, when monitoring events stimulated by a pulsed irradiation source, where the gate is only applied during the irradiation.

In both of these situations, the relative timing of the gate and the input pulse is critical; while such connections are often set up by the manufacturer, the ADC/MCA manual will specify what arrangement is required.

4.10 MULTIPLEXING AND MULTISCALING

These two processes have little in common and are included together here on the simple grounds of euphony, and a desire to dispel a possible cause of confusion.

Multiplexing is a means of sharing a single ADC with several separate counting chains. Multiplexers, also known as **mixer-routers**, are available for gamma spectrometry with 4, 8 and 16 inputs. The justification for using these is cost; one four-channel multiplexer is cheaper than three additional ADCs.

The multiplexer may be a separate unit or built into the MCA itself. For example, the ORTEC 919 Multichannel Buffers has four detector pulse inputs. As a pulse is detected at any of the inputs, it is allowed to claim the attention of the ADC and is converted and a count stored in a segment of memory (i.e. the buffer) corresponding to the input. In this way, four spectra can be acquired independently and simultaneously. During any conversion period, all four inputs are inactive. A disadvantage of multiplexing inputs in this way is that if any one of the inputs has a high count rate all inputs will experience a high dead time. An alternative method of multiplexing is to scan the inputs one by one on a regular basis, giving equal priority to each input. In such an instrument, it would be possible to have individual live time clocks for each input, but again very variable count rates on different inputs are likely to cause problems. In general, multiplexing is only worth considering for low count rate applications. One frequent use is in alpha spectrometry, where very low count rates are common.

Facilities on multiplexers vary – some allow simultaneous start and stop on all inputs while some can add together all inputs and produce one composite spectrum representing the combined output of a number of detectors. Clearly, setting up such a system, with each separate counter having identical gain and zero offset, would be critical.

Multiscaling, or multichannel scaling (MCS), is a function usually available on an MCA but very seldom used by spectrometrists. It is a means of measuring count rates as a function of time. 'Scalers' are pulse counters. The term derives from the very early days of radiometrics when pulses were counted with electromechanical registers. These could not handle even moderate count rates and so an electronic unit was introduced into the counting system to divide the count rate by 10, 100, 1000 , etc., to make it compatible with the register. This unit was called **scaler**. When new electronics displays became available and electronic counters were devised, the name stuck, and persists here in the term 'multiscaling'. It consists of taking the entire pulse stream, regardless of size (although within a range selected by discriminators), and counting them in a single channel of the MCA. After a certain time period, called the **dwell time**, the pulses are directed into the next channel, and so on. A 4096 channel MCA, for example, becomes 4096 individual counters. Dwell times can range from nanoseconds to days, depending upon the context.

An obvious application is, of course, the measurement of the half-life of a rapidly decaying nuclide. (For longer half-lives, multiple counts on a single counter would be adequate.) Such a use is facilitated by the ability of the MCA, or the program controlling it, to display channel counts on a logarithmic scale. Another common application is in Mossbauer spectrometry, where the change of position of the source is used to trigger the channel stepping rather than time.

4.11 DIGITAL PULSE PROCESSING SYSTEMS

We saw in Figure 4.1 that the conventional gamma spectrometer pulse handling system comprises the following:

- A preamplifier to collect the charge carriers.
- An amplifier, whose primary function is to extract the pulse height information from the preamplifier pulse by pulse shaping.
- An ADC to measure the height of the shaped pulse.
- A memory to store the numbers of counts.

Digital signal processing performs the same functions but on a digitized model of the preamplifier pulse. Such digitization demands an extremely fast ADC. Of all of the types of ADC, the *flash ADC* is the fastest. In principle, as a form of stacked SCAs, the complete pulse height measurement can be done almost instantaneously. Apart from their complexity in terms of numbers of components – an *n*-bit ADC needs $2n - 1$ individual SCAs – and the greater amount of power needed when compared to other types of ADC, the major disadvantage was the fact that resolution was limited to 8 to 10 bits, equivalent to 1–2k channels. Modern developments of the flash ADC have revolutionized pulse height measurement. They are now capable of providing 14 bit resolution, equivalent to 16k spectra, at a sampling rate of 10 Mhz – this is one pulse height measurement every 0.1 μs We saw earlier in this chapter that the rise time of the preamplifier output pulse is of the order of 0.5–1 μs, depending upon detector size, and its fall time might be 150–200 μs. The modern flash ADC can, therefore, measure the height of even a fast changing signal such as a preamplifier pulse.

Not only that. It is possible to digitize the whole of the rise and fall of the pulse. Having a digital representation of the pulse, it is then possible to perform mathematical operations on the digits to aid the spectrometry process – pulse shaping, pole-zero cancellation, baseline restoration, pile-up rejection and ballistic deficit correction are all achievable digitally. The liberating aspect of **Digital Spectrum Analysis** (Canberra – DSA), or **Digital Signal Processing** (Ortec – DSP) is that the digital mathematical operations employed are not restricted to those that can be achieved with analogue electronic circuitry. For example, the best analogue pulse filter achievable is semi-Gaussian, whereas, theoretically, cusp, triangular or true Gaussian filters would be better. A digital filter can emulate all of these filters. In practice, the digital systems commercially available utilize triangular shaping with a flat top to the filter, as shown in Figure 4.43. All of these systems will allow several options for rise time, fall time and flat-top width to allow precise matching to the detector/preamplifier system. Altering the flat-top width

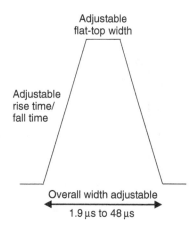

Figure 4.43 The triangular (trapezoidal) pulse filter used in digital signal analysers

and its slope is said to improve correction for ballistic deficit. ORTEC also include in their DSP a digital Low Frequency Rejection (LFR) procedure, working in conjunction with the trapezoidal filter, to remove electronic noise caused by ground loops and microphony. (Removing noise due to the latter has enabled ORTEC to design a complete handheld gamma spectrometer – the trans-SPEC – incorporating electrical cooling of the detector.)

The controls to a DSA/DSP system (software rather than mechanical, as one might expect) echo the controls on a traditional MCA. One difference in implementation concerns the LLD and ULD controls. In a digital system, this is performed digitally after digitization, meaning that electronic noise below the LLD will still contribute to spectrometer dead time even though not appearing in the spectrum. This particular unfortunate feature is causing the author some practical concern at the time of writing.

The manufacturers are, understandably, coy about the detail of their technology, but a research example of a digital system reported in the literature by Kim *et al.* (2003) gives some details of how the process works. In their system, a 100 MHz 12 bit flash ADC with a dual buffer was used. For each peak detected, this took 2000 samples over a time interval ranging from 5 μs before the pulse rise to 15 μs beyond it. Of that data, the first 400 samples were used to estimate the baseline to the pulse shape. After subtracting that baseline from every sample the data was compressed to 250 samples. The resulting smoothed pulse shape (Figure 4.44(a)) was then digitally filtered using a function containing terms taking into account pole-zero cancellation, giving the output pulse shape shown in Figure 4.44(b).

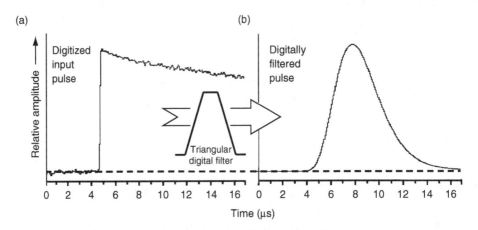

(a)

(b)

Relative amplitude

Digitized input pulse

Triangular digital filter

Digitally filtered pulse

0 2 4 6 8 10 12 14 16 0 2 4 6 8 10 12 14 16

Time (μs)

Figure 4.44 An example of the application of pulse filtering to a digitized preamplifier pulse

The advantages of digital signal processing when compared to analogue systems are as follows:

- Better temperature stability because, once the pulse is digitized, all of the subsequent operations are temperature independent.
- Higher throughput, because the digitization and filtering processes can be made faster than the alternative conventional ADC systems.
- Improved resolution stability at high count rate. Analogue systems are notorious for degraded resolution at high count rate. Digital systems offer much less resolution loss.
- Improved peak position stability. There is much less peak shift as count rate is increased in a digital system than in an analogue one.

With such an array of advantages, it is tempting to suggest that in time digital is bound to supplant analogue technology. Nevertheless, at the time of writing, there are special high specification analogue systems that can outperform digital spectrometers at the highest count rates. However, for all routine gamma spectrometry, digital is very attractive.

PRACTICAL POINTS

- NIM units are interchangeable. A prospective purchaser may take advantage of this by 'shopping around'; there is no need to buy all components of a system from the same manufacturer. This does not apply to the specialized high count rate systems described in Chapter 14.

- For many users, an 'all-in-one' system will fulfil all of their requirements.
- For high-resolution spectrometry systems, it is essential to use preamplifiers and amplifiers with a low noise specification.
- If the system is to be used routinely at high count rate, it might be appropriate to specify a preamplifier with a lower feedback resistor.
- If it is important that the detector system does not 'lock up', specify a transistor reset preamplifier. This may also provide better resolution.
- For best resolution, it is crucial that the pole-zero cancellation should be adjusted correctly. If high throughput is required, consider a trade-off between resolution and throughput.
- For routine high count rate operation, where short time constants are needed, a gated integrator amplifier should be used.
- Pile-up rejection is useful for reducing random summing at moderate to high count rates. At low count rates, it should be switched out.
- Choose an ADC conversion gain such that the FWHM is spread over about four channels. For many purposes, 4096 channels will be appropriate.
- If possible, operate at dead times of < 30 %; above this, qualitative identification should not be a problem, but quantitative measurements could become increasingly prone to error.
- There will be little difference in the performance of Wilkinson and Successive Approximation ADCs for most purposes. If high throughput is a criterion, Chapter 14, Section 14.5 may offer assistance.

- The system must have a well-engineered arrangement for dealing with dead time losses. This is crucial at high count rates and when the count rate changes significantly during the count period. Implement the manufacturer's instructions faithfully.

- A spectrum stabilizer will be useful if the system is subject to poorly controlled environmental conditions, especially temperature change, or is used at high count rates. For germanium detector systems, the digital stabilizer is recommended.

- A digital pulse processing system is certainly worth considering. At the time of writing, it would seem that they offer many advantages and, apart from cost, few disadvantages.

FURTHER READING

- The following is an excellent treatment of radiometric techniques in general and includes a non-mathematical coverage of the basic electronics:

Knoll, G.F. (2000). *Radiation Detector and Measurements*, 3rd Edn, John Wiley & Sons, Inc., New York, NY, USA.

- For a general appreciation of the transistor reset preamplifier, see:

Britton, C.L., Becker, T.H., Paulus, T.J. and Trammell, R.C. (1983). Characteristics of high-rate energy spectrometry systems using HPGe detectors and time-variant filters, presented at the *IEEE Nuclear Science Symposium*, San Francisco, CA, USA, IEEE, New York, NY, USA.

- The catalogues of ORTEC and Canberra have introductory sections which are worth reading for an outline of the electronic processes used in their equipment. Their websites are at: *http://www.ortec-online.com* and *http://www.canberra.com*, respectively.

- Those who have been baffled by the (necessary) electronic jargon might try the following for assistance: *http://www.maxmon.com/glossary.htm*.

- For information on CdTe and CZT detectors, see *Further Reading* for Chapter 3.

5

Statistics of Counting

5.1 INTRODUCTION

In this chapter, I will examine the statistical nature of radioactivity counting. Statistics is unavoidably mathematical in nature and many equations will emerge from the discussion. However, only as much general statistical mathematics will be introduced as is necessary to understand the relevant matters. I will go on to discuss the statistical aspects of peak area measurement, background subtraction, choosing optimum counting parameters and the often superficially understood critical limits and minimum detectable activity. I end with an examination of some special counting situations.

At its simplest, radioactivity counting involves a source, a suitable detector for the radiation emitted by the source, a means of counting those decay events that are detected and a timer. If we measure the rate of detection of events, we can directly relate this to the number of radioactive atoms present in the source. The basic premise is that the decay rate of the source (R) is proportional to the number of atoms of radioactive nuclide present (N), the proportionality constant being the **decay constant**, λ. Thus:

$$R = \frac{dN}{dt} = \lambda N \tag{5.1}$$

R is, of course, what would normally be referred to as the **activity** of the sample. In principle, therefore, if we count the number of events, C, detected by the detector in a fixed period of time, Δt, we can estimate the decay rate as follows:

$$R = \frac{C}{\varepsilon \Delta t} \tag{5.2}$$

where ε, in Equation (5.2), is the effective efficiency of counting, taking into account the source–detector geometry, the intrinsic detection efficiency for the particular

radiation and the probability of emission of the detected radiation.

While it is true to say that all scientific measurements are estimates of some unattainable true measurement, this is particularly true of radioactivity measurements because of the statistical nature of radioactive decay. Consider a collection of unstable atoms. We can be certain that all will eventually decay. We can expect that at any point in time the rate of decay will be that given by Equation (5.1). However, if we take any particular atom we can never know exactly when it will decay. It follows that we can never know exactly how many atoms will decay within our measurement period. Our measurement can, therefore, only be an *estimate* of the expected decay rate. If we were to make further measurements, these would provide more, slightly different, estimates. This fundamental uncertainty in the quantity we wish to measure, the decay rate, underlies all radioactivity measurements and is in addition to the usual uncertainties (random and systematic) imposed by the measurement process itself.

5.1.1 Statistical statements

At this point, it is appropriate to introduce a number of statistical relationships with which I can describe the distribution of a number of measurements. This section must necessarily be somewhat mathematical. However, textbooks on statistics will cover the theoretical basis of these parameters in much detail, and here I will content myself with a number of simple definitive statements. Later, these will become relevant to an understanding of counting statistics.

Let us assume we have m measurements, x_1, x_2, x_3, ... x_m, each of which is an estimate of some parameter. The nature of the parameter is not important: it might be a voltage, a length or, more relevantly, a number of events within a particular count period. The actual form, that is

Practical Gamma-ray Spectrometry – 2nd Edition   Gordon R. Gilmore
© 2008 John Wiley & Sons, Ltd

shape of the distribution of the measurements, need not concern us at the moment. The distribution will have a value, $E(x)$, which we can expect our measurements to have. Thus:

$$\text{Expected value} = E(x) \tag{5.3}$$

The difference between any particular value, x_j, and the expected value gives some idea of how good an estimate that particular measurement was. Taking the differences for all of the measurements into account would give an idea of the overall uncertainty of the measurements. However, some measurements will be below the expected value and others above; taking a simple sum of the differences is likely to give a result of precisely zero. To get around this, the sum of the square of the differences is used. The resulting factor is called the **variance**, so that:

$$\text{var}(x) = \text{expected value of } [x - E(x)]^2$$
$$\approx E\{[x - E(x)]^2\} \tag{5.4}$$

Note that the variance is not a function of x but a parameter of the distribution of x. A more convenient factor, which indicates the spread of the values about the $E(x)$, is the **standard deviation**, σ_x. This is simply the square root of the variance:

$$\sigma_x = \sqrt{\text{var}(x)} \tag{5.5}$$

Standard deviation is more meaningful in the sense that it has an obvious relationship to the expected value and the spread of the distribution. Variance will play a large part in this discussion. Variance is additive, standard deviations are not. Calculating the standard deviation relative to the expected value gives the **relative standard deviation**, r_x, sometimes referred to as the **coefficient of variation**, and often expressed as a percentage:

$$r_x = 100\sigma_x / E(x) \tag{5.6}$$

If we have the results of two measurements that we wish to combine, say x and y, then it is a straightforward matter to show that the following relationships hold:

$$E(x + y) = E(x) + E(y) \tag{5.7}$$

$$E(xy) = E(x)E(y) + \text{cov}(x, y) \tag{5.8}$$

The term $\text{cov}(x, y)$ is the **covariance** of x and y and is analogous to the variance:

$$\text{cov}(x, y) = E\{[x - E(x)][y - E(y)]\} \tag{5.9}$$

Covariance is a measure of the interrelation, or correlation, between x and y. When there is no correlation, as is likely to be in all the cases discussed here, then $\text{cov}(x, y) = 0$.

$$\text{var}(x + y) = \text{var}(x) + \text{var}(y) \tag{5.10}$$

$$\text{var}(x - y) = \text{var}(x) + \text{var}(y) \tag{5.11}$$

$$\text{var}(xy) \approx E(y)^2 \text{var}(x) + E(x)^2 \text{var}(y)$$
$$+ 2E(x)E(y)\text{cov}(x, y) \tag{5.12}$$

It can also be shown that, by making the covariance term negative, this relationship also holds for $\text{var}(x/y)$. More usefully, if as we expect $\text{cov}(x, y) = 0$, and using relative standard deviations, we can rearrange Equation (5.12) to:

$$r_{xy}{}^2 = r_{x/y}{}^2 = r_x{}^2 + r_y{}^2 \tag{5.13}$$

Finally, if k is a constant then:

$$\text{var}(k) = 0 \text{ and } \text{cov}(k, x) = 0 \tag{5.14}$$

$$E(kx) = kE(x) \text{ and } \text{var}(kx) = k^2\text{var}(x) \tag{5.15}$$

$$E(k + x) = k + E(x) \text{ and } \text{var}(k + x) = \text{var}(x) \tag{5.16}$$

These relationships are valid whatever the distribution of our measured values. When we make a radioactive count, our ultimate intention is to estimate the sample activity and a degree of confidence in that estimate of activity. Statistically we can achieve the former aim by identifying the measured count, C, as the expected number of decays, $E(n)$, and relating the confidence limit to the variance $\text{var}(n)$. Thus, in principle:

$$C \Rightarrow n \Rightarrow E(n) \Rightarrow \text{var}(n)$$

Both the expected value and the variance depend upon the form of the relevant statistical distribution and we can now move on to consider the particular case of the distribution of radioactive counts.

5.2 COUNTING DISTRIBUTIONS

5.2.1 The binomial distribution

In principle, the statistics of radioactive decay are binomial in nature. If we were to toss a handful of coins onto a table and then examine the arrangement, we would find coins in one of two dispositions – heads up or tails up. Similarly, if we could prepare a radioactive source and, during a particular period of time, monitor each individual atom we would see that each has only one of two possible fates – to decay or not decay.

Let us suppose that we could determine exactly which of the atoms, and how many, decayed during the count period. If we were able to repeat the experiment, we would find that different atoms and a different number of atoms decayed in the same period of time. We can regard each such measurement, each count, as a sample in the statistical sense, an attempt to estimate the true decay rate. We would expect the distribution of these counts to fit a **binomial distribution** (sometimes called a **Bernoulli distribution**). This distribution applies because:

- There are two possible states for each atom.
- The probability of an atom decaying during the count period is independent of how often we look.
- The decay of one particular atom does not affect the probability of other atoms decaying.

If we consider each atom in our source there is a certain probability, p, that the atom will decay during the period we choose to make our measurement. This probability is related to the decay constant of the atom and it is straightforward to demonstrate that:

$$p = (1 - e^{-\lambda \Delta t}) \tag{5.17}$$

where Δt is the count period and λ the decay constant. Since there are only two possible outcomes for each atom the probability that the atom will not decay must be $1 - p$. The binomial distribution predicts that, in any particular sample of N atoms the probability of n atoms decaying in a given time, $P(n)$, is:

$$P(n) = \frac{N!}{(N-n)!n!} p^n (1-p)^{N-n} \tag{5.18}$$

So if we have, say, 20 atoms and the probability of decay during the count is 0.1, Equation (5.18) predicts that on 9 occasions out of 100 we would find that 4 atoms decayed. This means that if our detection system were 100 % efficient in detecting decays then we would collect 4 counts on 9 out of 100 occasions. Figure 5.1 shows this probability distribution when the probability, p, is 0.1, 0.5 and 0.9. Unless the probability is close to 0.5, the probability distribution is skewed.

Regardless of the shape of the distribution, the most likely number of decays is given by Equation (5.19):

$$E(n) = pN \tag{5.19}$$

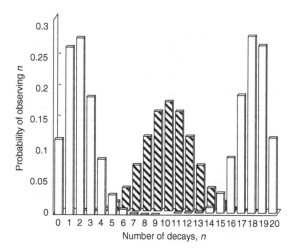

Figure 5.1 Binomial probability distributions for $p = 0.1$ (left), 0.5 (centre) and 0.9 (right)

In the specific cases plotted in Figure 5.1 the most likely counts would be 2, 10 and 18 for the three chosen probabilities. The variance of the distribution is:

$$\text{var}(n) = (1-p)E(n) = (1-p)pN \tag{5.20}$$

Taking the square root of the variance, we can calculate the standard deviation and, for the three specific cases, this would be 1.34, 2.24 and 1.34 decays (or counts, assuming 100 % efficiency). Equation (5.19) is interesting in that it predicts that as the probability becomes very small or very near to 1, the width of the distribution or, we might say, the uncertainty on the number of decays, tends to zero. This is not unreasonable. If $p = 1$, we can expect all atoms to decay and if $p = 0$ none to decay. In either case there is no uncertainty about the number of decays which would be observed.

To relate this to practice, suppose we have counted a sample on a detector with known efficiency, ε, and measured C counts in time Δt s. If the decay constant of the nuclide is known to be λ, then using Equation (5.17), p can be calculated. The overall probability of detection, as opposed to decay, is $p\varepsilon$ and the expected count could be:

$$E(C) = p\varepsilon N \tag{5.21}$$

If we take the measured count C as an estimate of the expected count then Equation (5.1) allows us to calculate the rate of decay, R, as:

$$R = \lambda N = \frac{\lambda C}{(1 - e^{-\lambda \Delta t})\varepsilon} \tag{5.22}$$

In most practical situations, the number of radioactive atoms present is exceedingly high and the probability of detection very small. This means that the number of decays detected (*n* decays or *C* counts) is very much smaller than the number of radioactive atoms present (*N*). (Exceptions to this general situation, when the efficiency of detection and probability of particle emission are very high and when the count period is comparable to the half-life of the nuclide, are discussed in Section 5.7.) In fact, if we assume the detection efficiency to be subsumed into *p*, it makes no difference to the statistics whether we consider number of decays or number of counts detected and from now on we can take *n* and *C* as equivalent. Under these circumstances, various mathematical approximations can be made to Equation (5.18) which lead to a new form for the probability distribution.

5.2.2 The Poisson and Gaussian distributions

The **Poisson distribution** is used in statistics whenever the total number of possible events, in our situation *N*, is unknown. The distribution is described by the equation:

$$P(n) = \frac{[E(n)]^n}{n!} e^{-E(n)} \tag{5.23}$$

As before, *P(n)* is the probability that a count of *n* will be observed given that the expected count is *E(n)*. This distribution has, as might be expected, some similar properties to the binomial distribution. For example, Equation (5.19) is still valid; however, because *p* << 1, Equation (5.20) approximates to:

$$\text{var}(n) = E(n) \tag{5.24}$$

Curiously, a strict consideration of the mathematics produces the conclusion that if we observe this count *n* then the expected value, *E(n)* is:

$$E(n) = n + 1 \tag{5.25}$$

This, at first, surprising statistical fact reminds us that if we were to detect no counts at all, the expected count need not be zero. In most situations, either *n* is large or is to be corrected for background and it is common practice to ignore this particular statistical fact and take *n* as a direct estimate of *E(n)*.

Figure 5.2 compares the binomial distribution and the Poisson distribution when both have *E(n) = 10*. The binomial case repeats the data in Figure 5.1 and represents 20 atoms and a probability of decay of 0.5. In the Poisson case, the number of atoms is unknown but large and *p* is very small. At such a low expected value, there are clear

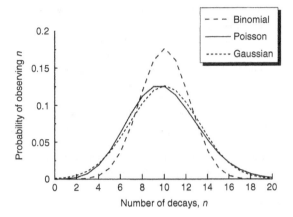

Figure 5.2   Comparison of the binomial, Poisson and Gaussian distributions for *E(n) = 10*

differences. The third distribution shown in Figure 5.2 is the **Gaussian** or **Normal distribution** for the specific case where the variance is equal to the expected value, again 10 counts. This is the distribution one would expect if the differences between the observed and expected counts were solely due to chance. The similarity between the Poisson and Normal distributions is not surprising. When the expected number of counts is greater than 100, then further mathematical approximations can be made to Equation (5.23) which yield the formula for a Gaussian distribution:

$$P(n) = \frac{1}{\sqrt{2\pi E(n)}} \exp\left\{\frac{-[n - E(n)]^2}{2E(n)}\right\} \tag{5.26}$$

To summarize, counting statistics are fundamentally binomial in nature. Under most counting circumstances, we can assume a Poisson distribution of counts. The exceptions to this general rule are:

- when the counting period is long compared to the half-life and the detection efficiency is high;
- when the total number of counts is very small.

These special situations will be discussed in Section 5.7.

5.3 SAMPLING STATISTICS

If we take a large number of measurements of the same parameter, we would find differences in the actual measured value from measurement to measurement. In effect, each measurement is a sample from the infinite number of possible measurements we could make. These measurements will have a distribution and, of course, an expected value and a variance. If the difference between each measured value and the expected value is due purely to chance, then there is considerable evidence to suggest that the distribution will be Gaussian, often referred to as a Normal distribution in this context. In this case, the equation will have a form similar to the special case in Equation (5.27):

$$P(n) = \frac{1}{\sigma\sqrt{2\pi}} \exp\left[\frac{-(x-\overline{x})^2}{2\sigma^2}\right] \qquad (5.27)$$

that is, the probability of measuring a value x given a particular expected value, $\overline{x}$, and a distribution with a standard deviation of σ (see Figure 5.3 below). Suppose, as I suggested earlier, we have m measurements, x_1, x_2, $x_3, \ldots x_m$. We can define the expected value, or **mean**, $\overline{x}$, of these measurements as:

$$\overline{x} = \frac{\sum x_i}{m} \qquad (5.28)$$

where the summation is understood to include all of the measurements x_1 to x_m. The mean is also referred to as the **average**. We can show that as m becomes larger, then $\overline{x}$ becomes a more precise estimate of the expected value. If the true, but unknown, value of the parameter is X, then:

in the limit as m increases: $\overline{x} \, E(x) = X$

It is worth emphasizing this point. The mean, $\overline{x}$, is not the true value of the parameter, only a better, more reliable estimate. The width of the distribution of measured values

gives an idea of the overall uncertainty of the measurements. The factor quantifying the width of a distribution is the variance, which is calculated as:

$$\text{var}(x) = \frac{\sum(x_i - \overline{x})^2}{m-1} = s^2 \qquad (5.29)$$

where s is an estimated standard deviation, not to be confused with the true standard deviation of the distribution, σ, from which we have taken our sample. The denominator of Equation (5.29), $m - 1$, is referred to as the number of **degrees of freedom**.

As with the mean, the more items taken together, the more precise the estimate of the standard deviation:

$$s^2 \, \sigma^2, \text{ as } m \text{ increases} \qquad (5.30)$$

It is becoming common to refer to **standard uncertainty**, that being the uncertainty on a value at the level of one standard deviation. This may seem an unnecessary addition to the vocabulary but the term does have the advantage of emphasizing that we are dealing with uncertain measurements. It is, perhaps, worth noting that in statistical texts it is more usual to discuss 'standard error' rather than 'standard uncertainty'. I shall keep to the latter usage as being descriptive of the actual situation, reserving the term 'error' for mistakes and the use of incorrect values (see also Section 5.8.1 relating to use of the terms 'accuracy' and 'precision').

5.3.1 Confidence limits

When we quote the result of an experimental measurement, whatever the technique used, it is essential that it is accompanied by a realistic estimate of the uncertainty of the measurement. If we refer again to the Normal distribution of all possible results of a particular measurement, then the uncertainty of the measurement must be related to the width of the distribution. Suppose then that we were to quote our result as, say, $a \pm s$, where s represents one standard uncertainty (a not uncommon procedure). This statement says that the true result (which we can never know) is most likely to be close to a and is less likely to lie below $a - s$ or above $a + s$. We can see from Figure 5.3, where the Normal distribution is plotted with the abscissa scaled in units of one standard deviation, that there is a great deal of scope for the true value to lie outside of these limits and still be 'within' the distribution of results.

To be more certain that our quoted limits encompassed the true value, perhaps we should quote two or three times the standard uncertainty. Whatever limits we choose, we still need to quantify the likelihood of the true value being

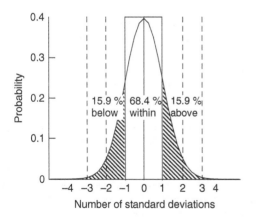

Figure 5.3 A Gaussian distribution with unit standard deviation

outside them. Or, to put it the other way about, we must quote our limits in such a way that we have a stated degree of confidence that the true value lies somewhere within them – hence the term '**confidence limits**'. This degree of confidence is related to the area of the Normal distribution lying within the limits and can be calculated precisely from the parameters of the Normal distribution. The number of times the standard uncertainty we decide to quote to achieve our desired degree of confidence is called the **coverage factor**. Table 5.1 lists the degree of confidence associated with various coverage factors. So, for example, if we wish to be 90 % confident that the true result lies between the quoted limits we might quote the result as follows, with a coverage factor of 1.645:

$$a \pm 1.645s \ (90\% \text{ confidence})$$

Table 5.1 Coverage factors and the associated degree of confidence[a]

Coverage factor	Area within confidence limits (%)
1.0	68.3
1.645	90.0
1.96	95.0
2.0	95.5
2.326	98.0
2.576	99.0
3.0	99.9

[a] Confidence limit = coverage factor × s.

The confidence limit quoted in this manner may be referred to as the **expanded uncertainty**. This particular result has confidence limits that are symmetrical about the mean because we have assumed that the distribution of the measurements is Normal. If the distribution were skewed in any way, or perhaps if we were aware that the measurement was possibly, for some reason, biased high (or low), then the lower and upper confidence limits would not be identical.

It is a common practice to quote confidence limits as a percentage of the value rather than as standard deviation. For example, Table 5.2 demonstrates a calculation of weighted mean (which will be explained in due course). If we take the first weighted mean result, we might quote it as:

10.33 (8.03 %, 1σ)

The advantage is that, expressed in this manner, the uncertainty of the result is immediately obvious whereas

Table 5.2 Illustration of weighted mean[a]

Set A

Count	Time (s)	cps	%RSD
102	10	10.2	9.90
53	5	10.6	13.74

Simple mean: 10.40 ± 0.28(2.72 %)
Weighted mean: 10.33 ± 0.83(8.03 %) pooled
　　　　　　　　 or ± 0.19(1.81 %) weighted

Set B

Count	Time (s)	cps	%RSD
1020	100	10.2	3.13
560	50	11.2	4.23

Simple mean: 10.70 ± 0.71(6.61 %)
Weighted mean: 10.51 ± 0.26(2.52 %) pooled
　　　　　　　　 or ± 0.46(4.41 %) weighted

[a] Figures are to be regarded as intermediate, un-rounded, values.

the alternative, 10.33 ± 0.83, needs a degree of mental arithmetic to appreciate whether the result is of good or poor quality.

While discussing the quoting of results it is, perhaps, appropriate to comment on the oft-abused matter of *rounding*. We might well have quoted the result above as 10.333 + 0.829. Taken at face value, that implies that we are able to determine the confidence limit to one part in 829. That is, not 0.828 nor 0.830, but 0.829. In fact, we were not able to determine the actual value to better than 829 in 10 333 (about one part in 12). In the light of this, is it reasonable to suggest such a high degree of precision for the estimate of the confidence limit? This is a prime example of *spurious accuracy*.

A 'rule-of-thumb' suggested in a code of practice published by the National Physical Laboratory (1973) is as follows:

- Take the confidence limit and round it *always upwards* so as to leave only one significant figure, e.g. 0.829 becomes 0.9.
- Round the result itself *up or down* according to the normal rule to the same degree of precision as the confidence limit, e.g. 10.333 becomes 10.3.

The result above rounded according to these rules would become 10.3 ± 0.9, a more honest statement of what was achieved by the measurement. The recommendation by UKAS, the United Kingdom Accreditation Service (1997), which is consistent with the broader advice in the NPL code, is to quote a 95 % confidence limit (coverage factor 1.96 – often rounded to 2) and to round to two significant figures. Thus, the result above would become 10.3 ± 1.7.

5.3.2 Combining the results from different measurements

Suppose that we have made two measurements of the same parameter and have calculated the uncertainty associated with them. For example, we might have taken measurements on two separate sub-samples of the same radioactive sample and calculated the activity, a_1 and a_2, in Becquerels per gram with confidence limits of s_1 and s_2, which for simplicity we will take as the 68.3 % confidence limit (one standard uncertainty). We will assume that these confidence limits include all sources of uncertainty, not only those due to counting uncertainty:

e.g. $a_1 \pm s_1$ and $a_2 \pm s_2$

Unless the variances of the two results are equal, it is not statistically valid to take a simple mean. This is not unreasonable. A simple mean accords equal importance to each result. A result with a larger variance is less precise and should not be taken as much notice of. The correct procedure is to calculate a weighted mean, $\bar{a}$:

$$\bar{a} = \frac{\sum a_i w_i}{\sum w_i} \qquad (5.31)$$

where w_i are weighting factors for each individual result and are simply the reciprocal of the variance of each result. (As usual, the summation is taken to mean the sum over all items.) For example, in the case suggested above: $w_1 = 1/s_1^2$ and $w_2 = 1/s_2^2$. The standard uncertainty of the combined result is calculated from the **pooled variance**:

$$\text{var}(a)_{\text{internal}} = \frac{1}{\sum w_i} \qquad (5.32)$$

Because this calculation takes into account only the individual sample uncertainties, implicitly assuming that the distribution about the mean is satisfactory, this is also known as the **internal variance**. Table 5.2 gives a couple of numerical examples to illustrate the difference between simple and weighted means. In Set A, the simple and weighted means are similar but the simple standard uncertainly does not reflect the fact that both measurements are of poor precision. The weighted mean and pooled standard uncertainty give a much more realistic assessment of the data.

What, however, if the quoted uncertainties do not take into account all sources of uncertainty? In Set B, count times are taken ten times longer. The data is such that the precision of each result is better but the actual results are further apart. In this case, the pooled precision, 2.52 %, is consistent with the precision of the individual results (as

it must be!) but does appear to be optimistic taking into account that the difference between the results is nearly 10 %.

Of course, such a large difference could happen by chance, by the statistical roll of the dice, but it is more likely that there are other sources of uncertainty in addition to that due to counting and not accounted for in the uncertainty quoted. We could, of course, simply ignore the uncertainties on the individual values and calculate a simple mean. That, however, would not take into account the relative degrees of reliability of the individual values. In such cases, a standard deviation derived from the **weighted variance** might be quoted, calculated as follows:

$$\text{var}(a)_{\text{external}} = \frac{\sum (a_i - \bar{a})^2 w_i}{\sum w_i (m - 1)} \qquad (5.33)$$

Because this takes into account the spread of the results about the mean, it is also known as the **external variance**. This is quoted in Table 5.2 as the **weighted uncertainty**. For Set B, the weighted uncertainty of 4.41 % is a more satisfactory estimate of the actual uncertainty than the pooled estimate. In practice, particularly if the work is done by computer, it would make sense to calculate both estimates and quote as the best result the weighted mean together with the larger of the two uncertainty estimates. There is no merit in underestimating uncertainties.

Calculating both has in any case diagnostic value. If experience of a particular measurement scheme shows that the pooled variance is always a significant underestimate of the actual variance, then the measurement process should be looked at in detail to track down the hidden sources of that extra uncertainty.

It should not be lost on us that a single radioactive count has an inherent uncertainty and this should be borne in mind when combining simple count data. A weighted mean should always be used. In fact, because the variance of a count is numerically equal to the count itself, simply combining the count data together will do just that as long as there are no significant sources of uncertainty other than counting uncertainty. As an example, take Set A data from Table 5.2. Simply adding together the counts $(102 + 53 = 155)$ and dividing by the sum of the count times $(10 + 5 = 15)$ provides the weighted mean result of 10.33 cps with an uncertainty (1σ) of 0.83 cps (i.e. $\sqrt{155}/15$), precisely the result shown in the table. Note, though, that applying the same procedure to the data in Set B would give an unsatisfactory result because of the extra, unknown uncertainties.

When calculating weighted means, it is important that the variances used only include those items of uncertainty that are different from measurement to measurement.

Common uncertainties should not be included; otherwise, correlations within the data are introduced.

5.3.3 Propagation of uncertainty

The previous section discussed combining the results of different measurements to obtain a better overall result. We noted that data Set B in Table 5.2 must have undisclosed sources of uncertainty. Let us suppose that it becomes apparent that the preparation of the sources had introduced an extra uncertainty of 6.5 % in the case of the first source and 5.3 % for the second. How can we include the information? The calculation of the uncertainty for each data item, using the example of Set B, is as follows:

- for the count of 1020: $\sqrt{(3.13^2 + 6.5^2)} = 7.21$ %;
- for the count of 560: $\sqrt{(4.23^2 + 5.3^2)} = 6.78$ %.

This would provide us with a weighted mean of 10.68 with a pooled uncertainty of 4.94 %, consistent with the actual spread of the data suggested by an external uncertainty of 4.68 %. This is an example of **propagation of uncertainty**. Because the source preparation factor is multiplicative, Equation (5.13) from Section 5.1.1 can be used to combine the uncertainties. The uncertainties are said to have been combined **in quadrature**. (We will meet this again later when discussing the factors that combine to create the width of gamma-ray peaks.)

In our example here, if the source preparation uncertainty were a fixed amount for the method it would be an item common to both sources. It should not, therefore, be included when the uncertainties on the individual results are calculated. It should be taken into account by adding in quadrature to the weighted mean result. If, in our example, the sample preparation uncertainty were 6.5 % for both samples, then the overall uncertainty of the weighted mean for Set B would be $\sqrt{(4.41^2 + 6.5^2)} = 7.85$ %. The weighted mean value would be unchanged.

In a radioactivity measurement, we may have several sources of uncertainty, all of which must be taken into account in our final uncertainty. For example, we might have:

$$r_T = \sqrt{r_A^2 + r_P^2 + r_S^2 + r_E^2} \qquad (5.34)$$

where the various factors are the relative standard deviations of, in order, the total, peak area measurement, source preparation, standard calibration and the efficiency estimate (which would, in turn, include uncertainties due to gamma-ray emission probability and half-life).

Equation (5.13) can only be used in this way when the various factors are multiplied together. If the factors contributing to the overall result are additive, then Equations (5.10) and (5.11) are relevant. For example, assume the result is calculated by an equation which includes additive and multiplicative factors, for example:

$$R = (C - B) \times Y/E$$

The process of combining uncertainties will have to be done in separate stages. In this example, the overall uncertainty of $(C - B)$ must be calculated by using Equation (5.11). This uncertainty, expressed in relative terms, can then be combined in quadrature with the relative uncertainties of Y and E. Combination of uncertainties will be discussed further in Section 5.8, where uncertainty budgets are discussed.

5.4 PEAK AREA MEASUREMENT

In Chapter 3, I explained that a gamma-ray spectrum consists of a large number of 'channels' in each of which are accumulated all of those counts which fall within a small energy range. We might have, for example, a 4096 channel spectrum covering an energy range of 2048 keV, the content of each channel representing the number of counts received within a 0.5 keV energy window. Successive channels represent increasing energy. Within such a spectrum, a gamma-ray appears as a distribution of counts, approximately Gaussian, about a central point which we can take to represent the gamma-ray energy (Figure 5.4). In principle, the actual distribution of counts in a peak is irrelevant; measurement of the peak area should require no more than a simple summation of the number of counts in each of those channels that we consider to be part of the peak and subtraction of an allowance for the background beneath the peak.

The background beneath gamma-ray spectrum peaks can arise from many sources. In most cases, the background will represent the Compton continuum from other gamma-ray interactions within the detector, within the sample itself and from general background radiation interaction with the shielding and the detector. Both background radionuclides and other radionuclides in the sample will contribute to this peak background. Unlike simple counting where total counts are accumulated, the measurement of natural background is of little use in estimating the continuum background beneath a peak. In some cases, those where the radionuclide we wish to measure can be detected in the natural background (^{60}Co, for example), allowance will have to be made for this additional peaked-background over and above the continuum background. This will be considered later, and for the time being we will make the assumption that the continuum beneath the peak is linear.

(a)

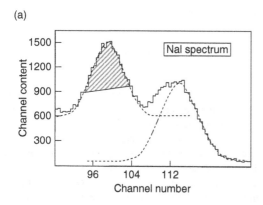

(b)

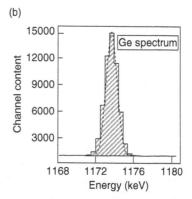

Figure 5.4 Portions of the spectrum of ^{60}Co gamma-rays measured on (a) NaI(Tl) scintillation, and (b) Ge semiconductor detectors (the dotted lines represent the underlying Gaussian distribution of counts)

Over the years, a number of simple algorithms for peak area calculation have been used. The *Covell method* was used in the early days of digital gamma-ray spectrometry for measuring peak areas in sodium iodide scintillation spectra. The procedure was to locate the highest channel in the peak and then to mark the peak limits an equal number of channels away from the centroid channel. When using low-resolution scintillation detectors, peak interference was frequent and it was often necessary to restrict the portion of the peak measured to minimize the effect of neighbouring, possibly overlapping, peaks (see Figure 5.4). The fact that not all of the peak area was taken into account was compensated for by ensuring that the same fraction of the total peak (i.e. the same measurement width) was used for all samples and standards.

With the advent of high-resolution detectors, peak interference became the exception rather than the rule and the peak limits were extended down the sides of the peak to the background continuum level. This, the **total peak area method**, is now the standard method for peak area estimation for single un-interfered peaks. Other methods for estimating peak background, such as the *Wasson* and *Quittner methods*, found limited favour but, except for a few special situations, these offered no overall advantages. As an example, the Quittner method, which involved fitting a polynomial function to the background channels either side of the peak, is more accurate when the peak sits on an obvious nonlinear background, such as the top of a Compton edge.

5.4.1 Simple peak integration

In both the Covell and total peak area methods, the background level is estimated by using the channel contents at the upper and lower edges of the peak region (Figure 5.5).

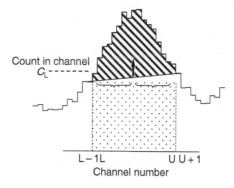

Figure 5.5 Calculation of peak area using the Covell method

If we take the first channel on each side of the peak beyond what we consider as being the peak region as representative of the background, then the gross (or integral) area of the peak is:

$$G = \sum_{i=L}^{U} C_i \qquad (5.35)$$

where C_i are the counts in the ith channel (see Figure 5.5). The background beneath the peak is estimated as:

$$B = n(C_{L-1} + C_{U+1})/2 \qquad (5.36)$$

where n is the number of channels within the peak region and C_{L-1} and C_{U+1} are the counts in the channels immediately beyond the lower and upper edge channels L and U. This background is, mathematically, the area of the background trapezium beneath the peak. It is more useful to think of this as the mean background count per channel beneath the peak, multiplied by the number of channels within the peak region.

The net peak area, A, is then:

$$A = G - B = \sum_{i=L}^{U} C_i - n(C_{L-1} + C_{U+1})/2 \qquad (5.37)$$

It is important to appreciate that while we can calculate precisely the number of counts within the peak region (G), we can only ever estimate the number of background counts beneath the peak. We can never know which counts within the peak region are due to background and which are the peak counts. In most spectra, the peak background continuum derives from the sample itself. Unlike simple total activity counting, such as Geiger–Müller counting, we cannot take away the sample to determine a precise background count. In certain circumstances, in particular, when small peaks lie on large backgrounds, the uncertainty on the background estimate can dominate the total uncertainty of the peak area measurement.

Background estimates can be made more precise (i.e. less uncertain) by using more channels to estimate the mean count per channel under the peak. Figure 5.6 shows the general principle. Instead of a single channel, m channels beyond each side of the peak region are used to estimate the background beneath the peak.

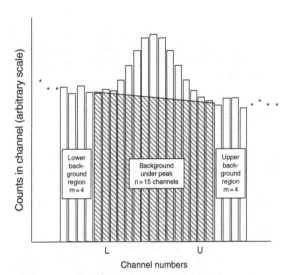

Figure 5.6 Calculation of peak area using extended background regions

Extending Equations (5.35) to (5.37) we find:

$$A = \sum_{i=L}^{U} C_i - n\left(\sum_{i=L-m}^{L-1} C_i + \sum_{i=U+1}^{U+m} C_i \right)/2m \qquad (5.38)$$

Again, the background term is the mean background count per channel, but now estimated using upper and lower background regions, m channels wide, multiplied by the number of channels within the peak region. There is little point in estimating a peak area unless the statistical uncertainty of that peak area is also calculated. If, as we have stated above, $A = G - B$, then, according to Equation (5.11), the variance of the net peak area is given by the sum of the variances of these two terms, giving:

$$\text{var}(A) = \text{var}(G) + \text{var}(B) \qquad (5.39)$$

Substituting for the individual variances and using Equation (5.15):

$$\text{var}(A) = \sum_{i=L}^{U} C_i + n^2\left(\sum_{i=L-m}^{L-1} C_i + \sum_{i=U+1}^{U+m} C_i \right)/4m^2 \qquad (5.40)$$

From this, we can calculate the standard deviation, σ_A.

This simple method described here assumes that the background is linear from the bottom to the top edge of the peak. In fact, examination of well-defined peaks shows that the background appears to have a 'step' beneath the peak (see Figure 9.6). Nevertheless, for most everyday purposes the method provides satisfactory results. The simple method cannot, of course, provide satisfactory results in cases where peaks are overlapped.

It is still possible to find incorrect expressions for the calculation of peak area uncertainty in the literature. The confusion arises because of a failure to appreciate that unlike single background counts where the variance of the count is numerically equal to the count itself, the variance of a peak background depends upon the number of background channels used. The offending expressions are variation of the form:

$$\sigma_A = \sqrt{(A + 2B)}, \text{ or } \sigma_A = \sqrt{(G + B)} \qquad (5.41)$$

These expressions are certainly correct for a single count plus background count, for example, from a simple beta counter. They are not valid for peak area calculations where Equation (5.40) must be used, resulting in the correct expression:

$$\sigma_A = \sqrt{[A + B(1 + n/2m)]} \qquad (5.42)$$

In the simple case of a single count plus background count, var(B), according to Poisson statistics is indeed equal to B and the expressions (5.39) and (5.40) are equivalent. In the peak area case, while var(G) is numerically equal to G, a sum of counts, the variance of the background estimate, var(B), depends upon the number of channels used to estimate (as opposed to measure!) the background as we saw earlier. Equation (5.41) does not take this into account and must, therefore, be generally incorrect. It is only true for the single case when $n = 2m$.

5.4.2 Peaked-background correction

So far, we have discussed only the situation when the background to the peak is a continuum. Measurements of radionuclides that are detectable in natural background must take that additional background component into account. In these cases, the backgrounds to the peaks will be peaks themselves and will be unavoidably included within the overall calculated peak areas. A background spectrum must be measured and the appropriate peak areas determined and subtracted from the sample peak areas. **Peaked-background** is most likely when dealing with environmental samples where, one hopes, the sample activity is near to background levels.

In some analysis programs, peaked-background correction is made after peak areas and the background contribution have been separately converted to nuclide activities. Since the calculation of activity necessarily introduces extra uncertainties, it makes sense to make the background correction at the earliest possible stage of the analysis process. Ideally, analysis programs should allow the correction to be made in terms of peak count rate in counts per second.

For example, if the peak area is A counts accumulated over Δt seconds of live time, the net peak area will be:

$$A_{NET} = A - B_{PBC} \times \Delta t \qquad (5.43)$$

where B_{PBC} is the background peak count rate in counts per second and, as in any background correction, the variance (from which the standard uncertainty and %RSD can be calculated) will be:

$$\text{var}(A_{NET}) = \text{var}(A) + \text{var}(B_{PBC} \times \Delta t)$$

$$\text{var}(A_{NET}) = A + (B_{PBC} \times r_{PBC} \times \Delta t)^2 \qquad (5.44)$$

where r_{PBC} is the uncertainty on B_{PBC}, expressed as a relative standard uncertainty (*not* as a percentage). Although commercial spectrum analysis programs will consider peaked-background correction, at least one, GammaVision™, takes no account of the uncertainty on

the peaked-background. The effect of that is to increase the number of false positive results when there is little or no nuclide present over and above natural background.

Apart from natural background, peaked-backgrounds can be experienced if the detector is used in an area where there is an enhanced neutron flux. Gamma spectrometry close to nuclear reactors and accelerators can be a problem in this respect. Although the neutron fluxes may not be significant from a safety aspect, activation of the materials of the detector system and prompt gamma-rays from neutron capture can sometimes be a problem. Appendix C lists the prompt gamma-ray from the activation of ^{114}Cd within a graded shield.

Leaving such special cases aside, it cannot be assumed that background is constant. An obvious example is the common background nuclide, ^{60}Co, which decays with a half-life of 5.27 years by about 1 % per month. Many peaks within the natural background spectrum originate in the uranium and thorium decay series. The degree of ventilation in a counting room might alter the amount of radon within the room and the amount of daughter nuclides in equilibrium with it. Even the external cosmic-ray background can change over a period of time. Background spectra should therefore be measured regularly. Because of this variation, it is advisable to collate the analysis of several backgrounds to establish a true uncertainty over time, rather than depend upon the measurement uncertainty of a single measurement.

There is further discussion of the sources of background in Chapter 13.

5.5 OPTIMIZING COUNTING CONDITIONS

5.5.1 Optimum background width

Equation (5.40) implies that the uncertainty of the estimate of the background must depend upon the number of background channels used. Since the more channels that are used the better the background estimate, it would appear that the more channels the better.

However, as one uses more channels there are decreasing returns and one must not overlook the possibility of neighbouring peaks causing a wide background region to be nonlinear. What is the optimum number of channels to use? This depends upon the circumstances.

Figure 5.7 summarizes the results of an assessment of the measurement of a particular ill-defined peak taken from an actual gamma-ray spectrum as a function of the width of the background region. It is apparent that the uncertainty on the peak area estimate (expressed as percentage relative standard deviation in Figure 5.7) decreases as the number of channels used to estimate the background increases. It is obvious that two channels is a

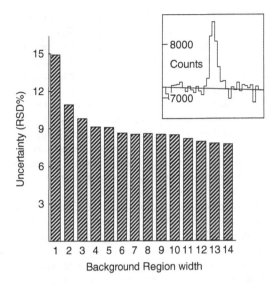

Figure 5.7 Variation of peak area uncertainty with background region width (the inset figure shows the actual peak measured)

considerable improvement on one, and three rather better than two are, but the reduction in uncertainty with each extra channel used gets smaller and smaller. There would be little extra value in using more than nine or ten channels and, in practice, the presence of neighbouring peaks may automatically limit the width of the background region.

If the peak is well defined and has a large area, then there may be little to be gained by using more than three or four channels. In such cases, the background uncertainty will have a much smaller effect on the uncertainty of the net peak area estimate. Note that the number of channels used for background estimate does not have any statistically significant effect on the net area, only on the uncertainty with which it is measured.

In an automatic spectrum analysis system, a compromise is usually made. Most commercial MCA and spectrum analysis programs use 3, 4 or 5 channels, depending upon the manufacturer and the situation. Note that there is no fundamental reason why the width of the background region should be the same above and below the peak region. If there were a potentially interfering neighbour above the peak, it would be sensible to use, say, three channels above and, perhaps, ten below. In such a case, the term $2m$ in Equations (5.38) and (5.42) would be replaced by $(m_L + m_U)$, where m_L and m_U are the lower and upper background region widths. When its 'Automatic' peak background width option is selected, GammaVision™ chooses 5, 3 or 1 channel widths on each side of the peak independently, depending upon whether

the channels are deemed to represent a flat portion of the background continuum.

5.5.2 Optimum spectrum size

How does the peak area uncertainty alter with the number of channels in the spectrum? Conventional advice is often to use as many channels as possible. If you have an 8192 channel MCA system, use 8192 channels, if 16 384 use 16 384. The argument is that as detector resolution increases with advances in detector manufacture, the number of channels within each peak becomes smaller at a constant energy range. From the point of view of the spectrum analysis program, it may be advantageous to have more, rather than fewer, channels in each peak. However, what is not always taken into account is that as the spectrum is spread over more channels, for a constant counting time, the numbers of counts within the channels decrease. In order to compensate we must increase the number of channels in the peak region and ought to increase the number of channels in the background regions. Unfortunately, while the former will be done automatically by the spectrum analysis software there may be no option of altering the background region width. Figure 5.8(a) demonstrates how the uncertainty of a peak area estimation deteriorates as the number of channels in the spectrum is increased without a corresponding increase in background width.

The curves were calculated for a peak of energy 1332 keV, measured with a resolution of 1.8 keV and for peak areas of 500, 1000 and 10 000 counts on a background of 1000 counts per keV with an overall range of 2048 keV. Figure 5.8(b) shows the appearance of the same 500 count peak in spectra ranging from 4096 to 32 768 channels. Although the overall peak area is unchanged, the uncertainty (i.e. the scatter from channel to channel) of the background is much greater because the counts are spread over more channels. The consequence is poorer precision for the area measurement. Even if the background region width is adjusted to suit the change in spectrum size by doubling the width for a doubling of spectrum size, there is no advantage, from the point of view of peak area precision, of using a larger spectrum size.

We should not forget though that peak width varies with energy. (The effect this might have on conversion gain was discussed in Chapter 4, Section 4.11.4.) If mainly low energy, and narrower, peaks are to be measured, then more channels per keV might be arranged either by increasing amplifier gain or doubling the spectrum size. If only high energy, and wider, peaks are of interest, then a smaller amplifier gain or spectrum size might be preferable.

(a)

(b)

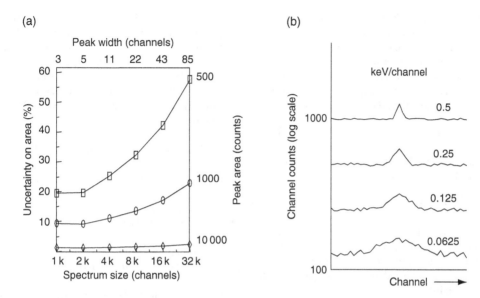

Figure 5.8 (a) Deterioration of peak area precision with increase in spectrum size. (b) Broadening of peak and increase in background 'scatter' with increase in spectrum size

Taking into account peak area measurement uncertainty, and the need for a reasonable number of channels within a peak (to facilitate peak searches and fitting), it would seem that spectrum sizes of 4096 and 8192 channels would be optimum. With current detectors and spectrum analysis software, there seems little point in seeking larger spectrum sizes.

5.5.3 Optimum counting time

In many laboratories, samples will not be submitted one by one for individual attention but in batches all to be counted within as little time as possible. Efficient use of counting equipment in terms of the time devoted to counting each sample can pay dividends when time and equipment are limited.

The first matter to be decided is the precision required of the final result. Let us suppose that, as an example, the reason for the count is to assess whether the ^{137}Cs in a sample of lamb is above or below some action limit. It might be that a poor precision result from a count of only five or six hundred seconds might answer the question for the majority of samples where the amount of ^{137}Cs was much lower, or indeed much greater, than the action limit. This would leave much more time available to achieve more precise results for those samples that are near to the action limit.

For a single sample with unlimited counting time available, the optimum count period is that which will provide

sufficient counts in the spectrum to allow a peak area estimate with the predetermined satisfactory precision or to achieve a stated upper limit on activity. All multichannel counting systems will allow count periods to be automatically terminated after a preset time – usually, live and real time presets are options. Some systems will allow preset maximum count or total count within a channel or spectrum region. While such options can be useful, their value is limited by the fact that the precision of small peak areas may depend largely on the background continuum level. If this varies greatly from sample to sample, then preset-maximum count or gross region count are unlikely to be of any real use. In this respect, the most useful systems are those which monitor the actual peak uncertainty (as %RSD) continuously during the count and allow acquisition to be terminated when the desired precision is achieved.

More thought is needed when there are a number of samples to be counted with differing activities. Unless the facility for continuous monitoring of precision is available, there seems to be few options. Either one can select a count period which is expected to be satisfactory for the majority of samples and accept that some will be 'over-counted' and some 'under-counted', or one might split the samples into groups of roughly equal activity and count each group under optimum conditions for that group.

A special case arises when a background spectrum is necessary. An example might be the measurement of low levels of ^{60}Co where a peaked-background correction must be made. Let us assume that we have a batch of samples to count, plus background, within a fixed overall time period. Is there an optimum way to split the available counting time between samples and background to achieve the best precision for the net count rates?

Taking simple single channel counts as an example, if we measure C counts in time Δt_C and measure a background count B in Δt_B, then the net count rate (R) is:

$$R = C/\Delta t_C - B/\Delta t_B \qquad (5.45)$$

and the variance of this net count, V, is, according to Equations (5.11) and (5.15):

$$V = C/\Delta t_C^2 + B/\Delta t_B^2 \qquad (5.46)$$

Now, if we have a fixed total count time, $\Delta t_C + \Delta t_B$, then the optimum sharing of the time will be found when the variance is at a minimum, i.e. when $dV/dR = 0$. If the mathematics is followed through, we find that this condition is obtained when:

$$\Delta t_C/\Delta t_B = \sqrt{(C/B)} \qquad (5.47)$$

Now because C, the total count, can never be less than B, then Δt_B, the time devoted to background counting, should never be greater than Δt_C, or otherwise the precision of sample measurement will suffer. For a sample of four times background, we would achieve the best precision if we counted the sample for two thirds of the available time and the background for one third.

This is counter to the instinct to devote more counting time to the background – on the basis that because the background correction is applied to all sample counts, it should therefore be of high precision. If the sample count is near to background, then both are equally important in terms of precision. If the sample count rate is higher than background, then the background is proportionally less important and can be counted for a shorter time. Ultimately, of course, as the sample activity becomes very large the background becomes insignificant and we might choose not to measure it at all. If there is more than one sample, the conclusions are still valid. If the activity of the samples is unknown, divide the counting time so as to give the same counting time for each sample and background. If the samples are known to be greater than background, then reduce the background time appropriately and share the saved time equally between the samples.

5.6 COUNTING DECISION LIMITS

There is a great deal of confusion about the meaning of such terms as 'limit of detection', 'minimum detectable

activity' and 'critical limit'. The terms are often treated as if interchangeable and there appears to be a considerable degree of freedom of choice in the manner in which they are calculated. Of these, minimum detectable activity (MDA) appears to be the most variable and I shall discuss this later. For now, I shall define a number of statistically determined levels that answer the following questions:

- **Critical limit** (L_C) – a decision level: '*Is the net count significant?*'
- **Upper limit** (L_U) – '*Given that this count is not statistically significant, what is the maximum statistically reasonable count?*'
- **Detection limit** (L_D) – '*What is the minimum number of counts I can be confident of detecting?*'
- **Determination limit** (L_Q) – '*How many counts would I have to have to achieve a particular statistical uncertainty?*'
- **Minimum detectable activity** (MDA) – '*What is the least amount of activity I can be confident of detecting?*'

These are considered in some detail by Currie (1968) and from a different perspective by Sumerling and Darby (1981). Note that, with exception of the MDA, the limits are calculated as a count rather than as an activity or other derived quantity. Note also that critical limit and upper limit relate to a measurement just made, whereas detection limit (and the associated MDA) and determination limit pose hypothetical 'what if' questions.

5.6.1 Critical limit (L_C)

'*Is the net count significant?*' After a peak area has been measured, it is important to establish its statistical significance. Since a peak becomes non-significant only by being 'lost' in the background, this cannot be done by reference to the peak area alone but must take into account the uncertainties of the background.

Let us suppose that a sample with no radioactivity at all in it was measured a large number of times. A series of counts – effectively background counts – would be obtained for which the mean net count above background was zero but distributed in a Gaussian fashion above and below zero (Figure 5.9). The spread, or standard deviation, of this distribution we will call σ_0.

How can we decide whether any particular measurement near to zero is truly zero or represents a true positive count? There must be some level, which we can call the **critical limit**, above which we can be confident, to a degree, that a net count is valid. We might decide that if the count, A, were above a certain number of standard

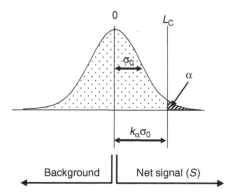

Figure 5.9 Definition of critical limit (the vertical axis represents the frequency of observing a particular count)

uncertainties of the distribution of counts we would be confident that the count existed, namely:

- if $A > k_\alpha \times \sigma_0$, the count is statistically significant;
- if $A \leq k_\alpha \times \sigma_0$, the count is not significant.

The factor k_α would be selected to provide a predetermined degree of confidence in the conclusion. For example, we may consider that it would be acceptable that if a count happened to be at the critical limit there would only be a 1 in 20, or 5 %, chance that we would judge the count to be present when in reality it was not. This is the same as saying that at the critical limit we would be 95 % certain that the count was not statistically significant. In this case, in statistical probability terms $\alpha = 0.05$ and, from one-tailed probability tables (Table 5.3), we find that k_α would be 1.645. (We use the one-tailed tables because we are only interested in the level being exceeded on one side, the higher, of the distribution.)

Table 5.3 k_α factors for particular probability intervals and the associated degrees of confidence

Probability interval, α	1-tailed confidence	2-tailed confidence	$k\alpha$ factor
0.1587	84.13	68.27	1.0
0.1	90.00	80.00	1.282
0.05	95.00	90.00	1.645
0.025	97.50	95.00	1.96
0.022 75	97.73	95.45	2.00
0.01	99.00	98.00	2.326
0.006 21	99.38	98.75	2.5
0.005	99.50	99.00	2.576
0.001 35	99.87	99.73	3.0

$$L_C = 1.645\sigma_0 \quad (95\ \%\ \text{confidence limit}) \quad (5.48)$$

Operationally, this is applied as follows. If the net count is above L_C, we can say that the activity has been detected and can legitimately quote a value together with an associated uncertainty (confidence limit). Otherwise, we must judge the count not significant and quote an upper limit (see below). There is, of course, nothing sacrosanct about the 95 % confidence level. A higher or lower level might be chosen with an appropriate change to the value of k_α. Whatever the value chosen, it should be a positive decision in the context of the overall measurement system rather than a selection by default.

In practice, we do not know σ_0, the standard deviation (or uncertainty) of the net background count distribution. All we do have are the sample and background estimates. Taking Equation (5.39) again and remembering that var(count) = count, we can deduce that:

$$\text{var(net count)} = \text{net count} + \text{background}$$
$$+ \text{var(background)} \quad (5.49)$$

This is true whether we are dealing with single counts or peak areas. In the case where single counts are measured, the total count is C, and the background the single count B. If the net count $C - B$ is N then:

$$\text{var}(N) = N + B + \text{var}(B) \quad (5.50)$$

Now σ_0^2 is the variance of N when $N = 0$ and, for a *single count*, var$(B) = B$. Therefore:

$$\text{var}(N = 0) = \sigma_0^2 = B + \text{var}(B) = 2B \quad (5.51)$$

and it follows from Equation (5.48) that the critical limit, for 95 % confidence, is given by:

$$L_C = 1.645\sqrt{(2B)} = 2.33\sqrt{B} \quad (5.52)$$

For peak area calculations, the situation is complicated by the fact that the uncertainty of the background estimate depends upon the numbers of channels in peak and background regions. The principles leading to Equation (5.50) are still valid but now B is not a single background count but an *estimate* of a background that has an uncertainty that is not numerically equal to B itself. If we return and consider Equation (5.39) again, we remember that the second term is in fact the variance of the background estimate. Taking the second term of Equation (5.38) as B then:

$$\text{var}(B) = nB/2m \quad (5.53)$$

For a peak area, the expression equivalent to Equation (5.50) is therefore:

$$\text{var}(A) = A + B + nB/2m \tag{5.54}$$

This is, in fact a restatement of Equation (5.42).

Taking the net peak area A as zero, and rearranging:

$$\text{var}(A = 0) = \sigma_0^2 = B(1 + n/2m) \tag{5.55}$$

and:

$$L_C = 1.645\sqrt{[B(1 + n/2m)]} \quad \text{for } \alpha = 0.05 \tag{5.56}$$

Note that when the total number of channels used for the background estimation equals the peak width ($n = 2m$), Equations (5.52) and (5.56) become identical. Equating the peak background and its variance is a common misconception that appears in some current analysis programs. The effect is to underestimate the critical limit. For example, in the case of a peak 21 channels wide, when background regions of three channels are used, the factor

used to calculate L_C should be 3.49 (Equation (5.56)) rather than 2.33 (Equation (5.52)) – an underestimate of 33 % which could lead to false positive identifications.

In most cases, the background to the peak will simply be the Compton continuum. However, if there is a peaked-background in addition, that must also be taken into account.

5.6.2 Upper limit (L_U)

'Given that this count is not statistically significant, what is the maximum statistically reasonable count?' The critical limit is used to assess the statistical validity of a calculated net count. If the net count, N, is below or equal to L_C, then the activity must be declared 'not detected' and an upper limit or 'less-than' level quoted. So we wish to define a level which we can be confident (to an appropriate degree) exceeds the actual peak area, if any. We can relate this to the notional distribution of counts we might obtain if we were to count the particular sample a large number of times (distribution (b) in Figure 5.10)

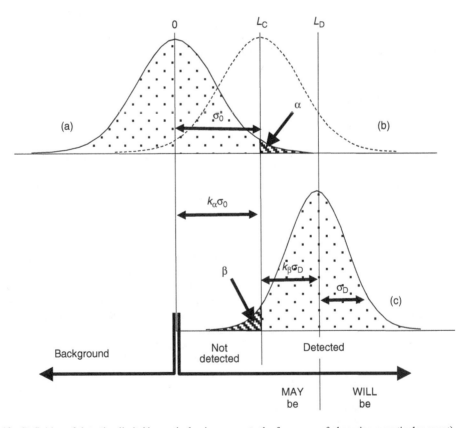

Figure 5.10　Definition of detection limit (the vertical axis represents the frequency of observing a particular count)

and define L_u accordingly. In fact for *any* distribution of counts, above or below the critical limit, we can say that:

$$L_U = N + k_\alpha \sigma_N \qquad (5.57)$$

where σ_N is the uncertainty of the actual measured value; k_α is again the one-sided confidence interval and if we take again as our confidence level 95 % ($\alpha = 0.05$), then we can be sure that there is only a 1 in 20 chance that the true activity is greater than L_U. If N is less than zero then, although statistically reasonable, it does not represent a true situation and N should not be included in the calculation of L_U. There is little point in underestimating the upper limit. For 95 % confidence then:

For a simple count:

$$L_U = N + 1.645\sqrt{(N+2B)} \qquad (5.58)$$

For a peak area:

$$L_U = A + 1.645\sqrt{[A + B(1 + n/2m)]} \qquad (5.59)$$

In both cases, the square root term is the standard deviation of the count, or of the estimated peak area, calculated in the normal manner, rather than of the background. You may notice that if N happens to be precisely zero, then Equations (5.58) and (5.59) reduce to the critical limit expressions (Equations (5.52) and (5.56)). Quite so. If the net count were zero, we would be 95 % certain that the true count were less than L_C – which is the definition of the critical limit. In spite of this, the upper and critical limits should not be used interchangeably.

5.6.3 Confidence limits

If a count, N, is found to be valid (i.e. greater than L_C), then the result may be quoted as a value with an appropriate confidence limit represented by k standard deviations of N, as explained in Section 5.3.1:

$$N \pm k_\alpha \sigma_N$$

The intention here is to state that the count or peak area we have measured lies, within a defined degree of confidence, between the two limits, $N - k_\alpha \sigma_N$ and $N + k_\alpha \sigma_N$. In this case, the factor for the two-tailed probability distribution should be used (see Table 5.3) and for 95 % confidence we might chose to present the result as:

$$N \pm 1.96 \sigma_N$$

In the case of a single count, σ_N is $\sqrt{(N+2B)}$ and for a peak area $\sqrt{[A + B(1 + n/2m)]}$.

5.6.4 Detection limit (L_D)

'*What is the minimum number of counts I can be confident of detecting?*' It is important to appreciate that the critical limit and upper limit are both *a posteriori* estimates based upon actual measured counts. They are statements of *what has been achieved* in the measurement. The detection limit answers the *a priori* question '*If you were to measure a sample, what would the count have to be for, say, 95 % certainty of detection?*' it is, therefore, a statement of *what might be* achieved. Detection limit is often confused with the critical limit. However, if the sample activity did happen to be exactly L_C (distribution (b) in Figure 5.10), statistically we would only be able to be sure (or 95 % sure!) of detection in 50 % of cases because the counts would be distributed symmetrically about L_C. It is clear that L_D must be some way above L_C (see distribution (c) in Figure 5.10).

Imagine that we have a sample with an activity that will provide a count precisely at our limit of detection. The distribution of counts, were we to measure the sample a large number of times, would have a standard deviation of σ_D. We wish to be certain, to a degree determined by k_β, that the chance of not detecting the activity when it is really there is only β, namely:

$$L_D = L_C + k_\beta \sigma_D = k_\alpha \sigma_0 + k_\beta \sigma_D \qquad (5.60)$$

If α and β are both taken to be 0.05 (although there is no reason, other than convenience, why they should be so), then $k_\beta = k_\alpha = 1.645$.

Taking the single count situation where the net count is equal to the detection limit (i.e. $N = C - B$ and $N = L_D$), we could make the following statements:

- the variance of the distribution of counts $= \sigma_D^2 = C + B$;
- at the detection limit, C must be $L_D + B$;
- for a single count, $\sigma_0^2 = 2B$;
- combining these, $\sigma_D^2 = L_D + \sigma_0^2$
- hence from Equation (5.60), $L_D = k_\alpha \sigma_0 + k_\alpha (L_D + \sigma_0^2)^{1/2}$.

Rearranging this equation produces the simple relationship:

$$L_D = k_\alpha^2 + 2k_\alpha \sigma_0 \qquad (5.61)$$

Putting $k_\alpha = 1.645$ and remembering from above that $\sigma_0^2 = 2B$ gives:

$$L_D = 2.71 + 4.65\sqrt{B} \qquad (5.62)$$

Although for the peak area case the expression for σ_0 is more complicated, the mathematics is identical except that the final expression becomes:

$$L_D = 2.71 + 3.29\sqrt{[B(1+n/2m)]} \qquad (5.63)$$

In practice, the calculation of L_D would be made once a background or spectrum, one which represented the particular situation for which detection limit is needed, had been measured. Again, although there are circumstances for which $L_D = L_C$, it is important to distinguish between these limits. Note that it is not essential to make $\alpha = \beta$. If they are not equal, the principle remains but the final expressions, derived from Equation (5.60) rather than Equation (5.61), will be more complicated.

From Figure 5.10, we can see that if the expected count due to a sample was below the critical limit we would almost certainly not detect the activity. If the expected count was above the critical limit but below the detection limit then we might detect the activity. If it was above the detection limit, then it is more likely than not that we would detect the activity. It is important to realize that it *is* possible to detect a count *below* the L_D – the detection limit. At first, this seems perverse. Consider a gamma spectrum. Is it not reasonable to suppose that, if a peak was detectable in 95 % of cases, it would be visible within the spectrum? Indeed, it is quite easy to show, by mathematically creating a continuum plus peaks, as in Figure 5.11, that if a peak contains a number of counts equivalent to L_D that it is indeed, in most cases,

visible by eye – and, one would hope, by the spectrum analysis software. Practical experience of visual examination of gamma-ray spectra over many years leads me to suggest that if a peak cannot be identified visually it is not there – regardless of what the spectrum analysis software decides.

Figure 5.11 shows that even peaks below L_D but above L_C can be seen. It follows that if a peak is not visible, and not detected, that the actual number of counts present, if any, must be less than L_D. However, the detection limit relates to a particular *confidence* of detection – in the equations derived above, 95 % confidence. Below L_D, detection will be less certain but will often still be possible.

5.6.5 Determination limit (L_Q)

'How many counts would I have to have to achieve a particular statistical uncertainty?' This limit is similar in concept to the detection limit and is also an *a priori* calculation but answers the question *'How many counts must there be to provide a result with, say, 10 % uncertainty?'* This implies that for a count, or peak area, equal to L_Q, the standard deviation σ_Q would be 10 % of L_Q or:

$$L_Q = k_Q\sigma_Q \qquad (5.64)$$

where k_Q is the inverse of the required relative standard deviation. Following the logic of the mathematics summarized above for calculating L_D, it can be shown that;

$$L_Q = k_Q(L_Q + \sigma_0^2) \qquad (5.65)$$

and the solution of this quadratic equation gives:

$$L_Q = k_Q^2[1 + (1 + 4\sigma_0^2/k_Q^2)^{1/2}]/2 \qquad (5.66)$$

For example, if the required precision is 10 %, then $k_Q = 10$ and:

$$L_Q = 50[1 + (1 + B/12.5)^{1/2}] \qquad (5.67)$$

for the simple count case; appropriate adjustment for σ_0 in the peak area case gives:

$$L_Q = 50\{1 + [1 + B(1+n/2m)/25]^{1/2}\} \qquad (5.68)$$

5.6.6 Other calculation options

Note that all these expressions are in terms of counts – the basic unit of uncertainty in radioactivity measurement – and assume equal count and background measurement

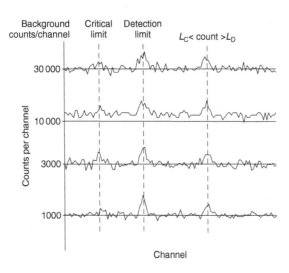

Background Critical Detection
counts/channel limit limit L_C< count >L_D

Figure 5.11 Peaks containing numbers of counts equivalent to L_C, L_D and midway between, on different levels of background continuum

times. If count times are not equal, then adjustment is needed to account for this by using count rates, bearing in mind that, if the count rate is $C/\Delta t$ then the variance of the single count rate is $C/\Delta t^2$. The adjustment is easily made by altering every occurrence of B in Equations (5.52), (5.58), (5.62) and (5.67) to $B\Delta t_C^2/\Delta t_B^2$, where t_C and t_B are the sample and background count periods, respectively. In the case of peak area calculation, the sample and background are derived from the same spectrum and the question does not arise.

In the mathematics above, k_α and k_β were set equal. There is no reason, other than convenience, why this should be so and, in general, these confidence levels can be set independently.

5.6.7 Minimum detectable activity (MDA)

'What is the least amount of activity I can be confident of measuring?' This is a term often used loosely without qualification or rigorous definition and different interpretations can be made. An acceptable general definition would be that, given the circumstances of the particular spectrum measurement, the MDA is the minimum amount of radioactive nuclide that we can be **confident** that we can detect. First, this limit is, then, an activity rather than a count limit. It is often equated to the activity equivalent of the detection limit, L_D. However, there is a problem. As we defined it above, the detection limit is that count which we can be 95 % certain of detecting in the particular spectrum. However, as we saw, the detection limit is some way above the critical limit. We could, therefore, have the situation where a peak area measurement gave a net area which was significant (i.e. above the critical limit) but below the detection limit. Our activity result would then be below the minimum detectable activity. In fact, the minimum detectable activity is *not* the minimum activity detectable! In Figure 5.11, there is visible proof that we can detect peaks that would give an activity below the MDA. How should we interpret that?

The problem stems from the fact that there is a general misunderstanding of the meaning of the limit of detection from which MDA is derived. L_D is that number of counts that we can expect to detect in 95 % of cases (assuming α is 0.05). From this, we can calculate the MDA, which then becomes the activity that we can expect to detect in 95 % of cases. It answers the *a priori* question *'How good is your method?'* This is what should be quoted on tenders or in documentation describing methods. *The MDA should not be quoted as an estimate of upper limit* when a peak is *not* detected. It is unfortunate that not all of the commercial spectrum analysis programs give the option of quoting anything but MDA when a peak is not detected.

If a peak is not detected, the client, the recipient of the information, will wish to know an upper limit on the activity in the particular sample, not what amount of activity the analyst would be 95 % confident of measuring in an arbitrary similar sample in future.

I recommend the following strategy for normal gamma spectrometry, which I should say, is counter to common practice:

- Examine the region in the spectrum where the peak is expected to be. Calculate the net peak area, its uncertainty and the uncertainty of the peak-background correction.
- Calculate the critical limit, L_C, and compare with the net peak area.
- If the peak area is greater than L_C, quote a result with an appropriate confidence limit.
- If the peak area is not significant ($A < L_C$), then calculate the upper count limit, L_U, and pass that value through the calculation to produce an eventual upper activity limit.

Note that no account is taken of whether the peak has been explicitly detected or not. If the net peak area is significant, we can imply that it would have been detected if sought. This procedure has the advantage that exactly the same calculation is performed, whether the peak is present or not. If the peak is not present, one can have some confidence that the upper limit is realistic.

If the question arises *'What is the performance of your method?'* – perhaps to satisfy the requirements of a tender to provide gamma spectrometry services – the procedure should be:

- Find a spectrum that adequately represents the analyses to be performed. For example, if tendering for measurements on soil, find a typical soil spectrum measured under the conditions, sample size, count period, etc., which are intended to be used. (One could also make a case for selecting a worst-case or a best-case sample spectrum.)
- Using data from the spectrum, calculate the L_D at the appropriate region of the spectrum and convert to activity. That is then is the MDA for 95 % confidence of detection.

If all those making tenders are using the same procedure, then comparing tenders will be realistic and fair. Unfortunately, that may not be the case. A major problem is that there is a variety of equations that are used, rightly or wrongly, to calculate the MDA. Those quoted in one tender might not be comparable to those in another. Even worse is the fact that in some software, even when the correct principle is applied, the equations properly relevant to the single count case are used for peak area measurement. Any spectrum analysis algorithms that equate the standard deviation of the background with the square root of the background are in error and will, in most cases, underestimate the limit. Further comments about the manner in which the MDA is calculated in spectrum analysis programs can be found in Chapter 9, Section 9.13.3.

In 1999, the UK Gamma Spectrometry Users Forum (now combined with the Alpha Spectrometry Users Forum as the Nuclear Spectrometry Users Forum) set itself the task of considering which were the 'correct' equations to use for MDA calculations. The intention was to make recommendations to users and software manufacturers alike. In the event, the scope was widened to encompass the correct use of the MDA and L_U. Recommendations were drawn up but, regrettably, have not to date been published. The reasons, apparently, are concerned with difficulties associated with recommendations about the uncertainties in the parameters used to calculate the MDA. To me, this seems to be an unnecessary delay, as I shall explain below.

5.6.8 Uncertainty of the (L_U) and MDA

The upper limit, in counts, is converted into an activity limit, A_U, by an equation of the form:

$$A_U = \frac{L_U}{m \times LT \times \varepsilon \times P_\gamma} \tag{5.69}$$

where m is the sample mass, LT the count period, ε the detector efficiency and P_γ the gamma emission probability. In some circumstances, there may be other factors involved.

L_U is estimated from the uncertainty on the peak-background continuum. If the sample were measured several times, the value of L_U would change statistically. However, *when determined from one particular spectrum*, it has no uncertainty – it is the number of counts below which we are 95 % confident that the true number of counts lies *in that spectrum*. On the other hand, all of the terms of the denominator have some degree of uncertainty, which could be propagated to an uncertainty on the value of A_U.

The unresolved, almost philosophical, question is whether the calculated A_U should be quoted 'as-is' or increased to take into account the uncertainties on the denominator. Suppose A_U were 100 Bq/g and the uncertainties on the denominator amount to, say, 5 %. Should we then quote $100 + 1.645 \times (A_U \times 5\%) = 108$ as the activity upper limit?

Replacing L_U in Equation (5.69) with L_D would allow calculation of the MDA and the same arguments apply. Again, there is an uncertainty on the MDA. Bearing in mind that the MDA is an *a priori* parameter, one could suggest that it should be determined from a number of measurements of an actual sample to estimate the true variability. In fact, the uncertainties on m and LT are very small and, for every measurement of a particular gamma-ray, the values used for ε and P_γ would be exactly the same; the variability of the MDA would be entirely due to the counting uncertainty on the background continuum. Experience shows, not surprisingly, that there is a much greater variability in the MDA between different samples than between different measurements on the same counting sample. Deriving a justifiable MDA is much more dependent on selecting an appropriate spectrum than on the details of the calculation.

Considering the fact that the uncertainties on m and LT are very small and, hopefully, those on ε and P_γ are also small, for the present, it seems reasonable to quote the MDA as calculated with no extra allowance for uncertainty. If the basis of reporting the MDA is to be altered, it is more important to achieve consistency between spectrum analysis programs than to worry about uncertainties on the MDA.

5.6.9 An example by way of summary

Statistics do not make light reading and the whole matter of decisions limits is undoubtedly confusing until one 'sees the light'. The following example might clarify matters. Consider the portion of a spectrum shown in Figure 5.12. The following have been calculated:

Gross counts in the
peak region, G:　　30 374　　($n = 15$ channels)

Sum of background
region counts, S:　　12 040　　($2m = 6$ channels)

Background
correction,
$B = nS/2m$:　　30 100　　(Equation (5.38))

Net peak area
(counts),
$A = G - B$:　　274

Critical limit
(counts), L_C:　　534　　(Equation (5.56))

Because A is less than L_C, we can conclude that the peak was not detected. We must therefore calculate an upper limit:

Upper limit (counts), L_U:　　809　　(Equation (5.59))

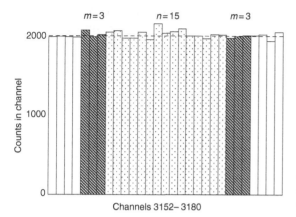

Figure 5.12 The peak used for the example calculations of critical, upper and detection limits (see text for further details)

The interpretation of this upper limit is that we are 95 % certain that the actual number of counts in the peak is less than 809. (In fact, because this is a test peak, we know that the actual number of counts in the peak is 250.) We may wish to determine the detection limit:

Detection limit (counts), L_D: 904   (Equation (5.65))

This means that, given the general level of background counts in this spectrum, if we were to measure another spectrum we could be 95 % certain of detecting a peak that had 904 counts in it. The difference between the upper and detection limits arises because we are asking different questions.

5.7 SPECIAL COUNTING SITUATIONS

In general, we assume that the statistics of counting can be adequately described by the Poisson distribution. When we calculated the various decision limits, we effectively assumed, for simplicity, the Normal distribution for the counts. We know, however, that Poisson statistics are only applicable when the probability of detection of the decay of any particular radioactive atom within the count period is small and when the statistical sample size is large. There are a number of circumstances when these conditions may not be met and we should consider whether the statistical treatment above is still valid.

5.7.1 Non-Poisson counting

If a sample is counted for a long time compared to the half life, the probability of decay within the count period is high. This condition is seldom met in most routine gamma spectrometry situations where half-lives are long, but may

be met frequently when dealing with short-lived radionuclides. In activation analysis, for example, the measurement of radionuclides with half-lives as short as a few seconds is commonplace and a typical count period could be a number of half-lives. Even in such circumstances, the day is saved because the efficiency of detection of gamma radiation is usually very small, partly due to the intrinsic detector efficiency and partly due to geometry factors, reducing the probability of detection.

However, if the count period were long and the detection efficiency high (perhaps a low-energy gamma-ray emitted with high probability and measured close to the detector or inside a well detector), then the assumptions underpinning our use of the Poisson distribution are no longer valid. It is then necessary to return to the binomial distribution. There is no place here for the mathematics involved but it can be shown that if we observe a count of n, then the expected true count is:

$$E(n) = n + 1 - p\varepsilon \qquad (5.70)$$

where p is the probability of decay, calculated using Equation (5.17), and ε is the known effective detector efficiency, taking into account the emission probability, the intrinsic detector efficiency and geometry factors. Similarly, the variance of n can be shown to be:

$$\mathrm{var}(n) = (n + 1)(1 - p\varepsilon) \qquad (5.71)$$

If $p\varepsilon$ is very small, then these relationships approximate to Equations (5.25) and (5.24) discussed above. If $p\varepsilon$ is large, then $\mathrm{var}(n)$ tends towards zero. This is not unreasonable; if we could detect every disintegration, then there would be no uncertainty associated with the number of counts detected.

Note, however, that even if the counts due to the measured species cannot be assumed to be Poisson distributed, it is most likely that the background count, be it a single count or a peak, will be. If this is so, then the gross count will also be Poisson distributed.

5.7.2 Low numbers of counts

If the number of counts accumulated is small, then, even though the count distribution will be Poisson, the approximation to a Normal distribution will not be valid. This means that the relationships to calculate the decision limits given above will not be valid. For number of counts of less than 25, we must resort to the Poisson distribution itself.

For example, if we wish to calculate the critical limit, L_C, we must consider the distribution of counts when the sample has zero activity – background count, in effect. Suppose that we have a background count B in time Δt, then the probability of accumulating n counts is given by Equation (5.23), thus:

$$P(n) = \frac{(B)^n}{n!}e^{-B} \qquad (5.72)$$

The probability that a blank sample could have a count greater than L_C, for a particular degree of confidence defined by α, is given by:

$$\sum_{i=n}^{\infty} \frac{[B]^n}{n!}e^{-B} \leq \alpha \qquad (5.73)$$

So, L_C is the minimum value of n for which this condition is satisfied. As an example, Table 5.4 lists factors taken from published tables of the Poisson distribution. If our background count were, say, 2 counts per counting period, then the critical level for 95 % confidence would be 6 counts. Any count below that would have to be interpreted as 'not detected'. Similar considerations apply to the calculation of the limit of detection, which, it turns out, depends only on the critical limit as long the background is well known. Again, the appropriate limit can be taken from tables. In the case above, where the critical limit was 6 counts, the detection limit would be 10.5 counts. If the background is not well known, as might be the case for a gamma-ray peak with very small numbers of counts, then it becomes more appropriate to consider the problem as a binomial one.

Table 5.4 Decision limits in low count situations (for 95 % confidence in each case)

Background (counts)[a]	L_C (counts)	L_D (counts)
0.1	2	3.0
0.2	2	3.0
0.4	3	6.3
1.0	4	7.8
2.0	6	10.5
4.0	9	14.5
10	16	23.0
20	27	36.0

[a] The average background count within the sampling time.

For small numbers of counts, the Poisson distribution is not symmetrical. This implies that confidence limits associated with a small count will not be symmetrical either. Again, appropriate limits can be tabulated and examples, taken from Sumerling and Darby (1981), are shown in Table 5.5. Taking as an example a gross count of 10 counts, and using Table 5.5, we could only say that there was a 95 % probability of the true count being somewhere between 4.8 and 18.39 counts.

Table 5.5 95 % confidence limits for low numbers of gross counts

Gross count	Confidence limit (counts)	
	Lower	Upper
0.0	0.0	0.025
1.0	0.025	5.57
2.0	0.242	7.22
4.0	1.09	10.24
10	4.80	18.39
20	12.22	30.89

A number of authors have commented on the failure of the Currie equations to cater adequately for the low count situation and proposed alternative equations for critical and detection limits for the single count situation. Strom and MacLellan (2001) compare eight equations for calculating the critical limit, looking at their tendency to allow false positive identifications. None of the rules appears to be completely satisfactory but the authors advocate the use of a 'Stapleton rule' rather than the usual Currie equation in situations where background counts are very low and α, as in Section 5.6.1, is 0.05 or less.

In gamma spectrometry, we can be comforted by the fact that when the number of background counts measured beneath our peaks is 100 or more, the Currie equations are valid. In practice, there will be few occasions when this is not so.

5.7.3 Non-Poisson statistics due to pile-up rejection and loss-free counting

It is not generally recognized that the uncertainty of counts in a spectrum may not be adequately described by Poisson statistics when there are non-random losses of counts, especially when counting rates are high. A good summary of the situation is given by Pomme *et al.* (2000).

Measurements under a fixed real count time may suffer from an increased uncertainty when count losses exceed 20 %. The increase could be a significant extra contribution to an uncertainty budget. When using a loss-free counting system, artificial counts are added to the spectrum to compensate for lost counts in a manner that is not random. That distorts the statistics to a small extent.

Finally, although Poisson statistics can be relied upon for almost all normal counting with an extending live time, when pile-up rejection is used at high count rate the counting uncertainty will be underestimated. Under these conditions, a significant proportion of the total counts might be rejected and the counts lost during the rejection dead time periods will not be truly random.

All spectrum analysis programs will make the assumption that count uncertainties are described by Poisson statistics. If the actual count situation is one of those described above, an extra uncertainty allowance would have to be made externally from the program.

5.8 UNCERTAINTY BUDGETS

5.8.1 Introduction

When we present the results of our gamma spectrometry, we have a duty to take care that the results are as accurate as possible and that the uncertainty that we quote is realistic. It must take into account all of the known sources of uncertainty within the measurement process. Identifying and quantifying those uncertainties provides us with an *uncertainty budget*. Laboratories that are seeking accreditation from bodies such as UKAS will have no choice but to create a satisfactory uncertainty budget.

The most obvious source of uncertainty is, of course, that due to the statistical nature of radioactivity counting. In many cases, it will be the major component of the total uncertainty and it is not uncommon for that alone to be the basis of the quoted confidence limits. However, that alone must underestimate the true uncertainty, especially when the counting uncertainties themselves are small and other sources of uncertainty become dominant. In setting up the budget, the uncertainty of all parameters which contribute to the final result must be assessed and quantified. Some may be justifiably found to be negligible. That conclusion in itself is part of the budget and should be documented.

Setting up an uncertainty budget is a valuable exercise, irrespective of the primary need to quote realistic uncertainties. While assessing in detail all of the relevant sources of uncertainty, it may be possible to remove some of them completely. For example, we might consider that positioning the sample on the detector would not make a significant difference to the result, even though there is potential for placing the sample in slightly different positions. To fulfil the requirements of our budget, we ought to check that out practically. However, if we provide positive sample location, preventing variable positioning, we remove that source of uncertainty completely. Similarly, preparing reference sources by mass, rather than by volume, will reduce the uncertainty on the result considerably.

It can be pointed out that matters such as errors caused by equipment malfunction or operator error do not form part of an uncertainty budget. Nevertheless, consideration of ways in which such problems can be avoided and, equally important, detected, should they happen, is a useful exercise that can only improve the robustness of the overall analytical procedure.

A good point at which to start setting up an uncertainty budget is to look at the way in which the result is calculated. The equation converting net peak counts, C, to activity per unit mass, A, might be:

$$A = \frac{(C - B_p) \times e^{-\lambda t} \times R \times S}{m \times LT \times \varepsilon \times P_\gamma} \tag{5.74}$$

where the parameters have the same meaning as in Equation (5.69), with the addition of $e^{-\lambda t}$, the decay correction, B_p, the peaked-background correction, R, a possible random summing correction and S, a possible self-absorption correction. Within those parameters, there may be others hidden from view. For example, in addition to the normal counting uncertainty there may be other factors affecting the count rate, such as sample positioning or differences in sample height. In the case of the decay correction, we must take into account the uncertainty on the correction due to the uncertainty of the half-life. In principle, we ought to include the uncertainty of the calculated decay period, t, but, under normal circumstances, this would be known very accurately.

When we consider the efficiency, ε, we find we have a separate uncertainty budget to consider taking into account all of the factors involved in preparing the efficiency calibration reference source and the measurement of the calibration data. This would take into account the uncertainty of the certified source from which the calibration source was prepared and the uncertainty of interpolation of the calibration curve.

Combining all of these different sources of uncertainty into a single value can be confusing. There is no other advice to be offered other than to work carefully through the factors one by one and quantify the uncertainty of each before combining them. One may be asked by a client or regulatory body for a single overall uncertainty value summarizing the budget. Frankly, in radioactivity measurements, this is not possible. The overall uncertainty on a result depends critically on the counting uncertainty, that, in turn, depending upon the magnitude of the background to the peak. Measurement of different samples with the same activity of a particular nuclide might give results with considerably different uncertainties if they have different amounts of other nuclides in them. It is possible, however, to give a reasonable uncertainty budget with the counting uncertainty listed as a separate, variable item.

5.8.2 Accuracy and precision

The performance of measurement systems has been traditionally defined in terms of *accuracy* and *precision*. **Accuracy** can be defined as a measure of how close a result is to the actual value and ***precision*** is thought of as the uncertainty of the result, which we could identify with the standard uncertainty. Modern usage in the context of quality of analytical results tends to avoid these terms. This is because there has been a more fundamental appreciation of the actual measurement process. For example, accuracy or, perhaps we should say, inaccuracy, involves bias within a measurement process as well as statistically determined factors that cause the result to be different from the true result. What, at one time, we would have blithely termed precision is now discussed as **repeatability**, the variability of a method when applied to measurements on a single sample within a laboratory, and **reproducibility**, which applies to measurements of that sample when applied by different laboratories using different instruments operated by different operators.

The IUPAC recommendations on this matter have been published by Currie (1995). On a day-to-day basis, there is little harm in applying the traditional usage. However, when producing formal documentation I would recommend that the IUPAC usage should be adopted.

5.8.3 Types of uncertainty

It was, at one time, conventional to identify uncertainties as 'random' or 'systematic'. Experience showed that it was not always possible to assign any particular uncertainty to one category or the other. For example, there may be sources of uncertainty with a Normal distribution – and therefore ought to be categorized as 'random' – which would be more understandable as 'systematic' in origin. Modern usage is to treat each source of uncertainty separately and calculate the standard uncertainty, taking into account the type of distribution involved, before combining with other uncertainties.

However, a new distinction has arisen – Type A and Type B uncertainties. Type A uncertainties are defined as those that have been determined by repeated measurements to assess the magnitude and distribution of the parameter. Type B uncertainties are those whose magnitude has been derived in any other manner. For example, the uncertainty on gamma-ray emission probability is

Type B because the data will have come from literature sources, as will the uncertainty on a certified source activity that will have been taken from the calibration certificate.

Counting uncertainties are a special case. Unlike all other measurements, the nature of radioactive decay, and a considerable body of theory and practice, means that we can establish the uncertainty of a count rate from a single measurement. Because of that, we regard counting uncertainty as Type A because we measure it, although not by repeated measurement. The designation Type A or Type B has no bearing at all on how the uncertainty is incorporated into the budget.

5.8.4 Types of distribution

So far, we have taken it as read that the parameters we are dealing with have a Normal distribution. In many cases, this will be so, but there may be exceptions. For example, suppose samples are placed manually on the cap of a detector for counting. The variation in sample position is likely to alter the sample count rate. What is the shape of the distribution of count rates when a large number are measured? One could argue that the most likely position would be the centre of the detector cap and that the count rates would, therefore, be distributed Normally. On the other hand, one might make an equally attractive case for saying that, taking into account the fact that different operators, with differing perceptions of where the centre was, would be placing samples on the detector with equal probability over a limited area of the detector face. In that case, there might be a rectangular distribution of count rates. Other situations might generate triangular or U-shaped distributions. How do we handle these different distributions? Fortunately, the answer is simple. It should be remembered that all of the statistical relationships and the equations for combining uncertainties are valid, whatever the shape of the distribution, as long as we are consistent in the use of standard uncertainty. It is only necessary to work out the standard uncertainty of the assumed distribution. This might be done by repeated practical measurements. However, if the extreme limits of the measurement are known, the appropriate factor in Table 5.6 may be used to convert the range between those extremes into a standard uncertainty.

5.8.5 Uncertainty on sample preparation

Ideally, samples presented for gamma spectrometry would be homogeneous. Unfortunately, in the real world samples are often far from homogeneous, much less representative. A 100 g sample of a decommissioning waste submitted

Table 5.6 Calculation of standard uncertainty[a] for different distributions

Distribution	Parameter	Divisor
Normal	68 % Confidence limit (1σ)	1
Normal	95 % Confidence limit (2σ)	2
Rectangular	Half-range	$\sqrt{3}$
Triangular	Half-range	$\sqrt{6}$
U-shaped	Half-range	$\sqrt{2}$

[a] Standard uncertainty = parameter/divisor

for gamma spectrometry can hardly be said to be representative of the tons of material to be disposed of, which might be crushed brick, concrete, soil or even floor sweepings. My personal feeling is that, because the gamma spectrometry laboratory has no control over the sampling procedure, the unrepresentative nature of the sample cannot be included as part of the uncertainty budget.

Once the sample has been received, however, it is a different matter. In principle, it is the duty of the analyst to provide an analysis that is representative of the sample provided. However, how far should the analyst go in achieving that? If the sample is clearly not homogeneous, steps need to be taken to make it so, especially if it must be sub-sampled. That may involve crushing or grinding to remove large lumps of material and perhaps segregation into clearly different portions of the sample. From a gamma spectrometry point of view, it may only be necessary to make the sample macroscopically, rather than microscopically, homogeneous – i.e. very small grain size is not essential. Although such procedures can be expected to reduce the uncertainty on the composition of the counting sample, whatever procedure is adopted, the laboratory should have some idea of the final uncertainty. How that is achieved is another matter. Homogeneity can only be properly assessed by a number of measurements on sub-samples. That may be acceptable in a research environment where any amount of time and effort can be devoted to a final high-quality measurement. However in a commercial environment the client is unlikely to wish to pay for anything other than a notional attempt at homogenization and a single measurement. One could, perhaps suggest that representative samples of a 'typical' matrix were homogenized and measured and several measurements made to assess the uncertainty on the composition of a sub-sample. However, practical experience suggests that in a commercial laboratory with a wide range of received samples there is little which can be described as 'typical'. I can only suggest that in such cases, prior to any work being done, an agreement should be reached with

the client on the procedure to be carried out on the sample to achieve assumed homogeneity. Having done that, one could reasonably exclude homogeneity from the uncertainty budget, unless one does indeed have a reasonable idea of the magnitude.

Having achieved a (notionally) homogeneous material, it must be weighed into a counting container. Two sources of uncertainty remain; the mass and height (shape) of the sample. The uncertainty on the mass is small and can easily be quantified in relation to the significance of the least significant digits on the balance display. Uncertainty of the count rate with height of the sample must be assessed experimentally. When samples are measured close to the detector, as is usually the case with low-activity samples, this uncertainty can be significant. Ideally, variation in sample height should be eliminated by using a plunger to lightly compress the sample to a standard height. Chapter 7, Section 7.6.6 describes an empirical correction to count rate for sample height. Determining the factors needed to make that correction would in itself give an idea of the variability of count rate with sample height.

5.8.6 Counting uncertainties

The very act of placing the sample in its counting position is uncertain unless there is a positive sample location. I would recommend that sample locators are always used.

Statistical counting uncertainties are always present, of course, but are always taken into account within the spectrum analysis program. Because these uncertainties vary from sample to sample, from nuclide to nuclide within the sample and even from peak to peak of each nuclide, it makes no sense to include counting uncertainty as part of the uncertainty budget, except to point out that it is variable and taken into account. If peaked-background corrections are involved, you should be aware that the spectrum analysis program may not take into account the uncertainty of the background correction itself.

In routine gamma spectrometry, uncertainty on the timing of the count can be ignored. Only if the count rate varies considerably during the count, for example, if measuring very short-lived nuclides, is there likely to be any live timing problem.

If counting losses due to random summing and/or self-absorption are corrected for, then these corrections will themselves have an uncertainty that must be accounted for. If these corrections are made by the spectrum analysis program, you should make sure, by reading the manual and by validation measurements, that the uncertainties assigned by the program are reasonable.

5.8.7 Calibration uncertainties

Nuclear data uncertainty

The value of the analysis result ultimately depends upon the value of the gamma-ray emission probability. Reputable nuclear data tables will provide you with an uncertainty on these values.

There is, of course, also an uncertainty on the half-life and, if a decay correction is made, the uncertainty on that should also be included. The commercial spectrum analysis program libraries may only allow a single nuclide uncertainty factor to be accounted. If so, it will be necessary to devise a single factor, taking into account the likely magnitude of any decay correction and the various emission probability uncertainties for each nuclide.

Should uncertainty on the nuclear data represent part of one's uncertainty budget? In an ideal world, every laboratory would use the same, well evaluated, set of nuclear data. For the nuclides within the DDEP database (see Chapter 15, Section 15.2 and Appendix B), the standard uncertainties are small; in general, less than 1 %. However, for other nuclides they may be much larger; for the 68.28 keV gamma-ray of ^{234}Th, the emission probability is quoted in the LARA database as 0.048 ± 0.006, a relative uncertainty of 12.2 %. This means that, if taken into account, no laboratory, however careful and skilful, can provide ^{234}Th results with an uncertainty of better than 12 %. Within an intercomparison exercise, where the intention of the measurements is to compare methods or laboratories, if everyone were using the same nuclear data the inclusion of the uncertainty on gamma emission probability would obscure underlying differences due to methodology. However, under normal circumstances (and from the point of view of the recipient) the nuclide data uncertainty does represent part of the overall uncertainty of the result and should be included.

Uncertainty on efficiency calibration standards

When purchased, the reference material from which the calibration sources are prepared will be accompanied by a calibration certificate. This will list, for each nuclide, the activity per unit mass and the overall uncertainty on that activity. These uncertainties should then be taken into account when the efficiency calibration curve is created (see Chapter 7, Sections 7.6 and 9.9). Ideally, they would be used to weight the corresponding points within the fitting process.

It is unlikely that the calibration points will lie exactly on the fitted calibration line. The degree of scatter of the calibration points around the line can be said to represent both the 'goodness of fit' of the calibration data and the uncertainty of estimating the efficiency obtained by calculation from the calibration equation. (We are ignoring here the effect of true coincidence summing, which would make the scatter worse.) This 'interpolation uncertainty' is the figure that one would wish to include in the uncertainty budget.

There will, of course, also be uncertainties introduced when preparing the calibration source. However, they will be a constant amount on each calibration point and it would not be useful to include them in the weighting process. In fact, little more than weighing will be involved in most cases and the extra uncertainty is likely to be small. Nevertheless, it should be accounted for by combining with the interpolation uncertainty. Note that the individual uncertainties on the amount of each nuclide in the reference material do not appear directly in the budget. These will contribute to the scatter on the calibration curve. On the other hand, if individual efficiencies, for particular gamma-rays of particular nuclides are used, the uncertainty on the amount of nuclide in the calibration source should be taken into account.

5.8.8 An example of an uncertainty budget

Table 5.7 shows a notional uncertainty budget based on Equation (5.74). The data in the table are quoted by way of examples; they should not be taken too seriously. Indeed, even the choice of items may not be relevant to other detector and analysis systems. The procedure one should follow is as follows:

- List all identified sources of uncertainty. It may help to group them into categories such as 'Source preparation', 'Calibration', 'Counting' (Column 1).
- For each assess, or measure, the magnitude of the uncertainty (Column 2).
- Decide what that magnitude means. Is it a standard uncertainty or is it a range? (Column 3).
- Decide what the probability distribution is (Column 4).
- Write down the divisor corresponding to that distribution taken from Table 5.6 (Column 5).
- Calculate the standard uncertainty by dividing the magnitude by the divisor (Column 6).
- Add all of the standard uncertainties in quadrature to give the overall standard uncertainty (at the foot of Column 6) and then multiply by the required coverage factor to give the final expanded uncertainty.

It is useful to consider where these various sources of uncertainty are taken into account, as in Column 7. In many cases, this will be within the spectrum analysis program, although it is possible that the program

Table 5.7 An example of an uncertainty budget based on Equation (5.74)[a]

Source of uncertainty	Magnitude (%)	Describing	Probability distribution	Divisor	Standard uncertainty (%)	Where taken into account
Sample homogeneity	Negligible	—	—	1	0	Sample preparation agreed with client is assumed to produce a homogeneous sample
Net counts: $(C - B)$	Variable	—	Normal	1	Variable	Automatic in spectrum analysis program
Sample position	Negligible	—	—	1	0	Sample locator used
Sample height	5, based on ±2 mm in sample height	Half-range	Rectangular	$\sqrt{3}$	2.89	'Additional random' in software
Decay[b] $e^{-\lambda t}$	0.1	Standard uncertainty	Normal	1	0.1	'Additional random' in software
Random summing correction: R	0.2	Standard uncertainty	Normal	1	0.2	Accounted post-analysis
Self-absorption correction: S	0.3	Standard uncertainty	Normal	1	0.3	Automatic in spectrum analysis program – empirical allowance
Mass: m	0.08 on 50.00 g	Full-range	Rectangular	$2 \times \sqrt{3}$	0.023	'Additional random' in software
Live time: LT	Negligible	—	—	1	0	—
Efficiency: ε	3.5 from calibration	2 × Standard uncertainty	Normal	2	1.75	'Additional random' in software
Emission probability: P_γ	3, typical from nuclear data	Standard uncertainty	Normal	1	3	Automatic in spectrum analysis program – defined in library

Overall distribution: Normal   4.53 %

Expanded uncertainty (95 % uncertainty; coverage factor = 2): 9.07 %

[a] The data in this table are intended as examples only and should not be taken to be representative of any particular or general circumstances.
[b] Corrected.

does not handle particular items correctly. In Table 5.7, this is indicated in the case of the uncertainty of the random summing correction where, in this case, a post-analysis adjustment to the uncertainty turned out to be necessary. The budget quoted referred to a laboratory when GammaVision™ was used for spectrum analysis. The items designated as being taken into account by 'Additional Random' refer to the box within 'GammaVision™' into which optional extra amounts of uncertainty, which would not otherwise be taken into account, can be specified. In the example of Table 5.7, an amount of 3.38 % would be specified, representing all the 'Additional Random' items summed in quadrature.

The example shown is incomplete in that it does not take into account the degrees of freedom of each of the uncertainty items. Unless the overall number of degrees of freedom is infinite, calculation of the expanded uncertainty by multiplying by the factors derived from Table 5.1 would not be valid. It would not be appropriate to go into such matters here and for a fuller explanation of uncertainty budgets other sources should be consulted (e.g. Bell, 2001 and UKAS, 1997). It should be said that, for the majority of uncertainty items in a gamma spectrometry budget, infinite degrees of freedom can be assumed.

PRACTICAL POINTS

- The basic distribution underlying counting statistics is binomial in nature.
- In most practical circumstances, it is appropriate to assume a Poisson distribution, which, if the number of counts is large, can be approximated by a Normal (Gaussian) distribution.
- For a Poisson distribution, the following is true: $\text{var}(n) = n$.
- The simple peak calculation area algorithms are:

$$A = \sum_{i=L}^{U} C_i - n \left(\sum_{i=L-m}^{L-1} C_i + \sum_{i=U+1}^{U+m} C_i \right) / 2m$$

$$\text{var}(A) = \sum_{i=L}^{U} C_i + n^2 \left(\sum_{i=L-m}^{L-1} C_i + \sum_{i=U+1}^{U+m} C_i \right) / 4m^2$$

- For lowest peak area uncertainty, the background region width (m) should be as large as possible under the particular circumstances. There would be little point in using more than about 10 channels.

- From the point of view of peak area uncertainty, the optimum spectrum size is 4096–8192 channels, depending upon the gamma-ray energy to be measured.
- Optimum sharing of counting time between samples and background is achieved when the ratio of count times equals the ratio of sample to background activity.
- Decision limits are calculated according to the following equations:

Limit	Single count	Peak area
L_C (95 %)	$2.33\sqrt{B}$	$1.645\sqrt{[B(1+n/2m)]}$
L_U (95 %)	$(N+)^*1.645$ $\sqrt{(N+2B)}$	$(A+)^*1.645$ $\sqrt{[A+B(1+n/2m)]}$

(*if $A < 0$ or $N < 0$, then that part of the equation in parentheses is ignored)

Confidence limits (95 %)	$1.96\sqrt{(N+2B)}$	$1.96\sqrt{[A+B(1+n/2m)]}$
L_D (95 %)	$2.71+4.65\sqrt{B}$	$2.71+3.29\sqrt{[B(1+n/2m)]}$
L_Q (95 %, >10 %)	$50\,(1+$ $\sqrt{[1+B/12.5)]}$	$50\,\{1+$ $\sqrt{[1+B(1+n/2m)/25]}\}$

- If the combined probability of decay and detection ($p\varepsilon$) is high, then Poisson statistics are inapplicable. The correct Binomial treatment provides the following – for a count n, the expected count is $E(n) = n + 1 - p\varepsilon$ and the variance, $\text{var}(n) = (n+1)(1-p\varepsilon)$.
- If the number of counts is small (< 25) then the decision limits cannot be calculated from the Normal distribution but must be taken from statistical tables of the Poisson distribution.
- All results should be accompanied by a realistic uncertainty, taking into account all sources of uncertainty in the measurement. This is arrived at by constructing a complete uncertainty budget.
- When counts are judged against the critical limit and found to be 'not significant', the upper limit should be quoted, *not* the MDA.
- The MDA should be used when assessing the performance, or expected performance, of a method.
- It is important to remember that the MDA is *not* the Minimum Activity Detectable.

FURTHER READING

- General statistics. A very good general introduction to statistics is:

Moroney, M.J. (1990). *Facts from Figures*, Penguin, London, UK.
Miller, J.C. and Miller, J.N. (1993). *Statistics for Analytical Chemistry*, Ellis Horwood, New York, NY, USA.

- Counting statistics. Although published some time ago, the following volume of the series on Nuclear Science is of value:

Stevenson, P.C. (1966). *Processing of Counting Data*, NASNS 3109, National Academy of Sciences – National Research Council, Washington, DC, USA.

- Nomenclature in the chemical measurement process:

Currie, L.A. (1999). Nomenclature in evaluation of analytical methods including detection and quantification capabilities (IUPAC Recommendations 1995), *Anal. Chim. Acta*, 391, 105–126.

- Experimental uncertainty and presentation of results:

Campion, P.J., Burns, J.E. and Williams, A. (1973). *A Code of Practice for the Detailed Statement of Accuracy*, National Physical Laboratory, HMSO, London, UK.

UKAS (1997). *The Expression of Uncertainty and Confidence in Measurement*, M2003, United Kingdom Accreditation Service, HMSO, London, UK. Also available at: http://www.ukas.com/Library/downloads/publications/M3003.pdf

- Decision limit (this article is essential reading):

Currie, L.A. (1968). Limits for qualitative detection and quantitative determination, *Anal. Chem.*, 40, 586–593.

- Statistics of small counts:

Sumerling, T.J. and Darby, S.C. (1981). *Statistical Aspects of the Interpretation of Counting Experiments Designed to Detect Low Levels of Radioactivity*, NRPB R113, National Radiological Protection Board, HMSO, London, UK.

Strom, D.J. and MacLellan, J.A. (2001). Evaluation of eight decision rules for low-level radioactivity counting, *Health Phys.*, 81, 27–34.

Currie, L.A. (2004). Detection and quantification limits: basic concepts, international harmonization and outstanding ('low-level') issues, *Appl. Radiat. Isotopes*, 61, 145–149.

- Non-Poisson statistics due to live time correction:

Pomme, S., Robouch, P., Anana, G., Eguskiza, M. and Maguregui, M.I. (2000). Is it safe to use Poisson statistics in nuclear spectrometry? *J. Radioanal. Nucl. Chem.*, 244, 501–506.

- Uncertainty and uncertainty budgets:

Bell, S. (2001). *A Beginner's Guide to Uncertainty of Measurement*, Issue 2, Good Practice Guide No.11, NPL, London, UK (this is a very worthwhile introduction to the subject with many references to more substantial documentation). This is now available at: http://www.npl.co.uk/server.php?show=nav.1177

6

Resolution: Origins and Control

6.1 INTRODUCTION

It is not an exaggeration to say that gamma spectroscopists are obsessed with **resolution**. In this chapter, I will explain what resolution is and the roots of that obsession. In the simplest terms, resolution is a measure of the width of the peaks in a gamma-ray spectrum – the smaller the width, the better the detector, the higher the resolution. The particular measure we use in gamma spectrometry is FWHM – the **F**ull **W**idth of the peak at **H**alf **M**aximum height, usually expressed in keV. I will explain later how this is defined and measured. There are two reasons why good resolution is prized:

- First, good resolution helps separate gamma-rays that are close in energy. When the centroids of peaks of good shape are $3 \times$ FWHM apart, then the individual peaks are clearly separated. (Figure 6.1) The factors in Chapter 5, Table 5.3 tell us that for such a separation, only 0.13 % of each peak will overlap its neighbour. Measurement of the peak area is straightforward, if not ideal. However, if the separation were only 1 FWHM, then simple measurement of the peak area is not possible. However, spectrum analysis programs should be able to resolve such a double peak into its components with good accuracy. As the peaks become closer, and the number of overlapped peaks becomes greater, the demands on the software become heavier. (See Chapter 9). Having narrower peaks, in the first place, helps.
- The second reason for seeking high resolution is less critical, but by no means unimportant. Most gamma-ray spectra contain many small peaks on an uncertain background. The better the resolution, the narrower the peak, and so what few counts are in the peak will be concentrated in a fewer channels. Those will then stand out more distinctly above the background continuum,

enabling more reliable detection and measurement. It is a signal-to-noise consideration. Figure 6.2 demonstrates this with an extreme example. This may be a sufficient reason for preferring germanium to sodium iodide detectors for low level measurements.

(a)

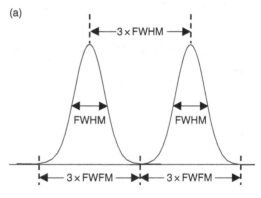

(b)

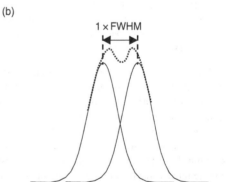

Figure 6.1 The influence of FWHM on the ease of discriminating between close energies: (a) centroids $>3 \times$ FWHM apart should pose no problem; (b) centroids $1 \times$ FWHM apart require deconvolution programs

Practical Gamma-ray Spectrometry – 2nd Edition   Gordon R. Gilmore
© 2008 John Wiley & Sons, Ltd

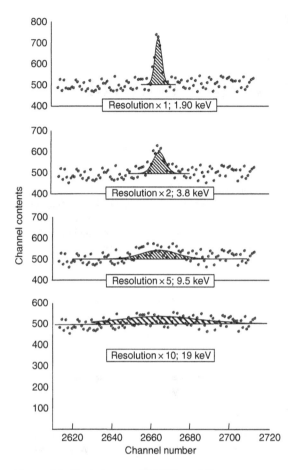

Figure 6.2 The influence of FWHM on ability to measure a small area peak on a statistically uncertain background. In both cases, the net peak area is 1000 counts and the mean background-continuum count per channel is 500

In an ideal world, every gamma-ray of the same energy detected would give rise to a count in the same channel of the gamma spectrometer. Clearly, they do not. Peaks are spread over several channels, with preponderance at a central point, which we can identify with the gamma-ray energy. The reason this is so is that there are uncertainties within the detection and measurement processes that cause nominally identical events to end up as counts in different spectrometer channels. While we are being neurotic about this, we might also wonder whether, in fact, all gamma-rays are emitted with exactly the same energy. Considering the emission and detection processes, the following sources of uncertainty can be identified and combined, in proper statistical fashion, in quadrature – thus:

$$\omega^2 = \omega_I^2 + \omega_P^2 + \omega_C^2 + \omega_E^2 \qquad (6.1)$$

where:

- ω = overall uncertainty in the energy measured by the spectrometer;
- ω_I = uncertainty on the energy of the gamma-ray – the intrinsic width;
- ω_P = uncertainty in the production of electron–hole pairs in the detector;
- ω_C = uncertainty in collecting the charge in the detector;
- ω_E = uncertainty introduced by electronic noise in processing the pulse.

In Chapter 1, Section 1.6.4, I explained how the uncertainty on the energy of a gamma-ray leaving the nucleus, referred to as the energy width, is determined by the sum of the widths of the two energy levels that gave rise to it. The magnitude of these widths is inversely related to the mean lifetime of the energy levels and the distribution of the energy levels is Lorentzian in shape (Section 1.7.4.). Consequently, ω_I also has a Lorenzian shape. The remaining terms in Equation (6.1) represent statistical processes which are expected to have a Gaussian distribution. Therefore, we expect peaks in our spectrum to have a shape described by a Voight function, a folding together of Lorentzian and Gaussian distributions. For a particular gamma-ray, we can calculate an approximate value for the intrinsic width, expressed as an FWHM, using a relationship derived from the Heisenberg Uncertainty Principle, $\omega_I = 10^{-15}/t_{1/2}$, where the half-life is in seconds. Comparing that with the overall measured peak width in a particular, typical, spectrum (Table 6.1), we can see that it will make a negligible contribution to the overall width, or shape, of the gamma-ray peak.

Table 6.1 Examples of gamma-ray energy widths (FWHM)

Nuclide	Energy (keV)	Half-life of level, $t_{1/2}$	Intrinsic width, ω_I (eV)[a]	HPGe spectrum peak width (eV)[b]
^{137}Cs	661.66	2.552 m	$\sim 6.5 \times 10^{-18}$	1420
^{60}Co	1332.50	0.91 ps	$\sim 1.1 \times 10^{-3}$	1850

[a] Lorentzian.
[b] Gaussian.

This means that, in practice, the measured shape of a gamma-ray peak in a germanium detector system is, to all intents and purposes, Gaussian (as are the peaks in Figure 6.1). It is important to bear in mind, though, that while this is true for gamma-rays, it is not necessarily the case for X-rays. Chapter 1, Table 1.4 demonstrated that the Lorentzian component of X-ray peaks greater than 100 keV could be significant.

It is obvious, then, that other sources of uncertainty are responsible for the width of the peaks in our spectra. I will discuss these in turn and to what extent their effects can be mitigated. As you will see, ω_P, the uncertainty on the number of electron–hole pairs created, is a matter of physics and is unalterable. It represents the ultimate resolution achievable. The uncertainty on collection of those charge carriers, ω_C, depends to some extent on detector design and, as long as we have a good optimized electronic system we can do little to reduce it. The electronic noise, ω_E, we do have a great deal of control over, although, once again, there will be an irreducible minimum we strive to achieve. Throughout, I will express these widths as an energy uncertainty in terms of FWHM.

6.2 CHARGE PRODUCTION – ω_P

We saw in Chapter 3, Section 3.2.3 that it takes, on average, 2.96 eV, which I will generalize as ε, to create an electron–hole pair in germanium, and therefore a gamma-ray of energy E eV can be expected to produce $n = E/\varepsilon$ electron–hole pairs. However, n is only an average number because ε is only an average, and the actual number will depend upon which particular energy levels the electron is promoted to within the conduction band (see Chapter 3, Figure 3.1) There is, therefore, an uncertainty on n, and that will lead to a distribution of pulse heights and hence to a broadening of the peak in the spectrum.

How can we quantify that uncertainty? Let us assume that the creation of each electron–hole pair is a normal statistical situation where the number of potential events is large and unknown and Poisson statistics can be applied. If that is so, then the expected uncertainty on n will be $\sqrt{n}$ – the same logic we applied when discussing radioactive decay in Chapter 5, Section 5.2.2.

$$\sigma_N = \sqrt{n} = \sqrt{(E/\varepsilon)}$$

or, in terms of uncertainty on the energy absorbed:

$$\sigma_E = \varepsilon \times \sqrt{n} = \sqrt{(E \times \varepsilon)} \qquad (6.2)$$

This can be converted to an FWHM in keV by multiplying by 2.355 and dividing by 1000:

$$\omega_P = 2.355 \times \sqrt{(E \times 1000 \times \varepsilon)}/1000 \qquad (6.3)$$

$$= 0.128 \times \sqrt{E}, \text{ if } E \text{ itself is in keV}$$

So, our irreducible minimum FWHM for a 1332.5 keV gamma-ray is 4.68 keV. Rather surprising when we consider that, on a day-to-day basis, gamma spectrometrists measure resolutions of around 1.85 keV. Clearly something is wrong! In fact, our basic assumption is faulty. Poisson statistics are only valid when each individual event is independent of all of the others. It would seem the creation of an electron–hole pair alters the local electron distribution within the crystal lattice. It would not be surprising if that affected the probability of creating another electron–hole pair within that local area. So, individual events are not independent and we cannot rely on Poisson statistics alone. In practice, the gap between theory and practice is bridged by introducing the **Fano factor, F**, defined as '*the observed variance in the number of electron–hole pairs created, divided by the variance predicted by Poisson statistic*'. In effect, this is a factor to convert the wrong answer to the right answer. (At school we would have called that a Fudge factor!) Equation (6.3) then becomes:

$$\sigma_E = \sqrt{(F \times E \times \varepsilon)} \qquad (6.4)$$

from which:

$$\omega_P = 0.128 \times \sqrt{(F \times E)} \qquad (6.5)$$

Measured values of F for germanium detectors range from 0.057 to 0.12, with a value of 0.058 often quoted. It does appear to be the case that the value varies with the particular charge carrier creation circumstances. Empirical estimates I have made, which are summarized in Section 6.5, suggest a value of 0.108. Using that value in Equation (6.5) gives a minimum FWHM of 1.51 keV for a 1332.5 keV gamma-ray peak. Using this value for the Fano factor, Equation (6.5) can be used to calculate how the charge production uncertainty varies with energy (Figure 6.3).

Equation (6.5) also allows us to compare detector materials. It is clear that detectors made from materials that have smaller F and/or smaller ε will provide spectra with narrower peaks, i.e. better resolution.

6.2.1 Germanium versus silicon

Table 6.2 compares the parameters for germanium and silicon used to calculate their resolution at 661.6 keV. These resolutions depend critically on the chosen value for the Fano factors – as noted above, there is quite a range to choose from for germanium. Often, it is assumed that the same value can be used for both materials. However,

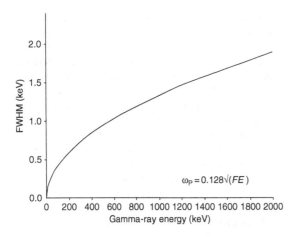

Figure 6.3   A calculated 'statistical' charge production component of FWHM for a germanium detector, assuming $E = 2.96\,\text{eV}$ and $F = 0.058$

Table 6.2   Comparison of ultimate resolution of Ge and Si detectors

Material	ε	F	FWHM at 661.67 keV	Resolution ratio	Source of data
Ge	2.96	0.058	0.794	0.74	Eberhardt (1970)
Si	3.76	0.084	1.077	0.74	Stroken *et al.* (1971)

there is some evidence to suggest that they are, in fact, different. The values in the table were taken from particular literature sources.

This table shows that at 661.67 keV, using germanium rather than silicon, is likely to create a detector with narrower peaks. This, of course, would depend upon the other factors – charge collection and electronic noise – being the same. This is one of the justifications for using germanium for ultra-low energy detectors rather than silicon. For example, the resolution of a 100 mm² Canberra Ultra-LEGe detector is quoted as '*less than 150 eV*' at 5.9 keV, while that of the best available comparable Si(Li) detectors is '*in excess of 175 eV*'.

6.2.2 Germanium versus sodium iodide

A similar comparison between germanium and sodium iodide (NaI(Tl)) is even more stark. For sodium iodide, both a higher Fano factor and a larger ε combine to give, at

Table 6.3   Comparison of ultimate resolution of Ge and NaI(Tl) detectors

Material	ε(eV)[a]	F	FWHM at 661.67 keV	Resolution ratio	Source of data
Ge	2.96	0.058	0.794	0.031	Eberhardt (1970)
NaI(Tl)	170	1	25.0	0.031	See Knoll (1989), p. 312

[a] ε is the energy needed to create an electron–hole pair in germanium, or the energy needed to produce a photoelectron at the photocathode of the photomultiplier of the NaI(Tl) detector.

the best, peaks 32 times wider than a germanium detector peak at the same energy (Table 6.3).

One gamma-ray of 661.67 keV energy completely absorbed in germanium can be expected to create 3.9 million electron–hole pairs. The same gamma-ray absorbed in sodium iodide will only give rise to 3.9 *thousand* photoelectrons. This thousand-fold difference in the number of charge carriers created is the most important reason for the poor resolution of sodium iodide detectors.

6.2.3 Temperature dependence of resolution

The energy needed to create an electron–hole pair is slightly temperature dependent. According to Pehl *et al.* (1968), ε decreases by 0.00075 eV per degree Kelvin, or $-0.0253\,\%$ per degree. This decrease in the number of electron–hole pairs would move a 1332.5 keV peak lower in its spectrum by 0.34 keV. The change in resolution would, however, be minute. Nevertheless, the peak shift is significant and it emphasizes the importance of allowing sufficient time for the detector to reach temperature equilibrium when cooling from room temperature.

6.3 CHARGE COLLECTION – ω_C

Having created a population of charge carriers, the detection process involves collecting them, within the preamplifier, and passing a voltage pulse to the amplifier. If the fraction of the electron–hole pairs produced in the detector that are collected varies, then resolution worsens. Charge collection is treated in detail in Chapter 3, Section 3.6.

Under normal conditions with a well designed and manufactured detector and optimized electronics, we can reasonably expect that the peaks in our spectrum will be Gaussian in shape. The manufacturers aim to provide systems that will collect at least 99 % of all charge within the integration time imposed by the amplifier shaping. Even under these ideal conditions, there will be some variability on the proportion of charge collected and this is responsible for our ω_C term in Equation (6.1). If, for any reason, some of the charge is lost or delayed so that it doesn't contribute to the height of the amplifier output pulse, that uncertainty will be increased.

Incomplete charge collection, no matter how caused, moves counts from the centre of the Gaussian distribution to lower channels, creating a low-energy tail to the peak. Unfortunately, there are several ways in which poor charge collection can be caused and in all cases the symptom is a low-energy tail. Not much help from a diagnostic point of view!

In all detectors, there are crystal imperfections and impurities which can act as trapping sites vary from one germanium crystal to another. Normally, the proportion of trapped charge carriers is small and we need not worry about it. Traps may be crudely categorized as 'deep' or 'shallow', depending on the energy binding the electron or hole to the site (see Chapter 3, Figure 3.13). An appropriately high electric field can release charge from the shallow traps within the rise-time of the preamplifier pulse, but not from deep traps where the charge may be considered as lost. It is, therefore, important that the detector be operated at the bias voltage recommended by the manufacturer. Operating at low voltage will effectively make shallow traps deep.

As detector size increases, the rise-time of the preamplifier pulse increases. The electrons and holes have further to travel to be collected and this must be taken into account when selecting the amplifier shaping time. The 'rule-of-thumb' suggests that the amplifier shaping time should be 10 times the longest preamplifier rise-time. With very large detectors, this may not be practicable – one's amplifier may not provide long enough shaping times, and even if it did, one would find inconvenient limitations on amplifier throughput. In such cases, it may be necessary to accept a small amount of tailing.

We should remember that irradiation of the detector can produce trapping centres within it. In particular, fast neutrons displace atoms into interstitial positions – called *Frenkel defects*. These are efficient traps for holes. If you are operating a detector close to a nuclear reactor or accelerator where a fast neutron flux might be found, it would be better to select an n-type detector for its greater resistance to such damage (Chapter 3, Section 3.6.6). In the event of neutron damage, annealing can help to repair the damaged detector.

Figure 6.4 actually shows the effect of radiation damage, but the peak shape is typical of incomplete charge collection in general. Detailed calculations of the effect that fast neutron damage has on peak shape have been made by Raudorf and Pehl (1987).

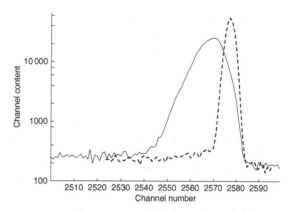

Figure 6.4 The 1332.5 keV peak of ^{60}Co from a radiation-damaged detector. The FWHM is >4 keV; vertical scale is logarithmic

It is worth reminding the reader that, before getting too excited about poor charge collection, under-compensated pole-zero cancellation will also cause a tail on the low-energy side of peaks. Whatever the circumstances, the FIRST diagnostic check to make should be pole-zero cancellation.

6.3.1 Mathematical form of ω_C

When it comes to mathematical modelling we have a problem; there is no simple means of expressing the uncertainties of charge collection as a function of energy. However, by taking a measured FWHM calibration, subtracting the calculated charge production uncertainty (using equations from Section 6.2) and the electronic noise, we can estimate the charge collection uncertainty at each calibration energy. The electronic noise can be estimated by using a pulse generator; pulses injected into the preamplifier at the test pulse input will not be subject to production and collection uncertainties and the width of the pulser peak in the spectrum will represent the electronic uncertainties. However, we can also estimate the electronic noise from the FWHM calibration itself. At zero energy, the charge production uncertainty is zero (Equation (6.5)) and the charge collection uncertainty must also be zero; there can be no uncertainty in collecting no charge carriers. The intercept of the FWHM calibration on the y-axis must represent the electronic noise alone. However, extrapolating the measured data depends upon the particular mathematical form chosen for the FWHM calibration. The data resulting from such a subtraction process are likely to have large uncertainties, but do seem to suggest a linear relationship to energy. As far as I am aware, there is no theoretical reason why that should be so, but for the purposes of modelling, as we will see in Section 6.5, it does appear to be satisfactory. The implication is that $\omega_C = cE$, where c is the proportionality constant.

6.4 ELECTRONIC NOISE – ω_E

All of our electrical pulses are sat on a baseline that may be nominally at zero volts or may be some way above or below that. That baseline itself will be variable. If the maximum amplifier output is 10 V and we assume that this would correspond to MCA channel 8192, then each channel is only 1.2 mV wide. Variability on the baseline, even if of only a few millivolts, will affect measurement of the pulse height. Such uncertainties that attach to an electronic signal as it is being processed by the preamplifier, amplifier and MCA are referred to as **electronic noise**. The most sensitive part of the system is the detector–preamplifier coupling; any uncertainties here are magnified within the amplifier. Note that the degree of electronic noise is independent of pulse height. Every pulse received the same amount of noise, which means that smaller pulses are affected more, in proportion, than larger pulses. The noise is, however, dependent on the shaping time set at the amplifier.

Noise is common to all electronic apparatus, not just gamma spectrometers, and the various sources have been identified in conventional terms. For example, **thermal noise** or **Johnson noise** is due to the random thermal vibrations of electrons, and **shot noise** is the variation in DC current in a diode caused by the statistical nature of the process generating the current, For our purposes, the electronic noise effects are usefully grouped into categories which reflect their coupling to the signal current flowing in the system. The three groups of *parallel*, *series* and *flicker* noise are of operational interest as they underpin the routine procedure for minimizing noise through choice of amplifier time constant.

6.4.1 Parallel noise

This noise is associated with the current flowing in the input circuit of the preamplifier which is in parallel with the detector. In particular, those sources that are integrated on the capacitor C_f (see Chapter 4, Section 4.3.1 and Figure 6.5). Because this gives discrete voltage steps, the noise is also known as **step noise**. It can be regarded as equivalent to a current generator at the input of the preamplifier.

Most parallel noise results from two sources:

- leakage currents at the detector element;
- thermal noise in the feedback resistor R_f.

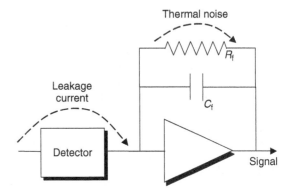

Figure 6.5 Representation of detector and preamplifier, showing contributions of parallel noise

Figure 6.5 indicates the 'parallel' nature of these uncertainties, and their magnitude varies as shown by the following expression, comprising a term related to detector current (leakage plus signal) and thermal excitation in the feedback resistor:

$$(\omega_{\text{parallel}})^2 \propto \left(I_D + \frac{2kT}{R_f} \right) \times T_S \qquad (6.6)$$

where:

- I_D = total detector current (signal + noise);
- T = temperature of the feedback resistor;
- T_S = shaping time of the signal out of the main amplifier.

It follows from Equation (6.6) that to minimize parallel noise we should:

- Use a low count rate (small I_D). This is not something we can always guarantee to do. Figure 6.7 below shows how with a larger count rate the parallel noise increases, particularly at high shaping times.
- Keep R_f cool (small T). Some preamplifier designs have the input circuit of the preamplifier within the cooled cryostat enclosure.
- Have a large value of the feedback resistor, R_f. The consequence of this is a long pulse width, which hampers throughput.
- Operate the amplifier with a short shaping time constant (small T_S). However, this may be in conflict with the 'rule-of-thumb' discussed above in relation to effective charge collection.

The second factor in Equation (6.6), the thermal noise term for the feedback resistor, can be eliminated altogether by removing it, i.e. by using an auto-reset mechanism. Thus, both transistor-reset (TRP) and pulsed-optical-reset (POR) preamplifiers can give better resolution than the resistive feedback preamplifier. (Transistor reset preamplifiers were discussed in Chapter 4, Section 4.3.2.) Note that parallel noise is independent of the detector capacitance.

6.4.2 Series noise

Series noise is those components of the noise that are considered to be in series with the detector signal. Shot noise in the preamplifier's FET is the prime source of series noise. It is equivalent to a voltage pulse generator in series with the signal at the input to an amplifying component. An alternative name is **delta noise**. The magnitude of series noise in the FET is described by:

$$(\omega_{\text{series}})^2 = C^2 \left(\frac{2kT}{g_\text{m} \times 2.1 \times T_\text{S}} \right) \tag{6.7}$$

where:

- C is the total capacitance at the preamplifier input (from detector, input of FET, feedback capacitor and detector–preamp connector);
- T is the temperature of the FET;
- g_m is the transconductance of the FET (i.e. its gain – current in divided by voltage out);
- T_S is the amplifier shaping time; the factor of 2.1 converts it to peaking time.

Equation (6.7) indicates that to minimize series noise, we should:

- Most importantly, minimize the capacitance of the detector (Chapter 3, Section 3.5) and of the detector–preamplifier connection. The increase in noise with increasing capacitance, as shown in Figure 6.6, is a symptom of series noise. It can be seen that detector capacitance can have a major influence on the overall noise of the system. Larger detectors have larger capacitances and hence larger values for series noise. The differences between lines (a) and (b) in the figure also demonstrate the importance of preamplifier design.
- Cool the FET. Manufacturers do cool FETs for low-noise applications; temperatures are close to liquid-nitrogen temperature and the detector plus preamplifier will be an integrated package.
- Select an FET with low noise characteristics and a high transconductance.

- Operate the main amplifier with a long time constant; note that this is in conflict with the parallel noise recommendation.

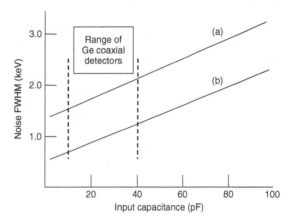

Figure 6.6 The FWHM noise due to capacitance effects in a charge-sensitive preamplifier: (a) a typical preamplifier, according to ORTEC; (b) a particularly low noise preamplifier, according to Canberra, who also provided the range of detector capacitances. Reproduced by permission of Canberra.

As already mentioned, the transistor reset preamplifier has no feedback resistor – hence, the parallel noise of Equation (6.8) is reduced. However, there could be a slight increase in series noise because the added capacitance of the reset transistor affects Equation (6.9), but this is usually negligible in comparison to the detector capacitance (the optical reset device avoids this transistor). These considerations mean that the TRP as well as being recommended for use in high throughput situations at low time constants, also has advantages of high resolution with long time constants.

6.4.3 Flicker noise

This is also called **1/f noise**. It is associated with variations of direct current in all active devices, such as carbon resistors. The magnitude of this source of uncertainty is a function of current through the detector and on the effective frequency of the signal. Fortunately, it is independent of amplifier shaping time and is small compared to series and parallel noise, but does increase at high count rates.

6.4.4 Total electronic noise and shaping time

To determine the overall magnitude of the electronic noise, the three noise components are added in quadrature in the usual manner:

$$\omega_\text{E}^2 = (\omega_{\text{parallel}})^2 + (\omega_{\text{series}})^2 + (\omega_{\text{flicker}})^2 \tag{6.8}$$

We saw above that parallel noise increases with shaping time constant, series noise increases, while flicker noise is independent of it. That being the case, we might expect that at some point there would be a trade-off between series and parallel noise with respect to shaping time. Figure 6.7 demonstrates that – there is an optimum shaping time at which the electronic noise component will be a minimum. The sum of the terms in Equation (6.8) as a function of shaping time is indicated together with the individual components. The minimum is at the point where $\omega_{parallel} = \omega_{series}$, and has been called the **noise corner**. The diagram is based on calculations using Equations (6.6)–(6.8), using typical values for the various parameters and a notional allowance for flicker noise. All data were then converted from coulombs to FWHM keV.

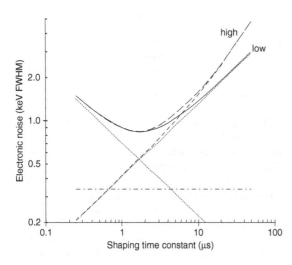

Figure 6.7 A schematic diagram showing how addition of various electronic noise contributions leads to a minimum total at a particular time constant

Because electronic noise is a major component of the total uncertainty comprising the spectrum peak width, it is important to operate the amplifier with the optimum shaping time. This is the background to the setting-up procedure in Chapter 11, Section 11.3.7, where the time constant is determined on the system as a whole.

Amplifier throughput is inversely proportional to the shaping time – the narrower the pulses, the more through the system per second. So, at any particular input rate there can be a trade-off between the throughput capability and resolution. If the optimum time constant is halved, then we expect twice as many counts to be processed before pile-up effects become a problem, at the cost of slight

increase of noise. The base of the noise/FWHM curve is broad and modest changes in shaping time constant will not have severe effects on overall resolution. With the arbitrary data of Figure 6.7, halving the shaping time would increase the electronic noise by about 11 %, but that would only be noticed at low energy. Above 200 keV, the degradation in resolution would be less than 5 %; this could be a price worth paying for what could be reflected in doubling productivity in terms of sample throughput.

6.5 RESOLVING THE PEAK WIDTH CALIBRATION

In the first edition of this book, I discussed at various points the best function to fit to FWHM calibration data. I compared a linear fit with the square root function recommended by Debertin and Helmer (see below). Ultimately, I suggested that a linear fit to FWHM calibration data was more satisfactory than any alternative and that this could mathematically be justified if one assumed that the charge collection uncertainty was linear with respect to energy. A little more thought on the matter revealed some inconsistency in the justification. I have looked at the matter again.

In Equation (6.1), I added in quadrature the factors I considered could affect peak width. Taking the intrinsic gamma-ray width as insignificant, gives:

$$\omega^2 = \omega_P^2 + \omega_C^2 + \omega_E^2 \qquad (6.9)$$

Each of these terms can be replaced by the mathematical representations introduced above. In Section 6.2, I showed that the width due to charge production is proportional to the square root of the gamma-ray energy. In Section 6.3, I suggested a linear relationship between collection uncertainty and energy, and in Section 6.4, I stated that the electronic contribution to peak width is independent of energy. Hence:

$$\omega^2 = p^2E + c^2E^2 + e^2 \quad \text{or, rearranged} \qquad (6.10)$$
$$\omega = \sqrt{(e^2 + p^2E + c^2E^2)}$$

where p, c and e are constants relating to production, collection and electronic noise, respectively. This equation is the square root of a quadratic in E.

The equation suggested by Debertin and Helmer is, in effect:

$$\omega = \sqrt{(AE + B)} \qquad (6.11)$$

We can equate B with our e^2 and A with our p^2. This function implies that either the charge collection uncertainty is negligible or that it has the same square root dependence on energy as the charge production, in which case $A = p^2 + c^2$.

To try and resolve the question of which function best fits FWHM calibrations, I took 22 routine calibrations from a number of sources:

- A number from different detectors and different source shapes from my own laboratory.
- Width calibrations derived from the NPL Test spectra (1997).
- Width calibrations from the IAEA Test spectra (1995).
- The example calibration distributed with Canberra's Genie 2000 analysis program.
- A width calibration derived from the Sanderson Test Spectra (1988).

The calibrations covered n- and p-type detectors, relative efficiencies from 11 to 45 % and differing sample geometry. The bulk of the measurements related to 45 % n-type detectors. I compared linear, Debertin and Helmer and quadratic fits to the calibrations. I found that about 50 % of calibrations seemed to be best fitted by a linear function, the rest by the Debertin and Helmer equation. There was no correlation between the best fit and type or size of detector, or with sample geometry. The problem is that routine FWHM estimation is inherently imprecise; the scatter of the calibration data makes it easy to fit any function at all with apparent justification. To achieve a more precise calibration, I pooled the data for all 22 width calibrations, scaling each to the mean gradient and intercept of linear calibrations. The scaled peak widths at each calibration point were combined as means. Gratifyingly, for each of the 13 calibration energies, the scatter of these means was less than 0.02 keV; on average 0.86 % of the width. That pooled data were then fitted to each of the functions listed in Table 6.4. In this table, I have taken the Debertin and Helmer parameters as squares so that they can be related to the relevant sources of uncertainty.

Figure 6.8 compares the pooled data fitted to linear, Debertin and Helmer and the square root quadratic function from Equation (6.10). Figure 6.9, by showing the differences between fitted and actual data, emphasizes the distinction between the different functions. From Table 6.4 and Figure 6.9, we can make the following observations:

- The best fit, in the sense of lowest RMS differences, is the square root quadratic.
- A simple quadratic fit is almost as good as the square root version.
- The quality of the square root quadratic fit was extremely good – an average difference between fit and data of 0.006 keV.
- The Genie 2000 fit is clearly not satisfactory.

Clearly, the square root quadratic function provides the best fit. (Hurtado *et al.* (2006) come to the same conclusion in a similar comparison.) The closeness of the fit gives some confidence to the speculative proposal that the charge collection uncertainty can be modelled by assuming it is linear with energy. We are no nearer to explaining why that should be so, but from a modelling point of view it is satisfactory. The estimated parameters for these particular data were:

Table 6.4 Results of fitting of pooled FWHM data to different mathematical functions[a]

Type of fit	Equation	RMS differences[b]	Fitted parameters			Estimated Fano factor
			a	b	c	
Linear	$a + bE$	0.020	0.992	7.14×10^{-4}	—	—
Quadratic	$a + bE + cE^2$	0.0065	0.968	8.23×10^{-4}	-6.52×10^{-8}	—
Genie 2000	$a + b\sqrt{E}$	0.055	0.664	0.035	—	—
Source of uncertainty related to parameter			Electronic	Production	Collection	
Debertin and Helmer	$\sqrt{(a^2 + b^2 E)}$	0.026	0.900	0.0473	—	0.137
Square root quadratic	$\sqrt{(a^2 + b^2 E + c^2 E^2)}$	0.0057	0.957	0.0421	5.29×10^{-4}	0.108

[a] Fitted points were derived from 22 width calibrations from a range of sources, of different detector types and sizes, and sample geometry.
[b] Square root of the sum of differences between pooled peak width and width deduced from the fit.

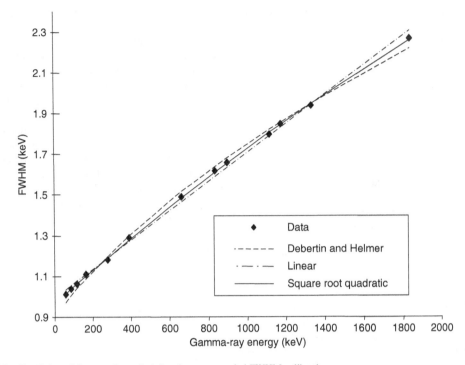

Figure 6.8 The fitting of three mathematical functions to a pooled FWHM calibration curve

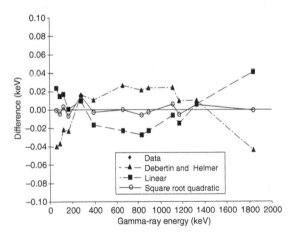

Figure 6.9 Differences between fitted functions and pooled measured values of FWHM

- electronic noise, $e = 0.956$ keV;
- charge collection parameter, $c = 5.28 \times 10^{-4}$;
- charge production parameter: $p = 0.0422$.

Parameter p in both square root quadratic and Debertin and Helmer equations can be related to Equation (6.5).

From that, I estimate the Fano factor to be 0.108, somewhat higher than the value of 0.058 generally used, but well within the range of reported values. The value from the Debertin and Helmer equation is somewhat higher, but that does, in effect, include an allowance for the uncertainty of charge collection.

Figure 6.10 resolves the pooled FWHM data into the three uncertainty components, production, collection and electronic, plotted separately. Note that below 500 keV, the peak width is dominated by the electronic noise contribution, underlining the importance of minimizing preamplifier noise. That is why it is necessary to use specially designed low noise preamplifiers and sophisticated amplifiers, both at high cost, of course. That expenditure is wasted unless proper attention is given to correct setting up of the spectrometer. When using low-energy detectors, one might consider using a pulsed optical resetting preamplifier to reduce electronic noise even further.

For scintillation detectors, the situation is very different. As we saw earlier, the charge production uncertainty is very large, compared to semiconductor detectors, and throughout the energy range will be much greater than the electronic noise. An electronic noise component of 1 keV, or so, is not very significant compared to the 25 keV

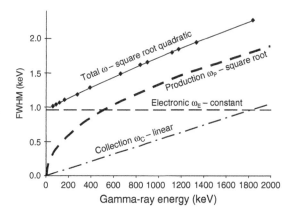

Figure 6.10 Resolution of the pooled FWHM curve into separate noise components, taking the Fano factor derived from a square root quadratic fit

charge production width at 661.6 keV (Table 6.3). One can justify compromising on preamplifier quality in scintillation systems.

PRACTICAL POINTS

- The resolution of peaks in a spectrum is much worse than any 'natural' spread in the gamma energy. The extra uncertainty is added in the processes of charge production, charge collection and electronic conversion. We can only influence the latter two.
- Incomplete charge collection can cause a low-energy tail to a peak. This may be due to too low a bias voltage or trapping. Check that the recommended working voltage is being applied. Trapping due to fast neutron damage can be annealed out in HPGe detectors.
- There are many influences on the overall electronic noise. This will be most influential at low energies. The minimum electronic noise should be determined at

the working count rate by varying the amplifier time constant.
- For general advice on improving poor resolution, see Chapter 12 on Troubleshooting.

FURTHER READING

- The following texts all have some treatment of noise in detectors and electronic components, and are listed roughly in order of increasing complexity:

Knoll, G. F. (1989). *Radiation Detection and Measurement*, 2nd Edn, John Wiley & Sons, Inc., New York, NY, USA.

ORTEC (1991/1992). *Detectors and Instruments for Nuclear Spectroscopy*, EG&G ORTEC, Oak Ridge, TN, USA

Canberra Reference 1 (1991). *Detector Basics*, 2nd Edn, Canberra Semiconductor NV, Olen, Belgium.

Dowding, B. (1988). *Principles of Electronics*, Prentice Hall, New York, NY, USA.

Nicholson, P.W. (1974). *Nuclear Electronics*, John Wiley & Sons, Ltd, London, UK.

Radeka, V. (1988). Low noise techniques in detectors, *Annu. Rev. Nucl. Particle Sci.*, **38**, 217–271.

- On the shape of FWHM calibration curves:

Hurtado, S., García-León, M. and García-Tenorio, R. (2006). A revision of energy and resolution calibration method of Ge detectors, *Nucl. Instr. Meth. Phys. Res., A*, **564**, 295–299.

REFERENCES

Eberhardt, J. E. (1970). Fano factor in silicon at 90 K, *Nucl. Instr. Meth. Phys. Res.*, **80**, 291–292.

Pehl, R.H. Goulding, F.S., Landis, D.A. and Lenzlinger, M. (1968). Accurate determination of the ionization energy in semiconductor detectors, *Nucl. Instr. Meth. Phys. Res.*, **59**, 45–55.

Raudorf, T.W. and Pehl, R.H. (1987). Effect of charge carrier trapping on germanium coaxial detector line shapes, *Nucl. Instr. Meth. Phys. Res., A*, **255**, 538–551.

Strokan, N., Ajdai, V. and Lalov, B. (1971). Measurements of the Fano factor in germanium, *Nucl. Instr. Meth. Phys. Res.*, **94**, 147–149.

7

Spectrometer Calibration

7.1 INTRODUCTION

A modern digital gamma-ray spectrum is in essence a list of numbers of pulses measured within small consecutive pulse height ranges. Detector calibration allows the gamma-ray spectrum to be interpreted in terms of energy, rather than channel number or voltage, and amount of radionuclide, rather than number of pulses. In addition, especially if computer analysis of the spectrum is contemplated, it may be necessary to provide information about the peak width variation with energy or channel number. There are, then, three main calibration tasks:

- Energy calibration – the relationship between channels and energy.
- Peak width calibration – the variation of peak width with energy.
- Efficiency calibration – the relationship between number of counts and disintegration rate.

Each of these is simple in principle, and I will discuss each in turn and the various factors that can render a calibration invalid and the means used to eliminate such difficulties. The potential sources of calibration error are:

- energy shifts caused by changing the source/detector orientation;
- anomalous peak widths;
- effect of source/detector distance;
- effect of sample density;
- pile-up losses (random summing);
- true coincidence summing;
- inaccurate decay corrections;
- live time correction errors.

Although I recognize that the majority of spectrum analysis will be performed by computer, I will defer the practical implementation of calibration algorithms until Chapter 9. In this chapter, I will concentrate on the principles of calibration. Unless we have an understanding of these, there is little chance of achieving the intuitive 'feel' for gamma spectrometry that allows the experienced practitioner to spot errors and inconsistencies in the output from the computer.

In practice, spectrometers are calibrated using appropriate measured gamma-ray spectra. It should go without saying that calibration spectra should be of high quality if the results are to be depended upon. In computing circles, there is a saying (oft quoted but nonetheless worth remembering) summarized in the acronym GIGO – garbage in, garbage out. This is equally applicable in gamma spectrometry. What is less obvious and less considered is the quality of the nuclear data used in the calibration and I shall start by considering that.

7.2 REFERENCE DATA FOR CALIBRATION

The data specifically needed for spectrometer calibration purposes are gamma-ray (and X-ray) energies, the probability of emission of these radiations and the half-life of the nuclide. It is not uncommon for it to be realized that the accepted value of an energy or half-life or, more often, a gamma emission probability which has been used for some time is in error. From time to time, there are re-evaluations of data to incorporate new values and to make judgments that are more informed on which of all available values should be taken into account. It is important, therefore, that the data used within a laboratory should be up to date.

There are many compilations which list such data and every laboratory has its favourite. It is important that reference data should have been validated in some way. Often, there is little to indicate the reliability of the various sources. In general, one should choose the most recent compilations and, in particular, one should be wary of limited data sets where the source of the information is not given. As a convenience, the commercial providers of spectrum analysis software will often also provide a nuclide library for use with the software. Such libraries should be treated with care. After all, the manufacturers' main concern is to provide up to date software, not nuclear data.

In the first edition of this book, I recommended that one's first-choice source of data should be the IAEA TECDOC-619 report, containing, at the time, the best-evaluated data for 35 nuclides. Since that time, there have been other efforts, involving international cooperation, to establish a single reliable body of data, including re-evaluation and extension of TECDOC-619. (In some circles, referred to as 'son-of-TECDOC'.) That was released in the spring of 2007 as IAEA XGAMMA. Appendix B contains data for the original 35 TECDOC-619 nuclides, updated to XGAMMA. BIPM, the international body that maintains the standards of weights and measures, has recommended that all laboratories should use the DDEP database. I discuss this, and its relationship to other databases in Appendix A. Nowadays, there is ready access to all of these databases via the internet.

7.3 SOURCES FOR CALIBRATION

As with the nuclear data, the radioactive sources used for calibration must be suitable for their purpose. For energy and peak width calibration, it is sufficient that the energies of the gamma-rays (or X-rays) it emits be known to a satisfactory degree of accuracy but the source strength need not be known. For efficiency calibration, it is essential that nuclides are used for which the gamma-ray emission probabilities are known accurately and that a source of known activity is used. Whenever possible, sources that have been certified as to their radioactive content should be used. I will discuss traceability in a later chapter but it is worthwhile remarking at this point that the value of a calibration is much reduced unless the activities of the sources used can be traced back to standards with international credibility.

As you will see in due course, the efficiency with which a source is counted depends upon its shape and density. The obvious conclusion is that calibration sources should be prepared in such a way as to have identical shape and density as the samples that are to be compared with them. Density differences are less critical than geometry differences and small differences can sometimes be tolerated. However, this should always be established by actual measurement and not just assumed.

The calibrations will not be greatly affected by source strength as long as it is not so high as to cause pile-up and other count rate problems. However, it is obviously convenient to use a source which will provide a spectrum with sufficient counts within a short period. Sources of 10 kBq, for calibrations close to the detector, to 100 kBq further away, would be reasonable. On the other hand, reference sources which are intended to check the analytical method by mimicking actual samples should be of similar activity. This poses certain problems for the preparation of environmental reference sources in that the several orders of magnitude dilution from the purchased calibrated source can only be done in stages. This loses traceability. For this reason, calibrated sources at near environmental levels are now being made available by the National Physical Laboratory (NPL) and other source manufacturers.

Commonly used commercial sources are the QCY and QCYK Mixed Nuclide sources provided in the UK by AEA Technology QSA and, in the USA, the calibrated mixture of ^{125}Sb, ^{154}Eu and ^{155}Eu available from the National Institute of Standards and Technology (NIST). In Germany the PTB (Physikalisch-Techniche Bundesanstalt) supply another suitable mixed radionuclide source. (The QCYK source, and its spectrum is examined in Chapter 8, Section 8.5.1.)

These sources are satisfactory for calibrations for normal decay gamma measurements; QCYK can be used up to the 1836.05 keV emission of ^{88}Y. If prompt gamma-ray measurements are to be undertaken, finding suitable calibrated sources that will provide gamma-rays up to 10 MeV is difficult. The energy range can be extended to 4800 keV by using ^{24}Na, ^{56}Co and ^{66}Ga, but beyond that it is necessary to generate prompt gamma-rays by neutron capture using americium/beryllium neutron sources.

7.4 ENERGY CALIBRATION

The object of energy calibration is to derive a relationship between peak position in the spectrum and the corresponding gamma-ray energy. This is normally performed before measurement, if only in a preliminary manner, but it is usual for spectrum analysis programs to include more sophisticated calibration options.

Energy calibration is accomplished by measuring the spectrum of a source emitting gamma-rays of precisely known energy and comparing the measured peak position with energy. It matters not whether the source contains a single nuclide or several nuclides. For many years, I used a ^{152}Eu source for routine energy calibration. Whatever source is used, it is wise to ensure that the calibration energies cover the entire range over which the spectrometer is to be used.

In practice, it is sufficient to measure the spectrum long enough to achieve good statistical precision for the peaks to be used for the calibration. The calibration process then involves providing a list of calibration peaks to be used and their true energy. The computer can then search for the peaks, measure the peak position to a fraction of a channel, and deduce the energy/channel relationship. If the spectrometer is completely un-calibrated, it might be necessary to perform a manual calibration first so that the computer has some idea where to find the peaks. (On older hardwired MCA systems, it would usually be necessary to type in the peak channel number. Under these circumstances, a certain amount of skill is needed to visually estimate the actual peak position between channels. If the MCA can only deal with integral channel numbers, the energy supplied by the operator must be adjusted to take that into account.)

and even some MCA Emulator programs might only allow a two point linear calibration:

$$E(\text{keV}) = I(\text{keV}) + G \times C(\text{channels}) \qquad (7.1)$$

where I and G are the intercept and gradient of the calibration line and C the channel position. Whether Equation (7.1) is adequate depends upon the integral linearity of the gamma spectrometry system and the use to which the information is to be put. The quoted integral linearity of spectroscopy amplifiers is 0.05 %, or so, and of ADCs 0.02 %. If we take the larger value, this implies that if we were to use a linear calibration the maximum energy error within a spectrum of 0 to 2000 keV would be 1 keV. Over most of the spectrum, the error would be smaller than this. The errors are most likely to occur at the extremities of the energy scale.

Experience suggests that the linearity of modern ADCs is extremely good and that errors of this magnitude are not found. It is, in fact, rather difficult to demonstrate non-linearity of modern ADCs. The example in Figure 7.2 is for an old ADC no longer in use. This degree of non-linearity would be unacceptable nowadays. However, even so, over most of the energy range, energies deduced from the linear calibration line would have been within 0.5 keV, sufficient for many purposes. An examination of the calibration line in Figure 7.1 reveals that in this case energies could be interpolated to within 0.15 keV. As it happens, using a two point energy calibration would have made little difference to the accuracy of the energy estimation. In 25 years of general gamma spectrometry, I found no need for other than a two point energy calibration.

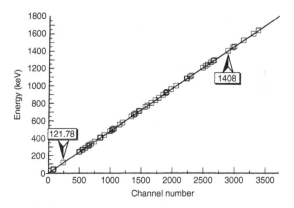

Figure 7.1 Energy calibration (55 points and the best fit straight line) using ^{152}Eu. The two marked points would be used for a two point calibration

Figure 7.1 shows an energy calibration using ^{152}Eu. Fifty-five points are plotted together with the best fit straight line. Fifty-five points is somewhat excessive and is by no means typical – 10 to 15 would be more usual. Normally, the spectrometer from which the data of Figure 7.1 were obtained would have been calibrated by using only the two points indicated. Although the data shown appears to fit a linear relationship very well, this would not take into account any integral non-linearity in the system. One might enquire *what is the most appropriate relationship to fit the data to?* There is no theoretical model to refer to, although the intention of the manufacturer is that the calibration should be linear, and in some cases, there is no choice. A hardwired analyzer

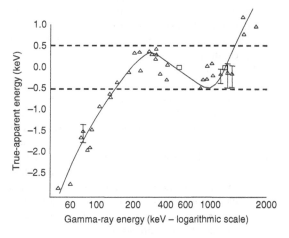

Figure 7.2 Integral non-linearity (data from an old ADC for the purposes of illustration)

While, from a general measurement point of view a two point calibration is probably acceptable, there are situations where this might not be so. At the nuclear level, energies are very precisely defined (see Chapter 1, Section 1.6.4), and studies of these levels might require a much greater degree of accuracy. A multi-point non-linear calibration would then be needed of a mathematical form chosen to reflect the 'shape' of the non-linearity. The most obvious first choice might be a quadratic function:

$$E(\text{keV}) = I(\text{keV}) + G \times C(\text{channels}) + Q \times C^2 \quad (7.2)$$

where Q is the quadratic factor. This is likely to fit a generally parabolic deviation from linearity but would not be adequate for the more extreme case shown in Figure 7.2. It should be remembered that where quadratic or higher order functions are provided, these are conveniences rather than representing any fundamental mathematical reality. If very high precision were required, it would be better to combine the results from more than one spectrum rather than agonize over the 'correct' polynomial to fit to the data.

If a quadratic, or higher order fit, is used, it is even more important than with a linear calibration that the calibration points should span the operating energy range. The reason for this is that small changes to the points within the central energy region can have a much greater effect on the curve outside of the calibration range. Extrapolation beyond the calibration points is always bad practice and might lead to errors.

7.4.1 Errors in peak energy determination

We generally assume that a sample will be presented to the detector on an axis normal to the face of the cap. Does this matter? Should it be necessary, could we get away with presenting the sample side onto the detector? In practice, the answer is yes, but there are implications which we ought to be aware of.

We saw in Chapter 2 that interaction of the gamma-ray with the detector produces fast electrons which then scatter within the detector, creating electron–hole pairs. These are then collected by applying an electric field. It follows, then, since that the primary fast electrons must travel through this electric field, their energy must be modified by it. Electrons moving down the field (i.e. towards the negative electrode) will be slowed down, and those travelling up the field will be accelerated. Only electrons travelling normal to the field direction will be unaffected. This change in primary electron energy, which is called the **field increment effect**, will be reflected in the number of electron–hole pairs produced and hence

on the position of the peak in the gamma spectrum. This appears then as an error in measured gamma-ray energy. For low-energy photons, the release of primary electrons after the interaction will be almost isotropic and the field increment effect will tend to cancel out. At higher energy, most of the primary electrons will tend to be moving in the general direction of the gamma-ray which produced them and this direction relative to the detector field will be important.

The effect is most clearly seen with detectors with a uniform field gradient, such as planar detectors. Fortunately, the error is relatively small. Helmer *et al.* (1975) reported a measured increase in energy of up to 250 eV at about 2700 keV and about 75 eV at the ^{60}Co 1332.5 keV peak when sources were measured in front of a planar detector rather than alongside (when there is no field increment).

Open-ended coaxial detectors can be expected to have no field increment error because, in the normal end-on source geometry the gamma-rays (and therefore primary electrons) will tend to travel perpendicular to the electric field. In closed-end detectors, the situation is complicated because the electric field will in some regions be perpendicular and in others parallel to the gamma-ray direction. Incidentally, gamma-rays which interact by pair production will show no field increment effect on that part of the energy absorbed (subsequent Compton scatterings may). This is because the field increment on the electron will be exactly counterbalanced by the field decrement on the positron (or vice versa, depending upon the field direction). Single and double escape peaks should, therefore, show little or no shift in energy.

There may also be differences in charge collection, depending upon where the gamma-ray interacts with the detector. Again, these will cause energy errors. The actual energy error in practice will be a combination of field increment and charge collection errors.

There can also be a noticeable difference in energy calibration due to changes in source-to-detector distance. Sources close to the detector will provide gamma-rays travelling through it in a wider range of directions than a source some distance away and this can manifest itself in energy errors. The work referred to above by Helmer *et al.* (1975) demonstrates the effect at 1489 keV for a number of detectors. The maximum energy error was about 0.1 keV. What is also apparent is that the effect depends very much on the individual detector.

In practice, as long as we energy calibrate for the geometry conditions we are to use, such matters need not concern us unless very precise energy measurement is needed. The magnitude of the energy error would seem to be less that 0.1 keV for gamma-rays of 1500 keV or less, similar to the uncertainties of gamma-energy estimation. Usually, such errors will be absorbed into the overall integral non-linearity of the system. However, very precise energy calibration may need to take them into account by modifying the calibration equation used. It is unlikely that a simple quadratic would then suffice.

7.5 PEAK WIDTH CALIBRATION

7.5.1 Factors affecting peak width

If a spectrometry system is used in simple fashion with peak areas derived from manually set regions-of-interest, there is no need for a peak width calibration. However, if the computer is used in any way in a calibration or analysis then it becomes necessary to tell the computer what the shape of a peak is. I will discuss this in more detail in Chapter 9 but in simple terms, the computer needs to be able to deduce the width (by convention, the full width at half maximum, FWHM) of a peak as a function of energy.

The procedure is much the same as for energy calibration and indeed may be done simultaneously with it, as when using the ORTEC Maestro-II MCA Emulator. Figure 7.3 shows the width of the 26 largest peaks in the ^{152}Eu spectrum used for the energy calibration in Figure 7.1. Again, a best fit straight line has been drawn through the points and once again we ought to consider whether this is appropriate. The factors responsible for the finite width of gamma-ray peaks were explained in detail in Chapter 6. I pointed out that the function that best fits the FWHM variation with energy is a square root quadratic. This is not an option in any of the commercial spectrum analysis programs and I will discuss what

options are available in Chapter 9. For now, I will content myself with explaining how FWHM is measured.

The scatter of the points in Figure 7.3 is noticeably greater than in the energy calibration. This demonstrates that there is an inherent uncertainty in peak width estimation from spectrum data. All but three of the peaks used for Figure 7.3 had uncertainties in their areas of less than 5 % and in most cases less than 2 %. It could well be that whether a width calibration is linear or conforms to Equation (7.4) (see below) depends upon the scatter of individual data points, particularly at the extremities of the curve.

For most practical purposes, the uncertainties in estimating the FWHM from spectrum data are much greater than the errors involved in assuming a linear FWHM to energy relationship. It is certainly true that, with routine width calibrations, the uncertainties on the individual width data points are often such that there is very little difference in the quality of the fit between different mathematical models.

7.5.2 Algorithms for peak width estimation

There are two simple methods of estimating peak width which are most easily explained by reference to Figure 7.4. The first, illustrated in Figure 7.4(a), is as follows:

(a) Estimate the peak height, C_T.
(b) Subtract the peak background, C_0, based upon the average background level above and below the peak.
(c) Divide this by two and add on the background level to get the expected count at half height (C_H).

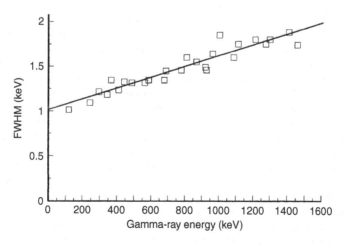

Figure 7.3 Peak width calibration (26 points and the best fit straight line)

(a)   (b)

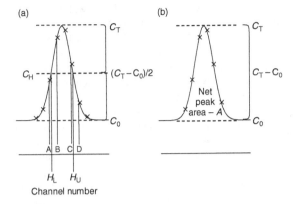

Channel number

Figure 7.4 Methods for estimating FWHM: (a) interpolation between channels; (b) using the area/height ratio

(d) On the low-energy side of the peak, find the channels with counts nearest below, A, and above, B, the expected level. If these are, say C_A and C_B, respectively, then the channel position of half maximum, H_L, below the peak is:

$$H_L = A + (C_H - C_A)/(C_B - C_A) \qquad (7.3)$$

(e) The position of half maximum on the high-energy side of the peak, H_U can be estimated in a similar manner:

$$H_U = C + (C_C - C_H)/(C_C - C_D) \qquad (7.4)$$

(f) The difference, $H_U - H_L$, is then the FWHM in channels.

There are four variations of this calculation, depending upon whether one interpolates from the lower or higher points, all of which will provide an identical result. However, the reliability of the final result depends on the uncertainty of the counts in the particular channels used. In some circumstances, other variations may give a lower uncertainty than the equations given above but in general they will provide an acceptable result.

Unless the peak is well defined, the uncertainty on the FWHM will be much greater than that on the area measurement. For example, for a peak of four channels FWHM containing 10 000 counts and an area uncertainty of 1 % (assuming no background continuum), C_H will be only a little over 1000 counts. A calculation propagating the uncertainties through Equations (7.3) and (7.4) gives an estimated FWHM uncertainty of about 2.3 %. If the peak background continuum is large, the FWHM uncertainty becomes much larger. A peak on a background

continuum the same height as the net peak height (i.e. when $C_0 = C_T/2$) will have an uncertainty about 4.5 times the peak area uncertainty.

If a very precise FWHM estimate is required, I would recommend accumulating at least 50 000 counts in the peak for a precision of 1 % in the FWHM. As far as the peak width calibration is concerned, where several peaks might be involved, it might be unrealistic to collect so many counts and one would have to rely on the smoothing effect of fitting the best line to the data. It is worthwhile remembering the uncertainty in estimation when interpreting FWHM values reported by computer programs. Obviously, FWHM values quoted for peaks with large 'scatter' should be treated with some scepticism.

The same principle can be used to estimate peak width at one tenth of full height but because the individual channel contents are even smaller the uncertainties on the final value will be even greater.

The algorithm described above makes no assumptions about peak shape. An alternative method of FWHM estimation (Figure 7.4(b)) makes the assumption that peaks are Gaussian in shape. For a good well-set-up detector system, such an assumption is not unreasonable and so the following simple equation can be used:

$$\text{FWHM} = 0.939 \times A/(C_T - C_0) \qquad (7.5)$$

where A is the area of the peak and the denominator is the full height of the peak corrected for underlying background. This formula can be derived very simply from the analytical expression for a Gaussian such as is given in Equation (7.6) below. Although much simpler, this calculation does not provide a result any more precise than the interpolation method. One could use the same equation with the appropriate factor to calculate the FWTM (or FW0.1M) (Full Width at Tenth Maximum). There would be little point in comparing it with the FWHM calculated in this manner because both estimates would be based on the same assumption of a Gaussian peak shape. (Chapter 11, Section 11.4.2 will explain why one should wish to compare those two estimates of peak width.)

Another difficulty with the second algorithm is the fact that in order to measure the peak area, we may need to have a FWHM relationship in order to define the peak limits – we have the equivalent of a circular argument. Nevertheless, for a manual estimation where peak limits would be set by eye, the second method is by far the more convenient.

A problem with both of the peak width estimation methods, as will be immediately obvious should the reader try to perform either calculation manually, is estimating the height of the peak. Unless the peak is centred on a channel, estimating the height needs either a little imagination (see Figure 7.5) or a cumbersome calculation (see below). For this reason, if the FWHM is being measured to check the system resolution I recommend that the amplifier gain be adjusted very slightly so that the centroid of the appropriate peak is centred on a channel, as in Figure 7.5(b).

(a)   (b)

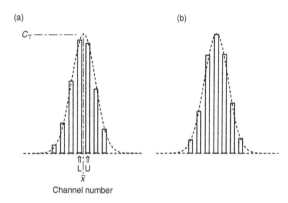

Channel number

Figure 7.5 Estimation of peak height from adjacent channels

7.5.3 Estimation of the peak height

In both of the peak width algorithms, it is necessary to estimate the height of the peak at a position which, more likely than not, is between two channels. A full Gaussian fitting over the peak region would provide this information but a simpler, and sufficiently accurate, approach can be used. The channel contents in a Gaussian peak should be related by the equation:

$$C_i = C_0 + C_T \exp\left[-\frac{(x-\bar{x})^2}{2\sigma^2}\right] \tag{7.6}$$

Substituting the data for the channels either side of the centroid, $\bar{x}$, which can be taken from the MCA, with contents C_L and C_U (see Figure 7.5), into this provides us with two equations which can be divided and rearranged to derive:

$$C_T = \exp\left[\frac{\ln(C_L - C_0) - F\ln(C_U - C_0)}{(1-F)}\right] \tag{7.7}$$

where $F = \left(\dfrac{L-\bar{x}}{U-\bar{x}}\right)^2$

An alternative, which slightly underestimates the peak height, is to assume a parabolic shape for the tip of the peak:

$$C_i = C_0 + C_T - k(x-\bar{x})^2 \tag{7.8}$$

which leads to Equation (7.9):

$$C_T = \left[\frac{(C_L - C_0) - F(C_U - C_0)}{(1-F)}\right] \tag{7.9}$$

7.5.4 Anomalous peak widths

We can leave aside, for the present, the obvious case of multiple overlapped peaks, often referred to as **multiplets**, which are expected to have an anomalous width. It is not always appreciated that spectrum peaks created with the involvement of a positron may be wider than excepted. The most obvious example is that of the 511 keV annihilation peak itself. This, we remember, is a consequence of the annihilation of a positron (whether created by a pair-production event or by β^+ decay) with an electron. Annihilation was discussed in Chapter 1, Section 1.2.2 and it was explained that for every annihilation photon emitted there is an extra uncertainty as to its energy. Extra uncertainty means wider peaks and the 511 keV peak is always wider than would be expected for a peak at that energy. The actual degree of Doppler broadening depends upon the environment of the annihilating electron and thus depends upon the material in which the positrons annihilate. From a normal gamma spectrometry point of view it is sufficient that we recognize that this peak is broadened and take appropriate action if its area is to be measured precisely. It would be convenient if our computer programs recognized that fact as well but, with few exceptions, they do not.

The anomalous width of the 511 keV peak is not the end of the story. Recall that the single escape peak in a spectrum is caused by the loss of one of the annihilation photons from the detector. If there is some uncertainty as to the precise energy of the annihilation radiation energy, then there will be a corresponding degree of uncertainty on the amount of energy lost. Single escape peaks are broader than would be expected for a full-energy peak of the same energy. The problem does not arise for double escape peaks, which we can expect to be of normal width, albeit slightly tailed to high energy. In this case, the full annihilation energy of 1022 keV is lost and it matters not how this is shared between the two photons.

Figure 7.6 is a plot of the peak width of a number of peaks in the spectra of various nuclides which provide higher than average gamma-ray energies. The single and

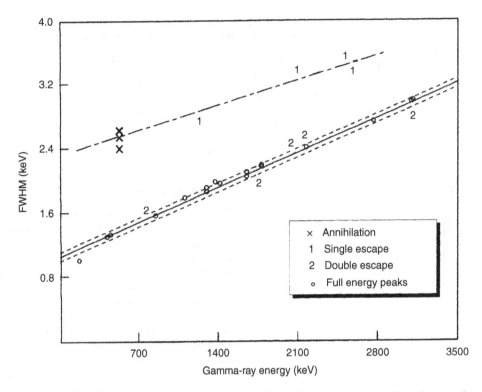

Figure 7.6 WHM of full energy, single escape, double escape and annihilation peaks as a function of energy, demonstrating the anomalous width of single escape and annihilation peaks (the detector used had a FWHM at 1332.5 keV of 1.88 keV)

double escape peaks and the 511 keV peak are marked. The figure confirms in every respect what we have discussed above. Doppler broadening contributes a little over 2 keV extra uncertainty to the annihilation and single escape peaks.

7.6 EFFICIENCY CALIBRATION

7.6.1 Which efficiency?

Before looking at this matter in detail, we should first establish what we mean by 'efficiency'. We can define it in a number of ways depending upon how we wish to use it:

- **Relative efficiency** is a general performance measure relating the efficiency of detection of the ^{60}Co gamma-ray at 1332 keV of the detector to that of a standard sodium iodide scintillation detector (this is discussed further in Chapter 11, Section 11.4.3).
- In gamma spectrometry, our intention is to relate the peak area in our spectrum to the amount of radioactivity

it represents. For this, we need the **absolute full energy peak efficiency**. This relates the peak area, at a particular energy, to the number of gamma-rays emitted by the source and must depend upon the geometrical arrangement of source and detector.

- **Absolute total efficiency** relates the number of gamma-rays emitted by the source to the number of counts detected anywhere in the spectrum. This takes into account the full energy peak and all incomplete absorptions represented by the Compton continuum.
- **Intrinsic efficiency** (full energy peak or total) relates the counts in the spectrum to the number of gamma-rays incident on the detector. This efficiency is a basic parameter of the detector and is independent of the source/detector geometry.

In this chapter, I will omit the word 'absolute' in referring to efficiency unless there is likely to be confusion with the intrinsic parameters. In previous chapters, we have seen that efficiency (however defined) varies with energy and a full calibration of a detector system needs the energy/efficiency relationship to be determined.

One might imagine that, knowing all we do about the interaction processes involved, the absorption coefficients of the detector material and attenuation within the encapsulation, it would be possible to calculate the detector efficiency from first principles. Unfortunately, there are limitations in the mathematical tools at our disposal and the lack of consistency with which detectors can be manufactured militate against such calculations. At the present time, efficiency calibrations are performed on actual gamma-ray spectra. There are, however, efforts being made towards provision by the manufacturers of theoretical calibration data with each detector supplied, so that the need for calibration by the user may diminish in the future. I will discuss some of these developments in Section 7.7.

7.6.2 Full-energy peak efficiency

This is the parameter of most significance in practical gamma spectrometry. (I will denote it by the symbol ε alone and subscript it with the letter 'T' (ε_T) to indicate the total efficiency when necessary.) The calculation of full-energy peak efficiency is straightforward; it is the ratio of the number of counts detected in a peak to the number emitted by the source:

$$\varepsilon = R/(S \times P_\gamma) \tag{7.10}$$

where R is the full-energy peak count rate in counts per second, S is the source strength in disintegrations per second (i.e. Becquerels) and P_γ is the probability of emission of the particular gamma-ray being measured. The source strength used in Equation (7.10) may need to be corrected for decay from the date of preparation. Once again, the values of half-life quoted even in some well-used and respected compilations, have been found to be significantly in error and I would again direct the reader to Appendix B.

It is conventional to construct an efficiency curve by measuring many gamma-rays and plotting efficiency against energy. Figure 7.7 shows such a plot for a p-type coaxial detector using logarithmic scales. Plotted thus, the relationship is approximately linear over much of the commonly used energy range, say 130 to 2000 keV. Below 130 keV, the efficiency falls due to absorption in the detector cap and dead layers. (This portion of the spectrum would be approximately linear and horizontal for an n-type detector.) At energies above 3000 keV, the efficiency falls more rapidly than a linear relationship would indicate.

If it is necessary to derive an equation for the efficiency calibration, once again we must consider whether

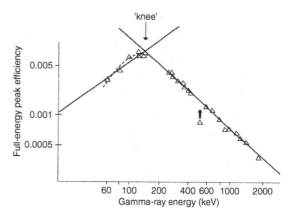

Figure 7.7 Efficiency curve for a p-type closed coaxial detector. The point lying below the line is that representing the 511 keV annihilation peak

a linear relationship on a log–log plot is satisfactory or whether a more complex relationship is needed. Commercial analysis programs offer various alternatives that will be discussed in Chapter 9, Section 9.9. Regardless of the form of equation chosen, it may be difficult to find a simple curve which will fit well over the whole energy range and it is convenient to split data, such as that given in Figure 7.7, into two parts, one above the 'knee' at about 130 keV and another below. For an n-type detector, the knee may be somewhat different in energy.

The data in Figure 7.7 include an efficiency value for the 511 keV annihilation peak deriving from ^{22}Na, which clearly lies below the calibration line. There are two possible reasons for this. First, unless the extra width of the annihilation peak is taken into account when estimating the peak area, a low result will be obtained. However, in the particular case plotted here the reason lies in the source environment. The ^{22}Na point source was positioned on a solid plastic shelf. Positrons from the sources scattering within the shelf would be annihilated close to the source. On the other hand, those leaving the source and travelling away from the shelf would tend to be annihilated some distance from the source and further away from the detector. As we shall see later in Section 7.6.4, this means that those annihilation photons will have a lower probability of being detected and a lower apparent efficiency. The solution is simple. Cover the source with sufficient solid material to ensure that all positrons are annihilated close to the source and a valid efficiency point can then be measured. This is a general point which applies whenever the 511 keV photon emitted by position-emitting sources is measured.

Having constructed the efficiency curve, or the equivalent mathematical equation, it can be interpolated to provide the efficiency data needed by the inverse of Equation (7.10) to convert peak area to activity.

Apart from noting that calibration spectra should be of high quality, we have not discussed experimental conditions under which the calibration sources should be measured or what form they should be in. Ideally, we would have point sources emitting single gamma-rays at low count rate and reasonably large source/detector distance. There are several reasons why such a calibration curve might not be relevant when confronted by a real sample, as follows:

- different source-to-detector distance;
- different shape of source;
- absorption within the source;
- random summing at high count rate;
- true coincidence summing at close geometry;
- decay of the source during counting;
- electronic timing problems.

7.6.3 Are efficiency calibration curves necessary?

Before going on to discuss the various factors that can affect their validity, it is worth considering whether in practice a complete calibration curve is necessary. In fields such as activation analysis, almost all measurements are made comparatively. Samples and standards irradiated together are measured under identical conditions and their spectrum peak areas compared directly. The calculation implicitly takes into account efficiency, including all of the factors likely to cause error. Again, environmental measurements involve a limited set of common radionuclides. There is a great deal of merit in making measurements relative to a reference standard for each nuclide rather than depending upon interpolation of a calibration curve. Such interpolation inevitably introduces extra uncertainties over and above those involved in producing a point on the curve. Indeed, because of one or more of the factors listed above the calibration curve may not be accurate in any case.

In practice, of course, one may be constrained by the tools at hand. If the spectrum analysis program provides no other option than using a calibration curve, then a curve must be created. However, it is worth bearing in mind the general point and the potential for error in using calibration curves.

7.6.4 The effect of source-to-detector distance

It is generally recognized that the gamma-ray intensity emanating from a source falls off with distance according to the inverse square law. This is certainly applicable to point sources of gamma radiation and point detectors. Does the inverse square law apply in gamma spectrometry? Could we use it to reconcile counts made at different distances?

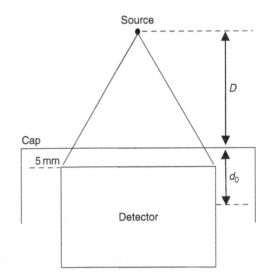

Figure 7.8 Geometric basis of the correction for source-to-detector distance

Figure 7.8 shows the general geometric arrangement. An immediate problem is that we cannot directly measure the true source-to-detector distance. Because the total absorption of gamma-rays often involves multiple scattering within the detector, the zero distance point must be somewhere within the body of the detector crystal. This point can be deduced experimentally. If we assume that the inverse square law is valid, then the count rate, R, must vary thus:

$$R \propto 1/d^2 \qquad (7.11)$$

Now the distance d is the sum of the known source-to-detector cap distance, D, and the unknown distance from the point-of-action within the detector to the detector cap, d_0:

$$d = D + d_0 \qquad (7.12)$$

or, by combining these two equations and rearranging:

$$1/R^{1/2} = kD + kd_0 \qquad (7.13)$$

where k is a constant. So, if the activity of a source is measured at different distances, D, and $1/R^{1/2}$ plotted against that distance, the intercept on the x-axis will be d_0. R could be taken as overall gamma-ray count rate or the count rate at a particular energy, as indicated by the peak area in a spectrum.

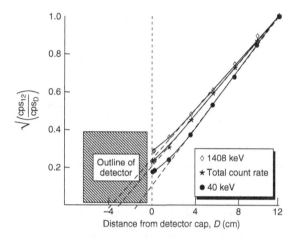

Figure 7.9   Inverse square root of source activity as a function of source-to-detector distance

In Figure 7.9, three sets of experimental data, using a point source of ^{152}Eu, are plotted. These represent the 40-keV peak area, the total count rate and the 1408-keV peak area. As we can see, the value of d_0 is not constant but depends upon the energy of the gamma-ray. The low-energy photons will be totally absorbed in the parts of the detector closest to the detector cap. As we expect, the effective point-of-action for higher-energy gamma-rays is deep within the detector. Apart from non-linearity close to the detector, it would appear that the inverse square law can be applied, as long as:

- the position of the point-of-action, d_0, has been determined for the particular gamma-ray energy;
- the source-to-detector distance is not too small.

The latter restriction is a consequence of lost counts due to true coincidence summing. I will discuss this fully in the next chapter. For a nuclide emitting a single gamma-ray, such a restriction would not necessarily apply. In general, the use of corrections of this type is best avoided. It is much more satisfactory to standardize on a small number of counting positions and create a separate efficiency calibration at each one. If it does become necessary to make mathematical corrections, it should be borne in mind that,

for an accurate correction, d_0 must be determined for each gamma-ray of each nuclide.

The idea of reducing the detector to a point is the basis of the **virtual point-detector** concept. The importance of this concept is that it allows approximations to be made that simplify, what would otherwise be, complicated mathematical calculations. Mahling *et al.* (2006) took the dimensions of 49 actual detectors and, using the Monte Carlo program MCNP, were able to derive an empirical equation for calculating d_0 based upon the radius and height of the detector and two energy dependent parameters, which have been quantified. The concept has been explored for planar and semi-planar detectors (Alfassi *et al.*, 2006). The rather curious conclusion was that the concept is valid, but the virtual point for small detectors can be outside the physical dimensions of the detector.

7.6.5   Calibration errors due to difference in sample geometry

At constant source-to-detector distance, distributing radioactive material within a volume of material, as opposed to concentrating it in a point source, must decrease the gamma-ray intensity at the detector. Figure 7.10 compares

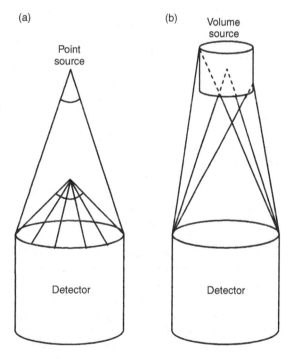

Figure 7.10   Illustration of defining solid angle subtended at the detector: (a) point sources at different distances; (b) a distributed source

point source and distributed source geometry. With a point source, calculating the solid angle subtended by the detector, which determines the incident gamma-ray intensity, is straightforward. For distributed sources, the calculation of an effective solid angle is complicated by the fact that every point within the source has a different aspect on the detector and therefore will contribute to the overall gamma-ray intensity to a different degree, as suggested in Figure 7.10(b).

For a thin disc, approximate equations have been derived and tables of factors have been published which can be used to correct the activity of a distributed source to a point source equivalent. (e.g. Faires and Boswell, 1981 and Debertin and Helmer, 1988). For volume sources, the integrations are more complicated and cannot be reduced to a simple expression for calculating a geometry correction factor. Abbas (2006) has reported a direct mathematical method of calculating solid angle subtended by a well-type detector for point sources, circular disc and cylindrical sources. The procedure can cope with sources inside and outside of the well. Presumably, application of the equations to a cylindrical, rather than a well-type detector, would be possible.

In practice, most laboratories will work with a small number of standardized sample geometries and I would suggest that by far the simplest way in which to relate samples of differing geometry is by means of empirical factors determined by actual measurement. However, you should be aware that again such factors will be individual for each gamma-ray of each nuclide to be measured.

For irregular objects, estimation of geometrical correction factors is more of a problem. Calculation, even with computer assistance, is difficult and an empirical comparison of different geometries might be thwarted by the unavailability of vessels of an appropriate shape. In a particular case where small (i.e. 5 cm across) irregularly shaped archaeological items were to be analysed by neutron activation analysis the problem of comparing the items with standard sources was solved by modelling (Warren 1973), in this case – literally. A clay compound, which happened to contain several of the elements of interest, was irradiated to activate them. Rough models of the objects were then made from the radioactive compound. After counting, the model could be re-formed into a regular cylinder and re-counted. The regular cylinder could be related to the standard geometry mathematically, which provided a chain of correction between the object and the standard. Modelling with radioactive materials is not a task to be undertaken lightly (and not necessarily to be recommended – a better solution might

be to activate the models themselves) but the general principle of relating non-standard geometries to standard ones is worth bearing in mind.

A more everyday geometry correction might be that necessary to correct for differences in sample height within standard cylindrical containers and is discussed below. However, in practice it is much better to avoid changes in sample height than to correct them. A routine procedure which can be recommended is to prepare counting samples in cylindrical containers filled to a standard height. That is, of course, straightforward for liquid samples but is more difficult with solids. Most powdered materials have some elasticity which allows for the container to be filled slightly beyond the standard height and then a plunger can be used to compress the sample to the standard height. For many materials, this will render the uncertainty due to variation in sample height negligible. For lumpy materials, there is likely to be some uncertainty which must be allowed for in the uncertainty budget.

7.6.6 An empirical correction for sample height

For small changes in the height of cylindrical sources, an empirical correction can easily be derived. We can extend the reasoning in Section 7.6.4 to include the height of the sample. A cylindrical source, such as that shown in Figure 7.11, can be considered as approximately equivalent to a point source placed at some distance within the volume of the source. If we assume that for small changes

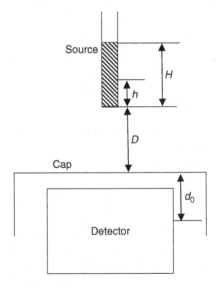

Figure 7.11 Geometric basis of the empirical correction for source height

in height, that position will be a constant fraction, f, of the full source height, H, we can say that:

$$1/R^{1/2} = kD + kd_0 + kh \tag{7.14}$$

where h is the effective height of the source. The first two terms of this equation will be constant for a constant source position. If we group these into a constant K, we can rewrite this as:

$$1/R^{1/2} = K + kfH \tag{7.15}$$

We can now place an amount of radioactive material in the container and measure the count rate as it is successively diluted. A linear regression of $1/R^{1/2}$ against the source height will allow K, the intercept, and kf, the gradient, to be determined. Using Equation (7.15) we can derive the following expression to correct activities measured with height, H, to a standard height, H_S:

$$R_S = R_h \left(\frac{1 + FH}{1 + FH_s} \right)^2 \tag{7.16}$$

where F is a composite factor, kf/K. This procedure has been used successfully over a number of years in an activation analysis laboratory to correct for the height of sources in $1\,cm^3$ and $3\,cm^3$ standard sample containers. Figure 7.12 summarizes a particular set of measurements for sources containing ^{241}Am, ^{137}Cs and ^{60}Co in $1\,cm^3$ sample tubes measured on the face of the detector. One point worthy of note is that on the particular detector used (18 % relative efficiency), a 1 mm change in source height leads to about a 2.5 % change in count rate. One might

wonder how many people control their source height to 1 mm. The case cited is in some respects a 'worst case' but it does underline the fact that small changes in geometry can have larger effects than is generally appreciated.

In deriving Equation (7.16), a number of simplifications have been made and one would not expect to be able to correct for gross changes in geometry in this manner. It should also be noted that changes in true coincidence summing are not accounted for explicitly although the empirical manner in which the correction factor is derived would tend to do this if the nuclide used were the same as that to be measured in practice.

A general point worth mentioning in connection with sample heights is that if samples are counted close to the detector the fall-off in detector efficiency with distance is such that there is little point in increasing the height of a sample beyond 20 mm or so (see Chapter 13, Figure 13.11). If a larger sample must be counted it is much better to increase the diameter of the source rather than height, at least up to the diameter of the detector crystal.

7.6.7 Effect of source density on efficiency

In Chapter 2, on the interactions of gamma radiation, I discussed absorption of gamma-rays. It is self-evident that there must be some degree of self-shielding within a large source with a high density. Because of this, it makes sense to try to arrange that samples are compared directly with standards prepared in the same geometry and density. If there are density variations from sample to sample, it is worthwhile considering distributing the source in a larger volume of inactive material. (Cellulose powder is useful for this purpose. Because its bulk density is low, it contributes little extra absorption.) There would still be density differences between such diluted samples but the differences may be small enough to ignore.

Some care is needed. While the mixed sample and diluent need not be homogeneous on a microscopic level, it would be sensible to reduce the sample to a reasonable grain size before mixing. With very dense materials (some separated geological minerals, for example) there is a possibility that if the sample is shaken during handling after preparation, the heavier grains will work their way to the bottom of the sample container, obviating the effect of the dilution.

If we consider making a mathematical self-absorption correction to the measured gamma spectrum peak count rate, R, we can use the following simple equation:

$$R_0 = \frac{R\mu t}{(1 - e^{-\mu t})} \tag{7.17}$$

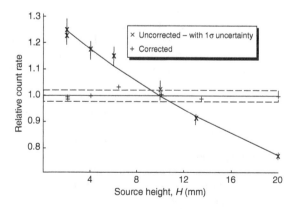

Figure 7.12 Empirical correction for variable height of a cylindrical source counted on the detector cap. The area enclosed by the horizontal dashed lines indicates one standard deviation of the corrected results

where t is the thickness of the sample and μ is the linear attenuation coefficient at the appropriate energy for the material of which the sample is made. Application of Equation (7.17) is simple enough when the attenuation coefficients are known (the 'Further Reading' list for Chapter 2 refers to useful compilations). If the sample matrix is not of a simple composition, there is an immediate problem in that the value of μ is likely to be unknown. For many matrices one could, perhaps, estimate a value from that for similar materials. If the composition of the sample is reasonably well established, another option would be to estimate the mass attenuation coefficient (μ/ρ) for the composite from the coefficients of the component parts, $(\mu/\rho)_i$, in the following way (where f_i is the fraction of each individual component and ρ the relevant density):

$$\frac{\mu}{\rho} = \sum f_i \left(\frac{\mu}{\rho}\right)_i \qquad (7.18)$$

An example of the use of this equation is provided by Oresugun *et al.* (1993).

Corrections based on estimated mass attenuation coefficients

Spectrum analysis programs may provide a correction for self-absorption. For example, GammaVision uses Equation (7.17) to make corrections based upon a database of linear or mass attenuation data. Facilities are provided to generate data by using Equation (7.18) for sample compositions not already in the database. This is useful, but would seem unwieldy when every sample might be of different composition and useless unless the sample composition is known. Fortunately, it happen to be the case that mass attenuation coefficients due to Compton scattering, which dominate the absorption of gamma-rays over much of the normal energy range, are almost independent of atomic number (see Equation (2.8)). This means that, even if the composition of the sample is only approximately known, a reasonable self-absorption correction will be made. This comfortable situation breaks down at low energy, where self-absorption is particularly severe, because of the strong dependence of photoelectric absorption on atomic number.

The use of mass attenuation coefficients involves a modification to Equation (7.17):

$$R_0 = \frac{R(\mu/\rho)t\rho}{(1 - e^{-(\mu\rho)t\rho})} \qquad (7.19)$$

where (μ/ρ) is the mass attenuation coefficient. The product, $t\rho$, has units of $g\,cm^2$ and for cylindrical sources

can be identified with the mass of the source divided by the area of the base of the source. GammaVision calls this the 'size' of the source. Note that, if a correction for the density of the calibration source was not made (Gamma Vision does not allow this), the mass used when making this correction would have to be the difference in mass between sample and calibration source.

This approach works well in practice and, in situations where the sample composition is undefined, even at low energy, where it cannot be expected to give an accurate correction, it is worth doing. Under those circumstances, even an approximate correction will produce results of greater value than if left uncorrected. In situations where a nuclide has peaks at low energy and high energy and, because of differing results, it is clear that the absorption correction has failed, there is scope for altering either the 'size' factor or creating a more appropriate mass-attenuation curve.

Empirical correction

In a routine service laboratory, there will inevitably be occasions when the sample is of unknown composition. A particular real, if uncommon, example is the measurement of ^{210}Pb in a lead-rich dust. Even without the presence of the lead, the low energy of the ^{210}Pb gamma-ray (45 keV) means that in any sample of reasonable size a self-absorption correction is needed. A procedure, which was followed in that case, is based upon an empirical determination of μt using an external source:

(1) The sample is placed in a standard container and slightly compressed by hand to a standard volume using a plunger (see Figure 7.13).
(2) The gamma spectrum is measured and the relevant peak area measured, say R cps. This count rate must be inaccurate (but hopefully is precise) and is to be corrected.
(3) A small point source of the nuclide to be measured is held just in contact with the surface of the sample. Another spectrum is measured. Now the peak count rate is R^+ and includes the contribution from the external source after absorption by the sample. The activity of this external source must be somewhat greater than the sample activity so that it can be measured in a short time.
(4) The point source is then measured at the same distance from the detector as it was when positioned on the sample. The peak area, R_S, is the point source count rate *sans* sample. (The arrangement shown in

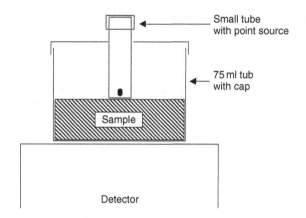

Figure 7.13 Source geometry used to allow correction for self-absorption within sources

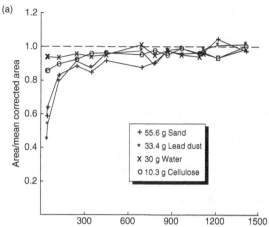

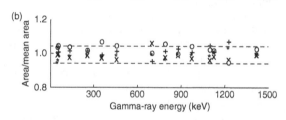

Figure 7.14 Empirical correction for self-absorption within sources: (a) uncorrected peak areas relative to the overall mean corrected peak area; (b) peak areas after correction

Figure 7.13 allows this to be achieved by replacing the sample container by an identical empty container.)

The second and third measurements allow an estimate of the degree of absorption by the sample that can be equated to the right-hand side of Equation (2.16) in Chapter 2, thus:

$$(R^+ - R_S) = R_S\,e^{-\mu t} \tag{7.20}$$

From this, we estimate the combined parameter μt. We need not bother measuring t at all. The factor can then be used to correct R by using Equation (7.17).

There are various theoretical objections to this procedure. Equation (2.16) (Chapter 2) is valid only for a collimated beam of radiation and we have ignored build-up. Differences in sample density can also have an effect on the degree of true coincidence summing. Nevertheless, the method does have the advantages of simplicity. It makes no assumptions about the state of the sample; it could even be applied to wet samples and does not depend upon literature data. Figure 7.14 shows the results of measurements of ^{152}Eu distributed in 30 cm^3 samples of four materials of very different density – cellulose, water, sand and a somewhat heterogeneous sample of chimney dust contaminated with lead. Figure 7.14(a) demonstrates the severe effect of absorption on the low-energy peaks, while Figure 7.14(b) shows the same results after correction by using the method described. For illustration, the mean count rate for the 40-keV peak of ^{152}Eu in the four spectra was 7.75 cps with an uncertainty of 28 %. The mean corrected count rate is 10.97 cps with an uncertainty of 4.78 %, rather larger than the counting uncertainties

of about 1.5 %, but a useful improvement, albeit at some cost in time and effort.

Mathematical tools for self-absorption correction

A simple mathematical tool for post-analysis self-absorption corrections for samples in cylindrical or Marinelli geometry, called *Gammatool*, is available from Isotrak (a division of Amersham QSA). The program must be supplied with the physical dimensions, density and composition of the sample, the position of the detector relative to the sample and information about the calibration source used to generate the efficiency curve. It then calculates a correction factor for each gamma-ray used.

Ideally, one would have an efficiency curve derived from measurement of a calibration source of the same composition and density as the samples to be measured. If a range of samples is to be measured, that might mean the preparation of a large number of sources – a time-consuming and by no means trivial task. A more acceptable solution might be a mathematical tool for creating efficiency data taking into account self-absorption.

A common program used for the calculation of efficiencies in volume sources is MCNP-Monte Carlo N-Particle transport code. I will describe this program further in Section 7.7. The program works by imagining a gamma-ray emitted in a random direction within the sample and follows its fate as is scatters through the sample and, if it happens to reach it, what happens to it within the detector to create the detector signal. It does this a very large number of times to generate an efficiency curve that takes into account the size and shape of the sample and self-absorption within it. The user must provide a considerable amount of detailed information about the detector, its constructional details and its mounting and a detailed description of the sample, including its composition. There are difficulties with the program (and others based on the same principle) when applied to volume rather than point sources and a considerable amount of setting-up is needed.

Saegusa *et al.* (2004) suggest a compromise procedure, that they refer to as the Representative Point (RP) method. Initially, they use MCNP to calculate point source efficiencies as a function of gamma-ray energy at some tens of thousands of points around the detector. Those points within the required source volume are integrated. They search the efficiency data for a point source efficiency curve that most closely matches the calculated volume source efficiency curve. The coordinates of that point define the Representative Point. A practical efficiency calibration can then be made by using a point source placed at the RP coordinates. Having a measured efficiency curve as a basis, the MCNP program can then be used to calculate an efficiency curve for a particular matrix, taking into account self-absorption. Only one measured efficiency curve is needed – all other efficiency curves being created by calculation. Saegusa *et al.* (2004). claim to achieve better than 4 % uncertainty on efficiency values over the range 22 to 1836 keV and that the results are less dependent on selecting appropriate values for the various detector parameters than normal MNCP procedures.

There are further details of MNCP procedures in Section 7.7.

7.6.8 Efficiency loss due to random summing (pile-up)

In Chapter 4, I discussed random summing in connection with the pile-up rejection circuitry in amplifiers. We came to the conclusion that even with pile-up rejection there must be some residual random coincidences. There is then, whether or not pile-up rejection is available, a need to be able to correct for random summing in high count rate spectra. In some circles, there seems to be an assumption that pile-up rejection is 100 % effective

in removing pile-up pulses. Any residual count losses as at high count rate are then attributed to limitations in the live time correction systems. Indeed, one test in the ANSI Calibration Standard (1999) to check for accurate live time correction is uncomfortably like the procedure described below to measure the correction for random summing loss. It is certainly true that the pile-up rejection circuits cannot remove all random summing – random sum peaks in our spectra are evidence of that. It seems sensible to take steps to correct for that before worrying about live time problems. It could be, of course, that the loss of accuracy in live timing has the same effect as random summing on peak areas – both are a function of random pulse coincidences. If that is so, then the correction described below is correcting for them in any case. There will be a point at which the correction ceases to be linear – this may be the point at which live time correction begins to fail.

A pulse will be involved in a summing whenever it is not preceded or followed by a certain period of time. This time, τ, is the resolution time of the electronic system. Using the Poisson distribution, it can easily be demonstrated that the probability of a random coincidence, p_C, within τ is:

$$p_C = 1 - e^{-2R}\tau \qquad (7.21)$$

where R is the mean count rate. We can equate this probability to the fractional loss in peak area in our spectrum. So, if A is the measured peak area and A_T is the true peak area:

$$(A_T - A)/A_T = p_C = 1 - e^{-2R}\tau \qquad (7.22)$$

If we rearrange this, we can derive a simple equation to correct peak areas for random summing:

$$A_T = Ae^{-2R}\tau \qquad (7.23)$$

Because the summing is random, this correction is applicable to all peaks in the spectrum – to apply it we need to know the resolution time, τ. Without pile-up rejection, we can expect it to be of the same order of magnitude as the shaping time constant of the amplifier, i.e. a few μs. It is best estimated by experiment.

If we take logarithms of Equation (7.22) and rearrange, we find that:

$$\ln A = \ln A_T - 2R\tau \qquad (7.24)$$

The factor we need, 2τ, is the gradient of a linear plot of $\ln A$ against the count rate R. Finding R in itself would seem to be a problem. Using a count rate meter at the amplifier output would not be satisfactory because it would have its own rather different pile-up problems. Using the input count rate (ICR) output on the amplifier derived from the fast discriminator (see Chapter 4, Section 4.9 and Figure 4.25) should provide a relatively loss-free pulse stream. This is fine if the correction is to be made manually. However, if the spectrum analysis program is to make the correction there will be no mechanism for it to read that externally measured count rate. In practice, the easiest measure of count rate is within the spectrum itself. All of the counts in the spectrum can be added together and divided by the live time to give an estimate of R. It will not be absolutely accurate because pulses below the lower threshold of the ADC and those outside of the linear gate window will not be taken into account. Nevertheless, it does provide a convenient measure of count rate, sufficient to provide an empirical estimate of the system resolution time.

The procedure, referred to as the **moving source method**, is as follows:

(1) Fix a source of ^{137}Cs in such a position near to the detector that the total count rate is, say, 2000 cps. It is important that this source should not be moved during any of the subsequent measurements (adhesive tape is useful here).

(2) Accumulate a spectrum for long enough to provide a peak area for the 661.6 keV-peak of ^{137}Cs, with a precision of better than 1 % or so.

(3) Measure the 661.6 keV-peak area and calculate the mean count rate.

(4) Place a source of a different nuclide which will not interfere with the 661.6-keV peak in such a position that the total count rate is doubled – ^{152}Eu is a convenient nuclide for this. The general 'shape' of the spectrum is not unlike the typical shape of a range of neutron-activated materials.

(5) Measure another spectrum and repeat the step.

(6) Move the other source closer to the detector so as to increase the count rate and repeat the measurement.

(7) Repeat step 6 until the total count rate exceeds the normal working range or the plotted data cease to be linear (in which case the maximum usable count rate has been exceeded).

(8) Plot ln (^{137}Cs peak area) against R and estimate the gradient – this is the correction factor.

Figure 7.15 shows the results of one such determination. In this particular case, the resulting correction factor, $3.81\,\mu s$ (2τ), was considered usable up to 40 000 cps when the peak area error would otherwise have been about 14 %.

In this example, the corrected results agreed to better than 1 %. At this count rate, on that particular detector with $6\,\mu s$ pulse shaping, the probability of two gamma-rays appearing within a single pulse width would be 75 %.

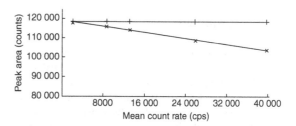

Figure 7.15 Empirical correction for pile-up (random summing): ($\times$), uncorrected; ($+$), corrected

Sometimes, Equation (7.22) is used in a simplified form. When the correction is small, $e^{-2R}\tau$ is approximately equal to $(1 - 2\tau R)$ and a linear form of Equation (7.21) results. With computers, calculators and spreadsheets so readily available there seems little point in using the simplfied form.

Section B.5.1 of the ANSI Calibration Standard (1999) describes a similar method of pile-up correction based upon the use of a series of ^{152}Eu sources of precisely known relative activity. The principle is the same – the only disadvantage is the need to prepare accurately related sources.

7.6.9 True coincidence summing

This source of error is a consequence of summing of gamma-rays emitted very nearly simultaneously from the nucleus. It is a potential source of error whenever nuclides with a complicated decay scheme with cascades of gamma-rays in it are measured. Unlike random summing, which is count rate dependent, true coincidence summing is geometry dependent and errors are particularly severe when sources are positioned very close to the detector. For this reason, multi-gamma-ray sources should not be used for close geometry efficiency calibrations.

True coincidence summing is of such great importance that Chapter 8 has been devoted in its entirety to the problem.

7.6.10 Corrections for radioactive decay

All measured activities must be related to a point in time. When generating calibration data, whether they be used as individual calibration points or contribute to a curve, the activity of the standard sources must be corrected for decay to a common point in time using the normal decay equation introduced in Chapter 1, which I reproduce here:

$$R_0 = R_t \exp\left(0.693\,t/t_{1/2}\right) \tag{7.25}$$

where R_t and R_0 are the disintegration rates at time t and at a reference time; $t_{1/2}$ is the half-life of the nuclide.

When setting up analysis systems, decay correction should never be left to the discretion of the individual. It is too easy when dealing with long-lived nuclides to assume that decay is negligible when it is not. For example, the half-life of ^{60}Co is 5.27 years and over a few days might be insignificant. However, over a 30-day period the activity decreases by 1 % and perhaps an error of this magnitude ought not to be ignored.

Activation analysis can involve the measurement of nuclides with half-lives as short as a few seconds, and half-lives of a few minutes to hours are common. As we saw in Chapter 5, under these circumstances the optimum count period is one to two half-lives. Obviously during the count period the activity of the source will decrease by a factor of somewhere between two and four and this must be accounted for.

Figure 7.16 shows the basis of the correction. The measured activity is as if a lower, but unchanging, activity

had been measured over the same period of time. The areas beneath the true decaying source curve and the apparent activity line must be equal. Expressing this mathematically and rearranging leads us to a correction:

$$R_t = R_{\mathrm{M}} \frac{\lambda \Delta t}{\left(1 - e^{-\lambda \Delta t}\right)} \tag{7.26}$$

where λ is the decay constant, R_t, is the activity at the start of the count and R_{M} the measured activity. In this context, the relevant time, Δt, is the overall count period, otherwise called the real time, clock time or true time of the count. As with the normal decay correction, it is unwise to leave the correction to the discretion of the operator.

It is sometimes advised that the correction for decay during counting can be neglected if the mid-point of the count, instead of the start time, is used when making the normal decay correction. If the count period is short compared to the half-life, the error introduced by doing this is indeed small. However, the error after a one half-life count period is 2 %. It would not seem worthwhile accepting an unnecessary error of this magnitude for the sake of a small amount of calculation. I would not advocate it.

7.6.11 Electronic timing problems

The main sources of error from this source arise when:

- Not all dead time sources are accounted for. This might arise, for example, if the dead time signal from the pile-up rejection circuit is not connected up properly.
- Short half-life nuclides are measured and the dead time is changing rapidly throughout the count period.

Such problems all relate to the accuracy of the correction for spectrometer dead time. These problems and their solutions are discussed in depth in Chapter 14, Section 14.7.

7.7 MATHEMATICAL EFFICIENCY CALIBRATION

Efficiency calibration in a laboratory environment is straightforward; take a reference source, or sources, containing known activities of nuclides for which well-defined nuclear data are available, measure a spectrum and, from the individual peak areas, calculate efficiencies over a range of gamma-ray energies. Such simplicity is not achievable when one needs a calibration relevant to measurement of a 220 L waste disposal drum, a transport

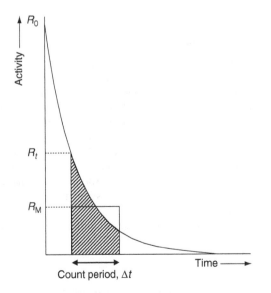

Figure 7.16 Correction for decay during counting. The area of the rectangle defined by R_{M} and Δt is equal to the shaded area of the decay curve

container or an area of land. How do you create an efficiency curve to cope with measurements in the field – of the field?

Consider all we know about the gamma-ray detection process: we understand the processes involved in absorption and scattering of the gamma-rays in the detector, the sample and the surroundings; we know the geometrical arrangement of sample, detector and shielding; we can refer to the decay scheme of the nuclides. It ought to be possible to create a computer program that will calculate an efficiency curve from first principles, no matter what the geometry of the source of the activity. Such programs exist. I referred to one of them above. Monte Carlo programs, such as GEANT and MCNP, are in the public domain and can be freely used, albeit with a considerable degree of understanding. There is a considerable literature on the use of both of these programs. What follows is not intended to be an exhaustive review, more a taster to encourage the reader to delve further.

These programs imagine a gamma-ray emitted from a position within the source, chosen at random and in a random direction, and follow its fate until it is totally absorbed or otherwise lost to the system. The program will consider interactions as it passes through the sample, through the detector enclosure, through the dead layer of the detector and finally as it scatters through the detector, giving up its energy until it is completely absorbed or it escapes from the detector. Each simulated event provides a count within the spectrum in a channel representing the amount of energy absorbed in the detector. At each stage, the program will consider the probability of interaction by various means. It will take account of gamma-rays that scatter within the detector and are then lost and those which would miss the detector altogether but backscatter from the shielding into the detector. It will take into account those gamma-rays that are absorbed within the sample itself. Repeating this process millions of times will create a spectrum that will be comparable with an actual measured spectrum from which an efficiency curve can be derived.

This mathematical process can cope with any type of detector, any type of source. In principle, it can generate efficiency curves relevant to anything from a point source measured on-axis to an infinite plane source of activity measured at an angle – all without radioactive sources. There are, however, drawbacks. Unless all samples are identical in shape, density and composition, a separate efficiency curve would need to be calculated for each sample. Depending upon the complexity of the model and the speed of the computer, the calculation could take from minutes to hours to complete.

More fundamental is the quality of the data available to the model. A common feature of reports dealing with the application of Monte Carlo codes to the simulation of gamma-ray detection is that the success of the model depends critically on the detector parameters supplied. In most cases, it would appear that the detector dimensions quoted by the manufacturer need to be adjusted in order to achieve satisfactory agreement between simulation and practice. The process of optimizing parameters is done by comparing simulation with actual measurements of reference sources in a simple geometry, such as an on-axis point source. In this way, Karamanis (2003) investigated the use of both GEANT and MNCP for simulation and found it necessary to increase the dead-layer thickness substantially for both p-type and n-type detectors and to adjust the detector-to-cap distance and detector diameter. Once characterized in this way, the model can then be expanded to take into account more difficult geometries. Lépy *et al.* (2001) examined a large number of mathematical methods, some of them empirical, some Monte Carlo, and the accuracy with which the efficiency curves created matched measured curves. They considered point sources at different distances and $88 \, cm^3$ volume sources of different density measured on the detector cap. Considering only those calculations where the parameters had been optimized, it is clear that, in general, some programs fared better than others, the further one departed from the reference geometry, the larger the errors became. Although on average, the errors were not large (typically 5 % and up to 9 %, depending on geometry and energy), the authors concluded that '. . . *up to now, the efficiency computation codes cannot be used directly for precise measurements. . . . However, the tested programs are operational for routine measurements, such as environmental or survey monitoring. . .* '. In the context of monitoring for waste disposal, as discussed in Chapter 17, Section 17.2, these programs would be acceptable. It is more difficult to establish how appropriate they would be for out-of-laboratory measurements.

7.7.1 ISOCS

Canberra market a program called ISOCS – *In-Situ* Object Counting System. This program calculates efficiency curves for user-defined source shapes. It can only be used in connection with a detector that has already been characterized by Canberra by MNCP. This is a neat way of reducing the computing time needed for creating the efficiency curve to a few seconds.

The cost of the software and the detector characterization is considerable but Canberra place great emphasis on the cost savings, in terms of purchase and measurement, by not using reference sources. This software is undoubtedly of great value when dealing with difficult geometries and measurements in, what we might call, field or industrial situations. Its value for laboratory calibrations is questionable. In one of their assessments, Canberra state that (Canberra, 1998), for a 1 L Marinelli beaker, ISOCS provides efficiencies to within 5 % of source-based measured values. That particular statement clearly satisfies the marketing department but hides the fact that it is a mean of deviations over a range of energies. In fact, results quoted by Canberra (1998) show that the deviations range from −9 % to +24 % – considerably more than the statement *'within 5 %'* would lead one to believe. Those figures should be compared with the uncertainties on the source-based calibration of 0.27 % to 1.04 %. The same source quotes similar results for other geometries, claiming that the overall agreement is *'well within the statistical uncertainty at one sigma'*. What the article does not quote is the uncertainty on the individual efficiency estimates.

It is difficult not to agree with the assessment of Lépy *et al.* (2001) quoted above. On the other hand, if ISOCS is to be used for field survey work there is no reason to believe that it is a useful tool and fit-for-purpose. Equally, I would have no hesitation in using it for waste monitoring of typical decommissioning samples, where the demands on accuracy are not as great.

7.7.2 LabSOCS

In 1999, the detector characterization facilities of ISOCS was improved to allow measurements at close geometry and, together with improved attenuation corrections, aimed at laboratory measurements. The program is now marketed as LabSOCS – Laboratory SOurceless Calibration Software. A report of a comparison of the use of LabSOCS, applied to a large number of source geometries used for measurements in nuclear power plant situations, with source-based efficiency estimations, is available on the Canberra website (Stewart and Groff, 2002). The comparison is favourable, although the statistical analysis in that particular report is flawed. Visual assessment of the graphs in the report suggests agreement within a few percent but it would appear that the uncertainties on either the LabSOCS estimates or the source-based measurements (or both) are overstated, making the conclusion *'LabSOCS... will agree with source-based efficiency*

calibrations' inevitable. LabSOCS estimates for a particular source quoted vary from 11 % lower than the source-based estimate to 9 % higher. The authors have calculated ratios of efficiencies, ISOCS/source-based, and state that in all cases the ratio = 1 point lies within the 95 % confidence limit. Statistically, we would expect 5 % of the 230 plotted efficiency estimates to be beyond the 95 % confidence limit. In fact, all are well within that confidence limit, suggesting that the confidence limit is too lenient. Considering the pattern of differences between LabSOCS and the source-based calibrations, it appears that LabSOCS underestimates at low energy and overestimates at high energy by up to 10 %. The uncertainties quoted for the LabSOCS efficiency estimates range from 10 % at low energy to 4 % at higher energy. These uncertainties are much greater than achievable with source-based calibration. Nevertheless, as a tool for low cost efficiency calibration when many complicated geometries are used, it has a deserved place. As with other mathematical tools, LabSOCS cannot yet replace conventional calibration for highest quality measurements, a conclusion shared by Bossus *et al.* (2006). After checking a detector calibrated for LabSOCS by the manufacturer, those authors suggest that *'... one should not have a blind confidence in factory-calibrated detector.'* They concluded that the efforts needed to check the manufacturer's calibration were almost as great as calibrating the detector themselves – at much less cost!

7.7.3 Other programs

Above, I have only mentioned a couple of programs. There are others available. EGS – Electron Gamma Shower – is another general-purpose package applicable to the measurement of detector efficiency curves. There is a considerable literature on the use of GESPECOR (GErmanium SPECtrometry CORrection) but, because much of that concentrates on its advantages for making corrections for true coincidence summing, I have left comment until the next chapter.

PRACTICAL POINTS

The points most worthy of special note in this chapter are:

- Ensure that your nuclear data is of good quality.
- Ensure that your efficiency calibration sources have the same source and density as your samples and are traceable to national standards.

There are many factors which can contribute to efficiency-calibration errors. Several of these can be eliminated empirically:

- sample height differences;
- source density;
- pile-up.

It is worthwhile considering whether the analysis can be performed comparatively rather than via an efficiency curve.

Mathematical generation of efficiency curves is becoming a possibility. At the present time, the tools are available and applicable for less critical applications. For highest quality efficiency curves, the source-based procedures are best, but that could alter as the efforts being put into improving the mathematical programs bear fruit.

FURTHER READING

- Sources of nuclear data – see Appendix A.

- For detailed information on the choice of radionuclides for energy calibration, see the following:

Debertin, K. and Helmer, R.G. (1988). *Gamma- and X-ray Spectrometry with Semiconductor Detectors*, North-Holland, Amsterdam, The Netherlands.

- Warren, S.E. (1973). Geometrical factors in the neutron activation analysis of archaeoligical specimens, *Archaeometry*, **15**, 115–122.

- For more information on the mathematical geometrical corrections, see:

Debertin and Helmer (as above).
Faires, R.A. and Boswell, G.G.J. (1981). *Radioisotope Laboratory Techniques*, Butterworths, London, UK.

- On pile-up corrections:

Wyttenbach, A. (1971). Coincidence losses in activation analysis, *J. Radioanal. Chem.*, **8**, 335–343.
American National Standards Institute (1999). *Calibration and use of germanium spectrometers for the measurement of gamma-ray emission rates of radio-nuclides*, ANSI N42. 14-1999, American National Standards Institute, New York, NY, USA (also available at *http://webstore.ansi.org*)

- On calculation (rather than measurement) of detector efficiency:

Moens, L., De Donder, J., Lin, X., De Corte, F, De Wispelaere, A. and Simonits, A. (1981). Calculation of the absolute peak efficiency of gamma-ray detectors for different counting geometries, *Nucl. Instr. Meth. Phys. Res.*, **187**, 451–472.

Canberra (1998). *ISOCS vs, Traditional IGe Efficiency Calibration Measurements For Routine Counting Room Geometries*, Canberra Chronicle, December, pp. 4 and 9.
Lépy, M.C., Altzitzoglou, T., Arnold, D., Bronson, F., Noy, R.C., Décombaz, M., De Corte, F., Edelmaier, R., Peraza, E.H., Klemola, S., Korun, M., Kralik, M., Neder, H., Plagnard, J., Pommé, S., de Sanoit, J., Sima, O., Ugletveit, F., Van Velzen, L. and Vidmar, T. (2001). Intercomparison of efficiency transfer software for gamma-ray spectrometry, *Appl. Radiat. Isotopes*, **55**, 493–503.
Stewart, J.P. and Groff, D. (2002). *LabSOCS™ vs. source-based gamma-ray detector efficiency comparisons for nuclear power plant geometries*, 48th Annual Radiobioassay and Radiochemical Measurements Conference, November 11–15, Knoxville, TN, USA.
Karamanis, D. (2003). Efficiency simulation of HPGe and Si(Li) detectors in g- and X-ray spectroscopy, *Nucl. Instr. Meth. Phys. Res., A*, **505**, 282–285.
Bossus, D.A.W., Swagten, J.J.J. and Kleinjans, P.A.M. (2006). Experience with a factory-calibrated HPGe detector, *Nucl. Instr. Meth. Phys. Res., A*, **564**, 650–654.

- On solid angle calculations:

Abbas, M.I. (2006). Analytical calculations of the solid angles subtended by a well-type detector at point and extended circular sources, *Appl. Radiat. Isotopes*, **64**, 1048–1056.

- On the virtual point-detector and representative point method:

Seagusa, J., Kawasaki, K., Mihara, A., Ito, M. and Yoshida, M. (2004). Determination of detection efficiency curves of HPGe detectors on radioactivity measurement of volume samples, *Appl. Radiat. Isotopes*, **61**, 1383–1390.
Mahling, S., Orion, I. and Alfassi, Z.B. (2006). The dependence of the virtual point-detector on the HPGe detector dimensions, *Nucl. Instr. Meth. Phys. Res., A*, **557**, 544–553.
Alfassi, Z.B., Pelled, O. and German, U. (2006). The virtual point detector concept for HPGe planar and semi-planar detectors, *Appl. Radiat. Isotopes*, **64**, 574–578.

- On the field increment error:

Helmer, R.G., Gehrke, R.J. and Greenwood, R.C. (1975). Peak position variation with source geometry in Ge(Li) detector spectra, *Nucl. Instr. Meth. Phys. Res.*, **123**, 51–59.

- Useful papers dealing with corrections for self-absorption are:

Bode, P., De Bruin, M. and Korthoven, P.J.M. (1981). A method for the correction of self-absorption of low energy photons for use in routine INAA, *J. Radioanal. Chem.*, **64**, 153–166.
Appleby, P.G., Richardson, N. and Nolan, P.J. (1992). Self-absorption corrections for well-type germanium detectors, *Nucl. Instr. Meth., Phys. Res., B*, **71**, 228–233.
Oresegun, M.O., Decker, K.M. and Sanderson, C.G. (1993). Determination of self-absorption corrections by computation in routine gamma-ray spectrometry for typical environmental samples, *Radioact. Radiochem.*, **4**, 38–45.

8

True Coincidence Summing

8.1 INTRODUCTION

In Chapter 7, I referred, without going into any detail, to the difficulties caused by true coincidence summing in achieving a valid efficiency calibration for close geometry measurements. The problem is not a new one (although as larger detectors become available, it becomes more significant) but appears often to be ignored in practice. The problem is important enough, and has been neglected long enough, to devote a chapter to it alone.

The problems caused by **true coincidence summing** (TCS) can be demonstrated by referring to the calibration curves for a 45 % HPGe detector shown in Figure 8.1. These were derived using ^{152}Eu near point sources of moderate activity such that the count rate in each case was about 7700 cps. The lower curve was measured with the source 115 mm from the detector, is smooth, consistent, and is, apparently, satisfactory.

The upper curve, measured with the source on the detector cap, is not at all satisfactory. The points do not lie on an orderly line and it would be difficult to draw an acceptable curve through them. The reason for this dramatic difference is TCS.

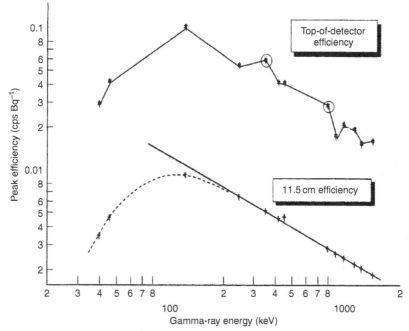

Figure 8.1 Efficiency curves on top of and at 115 mm from the detector endcap using ^{152}Eu

Practical Gamma-ray Spectrometry – 2nd Edition   Gordon R. Gilmore
© 2008 John Wiley & Sons, Ltd

8.2 THE ORIGIN OF SUMMING

Figure 8.2 shows a simplified decay scheme for [152]Eu. Atoms of this nuclide have a choice when they decay; they can emit a β^- particle and become [152]Gd or, more likely (on 72.08 % of occasions), undergo electron capture and become [152]Sm. Whatever the mode of decay, the daughter nucleus then de-excites by emitting a number of gamma-rays in one or other of the decay schemes. We must also remember that every electron capture decay to [152]Sm is likely to be accompanied by the emission of Sm X-rays.

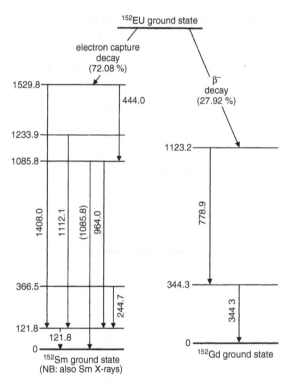

Figure 8.2 Simplified decay scheme for [152]Eu

The lifetimes of the individual nuclear levels are short, much shorter than the resolving time of the gamma-spectrometer system. From the point of view of the detector, every disintegration of a [152]Eu atom in the source will release a number of gamma-rays, and possibly X-rays, simultaneously and there is a certain probability that more than one of these will be detected together. If this happens, then a pulse will be recorded which represents the sum of the energies of the two individual photons. This is true coincidence summing; sometimes called **cascade summing**. It is the summing of two gamma-rays, or a gamma-ray and an X-ray, emitted in coincidence.

As with random summing, the event results in loss of counts from the full-energy gamma-ray peaks and a loss of efficiency. However, unlike the random summing that I discussed in Section 4.8, the summed pulse will not be misshapen and cannot be rejected by pile-up rejection circuitry.

8.3 SUMMING AND SOLID ANGLE

The degree of TCS depends upon the probability that two gamma-rays emitted simultaneously will be detected simultaneously. This is a function of geometry, of the solid angle subtended at the detector by the source. Figure 8.3 illustrates the geometrical arrangement under which the calibration curves in Figure 8.1 were measured. With the source on the detector cap, there is a 42 % chance that any gamma-ray will reach the detector and therefore a 17 % chance that two emitted together will both reach the detector. The further the source is from the detector, the less likely it is that the two gamma-rays will be detected together (1.5 % solid angle and 0.02 % chance of summing for our distant source). Note that at any source-to-detector distance there will be some degree of summing (Figure 8.4). However, beyond a certain distance, which

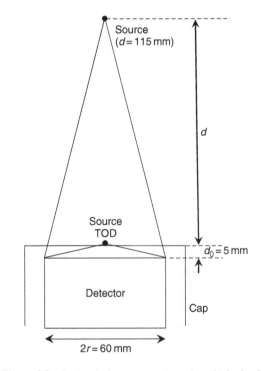

Figure 8.3 Geometrical arrangements used to obtain the data shown in Figure 8.1

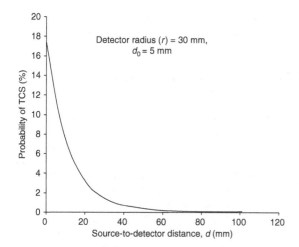

Figure 8.4 Calculated probability of summing as a function of source-to-detector distance: detector radius, $r = 30\,mm$; $d_0 = 5\,mm$

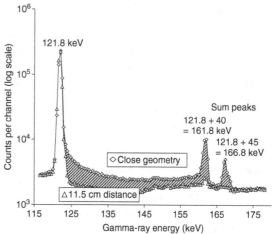

Figure 8.5 Sum peaks above the 121.78 keV peak of ^{152}Eu. Note also the raised level of the continuum at the high-energy side of the peak

depends upon the detector size, TCS losses will be negligible in practice.

A rough and ready estimate of the likelihood of summing of two photons emitted at the same time, P, can be estimated by using the following equation:

$$P = (R - D)^2 / (2R)^2 \qquad (8.1)$$

where R is the radius of the sphere into which the source emits and $D = d + d_0$, with d being the measured source-to-detector distance and d_0 the distance between detector face and the detector cap. If r is the detector radius, then $R = \sqrt{(r^2 + D^2)}$. The probability of one photon passing through the detector is the ratio of the solid angle subtended at the detector, $2\pi R(R - D)$, to the total area of the sphere, $4\pi R^2$. The probability of two photons emitted at the same time striking the detector is that probability squared.

It is worth noting that, for a given solid angle, the number of true coincidence summing events per second (but not the ratio of lost/total counts) will be directly proportional to the sample activity. On the other hand, random summing losses are a function of the *square* of the sample activity. In the situation in Figure 8.1, we can be quite sure that the problem with the 'top-of-detector' count is due to TCS, rather than random summing, because the measurements had almost the same count rate.

8.4 SPECTRAL EVIDENCE OF SUMMING

Since every de-excitation of ^{152}Sm, whatever gamma-rays are emitted, is likely to produce an X-ray these play a

prominent role in TCS. Figure 8.5 shows portions of the spectra from which the calibration curves of Figure 8.1 were calculated. Comparing the two spectra reveals the presence of pairs of small sum peaks approximately 39.9 and 45.3 keV higher in energy than the expected peaks. This is emphasized by the partial analysis of the low-energy portion of the close geometry spectrum reproduced in Table 8.1.

Table 8.1 Partial analysis of the close geometry spectrum of ^{152}Eu

Energy (keV)[a]	Area (cps)	RSD (%)	Attribution (energies in keV)
39.60	386.00	0.46	Sm Kα X-rays
45.23	109.60	1.00	Sm Kβ X-rays
121.83	469.20	0.30	—
161.62	19.40	3.20	Sum 121.78 + 39.91
167.16	6.00	8.64	Sum 121.78 + 45.4
244.83	63.80	0.98	—
284.70	2.60	15.01	Sum 244.70 + 39.91
290.17	1.40	26.54	Sum 244.70 + 45.4
296.09	3.00	10.34	—
344.44	225.10	0.41	—
367.33	10.20	4.85	Sum 121.78 + 244.70
411.28	13.20	2.70	—
444.12	14.40	2.52	—
488.79	1.80	13.51	—
564.29	2.60	13.32	Sum 121.78 + 443.97

[a] Energies here are as reported by the analysis program.

All the peaks not attributed in Table 8.1 are normal full-energy peaks of ^{152}Eu. The extra peaks due to summing with the X-rays are clearly identifiable, appearing after each major gamma-ray originating from the electron-capture decay. In addition, we can detect γ–γ coincidences between some of the ^{152}Sm gamma-rays with higher abundances. The independence of the two branches of the ^{152}Eu decay is underlined by the fact that there is no peak at 466.06 keV, which would indicate summing between the 121.78 and the 344.28 keV gamma-rays, because these originate in different cascades. Nor is there summing between the Sm branch X-rays and the Gd branch gamma-rays. However, elsewhere in the spectrum the (344.28 + 778.90) sum peak does appear as a consequence of summing within the beta decay branch.

Each sum peak represents only some of the counts lost from the main peaks – only some because there will also be a chance of summing in the detector with each and every gamma-ray in the cascade whether or not fully absorbed. In fact, since only a minority of gamma-rays are fully absorbed, the summing of a gamma-ray destined for a full-energy peak with an incompletely absorbed gamma-ray is more unlikely. As we shall see, these coincidences with partially absorbed gamma-rays must be taken into account if a TCS correction is to be computed.

Figure 8.5 also demonstrates a feature of summing that only occurs in p-type detectors. In the summed spectrum, the level of the background continuum at the high-energy side of the peaks is higher than that at the low energy side and the peak has a pronounced tail. Quite the reverse from the peak background we would normally expect, which is lower on the high-energy side (see Chapter 9, Figure 9.6). This phenomenon has been identified by Arnold and Sima (2004) as consequence of the electron capture X-rays generating a partial signal, in coincidence with the gamma-rays, within the dead layer of the detector, which may not be as 'dead' as thought. They suggest that the thick outer n+ contact layer diffuses some way into the high-purity germanium, creating a transition zone. That zone is capable of detecting the X-rays but with poor charge collection. These partial signals are then able to sum with the gamma-rays, creating the tail on the high-energy side of each peak. The effect does not happen with n-type detectors, where the contact layer is very thin, nor is there tailing on gamma-rays from β^- decay because there are no coincident X-rays. For example, elsewhere in the spectrum from which Figure 8.5 was taken, the 778.9 keV gamma-ray of ^{152}Eu, which is emitted following the β^- decay branch to ^{152}Gd, is not tailed.

A further possibility that I perhaps ought to mention is that in the case of beta decay, because the beta particle and the de-excitation gamma-rays are emitted almost at the same instant, a gamma-ray may sum with the bremsstrahlung produced as the beta-particle is slowed down.

8.5 VALIDITY OF CLOSE GEOMETRY CALIBRATIONS

Returning to the efficiency calibration data in Figure 8.1, I can say, without equivocation, that all the points in the close geometry set are invalid as far as constructing a calibration curve is concerned. That is not to say that they are altogether useless. Each point represents a valid calibration point for ^{152}Eu measured on the particular detector, in the particular source geometry on the detector cap. Although valid for ^{152}Eu, the points have no relevance to any other nuclide. For example, the point representing the 121.8 keV efficiency could not be used to estimate the activity of a ^{57}Co source via the 122 keV peak area. ^{57}Co will have its own, different, TCS problems.

While the lower curve in Figure 8.1, measured at 115 mm, appears to be satisfactory, we cannot say that there is no summing. All we can say is that the degree of TCS is negligible. (In fact, Figure 8.5 shows that even in the 115 mm spectrum there is a small sum peak indicating a small degree of summing.) For the same source measured at the same distance on a larger detector, the summing would not necessarily be negligible.

8.5.1 Efficiency calibration using QCYK mixed nuclide sources

In the United Kingdom, the readily available multi-nuclide reference materials QCY and QCYK, supplied by Isotrak, a subsidiary of AEA Technology QSA, are often used for gamma spectrometer efficiency calibration. Similar materials are available elsewhere, possibly with slightly different mixtures of nuclides. The QCYK mixed nuclide reference material contains twelve nuclides emitting gamma-rays from 59.54 keV (^{241}Am) to 1836.05 keV (^{88}Y). However, it seems not to be appreciated that these sources contain a number of nuclides, ^{57}Co, ^{60}Co, ^{88}Y and ^{139}Ce, which will exhibit true coincidence summing if measured close to the detector, significantly affecting the efficiency calibration, as demonstrated in Figure 8.6. In this figure, the dashed line represents the true calibration line passing through the non-summed nuclides. Table 8.2 lists the nuclides in QCYK and comments on various spectral features that became evident when a particular source was measured close to an n-type HPGe detector. These features are:

- ^{60}Co and ^{88}Y both emit two gamma-rays and summing between their gamma-rays is to be expected. That will reduce the peak areas of the full energy peaks of those nuclides and give rise to a sum peak in the spectrum corresponding to the sum of the gamma-ray energies. In Figure 8.6, these points can clearly be seen to lie below the corrected calibration line.

- Seven of the twelve nuclides decay by electron capture and, as we saw in Section 1.2.3, every decay will be accompanied by X-rays characteristic of the daughter atom. That means that we can expect summing between the gamma-rays and the X-rays. As Table 8.2 indicates, γ–X sum peaks from ^{57}Co, ^{139}Ce and ^{88}Y are very evident (see Figures 8.7(b) and 8.7(c)). Although similar summing in ^{241}Am and ^{65}Zn can be detected, it was barely significant in the particular spectrum analysed.
- ^{57}Co is a special case in that although the 122.06 keV peak sums out due to summing with the electron-capture X-rays and with the 14.41 keV gamma-ray, the 136.47 peak sums *in*. This is not obvious in the spectrum but is noticeable in the efficiency calibration where the 122.06 keV peak lies below the calibration line and the 136.47 above.
- A number of minor peaks were observed that could be identified as sums of gamma-rays with the germanium escaped electron capture X-ray (column 6 in Table 8.2).

Germanium escape peaks from the ^{241}Am gamma-ray were also detected (see Chapter 2, Section 2.2.1).
- Single and double escape peaks from the gamma-rays of greater than 1022 keV can be expected. Not all were observed, however.

We might expect γ–X summing from ^{109}Cd. After all, it is an electron capture nuclide and the low energy end of the spectrum is dominated by the intense X-ray peaks from the ^{109}Ag daughter. If that were so, there should be sum peaks at around $88.03 + 22$ keV – there are none. The gamma-ray we normally attribute to ^{109}Cd is actually emitted by the 39.8 s half-life metastable state ^{109m}Ag. Because of this, there is a delay in emission of the gamma-ray and no true coincidence between the silver X-ray, emitted at the moment of decay, and the 88.03 keV gamma-ray. ^{109}Cd is, therefore, not subject to summing. Interestingly, however, in the particular spectrum measured, with a 4.5 % dead time, we did observe minor random summing peaks between the major higher-energy peaks and the 22 Ag K X-rays. For the same reason, ^{113}Sn is not subject to TCS even though its electron capture X-rays are clearly visible in the spectrum. In this case, the 391.70 keV gamma-ray is, in fact, emitted by the 1.66 h half-life metastable state ^{113m}In.

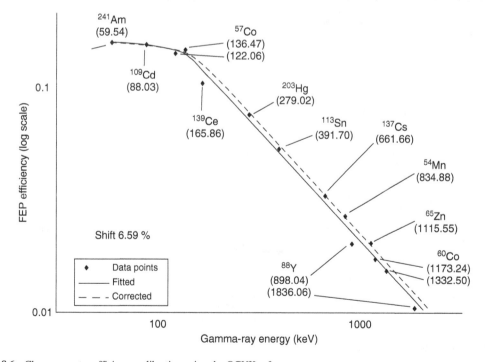

Figure 8.6 Close geometry efficiency calibration using the QCYK reference source

Table 8.2 Peaks observed within the QCYK spectrum measured at close geometry

Nuclide	Decay mode	X-rays	Gamma-ray energy (keV)	Summing[a]	Summing and Ge escape?	511 escape?[b]	Random-sum peak?[c]	Comments
^{241}Am	α	Np L	59.54	(γ–X)	—	—	—	Ge escape peaks
^{109}Cd	EC/IT	Ag K	88.03	None	—	—	—	No TCS – γ emission delayed by IT
^{57}Co	EC	Fe K	14.41	γ–γ	—	—	—	γ–X may be present but not resolved
			122.06	γ–γ and γ–X	✓	—	—	Sum with 14.41 with Ge escape
			136.47	γ–X	—	—	—	Sums in 122.06 + 14.41
^{139}Ce	EC	La K	165.86	γ–X	✓	—	—	—
^{203}Hg	β⁻	Tl K (IC)	279.02	None	—	—	—	—
^{113}Sn	EC/IT	In X	391.70	None	—	—	—	No TCS – γ emission delayed by IT
^{85}Sr	EC	Rb K	514.00	γ–X	✓	—	—	Close to 511 – difficult deconvolution
^{137}Cs	β⁻	Ba K (IC)	661.66	None	—	—	—	Possible SE from 1173.23 at 662.23
^{54}Mn	β⁻	—	834.84	None	—	—	—	—
^{88}Y	EC	Sr K	898.04	γ–X and γ–γ	✓	—	—	—
			1836.05	γ–X and γ–γ	✓	S, D	✓	—
^{65}Zn	EC	Cu K	1115.54	(γ–X)	—	—	✓	—
^{60}Co	β⁻	—	1173.23	}γ–γ	—	(S)	✓	—
			1332.49		—	D	✓	Close to SE from 1836.05 at 1325.05

[a] Parentheses indicates summing expected but not significant.
[b] S, single escape peak; D, double escape peak; Parentheses, indicate not observed.
[c] Random summing is with 22 keV Ag K X-rays from decay of ^{109}Cd.

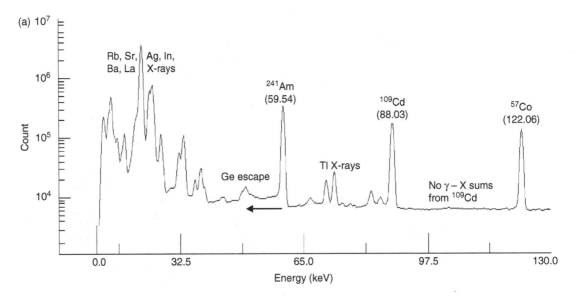

Figure 8.7 Parts of the QCYK spectrum at close geometry

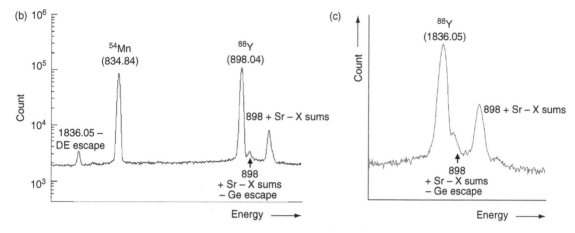

Figure 8.7   (Continued)

In the QCYK spectrum, there are a number of potential problems due to the close proximity of extraneous peaks to those to be measured. Particularly difficult is the measurement of the 514.00 keV peak of ^{85}Sr in the presence of the 511.00 keV annihilation peak. Unfortunately, most commercial software does not recognize the fact that the annihilation peak is Doppler broadened and deconvolution of the doublet may be questionable. A similar problem could arise when measuring the 1332.49 keV peak of ^{60}Co in the presence of the 1325.05 keV single escape peak of ^{88}Y, which will also be broadened. Fortunately, on any reasonable detector deconvolution will not be necessary. Attention can be drawn to the fact that the energy of the single escape peak of the ^{60}Co 1173.23 keV peak, 662.23 keV, is very close to the 661.66 keV peak of ^{137}Cs, although the intensity is very low and there is no problem in practice.

The effect of TCS in close geometry calibrations is to reduce the areas of the peaks due to ^{57}Co (the 136.47 keV peak is increased), ^{139}Ce, ^{88}Y and ^{60}Co. When the efficiency calibration is performed, the 'best fit' line will be somewhat lower than it should be (see Figure 8.6), meaning that activity estimates made using it will be a few percent in error. Nuclides that do not sum will be overestimated, while those that do sum will be underestimated by varying amounts. Whether that is acceptable or not depends upon the use to which the measured activities are to be put.

8.6 SUMMARY

I can summarize the essential points about TCS as follows:

- It usually results in lower full-energy peak areas (but see Section 8.8).

- It gets worse the closer the source is to the detector.
- It gets worse the larger the detector and is worst of all when using a well detector.
- It may be worse if a detector with a thin window is used because the X-rays that contribute to the summing will not be absorbed.
- It can be expected whenever nuclides with a complex decay scheme are measured.
- The degree of summing is not dependent upon count rate.

8.7 SUMMING IN ENVIRONMENTAL MEASUREMENTS

The last point in my summary is worth further discussion because of its importance in environmental measurements. Immediately after the Chernobyl accident, the gamma-spectrometric measurement of isotopes such as ^{137}Cs and ^{134}Cs was a major preoccupation for many laboratories. The sometimes heard opinion '*summing isn't a problem for us – we only work at low count rates*' is a dangerous simplification at best and probably utterly incorrect. ^{134}Cs has a relatively complex decay scheme and TCS is almost inevitable given that the environmental sample activities are low and must of necessity be measured close to the detector, often in Marinelli beakers. The extent to which TCS is ignored is best illustrated by Figure 8.8, which shows the results of an intercomparison arranged by the NPL in 1989 for the measurement of various radionuclides, including ^{134}Cs, at environmental levels. These broad features of the data were evident:

- Out of 58 ^{134}Cs results reported, only four were within the range expected by the NPL (i.e. within the shaded band in Figure 8.8.)

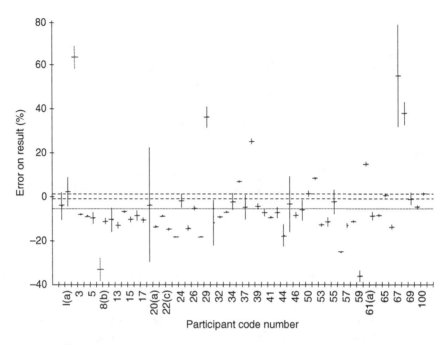

Figure 8.8   Summarized [134]Cs results from the 1989 NPL environmental radioactivity intercomparison exercise. (The band between the dashed lines represents the approximate 99 % confidence limit on the NPL source activity, the horizontal dashes mark the reported values and the vertical bars the reported 68 % confidence limits. Dashed vertical bars indicate that uncertainty was not reported)

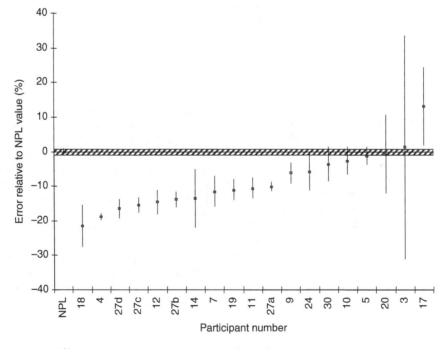

Figure 8.9   Summarized [134]Cs results from the 2002 NPL environmental radioactivity intercomparison exercise

- In only 11 cases did the true result lie within the 68% confidence limit reported by the measurement laboratory.
- It is obvious that the majority of the results were low; 64% of results reported were more than 5% below the expected value. (It is interesting to compare the magnitude of these errors with the calculated correction factors for ^{134}Cs given in Table 8.5 below.)

The most likely reason for this is that TCS had not been taken into account by the laboratories reporting these results. Bearing in mind that all the laboratories taking part in this intercomparison were reputable and believed that they were providing accurate results, Figure 8.8 was alarming. In fact, further NPL intercomparison exercises have been run regularly since then and, although some efforts do seem to be being made to correct for summing, the conclusion must be made that, in spite of the growing awareness of TCS, many laboratories still do not make appropriate corrections. Figure 8.9 shows similar data for measurements of ^{134}Cs taken from the 2002 NPL intercomparison. It is still the case that two thirds of the reported results are more than 5% low while 50% of the results can be judged as *significantly* low.

The message of this is obvious. True coincidence summing must be taken into account if accurate results are to be achieved. In defence of the gamma spectrometry community it should be said that until recently there were no easily usable tools for making TCS corrections and that in a busy laboratory it is not surprising that the time-consuming corrections were not made. A number of suggestions for making corrections and the modern software available are discussed below.

It is worth noting that, as important as it is in environmental measurement to achieve the lowest MDA, moving the sample a relatively small distance away from the detector will achieve a large decrease in TCS with only a small increase in MDA. The decrease in count rate due to larger source-to-detector distance is partly offset by fewer counts lost from peaks due to summing.

8.8 ACHIEVING VALID CLOSE GEOMETRY EFFICIENCY CALIBRATIONS

From what we have discussed it is evident that, leaving aside for the moment the possibility of making mathematical corrections, a valid calibration curve can only be measured under close geometry conditions by using radionuclides which do not suffer from TCS. Table 8.3

suggests a number of nuclides that might be used. The list includes several that emit only a single gamma-ray and therefore can be expected not to participate in summing. However, care should still be taken with those single gamma-ray nuclides that decay by electron capture (e.g. ^{51}Cr). A detector with a thin window, and especially n-type detectors, may allow significant summing with the X-rays due to the lower absorption in the entrance window.

Table 8.3 Radionuclides suitable for close geometry efficiency calibrations

Nuclide	Gamma-ray energy (keV)	Nuclide type[a]	Standard available[b]	Data in Appendix B
^{7}Be	477.60	S	Y	—
^{40}K	1460.82	S	(Y)	—
^{42}K	1524.67	M	Y	—
^{51}Cr	320.08	SX	Y	Y
^{54}Mn	834.84	SX	Y	Y
^{57}Co	122.06, 136.47	MX	Y	Y
^{64}Cu	1345.77	SX	Y	—
^{65}Zn	1115.54	S(X)	Y	Y
^{95}Zr	724.19, 756.73	M	—	—
^{95}Nb	765.80	S	Y	Y
^{103}Ru	497.08	S	—	—
^{109}Cd(^{109m}Ag)	88.03	S	Y	Y
^{113}Sn (^{113m}In)	391.70	M	Y	Y
^{131}I	364.49, 636.99	M	Y	—
^{137}Cs	661.66	S	Y	Y
^{139}Ce	165.86	SX	Y	Y
^{141}Ce	145.44	S	Y	—
^{144}Ce	133.52	M	Y	—
^{198}Au	411.80	M	Y	Y
^{203}Hg	279.20	S	Y	Y
^{210}Pb	46.54	S	—	—
^{241}Am	59.54	M	Y	Y

[a] S indicates a nuclide emitting a single gamma-ray; M indicates a nuclide for which the gamma-ray mentioned is the major one and has little coincidence summing; X indicates that summing with the accompanying X-rays (or other low energy gamma-rays may be a problem on thin-window or n-type detectors.
[b] Standards for these nuclides are available from radionuclide standard suppliers.

There are a number of nuclides which emit multiple gamma-rays but for which summing, of the specified gamma-ray, is usually negligible (e.g. ^{113}Sn and ^{131}I). Other gamma-rays are emitted with such a low abundance that summing, although possible, can be ignored. That might not be the case for a very large or a well detector.

Not all nuclides in the list are convenient to use in practice. The list includes some nuclides for which standardized sources may not be readily available. It also includes some sources with short half-lives that would not be appropriate for routine calibration. Nevertheless, with an appropriate selection of nuclides it is possible to create a valid close geometry efficiency calibration. There is, however, a scarcity of suitable nuclides providing gamma-rays above 1500 keV, thus preventing calibration at higher energy.

However, one must question what value such a calibration curve would have in practice. A large number of common nuclides have a complex decay scheme and are liable to exhibit the TCS problem. The carefully constructed TCS-free single gamma-ray efficiency curve will be irrelevant to the estimation of these nuclides. It will still be necessary to correct for the summing in the sample measurements. Overall, it is reasonable to suggest that the use of calibration curves for close geometry measurements is a waste of effort. The only value a summing-free close geometry calibration would have would be to help make the mathematical corrections described in Section 8.11.

In practice, we are usually constrained in the way that we handle our analysis by the software we use. It is pertinent to consider how the spectrum analysis program would treat calibration data such as that presented in Figure 8.1. Most likely it will blindly construct a 'best' (but completely invalid) fit, perhaps as shown in Figure 8.10. We know that all the points are likely to be lower than they should be and that the true calibration

curve should lie above them all. (This is not universally true. For example, in the case of ^{134}Cs, summing of the 569.33 keV (15.38 %) and 795.83 keV (85.5 %) gamma-rays increases the area of the 1365.19 keV peak. This is called a **crossover transition**. The phenomenon is referred to as **summing in**. In such a case, the true calibration curve could lie below the data point.)

8.9 TCS, GEOMETRY AND COMPOSITION

In Chapter 6, I discussed the effect of sample geometry on count rate. Once again, it would be useful to consider a practical example. Table 8.4 lists the peak areas measured when the same amount of ^{152}Eu was counted as a point source and when distributed in water and in sand. The distributed sources were 13 mm in diameter and 20 mm high and measured on the cap of a 45 % p-type HPGe detector. As one would expect, there is an obvious overall loss of count rate due to the lower effective solid angle of the distributed sources and a more pronounced loss of count rate in the low energy peaks.

Plotting the data graphically reveals more. Figure 8.11 shows the peaks areas for the distributed sources relative to the point source. The scatter of points, which cannot be attributed to peak area uncertainty, is at first unexpected but it is quite reproducible and is reminiscent of the scatter on the close geometry calibration. The explanation is straightforward. The ^{152}Eu in the point source is,

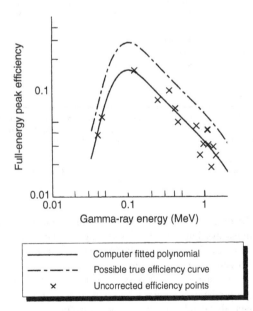

Figure 8.10 The efficiency curve constructed using a 'best fit' program

Table 8.4 Count rates of sources of different geometry[a]

| | Peak area (cps) | | |
Energy (keV)	Point	Aqueous	Sand
39.91[b]	386.0	243.1	174.5
45.75[b]	109.6	75.0	58.1
121.78	469.3	325.6	301.0
244.70	63.7	46.4	44.8
344.28	255.1	151.8	139.0
411.12	13.2	9.51	8.62
443.97	14.4	11.0	10.6
778.90	46.6	32.7	30.2
867.38	9.67	7.26	7.23
964.07	35.6	26.4	25.5
1085.84	24.8	16.9	16.3
1112.08	30.4	22.8	22.3
1408.01	38.0	27.4	27.2

[a] All sources have the same amount (to within 0.5 %) of ^{152}Eu. Aqueous and sand sources are 13 mm in diameter and 20 mm high. All were measured in contact with the detector end-cap.
[b] Apparent energy of unresolved X-ray doublet.

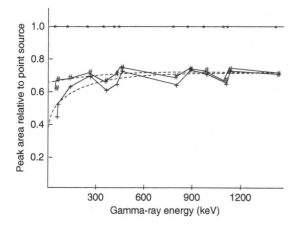

Figure 8.11 Relative peak areas for sources of different geometry and density: (∗) point source; (#) aqueous source; (+) sand source

on average, closer to the detector than in the distributed sources. The scatter represents the difference in TCS between the two source geometries. Moreover, the difference between the aqueous and sand matrix sources can, in addition to self-absorption, be attributed to the lower TCS in the sand source because of absorption of the Sm X-rays by the sand.

We can look further for evidence of the all-pervasive effects of TCS. Figure 7.9 (Chapter 6) relates the count rate of a source to its distance from the detector. The graphs confirm that the inverse square law can be applied to gamma spectrometry, at least in well-defined situations. A second look at the graphs reveals that the relationship becomes non-linear as the source nears the detector. The reason, of course, is the enhanced summing close to the detector.

The inescapable conclusion is that, unless sample and standard (or calibration) sources have identical shape and density, are in identical containers and are measured at the same distance, there will be differences in summing which will not be accounted for by the routine calibration process.

8.10 ACHIEVING 'SUMMING-FREE' MEASUREMENTS

It would seem that, if we must measure our samples close to the detector, our results would be in error due to TCS. It is important to remember that the degree of summing will be different for each nuclide and, if we are measuring more than one gamma-ray from a nuclide, different for each gamma-ray. Is there any way in which we can correct our results without resorting to the mathematical resolution described below in Section 8.11?

8.10.1 Using the 'interpolative fit' to correct for TCS

Both of the major gamma spectrum analysis programs allow a calibration option referred to as 'Interpolative Fit'. For gamma-rays not in the calibration set, this option interpolates between the adjacent calibration points to estimate an efficiency value. For most gamma-rays one is likely to measure, the value returned will be in error. However, for those nuclides in the calibration set, exactly the correct value, which takes into account summing, will be generated. There is scope there, using the interpolative function, for creating efficiency curves using all of the nuclides one is ever likely to measure which will have an automatic correction for summing built in – at the expense of a great deal of effort and a large number of certified reference sources.

As an example, take the measurement of the NORM nuclides (Chapter 16). Most of the nuclides to be measured, the ^{238}U and ^{232}Th decay series, have complicated decay schemes and suffer from TCS, seriously in some cases. If reference materials containing the relevant nuclides (IAEA RGU-1 and RGTh-1, come to mind) were to be used as calibration standards, then effective efficiency data could be acquired. The efficiency curve would not be a pretty sight, because of the TCS, but as long as the interpolative mode was used, the correct, TCS-accounted-for, efficiency data would be used when analysing the sample spectra. It would not be acceptable, though, to use that calibration for measurements of any nuclides other than those represented in the calibration data.

8.10.2 Comparative activity measurements

The intercomparison data in Figure 8.8 reveals a small number of points that lie within the uncertainty expected by the NPL. There is nothing remarkable about that. In at least two of these cases (one of which I have particular knowledge!), the reason for the close agreement is that the measurements were made *comparatively*. The sample was compared directly with a calibrated source of ^{134}Cs measured in the same geometry at the same distance from the detector. In that way, the TCS errors are the same for sample and standard and cancel out. There is no need for a calibration curve at all. This would seem to be the most direct way of avoiding calibration errors due to TCS and, unless there are specific reasons to do otherwise, I would recommend direct comparison with standards for measurements at close geometry. (This is, of course, in effect, what the interpolative fit referred to above is achieving.)

8.10.3 Using correction factors derived from efficiency calibration curves

Comparative analysis does pose logistic problems. Not least is the fact that the spectrum analysis program may insist that results are calculated by reference to a calibration curve. The only option then is to derive, by measuring reference sources, a set of correction factors for each gamma-ray to be measured that can be applied to the output from the program. (Indeed, in this situation, since the initial results are known to be wrong because of the inadequacy of the calibration curve, there is little point in measuring one at all. Any calibration data at all could be used as long as the correction factors were consistent with it.) Taking as an example the data in Figure 8.10, the procedure would be as follows:

- Construct a notional efficiency calibration.
- Measure calibrated sources of each of the nuclides to be determined, prepared in the standard geometry and of the same density as the samples to be measured.
- Use the computer program to calculate the activity of these sources based upon the notional calibration. The ratio between measured and actual activities is the correction factor to be applied to the sample measurements.

It is important to note that, if the final result is to be based upon more than one gamma-ray, it will be necessary to make a correction to the result from each individual gamma-ray before combining to achieve the final corrected result. If standardized reference sources are not available for a particular nuclide that must be measured, a direct measurement of the correction factor is not possible. In that case, it would be necessary to perform measurements at a large source-to-detector distance using an appropriate TCS-free calibration curve and compare these with the close geometry measurements.

8.10.4 Correction of results using 'bodged' nuclear data[1]

Having measured the correction factors as described above, it is then necessary to correct gamma-ray intensities for TCS, combining results for different gamma-rays

of the same nuclides, in order to achieve a satisfactory measurement. Although it is any easy task to set up in a spreadsheet, it is inconvenient for routine operation of a spectrometer system. It is much more acceptable if the system can be 'tweaked' so that the results from the program are correct. This can be achieved in a roundabout way.

Calculation of activity involves dividing the peak count rate by the gamma emission probability and by the efficiency. If the gamma emission probability in the nuclide library is adjusted by multiplying it by the TCS correction factor (i.e. the factor that the measured result would be multiplied by to achieve the correct result) then, because the efficiency is incorrect by that same amount, but in the opposite sense, then the calculated result must be correct. Note that this adjusted library must only be used for samples, not for the calibrations. There would have to be a separate adjusted library for each detector and each sample geometry. There is obvious scope for confusion. Nevertheless, this does appear to be the simplest means of coping with TCS corrections on a routine basis if activities must be calculated by reference to an efficiency curve.

8.11 MATHEMATICAL SUMMING CORRECTIONS

In principle, it is possible to correct for TCS errors mathematically. Take the simplest possible decay scheme in which we could expect TCS in Figure 8.12. The beta decay to one of two excited states is followed by the emission of the three gamma-rays shown. To simplify matters for the purposes of illustration, assume that the internal conversion coefficients for the gamma-rays are all zero. If the source activity were A Becquerels, in the absence of TCS, the count rate in the full-energy peak 1 would be:

$$n_1 = Ap_1\varepsilon_1 \tag{8.2}$$

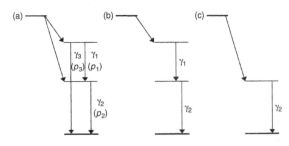

Figure 8.12   Simple illustrative decay schemes liable to true coincidence summing

---

[1] 'Bodge' is British slang for a clumsy, messy, inelegant or inadequate solution to a problem.

with p_1 and ε_1 being the gamma emission probability and the full-energy peak efficiency of detection of γ_1, respectively. Similar equations, with the appropriate p and ε, would be used to calculate the full-energy peak count rates for γ_2 and γ_3.

Now, we will lose counts from the γ_1 peak by summing with γ_2. We need not consider γ_3 as de-excitation of the upper level can only give γ_1 or γ_3 – not both – and we only need the partial decay scheme in Figure 8.12(b). The number of counts lost (per second) by summing can be calculated as the product of:

the number of atoms decaying (A)

$\times$ the probability of de-excitation producing γ_1 (p_1)
$\times$ the probability of γ_1 being detected and appearing in the full-energy peak (ε_1)
$\times$ the probability of γ_2 being detected and appearing anywhere in the spectrum (ε_{T2})

I noted elsewhere that we must account for all coincidences whether giving rise to a sum peak count or not and hence, the final term uses ε_{T2}, the total efficiency for the detection of γ_2. Therefore, the net peak area would be:

$$n_1' = Ap_1\varepsilon_1 - Ap_1\varepsilon_1\varepsilon_{T2} \qquad (8.3)$$

The ratio n_1/n_1' would then be used to correct for the TCS losses of the γ_1 peak area.

For γ_2, the situation is slightly different in that not all gamma-rays emanating from the intermediate energy level are a consequence of the de-excitation from the higher level. Some are preceded immediately by the β^- decay and cannot contribute to summing (see Figure 8.11(c)). The number of summing events is the product of:

the number of events giving rise to γ_2 (Ap_1)

$\times$ the probability of detection of γ_2 in the full-energy peak (ε_2)
$\times$ the probability of the detection of γ_1 anywhere in the spectrum (ε_{T1}).

and so the net area of peak 2 would be:

$$n_2' = Ap_2\varepsilon_2 - Ap_1\varepsilon_2\varepsilon_{T1} \qquad (8.4)$$

Every true summing event of completely absorbed gamma-rays will produce a count in a peak equivalent to the sum of the energies and so the peak corresponding to the crossover transition, γ_3, will be increased in area

rather than decreased. Following the same reasoning as above, the net count rate would be:

$$n_3' = Ap_3\varepsilon_3 + Ap_1\varepsilon_2\varepsilon_1 \qquad (8.5)$$

Often, it is the case that the crossover transition probability is small and, because the emission probabilities for the normal cascade transitions are high, the summing-in can be much greater that the direct emission. Unless taken into account, the error from using peak areas due to this transition would be large.

It must be emphasized that this is all a gross simplification. In general, we cannot depend upon the internal conversion coefficients being negligible and we have considered a much simpler decay scheme than we can normally expect. If we examine instead the ^{152}Eu decay scheme (which is more representative of the real situation), the enormity of the task of correction for TCS becomes apparent. Not only must we take into account every possible coincidence, every cascade on the ^{152}Sm side of the scheme is likely to be in coincidence with the Sm X-rays emitted.

There are further complications if the source emits positrons. The 511 keV annihilation quanta can appear in coincidence with the gamma-rays from de-excitation of the daughter nucleus. An analytical solution is to add a pseudo energy level to the decay scheme 511 keV above the level in which the positron emission leaves the daughter. It is then necessary to assign a conversion coefficient of -0.5 to the pseudo level to take into account both of the annihilation photons. We could also complicate matters even further by taking into account triple coincidences between the most intense gamma-ray and bremsstrahlung coincidences.

The mathematics become more complicated if we consider *real samples* that might have considerable size and may have a composition such that significant self-absorption within the sample occurs. We know that electron capture X-rays play a significant role in the summing process. Self-absorption will mean that those X-rays, and low energy gamma-rays, will be absorbed more than higher energy gamma-rays altering the correction factors. For accurate coincidence correction, either we need separate efficiency information for each sample geometry and composition or the mathematics must take into account self-absorption.

Of course, in order to perform these calculations at all we must have available a full-energy peak efficiency free of TCS errors *and* a total efficiency curve. The task is daunting; nevertheless, in principle, if we have available the detailed decay scheme, adequate full-energy peak efficiency data, adequate total efficiency data, complete

Table 8.5 Examples of calculated TCS correction factors for four detector/source arrangements (data abstracted from Debertin and Schötzig (1990))[a]

Energy (keV)	Correction factor (multiplier)			
	A2	B2	A4	C
^{152}Eu				
39.91	1.349	1.559	1.079	—
121.78	1.262	1.648	1.058	1.13
244.70	1.434	2.086	1.088	1.18
344.28	1.146	1.145	1.037	1.07
411.12	1.424	1.432	1.075	—
443.97	1.378	2.373	1.096	1.16
778.90	1.249	1.256	1.045	1.13
964.07	1.249	1.438	1.035	1.10
1085.84[b]	0.940	1.177	0.992	0.97
1112.08	1.182	1.709	1.035	1.07
1408.01	1.208	1.790	1.038	1.08
^{134}Cs				
604.72	1.252	1.249	1.063	1.13
795.83	1.265	1.258	1.055	1.13
1365.19[b]	0.839	0.761	0.975	0.85

[a] Detector – source geometry: A, 12.5 % Ge(Li); A2, point source on detector cap; A4, 1 L Marinelli beaker: B, 25 % n-type HPGe; B2, point source on detector cap: C, 30 % Ge(Li), 24 mm diameter, 24 mm high source.
[b] Crossover transitions.

conversion coefficient data and detailed knowledge of the shape and composition of the sample, then a mathematical correction is possible. Table 8.5 lists TCS correction factors for a number of major gamma-rays of ^{152}Eu and ^{134}Cs, calculated by Debertin and Schötzig (1990) using the principles described above. A number of points stand out:

- The magnitude of the factors is consistent with the scatter in the close geometry curve in Figure 8.1 and with many of the errors apparent in Figures 8.8 and 8.9.
- The larger ^{152}Eu factors for the n-type detector (B2) can be attributed to extra summing because of the lower absorption of the Sm X-rays in this type of detector.
- The enhancement of the crossover transitions (1085.84 keV in ^{152}Eu and 1365.19 keV in ^{134}Cs) is apparent by the fact that their correction factors are less than one.
- The larger factors for the point sources (A2, B2) compared to the distributed sources (A4, C) is as expected for sources that are effectively closer to the source.

The paper from which the data were taken and a similar paper by Sinkko and Abalone (1985) contain

useful compilations of correction factors which provide a general guide as to the likely importance of TCS for a large number of common nuclides. I must emphasize, however, that these correction factors should not be used to correct one's own data. They relate only to the particular detector and particular source geometries for which they were calculated. Correction factors for one's own data can only be determined by actual measurements or by reproducing the calculations described in the papers with the appropriate data for the detector system used. Unfortunately, while there are a number of well-known computer programs available for calculating TCS corrections (see the Reading List at the end of this chapter), these are not easily used in conjunction with the commonly used spectrum analysis programs.

One of the difficulties faced when collecting the information needed to perform reliable summing corrections is the reliability of the nuclear data in the literature. Of course, the quality of the data is worst for those nuclides for which summing is a particular problem. I can only remind the reader once more of the data given in Appendix B.

In Chapter 6, I mentioned the moves being made in the direction of providing detector calibrations based upon the dimensions and physical parameters of the detector and its encapsulation. Once this happy state of affairs has been attained, then the extension of that to include true coincidence corrections becomes possible. Again, one can envisage a situation where a new detector is delivered with a CD-ROM holding the theoretical calibration curve and the latest nuclear data for the nuclides of most common interest that would be accessed by spectrum-analysis programs for automatic summing corrections. For that, we must wait, but help does now appear to be on the horizon.

8.12 SOFTWARE FOR CORRECTION OF TCS

While the discussion above demonstrated the principle of calculating TCS corrections, in practice the situation is complex. It must be borne in mind that the corrections will depend upon sample distance, sample shape and, because of self-absorption, sample composition. Many attempts at making such corrections use Monte Carlo methods of calculation that consider the fate that many thousands of gamma-rays emitted from different parts of the source, impinging on different parts of the detector, might suffer, taking into account absorption within the source and in the detector. The program might construct an effective efficiency curve or may simply calculate correction factors to be applied to results from gamma-rays measured using the routine efficiency calibration. Both approaches have their advantages.

Monte Carlo methods demand a detailed knowledge of the detector geometry and construction. This information is not always available, that from the manufacturer being a nominal or estimated value. It is common, therefore, to read that Monte Carlo methods need 'fine tuning' with modification of parameters such as dead layer thickness and even detector diameter to make the model fit experimentally determined data. In some cases, people have resorted to X-raying their detector within its cooled encapsulation in order to measure the true detector size under operating conditions.

8.12.1 GESPECOR

GESPECOR (Germanium Spectrometry Correction Software) is one such Monte Carlo-based program which is commercially available. It was first created in 1996–1997 and has been under continuous development since, taking into account more and more minutiae of the detection process, including the partial dead layer phenomenon in p-type detectors referred to in Section 8.4. The program has been applied to all type of detector, including well types, and to all types of sample geometry – cylindrical, Marinelli and even 220 L waste drums. The program is undoubtedly worthwhile but it still not integrated into the spectrum acquisition process, meaning that there has to be some post-analysis correction to the result from each gamma-ray. In discussions with people who use GESPECOR and similar programs, they have warned that even minor inaccuracies in such factors as sample dimensions and composition can generate invalid results. Bearing in mind that in a routine gamma spectrometry laboratory, where often the composition of the sample is not well known and a separate calculation for each and every sample might be needed, this sensitivity is a drawback.

Normally one would not normally use sum peaks for efficiency calibration purposes. However, Arnold and Sima (20004) have shown how GESPECOR can be used to correct the sum peak intensities of ^{88}Y and ^{60}Co to extend the efficiency calibration energy range up to 2.7 MeV.

8.12.2 Calibrations using summing nuclides

I emphasized earlier that valid efficiency calibrations at close geometry can only be made directly using non-summing nuclides. For conventional calibrations, that is true, but Menno Blaauw (1993) has shown that it is possible to use the spectra of nuclides that are affected by summing to generate calibrations and TCS corrections.

The principle is as follows. Consider a decay scheme with, say, three excited states. There will be six possible gamma-rays emitted. For each of these gamma-rays, there will be an equation for the detection probability similar to Equations (8.2) to (8.5) above, albeit taking proper account of internal conversion coefficient, etc. Most of those equations will involve one full-energy peak efficiency and one total efficiency. Sum peak equations will have two full-energy peak efficiencies. We have six equations but much more than six unknown efficiencies. However, if we take into account the fact that on a log–log scale, the peak-to-total efficiency ratio is almost linear, it is possible to reduce the information needed to calculate total efficiencies to just two parameters. Even better, it is usually the case in complicated decay schemes that some of the crossover transitions (equivalent to the sum peak energies) have negligible emission probabilities. This leads to further simplification of the equations, such that we may have fewer unknowns than equations. It is possible to end up in a situation where not only the true full-energy peak efficiencies can be deduced, but also the parameters of the peak-to-total relationship and the activities of the source nuclides. We may not even need calibrated sources!

The solution of the equations is not trivial, needing an iterative fitting process. The method has been applied by Blauuw using ^{82}Br with some success and has been extended to deal with Marinelli beaker geometry and well-detectors. A development of the method introduces a third efficiency curve, called the 'linear-to-square' curve, in addition to full-energy peak and peak-to-total curves, which accounts for the variation of efficiency over the source volume due to self-absorption and scattering in the sample. This version is now incorporated into ORTEC's GammaVision, where it is assumed that the efficiency calibration source will include ^{134}Cs. It is, however, still necessary to create separate calibrations for each sample geometry and, presumably, sample composition.

8.12.3 TCS correction in spectrum analysis programs

When the first edition of this book was written, there were no facilities built into any of the readily available commercial spectrum analysis programs to make mathematical corrections for TCS. That situation has changed but it is not yet clear whether any great use is being made of these facilities or indeed whether they produce satisfactory results. While both GammaVision and Genie 2000 include facilities for making these corrections, the manuals provide little assistance to the inexperienced user on how to configure the programs.

Van Sluijs *et al.* (2000) evaluated the performance of three spectrum analysis programs in use in different European activation analysis laboratories from the point of view of TCS correction. None of the programs were easily available commercial programs and the conclusion was that inconsistencies between the programs needed investigation. Arnold *et al.* (2004) used the IAEA 2002 intercomparison spectra (see Chapter 15, Section 15.5.3) to evaluate seven commercially available spectrum analysis programs. Of these, only GammaVision and Genie 2000 could handle TCS corrections. They also used the external correction program, GESPECOR, to calculate correction factors. Where comparison between the three programs was possible, the correction factors were reasonably consistent. However, the results for GammaVision and Genie 2000 were not greatly better than for the other programs, especially for the spectra of uranium and thorium in equilibrium with their daughters. It would appear that there is no program available that can be trusted with TCS correction at the present time. It should be remembered that, depending on nuclide and source-to-detector geometry, TCS errors might be little more than 5%. If the correction program produces results with an extra 10% uncertainty, we would be little better off.

Even with a suitable program available, the TCS correction procedure will need some sort of extra calibration spectrum in order to derive a total efficiency versus energy relationship. The Genie 2000 system needs a number of nuclide sources, each emitting a single gamma-ray to cover the calibration range. The GammaVision calibration demands a single source containing a number of such nuclides and in addition, as mentioned above, a nuclide such as ^{134}Cs which suffers severe coincidence summing. From that single spectrum, GammaVision manages to derive all of the total efficiency data it needs. While it is easy, if expensive, to acquire appropriate single nuclide sources, there are few providers who can supply a traceable calibrated mixed source for the GammaVision calibration 'off-the-shelf'. However, recent work by Vidmar *et al.* (2005; 2006) may offer a solution. They have reported procedures for calculating both total efficiencies and the LS-curve for cylindrical samples, obviating the need for a practical measurement.

At the time of writing, facilities for TCS correction are beginning to become available but, as the results in Figures 8.8 and 8.9 suggest, are not being used routinely. Now that the process has started, perhaps we can hope that better, more user-friendly, programs, with adequate documentation, will be offered in the near future.

PRACTICAL POINTS

True coincidence summing:

- results in lower peak areas in general and possibly greater peak areas for crossover transitions;
- gets worse the closer the source is to the detector;
- gets worse the larger the detector and is worst of all when using a well-detector;
- may be worse if a detector with a thin window is used because more of the X-rays which might contribute to the summing will penetrate to the detector active volume;
- can be expected whenever nuclides with a complex decay scheme are measured;
- is not dependent upon count rate.

Accurate close geometry efficiency calibrations can only be achieved if nuclides are used which emit only one gamma-ray or if summing is accounted for. Whenever possible, close geometry calibration curves should be avoided by using direct comparison with an appropriate reference standard. If calibration curves cannot be avoided, correction factors must be derived for each gamma-ray of each nuclide used and the final results adjusted accordingly. It is important that that samples, standards and calibration sources have the same geometry and composition, or appropriate corrections are made.

In the last few years, a considerable amount of development effort has been put into programs that will allow TCS corrections to be made. In the near future, that effort may bear fruit.

FURTHER READING

- For a general understanding of summing errors, see:

Debertin, K. and Helmer, R.G. (1988). *Gamma- and X-ray Spectrometry with Semiconductor Detectors*, North Holland, Amsterdam, The Netherlands.

McFarland, R.C. (1993). Coincidence-summing considerations in the calibration of extended-range germanium detectors for filter-paper counting, *Radioact. Radiochem.*, **4**, 4–7.

- For a more detailed examination:

Sinkko, K. and Aaltonen, H. (1985). *Calculation of the true coincidence summing correction for different sample geometries in gamma-ray spectrometry*, Report STUK-B-VALO 40, Finnish Centre for Radiation and Nuclear Safety, Helsinki, Finland.

Debertin, K. and Schotig, U. (1990). *Bedeutung von Summationskrrektionen bei der Gammastrahlen-Spektrometne mit Germaniumdetektoren*, Report F Ra-24, Physikalisch-Technische Bundesanstalt, Braunschweig, Germany.

Debertin, K. and Schotzig, U (1979). Coincidence summing corrections in Ge(Li)-spectrometry at low source-to-detector distances, *Nucl. Instr. Meth. Phys. Res.*, **158**, 471–477.

- Calibrations using summing nuclides:

Blaauw, M. (1993). The use of sources emitting coincident γ-rays for determination of absolute efficiency curves of highly efficient Ge detectors, *Nucl. Instr. Meth. Phys. Res., A*, **332**, 493–500.

Blaauw, M. (1998). Calibration of the well-type germanium gamma-ray detector employing two gamma-ray spectra, *Nucl. Instr. Meth. Phys. Res., A*, **419**, 146–153.

Gelsema, S.J. (2001). *Advanced γ-ray spectrometry dealing with coincidences and attenuation effects*, phD Thesis, Interfaculty Reactor Institute, Delft University of Technology, The Netherlands.

Blaauw, M. and Gelsema, S.J. (2003). Cascade summing in gamma-ray spectrometry in Marinelli-beaker geometries: the third efficiency curve, *Nucl. Instr. Meth. Phys. Res., A*, **505**, 311–315.

- Evaluation of spectrum analysis programs:

van Sluijs, R., Bossus, D., Blaauw, M., Kennedy, G., De Wispelaere, A., van Lierde, S. and De Corte, F. (2000). Evaluation of three software programs for calculating true-coincidence summing correction factors, *J. Radioanal. Nucl. Chem.*, **244**, 675–680.

Arnold, D., Blaauw, M., Fazinic, S. and Kolotov, V.P. (2005). The 2002 IAEA intercomparison of software for low-level γ-ray spectrometry, *Nucl. Instr. Meth. Phys. Res., A*, **536**, 196–210.

- Assessments of some of the NPL intercomparisons are reported in:

Jerome, S. (1990). *Environmental radioactivity measurement intercomparison exercise*, Report RSA (EXT) 5, National Physical Laboratory, Teddington, UK.

Woods, D.H., Arinc, A., Dean, J.C.J., Pearce, A.K., Collins, S.M., Harms, A.V. and Stroak, A.J. (2003). *Environmental Radioactivity Comparison Exercise 2002*, NPL Report CAIR 1, National Physical Laboratory, HMSO, London, UK.

Arinc, A., Woods, D.H., Jerome, S.M., Collins, S.M., Pearce, A.K., Gilligan, C.R.D., Chari, K.V., Baker, M., Petrie, N.E., Stroak, A.J., Phillips, H.C. and Harms, A.V. (2004). *Environmental Radioactivity Comparison Exercise 2003*, NPL Report DQL-RN 001, National Physical Laboratory, HMSO, London, UK.

Some of these reports are available on the Internet via: http//publications.npl.co.uk/npl_web/search.htm

- Applications involving GESPECOR:

Sima, O. and Arnold, D. (2000). Accurate computation of coincidence summing corrections in low level gamma-ray spectrometry, *Appl. Radiat. Isotopes*, **53**, 51–56.

Arnold, D. and Sima, O. (2004). Application of GESPECOR software for the calculation of coincidence summing effects in special cases, *Appl. Radiat. Isotopes*, **60**, 167–172.

Arnold, D. and Sima, O. (2004). Extension of the efficiency calibration of germanium detectors using the GESPECOR software, *Appl. Radiat. Isotopes*, **61**, 117–121.

Sima, O., Cazan, I.L., Dinescu, L. and Arnold, D. (2004). Efficiency calibration of high volume samples using the GESPECOR software, *Appl. Radiat. Isotopes*, **61**, 123–127.

Information about the availability if GESPECOR can be found on the Internet at *http://www.matec-online.de*.

- On manual calculation of calibration curves:

Vidmar, T. and Likar, A. (2005). Calculation of total efficiencies of extended samples for HPGe detectors, *Nucl. Instr. Meth. Phys. Res., A*, **555**, 215–254.

Vidmar, T. and Korun, M. (2006). Calculation of LS-curves for coincidence summing corrections in gamma-ray spectrometry, *Nucl. Instr. Meth. Phys. Res., A*, **556**, 543–546.

- Test spectra that will allow TCS corrections to be tested:

Arnold, D., Blauuw, M., Fazinic, S. and Kolotov, V.P. (2005). The 2002 IAEA test spectra for low-level g-ray spectrometry software, *Nucl. Instr. Meth. Phys. Res., A*, **536**, 189–196.

9

Computer Analysis of Gamma-Ray Spectra

9.1 INTRODUCTION

I discussed the basic algorithms for spectrum analysis in Chapter 6 under 'Calibration' and in part in Chapter 5 under 'Statistics'. Although most of the useful information within a spectrum can be extracted manually, it is almost certain nowadays that a computer will be used to perform the spectrum analysis. In this chapter, I will examine the ways in which this is accomplished and discuss some of the pitfalls. To a large degree, computer programs follow (or should follow!) the principles I have already discussed.

At one time, the computer programs available could only be implemented on mainframe computers, but the increases in memory size and speed of the personal computer (PC) have put spectrum analysis on one's own desktop (as the advertising literature might have it) and nowadays there will be few people who do not use PC-based programs. Modern spectrum analysis programs are sophisticated packages that provide many facilities and I will refer to a number of commercial programs by way of example. These references are not intended to be endorsement of the programs, nor should any relative merit or demerit be implied. The programs will be referred to by name. Full details and attribution of these programs are given at the end of this chapter.

Before continuing, we perhaps ought to distinguish the different types of computer program available. The term 'software' will refer to the computer programs and 'hardware' to the computer on which they run. A hardwired MCA system is one in which the program is not loaded from an external source but is built into the wiring of the hardware. The distinction is somewhat blurred by the fact that some systems with a fixed program can be updated by installing a new version of the program, i.e. by loading new software:

- MCA emulators are concerned mainly with the acquisition of the spectrum data. They emulate the functions of the hardwired multichannel analyser. Examples are the *Canberra System 100* (no longer available) and ORTEC *Maestro-32*. These programs will provide for energy calibration but not necessarily width or efficiency calibration. Peak area measurement and peak search may be provided but this is unlikely to be as complex as a dedicated off-line analysis package. In our context, I will define 'online' as being work done interactively with data held and displayed in the MCA emulator. 'Off-line' would then be work done with spectra stored on computer disk.

- There are several 'off-line' programs which are intended to perform full calibration and analysis of spectra which have been acquired either by an MCA Emulator program or by a hardwired analyser and stored on disk. Examples of such programs are *Sampo 90*, *FitzPeaks* and *CompAct*.

- Since the first edition of this book, it is more likely that laboratory-based spectrum analysis will be achieved using one of the programs that combine the two functions, such as *GammaVision* and *Genie 2000*, and spectra may be analysed immediately after acquisition within the same program. Full energy, width and efficiency calibrations are provided in an interactive manner.

The programs within modern portable gamma spectrometers can, in some cases, rival that of the more sophisticated PC-based systems, especially where acquisition is controlled by a laptop PC. However, improvements have also been made to the PC-based software. In particular, and most laudable, is the introduction of true coincidence summing corrections into *GammaVision* and *Genie 2000*, even though, as I explained in Chapter 7, there is some room for improvement. It is disappointing that most of the changes to spectrum analysis programs have been cosmetic – more bells and whistles – rather than addressing known problems with the core algorithms that remain unchanged since the first edition of this book.

More often than not, spectrum analysis software is purchased with the hardware, if only to be certain of compatibility between them, and the market is dominated by software produced by the major gamma spectrometry equipment manufacturers. There are, however, a small number of software providers, independent of the manufacturers, providing reputable programs at a much lower cost; these are worthy of consideration. Such companies may be more amenable than the major players to rapid 'bug-fixes' when problems in their programs are identified and may be prepared to customize the program to suit a particular buyer.

It is evident that there is often an unwarranted faith in the power of computer programs which leads to a 'black-box' mentality. A spectrum is inserted into one end of the 'black-box' and out of the other end come results which are accepted uncritically as 'the truth'. Anyone who understands computers and has experienced the joys of programming is likely to have a much more jaundiced view of computer output. There are sound philosophical and logical arguments to suggest that it is impossible to prove without doubt that a computer program will work in every possible situation. Bearing that in mind, we would do well to keep an open mind about computer-generated spectrum analysis results. At the least, an understanding of the processes of calculation – the algorithms – used by the computer program to perform the analysis will help to alert the gamma spectrometrist to potential problems. Modern spectrum analysis packages will provide many analysis options and setting up the program may not be a trivial task. The user owes it to himself or herself to ensure that whenever an option is chosen it is a considered and informed choice rather than a blind reliance on conventional wisdom or the default option predetermined by the computer programmer.

In this chapter, I will consider in a general way the principles behind spectrum analysis. I shall leave aside the detailed implementation of those principles because that will vary from program to program. In Chapter 15, I will discuss the testing and validation of spectrum-analysis software. There are a large number of tasks that an ideal gamma spectrum analysis program might be asked to perform. As a minimum, the program should:

- Determine the position of peaks in the spectrum.
- Estimate the areas of the peaks in the spectrum, together with uncertainties.

- Calculate the energy of the gamma-ray each peak represents.
- Correct for counting losses due to dead time and random summing.
- Make corrections for decay from a reference time and, when necessary, decay during the count interval.
- Convert peak areas to activity (or concentration depending upon usage), either by reference to an efficiency function or by direct comparison with a reference spectrum.

It would be an advantage if the program could also:

- Create efficiency, peak width and/or efficiency calibration curves.
- Resolve multiplet peaks, either by peak stripping or by deconvolution.
- Make corrections for irradiation or sample collection time where appropriate.
- Estimate an upper limit activity when appropriate peaks are not detected.
- Identify nuclides in the spectrum.
- Make corrections for gamma-ray absorption within the sample and/or between source and detector.
- Make corrections for true coincidence losses.
- Make a full account of all sources of uncertainty within the measurement process.

Not all programs will perform all tasks and not all tasks will be relevant for all analyses but one might expect a typical commercial analysis program to be able to cope with the majority. In this chapter, I will discuss all of these items individually with the exception of true coincidence summing, which was discussed separately in Chapter 7. In general, a full computer spectrum analysis will consist of three phases:

- Set up data libraries for energy, peak width and efficiency calibration and for sample analysis. Different libraries might be needed for each phase of the analysis.
- Use spectra of reference sources to generate energy, width and efficiency calibration data files.
- Analyse sample spectra by referring to those data libraries and calibration files.

The processes involved are shown in flow chart format in Figures 9.4, 9.9, 9.12 and 9.13 and I will consider the various parts of these in due course but first we must consider how the program locates peaks in the gamma-ray spectrum.

9.2 METHODS OF LOCATING PEAKS
IN THE SPECTRUM

9.2.1 Using regions-of-interest

At the root of all spectrum analysis programs is the ability
to locate gamma-ray peaks. However, it is well to bear
in mind that our objective is to measure the number of
gamma-rays detected within a particular energy range.
That objective does not determine or demand any particular energy distribution of the gamma-rays. There is,
therefore, nothing reprehensible or shameful in ignoring
the sophisticated peak search facilities and simply telling
the program where the peaks are by defining regions-of-
interest (ROIs). This is more likely to be appropriate when
using an MCA emulator than a full spectrum analysis
program. Using ROIs, such programs can calculate, and
subsequently print out, peak areas corrected for the background continuum and a peak area uncertainty. The use of
ROIs is probably an under-utilized option in gamma-ray
spectrometry. Many laboratories will measure the same
few gamma-ray peaks day in and day out. If the spectrometer is free from gain drift, there is no harm and there
are potential benefits in using such a simple approach. It
is a pity that the MCA systems only allow ROIs to be set
up in terms of channel number, rather than energy. If the
latter were possible, a certain degree of gain shift would
be tolerable.

In most spectrum analysis, an active peak search will
be involved. Although the energy calibration procedure
for an MCA emulator usually requires the operator to set
up ROIs about the reference peaks, even then an active
search will be performed within these regions to determine
the exact peak position.

9.2.2 Locating peaks using channel differences

Peak search is not as straightforward as one might
imagine. Figure 9.1 demonstrates the essential problem.
Figure 9.1(a) shows a moderately well-defined peak. A
simple, intuitive peak search method might scan through
the spectrum seeking a number of channels each successively significantly greater, in the statistical sense, than its
earlier neighbour. Having established a consistent rise, a
corresponding series of channels with significantly falling
contents might be sought. The pattern of rise and fall
would then indicate the presence of a peak, and the onset
of the rise and cessation of fall could be used to determine
the peak limits. Such a naïve algorithm fails, however,
when presented with the data in Figure 9.1(b) where few
channels are statistically different from their neighbours
and there is certainly no consistent sequence of statistically significant differences. Yet, the human eye and brain

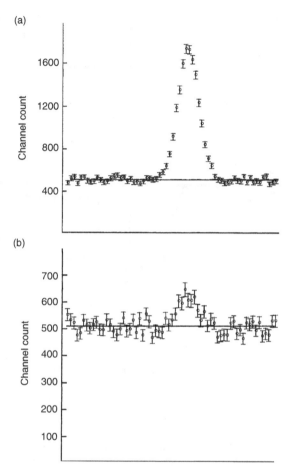

Figure 9.1 Well-defined (a) and poorly defined (b) peaks to
illustrate the limitations of a simple difference search. The circles
represent the measured channel content and the bars the 68 %
confidence intervals (*x*-axes represent channels)

detect the peak with ease. In effect, the brain suppresses
the statistical uncertainties (it smoothes the data) and
detects the underlying structure. That, in essence, is the
modus operandi of practical peak search programs.

9.2.3 Derivative peak searches

The most commonly used method for peak search is that
attributed to Mariscotti and, in general terms, might be
called a *derivative method*. Many years ago, this was
incorporated into a mainframe computer program called
SAMPO which has since formed the basis for many other
spectrum analysis programs, including those still in use
today. Figure 9.2 demonstrates the principle. Figure 9.2(a)
shows a basic Gaussian shape on a horizontal straight

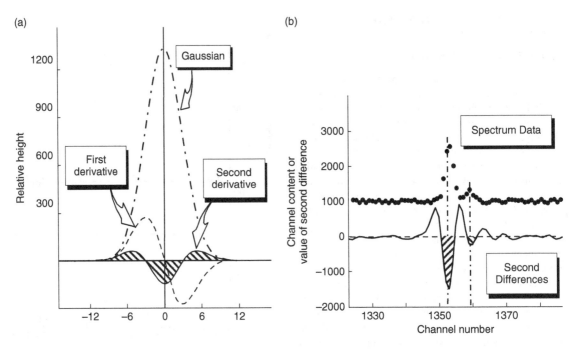

Figure 9.2 (a) The first and second derivation functions of a Gaussian peak. (b) Spectrum data and the corresponding smoothed second difference function (negative regions indicate the presence of peaks)

line (the model for our spectrum peak) with the first and second differentials of that Gaussian, also known as the first and second derivatives. Both of these curves have features which can in principle be used to detect the presence of peaks. For example, the first derivative changes sign as it crosses the peak centroid, the second derivative reaches a minimum at the centroid and so on. Third and even fourth derivatives have been discussed as potential search functions. Of course, gamma-ray peaks are not mathematical Gaussian curves; they are histograms that approximate a Gaussian curve. Because this is so, we cannot calculate a differential as such but must use the *differences* between channels as an approximation to the gradient.

The commercial programs *Genie-2000*, *GammaVision* and *Fitzpeaks* are all based upon the Mariscotti algorithms using the second difference method, as in the original mainframe program SAMPO. The actual implementation of the algorithm means that several channels are taken into account at any one time, smoothing out the statistical uncertainty in the data. Figure 9.2(b) shows such an algorithm applied to actual spectrum data. The two larger negative excursions of the second difference function (cross-hatching) indicate the position of the two gamma-ray peaks. In this particular example, the peaks are two FWHM apart. Clearly, as the separation between the peaks gets smaller, it becomes more difficult to resolve them. In general, two peaks of similar size are likely to be detected individually as long as the separation between channels is more than about one FWHM. Small peaks will tend to be lost in the presence of larger peaks at separations smaller than this.

9.2.4 Peak searches using correlation methods

Another less used, but useful, method of peak location is based upon cross-correlation. This is demonstrated in Figure 9.3. A search function, here shown as a Gaussian, is scanned across the spectrum. Over the width of the search function, each spectrum count is multiplied by the corresponding value of the search function. The sum of these products is then a point on the correlation spectrum. After applying an appropriate bias to take into account the underlying continuum, any channels in the correlation spectrum which are greater than zero represent channels within a peak. This author's experience is that correlation peak search works well and is easily set up to avoid spurious peak detection. A correlation search is used in the *CompAct* spectrum analysis program. As with the derivative search methods, because several channels are taken into account at one time, statistical scatter is effectively smoothed out without the need for a separate smoothing operation.

The correlation method is quite general and need not use a Gaussian search function, although that does have

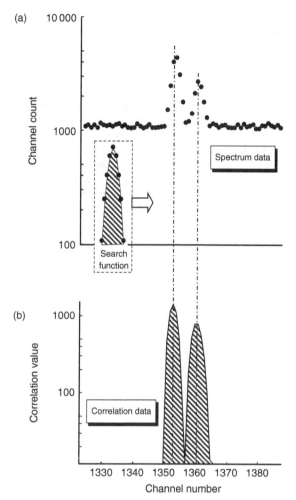

this. When applied to a background, the correlation spectrum must automatically be zero, obviating the need to make an allowance for background. The smoothing function used in the SAMPO-like programs is also a zero-area function and since the practical implementation of correlation methods and the Mariscotti second derivative method are very similar, there is a possible overlap of terminology.

9.2.5 Checking the acceptability of peaks

Whichever method is used to locate the peaks, the magnitude of the peak search function will be compared with some parameter related to a user set sensitivity value. Unless the function exceeds this value, a peak will not be detected. The sensitivity parameter might be related to peak area uncertainty – peaks with an uncertainty greater than the threshold being rejected, or some sort of empirical threshold factor might be used.

Whichever search method your chosen software uses, it is important that the peak location sensitivity criteria are set up properly; too high a sensitivity (which is very likely to mean a smaller sensitivity factor within the program) would mean that many spurious peaks are reported, while too low a sensitivity would mean small, but real, peaks are ignored. Having located a peak, the program may make some attempt to check that the peak shape is acceptable. Tests may be included to discriminate between true peaks and features such as Compton edges, backscatter peaks and the like. Sometimes, such peak shape tests are self-defeating when applied to ill-defined peaks where the estimation of FWHM is barely valid, resulting in rejection of real peaks.

Figure 9.3   (a) Spectrum data together with the Gaussian search function. (b) The resulting correlation spectrum (positive regions indicate the presence of peaks)

9.3 LIBRARY DIRECTED PEAK SEARCHES

There are two ways of approaching spectrum analysis. One might search the spectrum, take all of the peaks detected and measure their areas and then assign them to nuclides. Alternatively, one might specify which peaks of which nuclides to measure and then perform a limited peak search and measurement within the expected peak regions. This latter, library directed, method is reputed to be more sensitive than the former open search methods and there is some evidence to support that assertion. Certainly if a particular nuclide must be determined, even if only as an upper limit, then a library directed search must be used – otherwise nothing would be reported at all if the appropriate peak were not detected. Of course, unless a library directed search has associated with it a general search, the user will not be alerted if unexpected nuclides are detected. There is a need for both facilities within the same program and most software will provide them.

a certain logical appeal. The function could be modified to suit the data structure being sought. It is also possible to simplify the calculations by using a rectangular search function with integral values (ideally 0 and 1). The advantage of this is that the calculation of the correlation function reduces to integer summing rather than floating-point multiplication, giving a faster algorithm in the hands of a competent programmer. (This was applied to good effect in an MCA emulator program for the BBC Microcomputer once marketed by ORTEC which, in its day, out-performed PC alternatives.)

Oxford Instruments described the search algorithm used in *GammaTrac* (no longer available) as a 'correlation method using a zero-area correlation function' instead of a Gaussian search function. There are practical advantages to

9.4 ENERGY CALIBRATION

Energy calibration will often be performed before acquisition of the spectrum as part of the setting up procedure. It is included here because, in some cases, the MCA calibration may only be rudimentary (especially if a hardwired MCA is used) and a more precise calibration might be required before analysis of the spectra. Energy calibration involves the following steps (see Figure 9.4):

- Measure the spectrum of a radioactive source which emits gamma-rays at precisely known energies.
- Tell the system, whether hardwired or software, which peaks to measure.
- Supply the precise energies of the selected peaks.

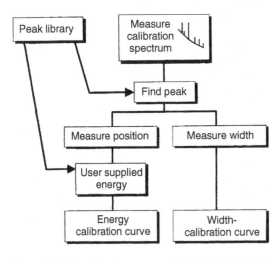

Figure 9.4 Flow chart for energy and peak width calibration

The system can then search for the required peaks, find their centroids and fit an appropriate function to the pairs of position/energy data. In most cases, the choice of 'appropriate' function will be determined by the system itself. A hardwired analyser is unlikely to allow more than a two point calibration. Even in an MCA emulator system such as *Maestro-II*, users are limited to a two point linear calibration. Using the companion full spectrum analysis program, *GammaVision*, this initial calibration can be replaced by a multi-point calibration fitted to a quadratic equation:

$$\text{Energy} = \text{Intercept} + \text{Factor1} \times \text{Channel}$$
$$+ \text{Factor2} \times \text{Channel}^2 \quad (9.1)$$

In some programs, a higher-order polynomial fit may be allowed and could provide a more precise fit. However, care should be taken. We have no theoretical reason to suppose that the energy/voltage relationship of any amplifier/ADC combination is quadratic. In fact, they are designed to be linear. The specification sheets for our amplifiers and ADCs tell us to expect serious non-linearity only at the extremities of the linear range. Without a theoretical basis, curve fitting is merely a mathematical game to achieve a more precise fit to the experimental data. It is not necessarily the case that a higher-order fit will provide a more satisfactory fit in a practical sense. The peak position data have uncertainties associated with them. Are they taken into account in the curve fitting? (Most likely not!) It may be the case that a higher-order fit is simply 'bending' its way around the uncertainties. A particular, and possibly non-typical, example is illustrated in the Figure 9.5(a). This shows the difference between actual and derived energies using a manual linear fit and a 13 point quadratic fit for the calibration spectrum associated with the Sanderson test spectra (see Chapter 15). It is quite clear that, although the errors are smaller at low energy, overall the quadratic fit gives larger errors than the linear. Evidently, a perfect fit to the data would require something more sophisticated than a quadratic.

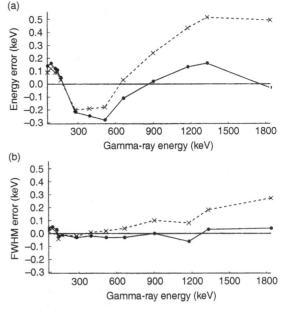

Figure 9.5 Comparison of linear and quadratic fits to energy calibration data (a) and peak width data (FWHM calibration) (b). In each case, the dots represent the linear fit and crosses the quadratic fit

A typical modern ADC might claim integral non-linearity of better than 0.05 % over the top 99 % of the channels. That implies a maximum error on an energy estimate of 1 keV, assuming an energy range of 0 to 2000 keV. Experience shows that the actual non-linearities are rather lower than this except, perhaps, at the extremities of the energy scale.

At the acquisition stage, there is probably little need to use anything more than a two point (and therefore) linear calibration. Whether subsequently there is need for a multi-point, and perhaps, non-linear calibration may depend to some extent on the use to which the information is to be put. At one extreme, if the MCA system is stable and always used to measure the same gamma-ray peaks using preset ROIs then a precise measurement of energy becomes irrelevant. At the other extreme, studies of the energy levels within nuclei would need very high-precision calibrations and it could be that a commercial program would not provide sufficient flexibility to do this directly.

Whichever particular mathematical function is chosen, the energy calibration data should span the intended measurement energy range. Extrapolation beyond the calibrated energy range should be done with caution. This is particularly important if polynomial fits are used. Higher order polynomials will certainly provide a better fit to all the nuances of the data but may deviate far from the 'true' curve beyond the first and last data points. Small changes in the data within the data can result in the ends of the curve 'waggling' in an unexpected way.

It is common practice for interactive spectrum analysis programs to provide a visual display of the energy-calibration data, together with the fitted line. This is always so nearly a straight line that it is impossible to distinguish between different fits visually. (If it were not so, one would probably have a serious instrumental problem.) It is usually necessary to examine carefully the printed-out results of the calibration. I look forward to the program that provides a graphical display of the differences between measured and fitted positions together with a goodness-of-fit factor to aid comparison between different types of it.

9.5 ESTIMATION OF THE PEAK CENTROID

It is unlikely that the true centroid of a peak will coincide with a channel number. If we are to estimate the gamma-ray energy represented by the peak, then we must have a means of determining the position of the peak centroid to within a fraction of a channel. An almost universal algorithm for this is to calculate the following:

$$\text{Centroid} = \sum C_i i / \sum C_i \qquad (9.2)$$

where C_i is the count in the ith channel, usually corrected for the underlying continuum background. After an active peak search, it is more likely that the value of the search function will be used for this calculation rather than the actual spectrum data. After a derivative peak search, the summation might be taken over the region where the derivative is negative and for a correlation method where the correlation function is positive. Alternatively, the summation width might be equivalent to a full peak width.

9.6 PEAK WIDTH CALIBRATION

Knowledge of peak width is fundamental for the measurement of peak area and for peak fitting. As we will see later, the computer program will, at some stage, need to know this. The gamma-ray peaks measured on a good modern detector free of neutron damage and other charge collection problems can usually be approximated by a Gaussian shape. In fact, as Figure 9.6 demonstrates, the actual shape is much more complicated. Even a cursory visual examination of a well-defined gamma-ray peak reveals that the continuum level on the low-energy side of the peak is higher than that on the high-energy side. In principle, this 'step', which is attributed to the loss of primary or secondary electrons from the sensitive volume of the detector, should be accounted for in some way. In

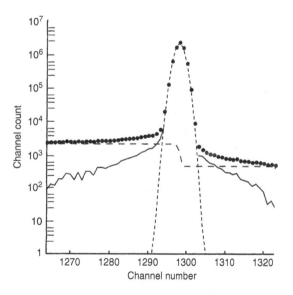

Figure 9.6 A gamma-ray peak resolved into a Gaussian (small dashes) with an underlying step function (large dashes) and low- and high-energy tails (full lines). The dots represent the channel's contents

the particular case in Figure 9.6, subtraction of the Gaussian peak and a step function from the channel counts reveals that there are also 'tails' on both sides of the peak. Again, a rigorous definition of the peak shape ought to include a description of these.

In practice, assuming that the detector is set up correctly and there are no gross problems which might affect the peak shape adversely, these underlying features are only significant when large well-defined peaks are analysed. In Figure 9.6, the departure from Gaussian only becomes significant below 1/400th of the full peak height. For the majority of peaks in routine spectra there is little to be lost by neglecting deviations from Gaussian. Nevertheless, some spectrum analysis programs will allow tails and/or step functions to be included in the peak description. For example, the current version of *Genie-2000* will allow low-energy tails to be included. *Fitzpeaks*, and possibly future versions of *Genie-2000*, will take into account low- and high-energy tails. *GammaVision* does not take into account tailing but will, under some circumstances, provide a stepped continuum.

The actual profile of the step beneath a peak does not appear to have been defined theoretically. Debertin and Helmer list eight functions which have been suggested in the literature. In practice, for peaks where the step is a noticeable feature, the measured peak area would be unlikely to alter significantly if the step function were changed as long as it remained symmetrical about the peak centroid. Sometimes, there are occasions where the peak background is obviously not linear. Typically, this would be the case for peaks on the low-energy side of a backscatter peaks. To cope with this, *GammaVision* will use a parabolic background function. *FitzPeaks* is able to use a number of different background functions based on polynomials, including the parabolic function.

In spite of all such qualifications, for many purposes the peak shape can be indicated to the program by supplying a single Gaussian width parameter – either the standard deviation or FWHM. The width of a peak increases with energy and it is necessary, therefore, to provide the parameters of a width equation similar to the energy calibration.

The arrangements for peak width calibration may not be as explicit as for energy calibration. For example, the ORTEC *Maestro-II* MCA emulator automatically performs a two point peak width calibration simultaneously with the energy calibration and stores both sets of information within the spectrum file. The equivalent program from Canberra, the *System 100*, makes no attempt at peak width calibration. *GammaVision* also does its multi-point peak width calibration at the same time as energy calibration.

Of the methods for FWHM estimation discussed in Chapter 7, the interpolation method (Section 7.5.2) is usually used. In some cases, (programs based on SAMPO for example) the peak width is deduced within the non-linear least squares peak fitting process which derives width, position and peak area simultaneously. It is well to remember that unless the peak is well defined any estimate of the peak width is likely to be of dubious value.

The analysis program must construct a mathematical relationship between FWHM and energy (or channel number). Unlike energy and efficiency calibration, it is possible to suggest a theoretical form for the FWHW/energy relationship as expressed by Equations (6.10) in Section 6.5. None of the available spectrum analysis programs allows a fit to the square root quadratic. Most programs assume a simple quadratic or other polynomial relationship. *Genie-2000* and *SAMPO 90* programs use Equation (9.3), perhaps believing that it is equivalent to Equation (6.11). It is not.

$$FWHM = F_1 + F_2 \times E_\gamma^{1/2} \qquad (9.3)$$

Figure 9.7 shows a typical FWHM calibration derived from a *Genie 2000* calibration fitted to Equation (9.3). (Spectrum *nbsstd.cnf*, distributed with the program.) Clearly, the fit is not satisfactory.

In fact, there is no theoretical justification for any of these alternative equations. The quadratic has a particular failing in that the statistical scatter of points can lead to an FWHM curve that curls upwards rather than downwards. This clearly flies in the face of physical reality. It would be useful if, whatever other FWHM fit options were

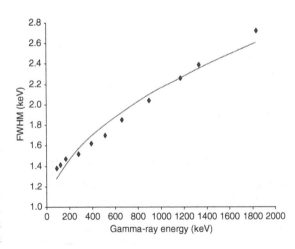

Figure 9.7   Peak width data as a function of gamma-ray energy, together with a calibration curve fitted by *Genie 2000*. The fit is clearly inadequate

provided, analysis programs always allowed a linear fit, which experience shows is a reasonable practical approximation to the theoretical square root quadratic equation.

Figure 9.8 show the data of Figure 9.7, together with the curves derived from fits to five mathematical equations: linear, quadratic, *Genie-2000* (Equation (9.3)), Debertin and Helmer and the square root quadratic equations. For this particular data, all the relationships but the Genie-2000 equation give similar root mean square differences between measured and fitted widths. Over most of the calibrated energy range, any of them would provide a reasonable estimate of FWHM. However, the figure does suggest that Equation (9.3) would seriously underestimate the FWHM at low energy and this is borne out by other similar calibrations. The reader should also remember the fact that X-ray peaks are unlikely to have the same width response to energy as gamma-rays (see Chapter 1, Section 1.7.4) and are likely to be wider than expected at low energy.

As with the energy calibration, a quadratic can produce a poorer fit than a linear relationship, as shown in Figure 9.5(b) above where the differences between measured and calculated FWHM using linear and quadratic fits are compared. It is up to the analyst, if there is a choice, to select the most realistic and sensible fit rather than the mathematical best fit.

In worrying about the 'true' mathematical width/energy relationship, perhaps we should consider the use to which the width calibration is put. Values interpolated from the curve are most likely to be used for defining the peak-integration limits, in which case there will be a rounding to the nearest channel, or, by comparing the value with its width, to establish whether a peak is a singlet or a multiplet. In either case, the errors caused by fitting the 'wrong' function (with the possible exception of the *Genie 2000* function at the ends of the calibration range) are not likely to be serious enough to affect the outcome of the analysis.

9.7 DETERMINATION OF THE PEAK LIMITS

Let us assume that we have established the position of a peak in the spectrum (or at least its expected position if it has not been detected). Before the peak area can be measured, making use of the algorithms discussed in Chapter 5, a decision must be made as to the limits of

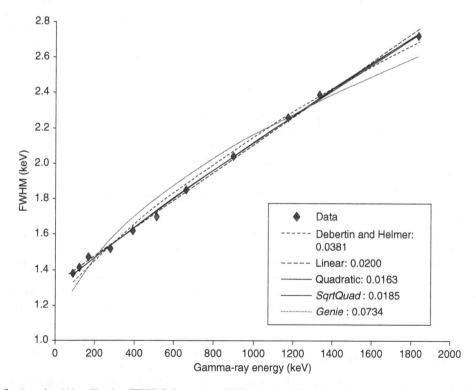

Figure 9.8 A peak width calibration (FWHM) for a p-type HPGe detector. The diamonds represent the measured points and the lines various computer fitted functions. The quadratic and square root quadratic fits are indistinguishable in this particular case

the peak region. Limits too far from the centroid will tend to include channels which are really background and will tend to render the area measurement vulnerable to the influence of near neighbour peaks. A peak region which is too narrow will result in an underestimate of the true peak area. (However, as suggested in Chapter 5, that would not necessarily be an overwhelming problem.)

A sensible strategy is to define what proportion of the total peak area we wish to measure and set the limits accordingly. Let us say that we would be content to measure 99.7 % of the peak area. If we assume that the peak is Gaussian, taking the data from Chapter 4, Table 4.3 we can see that that would imply a peak region of twice 3 standard deviations width (i.e. both tails of the distribution) equivalent to a total peak width of about 2.5 FWHM. We can then fix the peak integration limits at 1.25 FWHM below and 1.25 FWHM about the centroid.

Lars-Erik de Geer (De Geer 2005) argues that the minimum peak area uncertainty is achieved if the total integration width is restricted to 1.25 times the FWHM, which encompasses only 85.9 % of the total area. He has used that principle within the peak search algorithms used in CTBTO monitoring stations, where a high degree of confidence in the detection of small peaks is demanded (see Chapter 17, Section 17.1).

9.7.1 Using the width calibration

Since the program does have access to a peak width calibration function, the FWHM can be calculated for each individual peak by using its measured energy. This approach has the advantage of consistency; every time the same peak is measured the same peak limits will be selected. Peak area estimates measured by using limits selected in this way must be unbiased since no account is taken of the detail of the spectrum scatter. However, there are limitations. Unless special arrangements are made, incorrect peak limits will be selected for peaks which have anomalous widths, such as the annihilation peak at 511 keV and single escape peaks (see Chapter 7, Section 7.5.4). Since these are all wider than normal gamma-ray peaks at the same energy, their peak areas will be underestimated. However, such anomalous peaks would not normally be selected for nuclide activity measurement and so the problem is academic.

9.7.2 Individual peak width estimation

An alternative is to measure the FWHM of each peak as it is located and use that value to define the peak limits. That steps around the problem of anomalous peak widths

but there are disadvantages when the peak statistics are poor. Estimates of the FWHM of poorly defined peaks are inevitably uncertain and this means that the peak limits for the same peak in different spectra may appear to be different. Whatever the size of a peak or the particular statistical scatter of the points, the physical reality is that peaks at a particular energy measured on a particular detector system should have a constant width and position (leaving aside problems such as gain drift and count rate associated peak shape changes). It is possible that choosing peak limits in this manner could be biased. For example, if the FWHM algorithm were to consistently underestimate the width of peaks with large uncertainty then small peak areas would also tend to be underestimated.

9.7.3 Limits determined by a moving average minimum

Both *Genie 2000* and *GammaVision* use a method of peak-limit detection which uses a *moving* five point average to search for a minimum on each side of the peak. This is again subject to the same qualification noted above that the position of the peak limits might alter from spectrum to spectrum. More importantly, such a method must, in principle, be biased to some extent. If the position of the peak limits is always chosen to be at minimum points, then the estimate of the peak background underlying the peak must also be minimized, and the net peak area must therefore tend to be biased high. Although myself, and others, have confirmed that there is a positive bias in *GammaVision* results, it is only barely significant for very small peak areas, and not at all significant otherwise. Presumably the five point averaging almost removes the bias.

9.8 MEASUREMENTS OF PEAK AREA

Having detected the peak, determined its centroid (and therefore energy) and set peak limits, the peak-area measurement can proceed by peak integration (as described in Chapter 5, Section 5.4) as long as the peak is a singlet. Indeed, many programs do use the algorithms defined there. Exceptions are SAMPO-based programs (e.g. *Genie 2000* and *Fitzpeaks*) which perform a non-linear least squares fit on all of the peaks, including singlets. In such a fit, the peak area, width and position are determined simultaneously. It has been shown that, for a singlet peak, there is no advantage to be gained, in terms of measurement uncertainty, by fitting a peak instead of using peak integration. Peak-Fitting will be discussed later in Section 9.10.

Section 5.5.1 (Chapter 5) discussed in some detail the optimum number of channels to use for peak background estimation. In view of the significant influence of this, it is important that users of automatic spectrum analysis software be aware of how the program assigns the background channels. Many programs will simply use a fixed number of channels, say three, regardless of the peak situation. This is particularly so for MCA emulator programs where analysis facilities are more rudimentary. For example, in *Maestro-32* three channels are chosen and in the *Canberra System 100* four. This must be taken into account when setting up ROIs; the appropriate number of channels must be provided, *within the ROI*, to represent background. More sophisticated programs may make a more intelligent selection of background channels. *Genie-2000* will make sure that at least five channels are available beyond the peak limits to allow a valid background estimate. If there is a peak too close to the current peak, the pair will be treated as a doublet and deconvoluted together. *GammaVision* will look for five consistent background channels on each side of the peak. If necessary, the number will be reduced to three or even one channel if the neighbouring peak is too close. This could mean different background widths on each side of a peak. This is a sensible approach which mirrors that which one might take if calculating the peak area of a particularly awkward peak manually. There is after all no fundamental objection to using different background region widths below and above a peak, as long at the calculations take that into account.

9.9 FULL ENERGY PEAK EFFICIENCY CALIBRATION

The construction of a full energy peak efficiency curve was discussed in Chapter 7, Section 7.6 and is shown as a flowchart in Figure 9.9. I shall discuss here only the facilities available in spectrum analysis programs. The advice given earlier in that section about the quality of data is worth re-iterating. Only singlet peaks with good precision in conjunction with good quality nuclear data should be used to construct an efficiency calibration.

The option of efficiency calibration is usually not available on simple emulator programs and until recently was restricted to off-line analysis packages. More recent software, such as *Genie 2000*, *GammaVision* and *Fitzpeaks*, provide online efficiency calibration facilities within the overall acquisition/analysis package. Regardless of which of the many mathematical functions the analyst might choose to fit to the data, the spectrum analyst is constrained by the choice of functions provided within

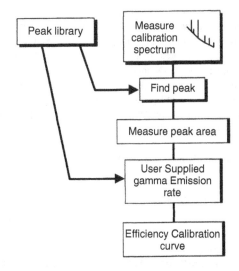

Figure 9.9 Flow chart for the measurement of an efficiency calibration point

his/her system. These might be one or more of the following:

- Polynomials in log (ε) against log (E_γ):

$$\log (\varepsilon) = a_0 + a_1 \log (E_\gamma) + a_2 [\log (E_\gamma)]^2$$
$$+ \cdots + a_n [\log (E_\gamma)]^n \qquad (9.4)$$

where E_γ is the gamma-ray energy, ε the full energy peak efficiency and a_0, a_1, etc. are coefficients determined by the fitting algorithm. The order, n, may be as high as 8. It is sometimes possible to take only the first two terms to provide a linear relationship, which satisfies the initial impression of efficiency calibration data that, above the knee, the data are approximately linear when plotted on log–log scales. Taking the first three terms of Equation (9.4) provides a quadratic – again a common option. Because of the two distinct regions of the efficiency curve, programs may allow the data above and below the knee to be fitted separately, maybe using different functions, in a way that ensures that the two curves meet reasonably convincingly.

Gunnink (1990) described polynomial equations of this type for calculating intrinsic efficiency, from which absolute efficiency can be calculated. By examining the efficiency calibrations of a large number of detectors, he was able to relate some of the parameters of these equations to the dimensions of the detector and other details of the detector system. Different equations were used from 50–90 keV, 90–200 keV (second order polynomial) and above 200 keV (sixth order polynomial). At first sight, this would seem to be a very reasonable way to approach efficiency calibration, but applying the idea in practice seems to be problematic and it does not seem to have been widely adopted. In particular, it would appear that attempting to apply the method to other than point sources is likely to obscure the relationship between parameters of the equations and physical properties of the detector. None of the easily available spectrum analysis programs allow a Gunnink calibration.

- Polynomials in log (ε) against E_γ:

$$\log (\varepsilon) = a_1 E_\gamma + a_2 + a_3 E_\gamma^{-1} + a_4 E_\gamma^{-2} \cdots + a_n E_\gamma^{-n}$$
(9.5)

up to, perhaps, terms in E_γ^{-4}. This function, the 'polynomial' in Figure 9.10 (see below), would usually be applied as a single equation covering the whole energy range. This is an option in *GammaVision*, where it is described as being optimized for p-type detectors but is not satisfactory for n-type detector efficiency curves, particularly below 60 keV.

- Polynomials in log (ε) against log ($1/E_\gamma$):

$$\log (\varepsilon) = a_0 + a_1 \log (c/E_\gamma) + a_2 [\log (c/E_\gamma)]^2$$
$$+ \cdots + a_n [c/\log (E_\gamma)]^n \qquad (9.6)$$

This function is available in *Genie 2000* and is referred to as the 'Empirical function' for, as the manual says (obscurely), 'historical reasons'.

- The inverse exponential, quoted as being particularly suitable for HPGe calibrations, was provided in the spectrum analysis program *GammaTrac*:

$$\varepsilon = 1/(a E_\gamma^{-x} + b E_\gamma^{y}) \qquad (9.7)$$

- A number of other functions are suggested in Debertin and Helmer (1988) and examined in a paper by Kis *et al.* (1998).
- An interpolative efficiency curve is an option within *GammaVision* which is intended for situations where the actual efficiency curve is known to be complicated and unlikely to conform to the other options available. In effect, the pairs of energy/efficiency data are simply stored as provided. Efficiencies at points between the calibration energies are then estimated by interpolation between an appropriate pair of data points. In general, this is of dubious value – if the data is too

scattered to provide a proper efficiency curve, do interpolated values have any meaning? However, I will discuss in Section 9.12.4 how such an option might be used in situations where true coincidence summing is a problem.

It is important to recognize that none of these equations has any theoretical basis. They are all simply empirical relationships – mathematical games – which may fit the experimental data to a greater or lesser degree. The actual choice of function from the limited options provided within a particular program can only be made on the basis of experience. A function that performs well on a p-type detector may not be satisfactory for an n-type where the low-energy behaviour of the efficiency curve differs. Where a choice in the order of the fitting polynomial may be exercised, care should be taken. A higher-order polynomial will certainly be able to fit all of the slight variations in the data better than a lower one and will give a better 'goodness-of-fit' factor (assuming this is displayed by the program). However, if these extra 'wiggles' in the data are simply due to statistical uncertainty, then the higher-order fit may be further from the 'true' curve (whatever that may be) than a lower-order fit. Beware of removing points from the data set on the basis that they 'don't fit' without finding out why. Certainly, if true coincidence summing is a problem, it is conceivable that a single odd point could be the only correct point.

It is desirable that the fitting process should take into account the uncertainty of each point so that less reliable points have less influence on the fit. This would be described as a weighted fit. Ideally, the weighting factors would include not only the counting uncertainty but the calibration source strength uncertainty (which should, in turn, include the uncertainty on the decay correction) and the uncertainty on gamma emission probability. At the present time, not every program available weights the fit and no program takes into account all sources of uncertainty.

Figure 9.10 shows an efficiency calibration using a source of [152]Eu (under conditions where summing was not significant) to which three particular functions have been fitted; double log–linear, double log–quadratic and a single linear polynomial with six terms. Within the data points above the knee of the curve, there is little to choose between the three relationships. The goodness-of-fit improves as the complexity increases but as we have already discussed above unless the fit is weighted to take into account the statistical uncertainty of the data points such improvement may be more apparent than real.

The data below the knee point is clearly unsatisfactory and is shown here merely as an example of how a

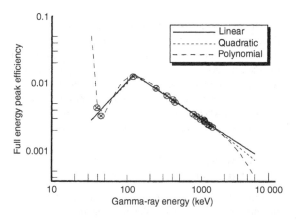

Figure 9.10 Efficiency calibration data for a p-type HPGe detector. The circles represent the experimental points and the lines various computer fitted functions. ('Linear' and 'quadratic' fits refer to Equation (9.4) with two and three terms, respectively. 'Polynomial' refers to the *GammaVision* fit using Equation (9.5))

complex fit can be led astray by inconsistencies within the data. The two low-energy points appear to be 'the wrong way round', which probably indicates a problem somewhere in the peak-area measurement, and this causes the polynomial curve to behave in an unreal manner at low energy. Had this particular data been intended to be used below the knee point, more low-energy data points would have been needed to define the curve below the knee. In this case, since the curve was intended only to be used above the knee, perhaps it would have been better to omit the low-energy points altogether to avoid any distortion of the high-energy part of the curve.

The problems of efficiency curve fitting are greater the longer the energy range. The energy range necessary for prompt gamma-ray measurements must extend to 10–12 MeV. Beyond 3 MeV, the log–log HPGe efficiency plot curves downwards after being approximately linear below that energy down to around 200 keV. Kis *et al.* (1998) compared a number of functions with a straightforward polynomial of log (ε) against log (E_γ) as implemented in the *Hypermet-PC* program. They found all functions, other than the *Hypermet-PC*'s 9th-order polynomial, to be wanting. In many cases, the alternative functions became unstable as the number of terms increased. This polynomial provided an acceptable fit up to 10 MeV.

9.10 MULTIPLET PEAK RESOLUTION BY DECONVOLUTION

Deconvolution is the term given to the process of extracting peak area information from a composite (multiplet) peak. The results of deconvolution should be treated with caution. Most other spectrum analysis calculations can be checked manually. That is not the case with deconvolution and the analyst is in the hands of the computer programmer.

In principle, each peak within a multiplet is described by three terms:

- its area, which, of course, we wish to determine;
- its position;
- its shape.

If we know any two of these factors the third can be easily estimated – give or take a little matrix arithmetic. (Some programs work with the height of a peak rather than its area. The two are linked, of course, but since height depends upon both area and width, using the area alone would seem to be more appropriate.) The shape of the peak we know (Section 9.6). The program may simplify matters by using a Gaussian approximation utilizing the peak width calibration to estimate the standard deviation.

The position can be determined either from information derived from the peak location or by taking account of library data which indicate which peaks to expect within the particular multiplet. Both have their advantages and disadvantages. Obviously, unless a gamma-ray is in a library then it will not be taken account of and so a simple library directed approach cannot cope with the unexpected. On the other hand, small peaks within a multiplet and very close multiples may not be resolved by the peak search and incorrect peak areas may again result.

Having reasonable estimates of the shape and position, we can say that for each channel over the region of the spectrum covered by the multiplet that the count, C_i, in each channel i can be represented by:

$$C_i = \sum a_j A_j g(i, s_j) + R_i \qquad (9.8)$$

where the summation is taken over the j components of the multiplet, a_j is the fitting factor (i.e. the proportion of the nuclide in the multiplet. This, in due course, would give us our nuclide activity), A_j is the peak area of the jth component and $g(i, s_j)$ the mathematical function describing the peak, with s_j being the peak standard deviation. R_i represents an unknown adjustment to the channel count due to statistical factors. Equation (9.8) represents a set of simultaneous equations which can be solved by simple matrix operations in such a way that the sum of the squares of the differences between each actual channel count and an estimate of the channel count from the fit is a minimum. (Which is why it is referred to as a least-squares fit. For a simple implementation, see Gilmore, 1979.)

As it stands, the fit would take equal account of every channel within the composite peak. Clearly, from what we know of counting statistics, channels with fewer counts have a greater uncertainty and should not be allowed an equal say in the fitting process. To take this into account, the equations generated by Equation (9.8) are weighted by the inverse of the variance of the corresponding channel count, C_i. Each weighting factor, w_i, is:

$$w_i = 1/\mathrm{var}(C_i) = 1/C_i \qquad (9.9)$$

Non-linear fits treat the data in the same sort of way but do not assume that the position and peak width are constant but deduce them together with the peak area (or height) in an iterative process.

If a library directed fit is chosen, it is essential that the library is tailored specifically to the job in hand. Libraries for this purpose should not contain 'just-in-case' entries. While the least squares fitting program would normally be written so as to reject any non-significant or negative components, there is often sufficient slack within the statistical scatter to fit another component. If your library tells the software to expect four components when there are really only three, you should not be surprised to find results for four components as the rule rather than the exception. This tendency to self-fulfilling prophecy is exacerbated when statistical scatter is large or when small uncorrected peak shifts are encountered.

While there is good reason to use the library directed approach, unless it is known in advance which interfering gamma-rays will be present, the open peak fitting approach must be used. Whichever approach is used, the results of deconvolution should be treated with caution. Published reports by the software vendors (see Reading List) on the analysis of artificially generated doublet peaks demonstrate that, unless the peak separation is greater than 1 FWHM, substantial errors in the deconvoluted peak areas are likely. This is especially so when the peak area ratio is high. Small peaks may be consistently under-represented in the peak fit. (Although it may be claimed that the error is within the statistical uncertainty of the individual peak area estimations, there would still be a bias on the final results. Whether this is significant or not depends upon the context of the analysis.) The moral is clear – view with some suspicion the results reported for the minor components of multiplets especially if the separation from their neighbours is less than 1 FWHM or so.

Blaauw *et al.* (1999), using the 1994 IAEA reference spectra, compared the performance of three programs that used different methods of peak area determination. The program *Apollo* measures peak areas by a simple peak-integration method, *Hypermet-PC* uses iterative fitting to derive peak areas and *GammaVision* uses a library-directed approach. The authors commented on the sensitivity of *GammaVision* with respect to the quality of the data in the nuclide library. The *Apollo* program was reported to provide the best results for doublets, but at the expense of manual intervention. *Hypermet-PC* came into its own when good library data was not available and when the highest deconvolution power was needed.

9.11 PEAK STRIPPING AS A MEANS OF AVOIDING DECONVOLUTION

Bearing in mind the uncertainties in deconvolution and the difficulty of checking the performance, it is worth considering whether it can be avoided altogether. In fact, for many routine measurements gamma-ray peaks can be selected that will never be interfered with by other gamma-rays under normal conditions. It is worth looking carefully at one's analysis libraries with a view to removing peaks which would need deconvolution. It may be better to use a gamma-ray with a lower emission probability which is unhindered than one of higher emission probability which is. What you lose in terms of counts, you may gain in precision and accuracy of measurement by not having to deconvolute.

If one component of a doublet must be measured it is worth considering whether a simple peak strip could be used. The procedure is demonstrated in Figure 9.11. The peak to be measured is labelled A and the interfering peak B_1. Nuclide B is known to have one or more

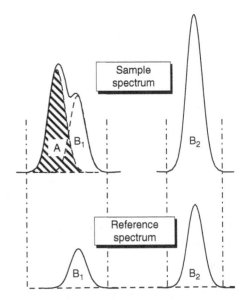

Figure 9.11   Doublet peak resolution by peak stripping

other gamma-rays which can be measured (e.g. B_2 in the figure). If a source of pure nuclide B is measured, then analysis of its spectrum will allow a peak area ratio, B_1/B_2, to be calculated. (Note that this measurement must be under identical conditions of geometry to the sample measurement or otherwise differences in true coincidence summing might alter the peak ratio.) It is, of course, possible to calculate the peak ratio by using gamma-emission probabilities and the detector efficiencies for the two gamma-rays. (Which is how the peak stripping operation within *GammaVision* works?) However, that ratio will be in error if true coincidence summing is involved.

In subsequent sample analyses, the area of peak B_2 can be multiplied by this empirical correction factor and subtracted from the total peak area of the doublet $A + B_1$. While this all seems very straightforward, it is not always as useful as expected. An oft-quoted example of the use of peak stripping is the resolution of the 186 keV peak in the spectra of naturally occurring radionuclides. This is a composite of the 185.72 keV peak from ^{235}U and the 186.21 keV from ^{226}Ra. ^{235}U also emits a 143.76 keV gamma-ray and, in principle, this can be used to calculate the proportion of the 186 keV peak due to ^{235}U. However, the 143.76 keV gamma-ray has a much smaller emission probability than the 185.72 (see Chapter 16, Table 16.1) and there will be a much greater uncertainty on its measurement, which will have a large effect on the uncertainty of the estimation of ^{226}Ra. The method is of limited use unless the correction peak (143.76 keV, in this case) can be measured with good statistics.

Although the procedure is simple and direct it will not necessarily be available within a commercial spectrum-analysis program. Programs which do are *GammaVision* and the comparative analysis program *CompAct*. In the absence of facilities within the spectrum analysis program, a simple spreadsheet could be used to perform an off-line peak strip on the output from the analysis program.

We should perhaps make clear that peak strip is not the same as the spectrum strip option often provided on old MCA systems. That would subtract one spectrum, or a proportion of it, from another on a channel-by-channel basis. Spectrum stripping permanently alters the spectrum data and is not recommended. One particular problem with it is that the statistical scatter of the stripped spectrum is not representative of the actual data.

9.12 THE ANALYSIS OF THE SAMPLE SPECTRUM

Having constructed energy, peak width and efficiency calibration curves, the way is clear to performing a full

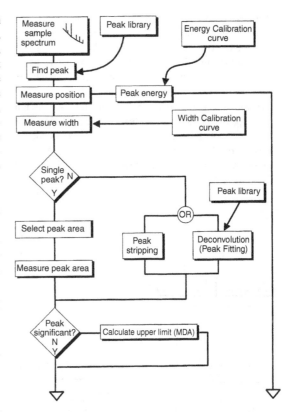

Figure 9.12 Flow chart for sample spectrum peak search and measurement

spectrum analysis to detect and determine the nuclides represented in the spectrum. The overall procedure for a full spectrum analysis might follow the flow chart shown in Figure 9.12, with a continuation of this figure presented in Figure 9.13. Not every part of this will be followed by any one program and not every part of it would be necessary for all situations. During calibration, peak detection and measurement is straightforward because (ideally) only singlet well-defined peaks are involved. In handling sample spectra, the algorithms are tested to their limits. They must be able to cope with poorly defined peaks which may be subject to interference from other peaks.

There are two general ways in which we may approach sample spectrum analysis:

- Select a limited number of gamma-rays for each nuclide of interest and search the spectrum only for these peaks. This would be done via a nuclide library that might also contain information to allow deconvolution if necessary. This approach has the advantage of speed and simplicity but would not alert one to the presence of unexpected nuclides. It would be more usual to combine this with a general spectrum search which would provide additional information from which the presence of other nuclides could be inferred.

- Perform a general spectrum search and then assign every peak (as far as possible) to nuclides using data held in a general library. As we will see, nuclide identification can be a complicated matter and in many cases an overall search and identification such as this would seem to be 'over the top'.

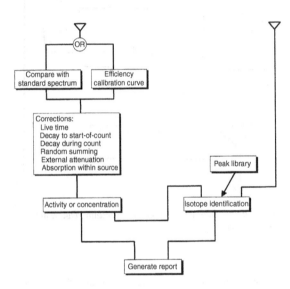

Figure 9.13 Flow chart for the conversion of peak areas into nuclide activities (continuation of Figure 9.12)

Whichever approach is used, there are common elements in the analysis which are worth examining in detail.

9.12.1 Peak location and measurement

Having located a peak, using the methods described above in Section 9.2, the program must determine whether the peak is a singlet or multiplet. This might be done on the basis of peak width – a peak wider than expected might be a multiplet – or on the basis of proximity of neighbouring peaks. For example, if *Genie 2000* detects another peak centroid within five FWHM of the peak, a multiplet is assumed and eventually deconvoluted.

9.12.2 Corrections to the peak area for peaked background

If a peak overlies a peak in natural background (for example, the measurement of ^{137}Cs or ^{60}Co at low levels) a peaked background correction (see Chapter 5, Section 5.4.2) would be necessary and may as well be done at this stage. The uncertainty on the peak area should also be adjusted to take into account uncertainty of the background correction. At the present time, not all spectrum analysis programs incorporate peaked background correction, and even those that do may not make proper allowance for the uncertainty on the correction. For many analysts that would seldom be a problem. For near background counting and measurements on naturally occurring nuclides, it is essential if false positive peak identifications are to be avoided.

9.12.3 Upper limits and minimum detectable activity

In principle, the net peak area should be assessed for significance using a critical limit, as discussed in Chapter 5, Section 5.6.1. If the peak area exceeds the critical limit, the peak area can be legitimately passed on to the activity calculation. If not, the peak should be declared 'NOT DETECTED' and an upper limit to the peak area calculated (Chapter 5, Section 5.6.2). That upper limit should then be passed through the calculation as any other peak area and reported as an activity upper limit – that activity which we are 95 % certain exceeds the actual activity.

In practice, all of the analysis programs carry through the net peak area without any significance test, together with a possibly outrageously large uncertainty. This is then tested later and may then be used to calculate the MDA. (*Genie 2000* does allow a critical limit test to be applied. Unfortunately, if a peak fails the test it is then completely ignored! This seems somewhat drastic. It is possible for the peak search to detect a genuine peak whose area turns out to be insignificant. In that case, what is needed is an upper limit.) In Chapter 5, Section 5.6.7, I explained that MDA is not the Minimum Activity Detectable and should *not* be quoted in place of the activity upper limit.

One aspect of computer calculation of MDA is worthy of comment. Let us assume that a peak area is measured by the normal peak integration method (Chapter 5, Section 5.4), the peak background being estimated using a few channels beyond the peak limits on either side of the peak. It is a feature of some programs that if the net peak area fails the critical limit (or similar) test, the upper limit (or MDA) is then estimated from the channel counts *within* the peak region, using them as an estimate of the background. This is clearly perverse. Figure 5.11 demonstrated that if there were counts in the peak region equivalent to the MDA, a peak would be visible and detectable (95 % of the time). In that case, we would have to use channel contents on either side of the peak to estimate the peak area. If we cannot use the channels within the peak region when we measure the peak background, how do we justify using them to estimate the MDA? MDA should be estimated from the uncertainty on the background as measured.

As I indicated in Chapter 5 and suggest above, there is a great deal of confusion over statistical decision levels and users of commercial programs should be aware of this. Bowing to this confusion, *GammaVision* provides twelve choices for MDA calculation so as to satisfy as many users as possible. Unfortunately, the equations I advocate in Chapter 5, Section 5.6 are not included in these choices. (The *GammaVision* MDA option 'Nureg 4.16 Method' may be acceptable. The manual quotes a valid equation for this option, but it unclear how the uncertainty on the background is calculated.)

9.12.4 Comparative activity estimations

It is unfortunate that spectrum analysis programs do not recognize that not every spectrometrist will want to analyse his or her spectra in the same way. All of the more sophisticated spectrum analysis programs assume that the analysis will be made by reference to an efficiency curve. Only one of the programs I have referred to, *CompAct*, makes specific provision for direct comparative analysis (as opposed to an analysis where both standards and samples are measured relative to an efficiency curve and the results compared afterwards to deduce the correct activity or concentration.)

In particular, the neutron activation analyst, to whom efficiency curves may be an irrelevance, is not well served by most spectrum analysis packages. As far as activation analysis is concerned, there is much evidence to show that absolute analysis, calculating concentrations from first principles, is much less accurate than comparative analysis. Apart from all of the problems which derive from having to use efficiency calibration curves, there are specific problems associated with defining and measuring neutron fluxes and cross-sections which make absolute analysis not worthwhile, in my opinion (although there are those who have devoted a considerable amount of effort into developing absolute neutron activation analysis procedures who would dispute that). For that reason, almost every activation analysis involves irradiation of samples and standards. A direct comparison between them is the simplest solution.

In Chapter 7, where the problems caused by true coincidence summing in close geometry measurements were discussed, we saw how TCS can make nonsense of an efficiency curve and strongly recommended comparative analysis, particularly for environmental measurements.

In a comparative analysis, the sample peak count rates would simply be compared with those of a standard:

$$C_{\text{sample}} = R_{\text{sample}} \times C_{\text{standard}} / R_{\text{standard}} \qquad (9.10)$$

where the Cs may represent activity or concentration and the Rs the appropriate peak count rates, ignoring for the time being decay and other corrections to be discussed below. The uncertainty of the final result would need to take into account the uncertainties of the individual items in Equation (9.10).

GammaVision provides a limited means for performing a comparative analysis in that there is an option to provide an 'interpolative' efficiency curve. If the efficiency calibration data is provided for each gamma-ray of each nuclide to be measured, then each request for an efficiency value would return the actual calibration data derived from the standard spectrum. If true coincidence summing were a problem, then as long as the standard spectra were of the same nuclides measured under the same conditions as the sample spectra, the summing errors would cancel out. The procedure is not elegant, but as far as I can see, should work satisfactorily.

9.12.5 Activity estimations using efficiency curves

Apart from *CompAct*, all of the programs referred to use the inverse of Equation (6.13) to convert peak count rate, R, to source strength, S, i.e. sample activity, calculating the efficiency, ε, from the calibrated efficiency function. Ideally, the program would also fold into the uncertainty of the peak area measurement the additional uncertainties due to interpolation of the calibration curve.

$$S = R/(\varepsilon \times P_{\gamma}) \qquad (9.11)$$

9.12.6 Corrections independent of the spectrometer

The activity estimate at this point may need correction for a number of counting losses which were covered in Chapter 6. Only the following are routinely catered for in spectrum analysis programs (numbers in parenthesis refer to the equation used to make the correction):

- decay from (or to) a reference time (7.25);
- decay during counting (7.26);

- self-absorption within the sample (7.17);
- random summing (7.24).

These corrections can usually be selected or not as required for a particular analysis. The correction for decay during counting may have a subtle problem associated with it. Chapter 6, Equation (6.30) involves the ratio of two factors that, if the count period (that is the real count time) is very short compared to the half-life, may both be very small and almost equal. If these factors are too small, round-off errors associated with the number of binary bits used to represent floating point numbers within the computer can result in a correction that is very large instead of negligibly small. This should be taken care of within the program but it makes sense not to enable the decay-during-counting correction if the count period is less than one thousandth of a half-life. At that point, the correction would be only 0.03 % in any case.

The random summing correction may not be implemented as implied by Chapter 7, Equation (7.26). Under most conditions, this correction is small and an approximation is made which results in a correction of the type:

$$A_0 = A(1 + 2R\tau) \tag{9.12}$$

where R is the mean count rate over the count period (determined by summing the whole spectrum counts) and 2τ is an empirical correction factor estimated using the method described in Chapter 7, Section 7.6.8.

Various other correction options of less general use may be provided. For example, *Genie 2000* allows corrections for irradiation time (or sample collection time, as appropriate) for the situation when that is long in comparison to the half-life of the nuclide measured. *GammaVision* makes provision for a correction for geometry differences between sample and calibration. This is in effect via a look-up table of empirical factors. This should be used with care. We saw in Chapter 8, Section 8.9 that if true coincidence summing is not negligible, count rate differences due to geometry are not a simple function of energy. Factors would have to be known for each gamma-ray for the specific nuclides being measured.

All of the spectrum analysis programs will allow corrections to be applied to take account of sample weight or volume and various empirical constant scaling factors. Sometimes, additional allowances can be included to take into account sources of uncertainty known to the analyst but not to the program. I would encourage the use of such facilities. There is little point in underestimating the overall uncertainty of an analysis.

9.13 NUCLIDE IDENTIFICATION

Nuclide identification is included in several of the common analysis programs in one way or another. There are various approaches ranging from the naive look-up approach to very sophisticated programs which fit gamma-rays to the pattern expected from the known gamma-ray emission probabilities. Whichever is used, identifications should be treated with caution, especially when an unexpected nuclide is reported. My advice would be that in such a situation the spectra and results should be examined critically by an experienced spectrometrist.

9.13.1 Simple use of look-up tables

The simplest possible approach is, having found a peak, to compare the energy with a simple look-up table of energies with corresponding nuclide names. For example, both the MCA emulator *Maestro-32* and *GammaVision* indicate on the computer screen a suggested nuclide by reporting the nearest library entry to the measured position. This is a very crude procedure and will often draw attention to library entries which are nowhere near the current peak. If there are two close entries, only the nearest will be given. For example, a peak at 122 keV might be reported as [152]Eu (121.78 keV) or [57]Co (122 keV) depending upon the energy calibration. A more useful approach would be to report all table entries within a specified energy tolerance of, say, 0.5 keV, with nothing at all reported if library entries are further away than that.

9.13.2 Taking into account other peaks

The simple look-up procedure cannot necessarily provide a definite identification and for that, some account must be taken of the presence or absence of other peaks in the spectrum and at least some account of their emission probabilities. For example *Genie 2000* subjects all preliminary nuclide identifications to three tests, each of which reduces a factor which indicates the degree of confidence in the nuclide assignment. These are based upon closeness of the energies of the peaks to the library energies, the proportion of the expected gamma-rays actually detected and a half-life criterion (nuclides with large decay factors are given less weight). Nuclides which still have a confidence factor greater than a user selected threshold are then declared as detected.

9.14 THE FINAL REPORT

The final report, a list of nuclide activities, optionally with individual peak areas, is the object of all of the preceding analysis. It is very unlikely that the output format provided by the software vendor will satisfy exactly every user's needs. The output might be too verbose, or not verbose enough, or not provide what the user regards as vital information. Fortunately, most general spectrum analysis packages provide the means to customize the output format to solve that particular problem.

For each nuclide, there may be several gamma-ray peaks measured, each with a different uncertainty depending upon its size and whether or not it has had to be deconvoluted or stripped. These must be combined in some way to produce a single result for each nuclide. It is worth consulting the software manual to establish how the final nuclide activity is calculated and just what is included in the quoted uncertainties.

All programs purport to provide a weighted mean of all of the peaks associated with a particular nuclide. However, not all do this in the same manner. The obvious way is to weight each individual result by the inverse of its variance. This is statistically sound and is often used. An exception to this is *GammaVision* where the individual calculated activities are completely ignored and a final result calculated as the sum of the individual peak counts for each gamma-ray of the nuclide divided by the sum of the emission probabilities. This does *not* provide a weighted mean. In fact, it gives equal weight to all of the counts whether they are derived from a poorly defined peak or a well-defined peak.

In situations where one or more peaks of a nuclide have an unresolved interference, there is a problem. If included in a normal weighted mean, they will affect the overall result for that nuclide adversely. It is common practice to quote a final uncertainty based only on the internal (or pooled) variance of the values (see Chapter 5, Section 5.3.2). It would be useful if programs also calculated the weighted (or external) variance and indicated when it exceeded the internal variance to alert the analyst to the fact that some other source of uncertainty was present.

More complicated is the situation where several nuclides have mutually interfering peaks. *Genie 2000* and *Sampo 90* use what the manual refers to as a 'Common Algorithm Nuclide Identification'. That identifies unresolved mutual interferences by a process of least squares minimization of a set of simultaneous equations, one for each nuclide, involving all of the peaks measured. The process is a more general treatment of the peak stripping explained in Section 9.11.

There are programs which use the concept of a 'key peak'. This might be the first entry for each nuclide in the nuclide library or be indicated by a flag within the library. A judgement may be made by the program on whether to include individual results in the calculation of the mean on the basis of their agreement with the key peak. A result not statistically consistent with the key peak value is rejected. Obviously, the selection of the key peak is critical. It must be able to be measured accurately and, hopefully, precisely, under all circumstance. (I have personal experience of a situation where inappropriate selection of the key peak could lead to mis-identification. The summing of the 554.3 keV and 618.7 keV gamma-rays of ^{82}Br produces a peak certain to be confused with the 1173.2 keV peak of ^{60}Co. In this case, the 1332.5 keV would be the better key peak.) When nuclides subject to coincidence summing are measured using *GammaVision*, it is quite common for all but the key peak to be rejected because they give significantly different activities – another factor to consider when choosing the key peak. In such circumstances, it is possible that it is the key peak that is most in error, rather than the less-intense peaks.

9.15 SETTING UP NUCLIDE AND GAMMA-RAY LIBRARIES

Throughout this chapter, nuclide and gamma-ray libraries have been mentioned in various contexts and it is evident that for a complete spectrum analysis several libraries may be needed – perhaps calibration libraries, nuclide-identification libraries, peak interference libraries and the like. It is worth reiterating the point made above that libraries should be tailored to their purpose. Putting in peaks 'just-in-case' is likely to cause trouble.

Libraries that include gamma-ray abundances and half-lives should contain the best available data. Again, Appendix B contains a highly recommended set of evaluated data for many common nuclides. Do not use the data in libraries provided by the software vendors without checking. They may be old and inaccurate. One could be more relaxed about the accuracy of gamma-ray energy data because of the tolerance within the peak-identification procedures but there seems little point when accurate information, at least for all common nuclides, is available. However, as reported by Blaauw *et al.* (1999), accurate deconvolution does demand accurate gamma-ray energies.

All of the data within a nuclide library – energies, emission probabilities and half-lives – have an uncertainty and, ideally, it should be possible to incorporate all of those uncertainties into the library. This is not always the case and it may then be necessary to account for nuclear-data uncertainties by increasing the uncertainty on the final result by an appropriate amount.

9.16 BUYING SPECTRUM ANALYSIS SOFTWARE

It is not my intention here to suggest a 'best buy'. Most programs readily available at the present time provide much the same sort of facilities and performance. Software is complex and not only must the algorithms and facilities within the software be considered but also the user interface. There is little point in having perfect algorithms within a program which is grossly inconvenient to use and cannot be tailored to one's needs. A few general points are worth bearing in mind when shopping around:

- Try to arrange a 'hands-on' demonstration of the software, preferably with your own spectra and a real analysis to perform. That will immediately show how easy or otherwise a program is to use.
- Try to persuade the vendor to analyse standard spectra (see Chapter 15) using the program, preferably while you observe. The vendors will not thank you for this, but their response to your request will be instructive.
- If you need to run automatic count/analyse sequences, find out how easy it is to set these up. (These are often referred to as batch jobs because they run under the computer batch file system.) Some MCA emulator sequences are particularly inconvenient to set up. *Maestro 32* must have a text list of instructions specially compiled to a *Maestro 32* readable form. *GammaVision* will allow you to control a number of detectors at the same time, but will only allow you one automatic sequence.
- Find out how easy it is to set up the necessary nuclide libraries. Libraries which are stored in text, rather than coded, form provide much more flexibility from the point of view of editing. On the other hand, plain text files are much more easily tampered with and, from a security point of view, a system with coded libraries might be deemed more appropriate.
- Don't get carried away by the multiplicity of fringe benefits. As with word processors and spreadsheets, most users use only a small proportion of the facilities available. Just make sure that the program does its *core business* well.
- Try to borrow a copy of the software manual so that you can study the algorithms. If you can't understand the manual, will you be able to understand the program?
- If you are buying a program to run on your existing computer, is it compatible in terms of both hardware and operating system?
- If you are not buying software from the manufacturer of your hardware, can the software handle your spectrum files easily?

9.17 THE SPECTRUM ANALYSIS PROGRAMS REFERRED TO IN THE TEXT

The MCA emulator and spectrum analysis programs referred to in this chapter were as follows:

- From ORTEC (*http://www.ortec-online.com/software/software-available.htm*)

Maestro-32
MCA emulator used in conjunction with ORTEC multi-channel buffer modules.
GammaVision
Combined MCA emulator and full spectrum analysis used in conjunction with ORTEC multichannel buffer modules. Runs under Microsoft Windows. Comments in this book refer to version 6.01.
ScintiVision
Spectrum analysis system tailored to scintillation spectrometry (not referred to in the text).

- From **Canberra** (*http://www.canberra.com/products/831.asp*)

System 100
MCA emulator system interfaced to Canberra ADCs. Runs under Microsoft Windows (no longer available).
Genie 2000 (and other variants)
MCA emulator and full spectrum analysis interfaces to the Canberra ADCs. Full multi-tasking facilities. Comments in this book refer to version 2.0.
SAMPO 90
Full off-line spectrum analysis facilities. Runs under Microsoft Windows.

- **Hpermet-PC**
Off-line spectrum analysis program (*http://www.iki.kfki.hu/nuclear/hypc/index.html*).
- From *JF Computing*
FitzPeaks
Off-line spectrum analysis package (*http://www.jimfitz.co.uk/*).
- From **Nuclear Training Services Ltd**
CompAct
Off-line program for activity estimation. Intended for neutron activation analysis and general comparative analysis. Efficiency curves are not used (*http://www.gammaspectrometry.co.uk/compact*).

PRACTICAL POINTS

- It is advisable to understand the way in which the spectrum analysis program handles the data.

- Whenever you must select options within the software, you should make an informed choice. Don't accept the default uncritically.
- When buying software, find out as much about it beforehand as possible. Ask for test spectra to be analysed. Ask for a hands-on demonstration – yours, *not* theirs!
- Whenever a new analysis situation is encountered, the analysis options should be re-appraised. Don't accept unexpected computer results uncritically.
- When, in a routine analysis, unexpected nuclides are identified or odd results obtained, the analysis should be assessed by someone who understands the spectrum-analysis program in depth.

FURTHER READING

- General:

The manufacturers' software manuals (essential).

Gilmore, G.R. (1979). A least squares spectrum fitting method for the measurement of Ge(Li) gamma-ray peak areas, *J. Radioanal. Chem.*, **48**, 91–104.

Brown, R.C. and Troyer, G.L. (1990). A directed-fit approach to estimate the lower limit of detection in a gamma-ray spectrometry system, *Radioact. Radiochem.*, **1**, 23–36.

The above should be read in conjunction with the following letter: Seymour, R. (1991). *Radioact. Radiochem.*, **2**, 4–5.

Seymour, R.S. and Cox, J.E. (1991). Library-Directed versus Peak-Search-Based Gamma-Ray Analysis, *Radioact. Radiochem.*, **2**, 10–18.

- Deconvolution performance:

Koskelo, M.J. and Mercier, M.T. (1990). Verification of gamma spectroscopy programs: a standardized approach, *Nucl. Instr. Meth. Phys. Res., A*, **299**, 318–321.

Blaaw, M. (1993). Multiplet deconvolution as a cause of unstable results in gamma-ray spectrometry for INAA, *Nucl. Instr. Meth. Phys. Res., A*, **333**, 548–552.

- Efficiency curves:

Gunnink, R. (1990). New method for calibrating a Ge detector by using only zero to four efficiency point, *Nucl. Instr. Meth. Phys. Res., A*, **385**, 372–376.

McFarland, R.C. (1991). Behavior of several germanium detector full-energy-peak efficiency curve-fitting functions, *Radioact. Radiochem.*, **2**, 4–10.

Kis, Z., Fazekas, B., Östör, J., Révay, Zs., Belgya, T., Molnár, G.L., and Koltay, L. (1998). Comparison of efficiency functions for Ge gamma-ray detectors in a wide energy range., *Nucl. Instr. Meth. Phys. Res., A*, **418**, 374–386.

Fazekas, B., Révay, Zs., Östör, J., Belgya, T., Molnár, G.L., and Simonits, A. (1999). A new method for determination of gamma-ray spectrometer non-linearity, *Nucl. Instr. Meth. Phys. Res., A*, **422**, 469–473.

Molnár, G.L., Révay, Zs. and Belgya, T. (2002). Wide energy range efficiency calibration method for Ge Detectors, *Nucl. Instr. Meth. Phys. Res., A*, **489**, 140–159.

- Peak search algorithms:

Black, W.W. (1969). Application of correlation techniques to isolate structure from experimental data, *Nucl. Instr. Meth. Phys. Res., A*, **71**, 317–327.

Routti, J.T. and Prussin, S.G. (1969). Photopeak method for the computer analysis of gamma-ray spectra from semiconductor detectors, *Nucl. Instr. Meth. Phys. Res., A*, **72**, 125–142.

Mariscotti, M.A. (1967). A method for automatic identification of peaks in the presence of background and its application to spectrum analysis, *Nucl. Instr. Meth. Phys. Res., A*, **50**, 309–320.

De Geer, L-E. (2005). *A Decent Currie at the PTS*, CTBT/PTS/TP/2005-1, CTBTO Preparatory Commission, Vienna, Austria.

- For program intercomparisons and other performance reports: *Further Reading* for Chapter 15.

Blaauw, M., Keyser, R.K. and Fazekas, B. (1999). Comparison of alternative methods of multiplet deconvolution in the analysis of gamma-ray spectra, *Nucl. Instr. Meth. Phys. Res., A*, **432**, 77–89.

10

Scintillation Spectrometry

10.1 INTRODUCTION

Scintillation detection has been used since the earliest days of radioactivity and is still today employed to measure the whole range of radioactive emissions – alpha- and beta-particles, gamma-rays, neutrons and the more exotic leptons and mesons. In this chapter, I will restrict myself exclusively to scintillation as applied to gamma-ray measurements.

Until the commercial advent of the semiconductor detector, scintillation detectors, in the main based on sodium iodide, were standard for gamma spectrometry. Indeed, even now their influence on gamma spectrometry is apparent in the fact that we still relate semiconductor detector efficiency to sodium iodide. There is a general feeling abroad that scintillation detectors are a thing of the past; an attitude which is not sustainable in practice. It is certainly true that where spectrometry of many gamma-rays, as opposed to simple measurement of one or two, is involved the advantages of the semiconductor detector far outweigh those of the scintillation detector. Nevertheless, scintillation spectrometry still has a number of valuable roles to play in gamma-ray measurement and, with recent developments in lanthanum halide detectors, its scope may broaden. There are circumstances where space limitations or a hostile environment preclude the use of semiconductor detectors. (An example which comes to mind is the use of a sodium iodide spectrometer for measurement of nuclear fuel burn-up at the bottom of a storage pond.) In addition to these admittedly special situations, scintillation detectors have an important part to play in the construction of active shielding for high-resolution detectors (see Chapter 13, Section 13.5.5).

10.2 THE SCINTILLATION PROCESS

In Chapter 3, I introduced the idea that primary electrons produced by gamma-ray interaction lose their energy by creating secondary electron–hole pairs and that the function of a detector was to collect these and produce an electrical signal. In a semiconductor, this is effected by means of an electric field. In scintillation detectors, the primary ionization of the detector medium is broadly the same but the collection process differs somewhat because scintillators are insulators.

According to our band structure model (Chapter 3, Section 3.2), the primary electrons produced by the gamma-ray interaction raise secondary electrons to the conduction band, leaving holes in the valence band. In some cases, the energy given to the electron may not be quite sufficient to raise it to the conduction band. Then, the electron and hole could remain electrostatically attracted to each other as an entity called an *exciton*. In terms of the band structure model, this represents elevation to an extra band just below, but continuous with, the conduction band, as illustrated in Figure 10.1.

If the electrons are allowed to de-excite by falling back to the valence band, they will emit electromagnetic radiation. If this radiation is in, or near, optical wavelengths, it can be detected by a photomultiplier or other light-measuring device to provide the detector signal. This is the basis of the scintillation detector.

If we are to construct a scintillation detector for gamma-ray detection and spectrometry, the scintillator material must have a number of particular properties:

- there must be a reasonable number of electron–hole pairs produced per unit of gamma-ray energy;
- it would be very desirable for the material to have a high stopping power for gamma radiation (which means, in practice, high density and atomic number);
- for spectrometry, the response must be proportional to energy;
- the scintillator must be transparent to the emitted light;

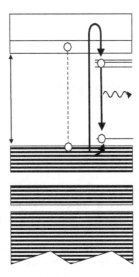

Figure 10.1 Band gap structure in a scintillator

- the decay time of the excited state must be short to allow high count rates;
- the material should be available in optical quality in reasonable amounts at reasonable cost;
- the refractive index of the material should be near to that of glass (ca. 1.5) to permit efficient coupling to photomultipliers.

The materials that have found particular application for gamma-ray measurements are all inorganic crystals: sodium iodide (NaI), caesium iodide (CsI), calcium fluoride (CaF_2), bismuth germanate (BGO) and, recently, lanthanum halides. Of these, the first is the most important and the last are materials rapidly gaining in importance.

10.3 SCINTILLATION ACTIVATORS

As it happens, the band gap of sodium iodide is large and photons emitted by de-excitation of electrons directly from the conduction band would be far outside of the visible range. This makes detection of the light difficult. Not only that, the bulk of the material absorbs the emitted photons before they reach the photomultiplier. Both problems are solved by using an **activator**. In the case of NaI, this would be thallium and for CsI it is thallium or sodium. The shorthand descriptions for these activated scintillators are NaI(Tl), CsI(Tl) and CsI(Na).

The introduction of about 10^{-3} mol fraction of the impurity produces defect lattice sites which give rise to extra levels within the forbidden band between the valence and conduction bands (see Figure 10.1). The ground state

of these activator sites lies just above the valence band and the excited states somewhat below the conduction band. When an electron–hole pair is formed, the hole may migrate to a nearby activator site. Electrons in the conduction band and within the exciton band will tend to be captured by the excited activator states. This means that the photon energy released when these levels de-excite will be lower and the electromagnetic radiation will be of a longer wavelength, perhaps in the visible range. It also means that the emission wavelength will no longer match the absorption characteristics of the scintillator and so much less light will be lost before measurement by the photomultiplier. To summarize the process:

- the gamma-ray is absorbed and primary electrons are formed;
- the primary electrons create electron–hole pairs;
- excitons are formed and electrons are also raised to the conduction band, leaving holes in the valence band;
- activator levels are populated by capture of electrons, holes and excitons;
- activator levels de-excite emitting light;
- light is collected and measured by a photomultiplier to produce an electrical signal.

Not all of the energy absorbed from the gamma-ray will be re-emitted as scintillation photons. NaI(Tl), the most efficient in terms of light output, will only release about 12 % of the total. The rest of the energy is retained as lattice vibrations or heat. The size of the actual output signal from the detector will also depend upon how well the response of the light detector matches the scintillation-light spectrum.

10.4 LIFETIME OF EXCITED STATES

The mean life, or lifetime, of an excited activator state is very short – of the order of 0.1 μs. This direct emission is termed **luminescence**. The short decay time means that very short detector pulses are possible. (Figure 10.2 shows schematically the shape of the light pulse.) In most cases, only one excited state is significantly populated but in others a more complex decay is evident. For example, the decay of the luminescence from bismuth germanate is characterized by two components of 60 ns and 300 ns lifetimes.

Transitions from some of the excited states to the ground state may be forbidden. An electron finding itself in one of these states must first be promoted by thermal excitation to a level which is not forbidden before it can de-excite. The lifetime of these states can be much longer that the normal excited states. The slowly decaying

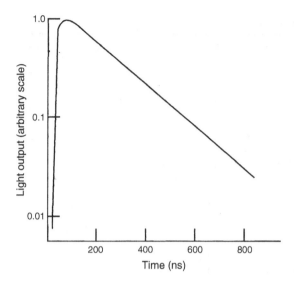

Figure 10.2 Light output from a NaI(Tl) scintillator as a function of time

proportion of light emitted by the de-excitation of these long-lived states is called **phosphorescence**, also referred to as **afterglow**. This can be a problem in that it causes an increase in the background to the normal pulses. In NaI(Tl), for example, where the primary scintillation lifetime is 230 ns, 9 % of the total light is emitted as phosphorescence with a decay time of 0.15 s.

10.5 TEMPERATURE VARIATION OF THE SCINTILLATOR RESPONSE

Scintillation detectors have an advantage over HPGe semiconductor detectors in that they can be operated at room temperature, and indeed the light output from NaI(Tl) is at a maximum at room temperature. The light output is reasonably constant over the normal range of room temperature but decreases beyond that. Other materials behave in a similar manner but the temperature of maximum response differs from material to material.

Even though the response of the detector crystal may be reasonably constant within the normal operating temperature range, the stability of the electronic system may not be so forgiving.

10.6 SCINTILLATOR DETECTOR MATERIALS

The relevant physical parameters for a number of materials in use for gamma-ray measurement are listed in Table 10.1. A word of explanation is needed about the measures of light emission quoted in the literature. In absolute terms, the light output of a scintillator could be expressed as the number of photons emitted per unit of gamma-ray energy absorbed. For NaI(Tl), commonly regarded until recently as the most efficient inorganic scintillator, the value would be 38 000 photons per MeV. Often, scintillators are given an efficiency measure relative to NaI(Tl), for example, CsI(Na) might be quoted as 85 %. The factor is variously referred to as 'relative conversion efficiency', 'scintillation conversion efficiency', 'relative scintillation efficiency', 'photoelectron yield', or even 'relative light output'. It is important to appreciate what these relative efficiencies relate to. The factor indicates the effective detector response, assuming that a particular photocathode, usually the bialkali type, is used with the scintillator. It takes into account the efficiency of conversion of gamma-ray energy into light by the scintillator and the efficiency of conversion of that light energy to photoelectron energy by the photocathode. The factor can only be quoted for specific scintillator/photocathode combinations. Since the number of photoelectrons is related directory to the final detector signal, it can be used as a measure of overall output pulse height ratio. Here I shall adopt the term **relative conversion efficiency** as being the most descriptive and least ambiguous of those in common use.

As long as the emission spectrum of the scintillator is similar to NaI(Tl), then the comparison is valid. In the case of CsI(Tl) in particular, the relative light output of 45 % quoted in Table 10.1 is misleading. The actual amount of light emitted by a CsI(Tl) scintillator is 52 000 photons per MeV of gamma-ray energy absorbed, compared to 38 000 for NaI(Tl). In fact, CsI(Tl), not NaI(Tl), has the highest light output of the traditional inorganic scintillators, surpassed only by the new lanthanum halide materials. The discrepancy arises because the spectrum of light emitted by CsI(Tl), which peaks at about 550nm, is not well matched to the bialkali photocathode. Figure 10.3 compares the emission spectrum of various scintillator materials with the photocathode response of two common photomultipliers. In some cases, scintillators produce more than one component of fluorescence. In Table 10.1, the data for wavelength of maximum emission and the decay times are given for the major components only.

10.6.1 Sodium iodide – NaI(Tl)

This is the most commonly used scintillator material. It is cheap and readily available. Detectors up to 0.75 m diameter have been produced. More typically, the 76 mm diameter by 76 mm high cylindrical sodium iodide detector was for many years the standard gamma-ray spectrometer

Table 10.1   Properties of scintillator materials for gamma-ray detection[a]

Scintillator	Activator	Abbreviation	Density $(g\,cm^{-3})$	WL (nm)[b]	DCT (ns)[c]	RI[d]	RCE[e] (%)	After glow (%)[f]	Hyg[g]	FWHM[h] (%)
Sodium iodide	Tl	NaI(Tl)	3.67	415	230	1.85	100	0.3–5.0	Y	7.0
Caesium iodide	Tl	CsI(Tl)	4.51	550	1000	1.79	45	0.5–5.0	Y	—
	Na	CsI(Na)	4.51	420	630	1.84	85	0.5–5.0	S	7.5
	—	CsI	4.51	315	16	1.95	4–6	—	Y	—
Caesium fluoride	—	CsF	4.64	390	3–5	1.48	5–7	0.003–0.060	S	—
Calcium fluoride	Eu	$CaF_2(Eu)$	3.18	435	940	1.47	50	< 0.30	Y	—
Barium fluoride	—	BaF_2	4.88	310	630 + 0.6	1.50	16+5	—	N	—
Bismuth germanate	—	BGO	7.13	480	300 + 60	2.15	15–20	0.005	N	> 10
Cadmium tungstate	—	$CdWO_4$	7.90	540	5000	2.3	40	0.10	N	—
Lanthanum Chloride	Ce	$LaCl_3(Ce)$	3.79	350	28	~1.9	130	—	Y	3.8
Lanthanum Bromide	Ce	$LaBr_3(Ce)$	5.29	380	16	~1.9	160	—	Y	2.7
Gadolinium silicate	Ce	GSO	7.13	430	30–60	1.85	20	—	N	—
Lutecium silicate	Ce	LSO	7.4	420	40	1.82	40–75	—	N	—
Yttrium aluminium perovskite	Ce	YAP	5.37	347	28	1.94	40	—	N	—

[a] Data are taken from the Harshaw QS Scintillation Detector Catalogue (March 1992) and Saint-Gobain Ceramics and Plastics Inc. Internet sources.
[b] Wavelength at maximum emission.
[c] Decay time.
[d] Refractive index.
[e] Relative conversion efficiency (relative to NaI(Te)), i.e. net detector output using a bialkali photomultiplier tube (PMT).
[f] Fraction of light emitted more than 6 ms after the initial fluorescence.
[g] Hygroscopic?: Y, yes; N, no; S, slightly.
[h] Quoted resolutions are quoted at 661.6 keV and are typical rather than definitive.

detector. (At the time, this was referred to as a '3-by-3 detector' – 3 in. × 3 in.)

The iodide atom of the NaI(Tl) provides a high gamma-ray absorption coefficient and, therefore, high intrinsic efficiency. At low energy, there is a high probability of complete absorption (see Figure 10.4 below for data for a 38 mm × 38 mm cylindrical detector). Because NaI(Tl) provides the greatest light output of all of the traditional inorganic scintillators using standard photomultipliers, it also has the best energy resolution (see Section 10.12).

Notwithstanding its excellent performance compared to other materials, NaI(Tl) does has several drawbacks. It is brittle, sensitive to thermal gradients and thermal shock. It is hygroscopic and must be encapsulated at all times. It also exhibits the long afterglow referred to above. At low count rates, this is not necessarily a problem, in that pulses due to the phosphorescence can be eliminated electronically, but at high count rate they tend to pile up and limit high count rate performance.

Potassium is chemically similar to sodium and is a very likely impurity in sodium salts. Because a small proportion of natural potassium is the radioactive ^{40}K, it is

important that sodium iodide used for detector manufacture is of high purity. Otherwise, the detector background would be higher than necessary. Typically, a manufacturer might specify the sodium iodide purity as less than 0.5 ppm of potassium.

10.6.2 Bismuth germanate – BGO

BGO, with the chemical formula $Bi_4(GeO_4)_3$, is a relatively new material in scintillation spectrometry. It is grown from a molten mixture of bismuth and germanium oxides. Although its light output is low compared to NaI(Tl), its much greater density gives it a much greater stopping power and makes it ideal for active shielding systems. Figure 10.4 compares the intrinsic peak efficiency for the two materials. This, of course, does not take into account incomplete absorption interactions which would enhance the efficiency of BGO for background suppression even further. A 150 keV photon is 90 % absorbed by only 2.3 mm of BGO. In simple terms, a BGO detector of only 1/16th the size of a NaI(Tl) detector would have about the same efficiency. However, because the light output is much smaller, the resolution of BGO

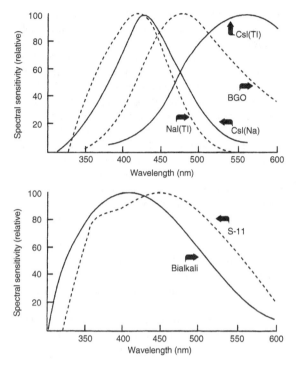

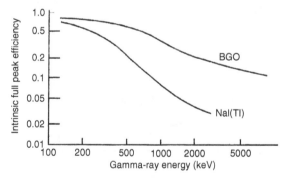

Figure 10.3 (a) Emission spectra from common scintillators. (b) Spectral response (sensitivity) of common photomultipliers

Figure 10.4 Intrinsic full peak efficiency of identically sized NaI(Tl) and BGO detectors (38 mm × 38 mm)

is worse, which makes it not the material of choice for spectrometry as distinct from detection.

Unlike NaI(Tl), BGO is used without activator. Since there is a large shift between the optical and the emission spectra of the Bi^{3+} states, relatively little self-absorption takes place, and the crystal is transparent to its own emission. Even so, the relative efficiency is only 15–20 % that of NaI(Tl). BGO is inert and is not hygroscopic and need not be hermetically sealed into its housing.

10.6.3 Caesium iodide – CsI(Tl) and CsI(Na)

The density of CsI is somewhat greater than that of NaI and has a correspondingly greater absorption coefficient. It is less brittle and more resistant to thermal and mechanical shock and for that reason has found applications in space instrumentation. It is less hygroscopic than NaI but must still be sealed within its container.

The light output of CsI(Tl) peaks at about 565 nm, well beyond the optimum sensitivity for the normal photomultiplier tubes. For this reason, the light output is effectively low. To alleviate this problem, a photodiode detector can be used and the fact that that the light output is greater than that for NaI(Tl) can be capitalized on. CsI(Tl) detectors mounted on photodiodes are now produced commercially but their size is limited by the size of the photodiodes currently available. The small size and rugged nature of such devices makes them ideal for medical and other applications where space is at a premium.

The alternative material, activated with sodium, CsI(Na), has a similar emission spectrum to NaI(Tl) and a similar efficiency with the advantage of higher absorption coefficients. The main problem with this material is that the decay time of the luminescence is long and, as with NaI(Tl), phosphorescence is also observed.

10.6.4 Undoped caesium iodide – CsI

Caesium iodide can also be used without an activator but at the expense of a much reduced relative conversion efficiency – 4–6% compared to 45 % and 85 % for the doped materials. The emission maximum at 315 nm means that better output would be obtained when using quartz-windowed photomultipliers.

The advantage of the undoped material is that the decay time of the fast component, 16 ns, is much shorter than most other scintillators, which gives it potential in timing applications. However, 15–20 % of the light output is due to a longer-lived fluorescent component with a decay time of 1000 ns.

10.6.5 Barium fluoride – BaF$_2$

As with BGO, barium fluoride needs no activator. At first sight, the material appears to offer little advantage over other scintillators. The low light output means that the resolution of barium fluoride detectors is particularly poor. Its only real advantage lies in the fact that the luminescence has two components, a rather slow one with a decay time of 630 ns and an extremely fast one with only 0.6 ns lifetime. The fast emission is in the ultraviolet and needs an appropriately sensitive light detector. Barium fluoride is the only practical scintillator with a luminescence lifetime below 1 ns and finds applications where timing is more important than resolution.

10.6.6 Caesium fluoride – CsF

As with barium fluoride, there are few advantages of this material when compared to the more popular scintillators, except for the short lifetime of the light output. Although much longer than for barium fluoride, the 4.4 ns scintillation lifetime is considerably shorter than that of the fluorescence in more conventional materials. Yet again, the low light output means poor resolution.

10.6.7 Lanthanum halides – LaCl$_3$(Ce) and LaBr$_3$(Ce)

Two materials since the first edition of this book, which are becoming available in detector-sized amounts, are the chloride and bromide of lanthanum. Both materials are activated with cerium and are designated as LaCl$_3$(Ce) and LaBr$_3$(Ce), respectively. Both offer advantages over NaI(Tl) and could, in time, replace it as the scintillators of choice for some applications. LaCl$_3$(Ce) has a similar density, refractive index and relative conversion efficiency to NaI(Tl) but has a much shorter decay time, offering improved timing resolution. More importantly, from a spectrometry point of view, detectors have been constructed with resolutions of 3.8 % at 661.6 keV, for a 1 in. × 1 in. detector, compared to the 7 % of NaI(Tl). The light output per keV of energy absorbed is 25 % greater than NaI(Tl) but its emission wavelength is lower, resulting in a similar relative conversion efficiency.

Early in 2007, the sole manufacturer, Saint-Gobain, announced the release of these lanthanum-based scintillators under the trade name, *BriLanCe®*; B350 is LaCl$_3$ (10 % Ce) and B380 LaBr$_3$ (5 % Ce). Their manufacturing technique has improved to the extent that, at the time of writing, 3 × 3 (i.e. 3 in. diameter by 3 in. long) crystals are being produced. The white paper available on their website discusses the advantages of B380 as follows:

- 60 % greater light output than NaI(Tl).
- Better resolution – 2.9 % compared to 7 % for NaI(Tl). Example spectra demonstrate complete resolution of the two ^{60}Co peaks. At 1332.5 keV, the resolution is 28 keV for a 3 × 3 detector.
- More efficient than NaI(Tl) because of its higher density, meaning more counts in the full energy peaks.
- The scintillation decay time much shorter than for NaI(Tl), meaning that it can be used at much higher count rates. The decay times are 16 ns and 28 ns for B380 and B350, respectively.

A single disadvantage of these materials is their inherent radioactive impurity content. Lanthanum has a naturally occurring radioisotope, ^{138}La. That decays 66.4 % by electron capture to ^{138}Ba, which emits a 1435.80 keV gamma-ray and the inevitable Ba X-rays. The remaining 33.6 % decays by β$^-$ emission to ^{138}Ce, releasing a 788.74 keV gamma-ray. Background spectra of these detectors show a bremsstrahlung continuum extending to the beta end point at 255 keV (Figure 10.5). Because the lanthanum atoms are decaying inside the detector, neither of the full-energy gamma-rays appear in the spectra as expected; both are summed – the 788.74 keV with its accompanying beta particles and the 1435.80 keV with the electron capture X-rays. At higher energies, there is evidence of low level alpha emitting contamination due to ^{227}Ac contamination. While the latter can be controlled, the ^{138}La is unavoidable. This background activity, which seems to be about 1 Bq cm^{-3}, would limit the use of these detectors for low background applications. In principle, it would be possible to remove the background features from a spectrum by stripping off a background spectrum. However, as I have indicated elsewhere, that has consequences in terms of disturbing the relationship between channel counts and their uncertainties.

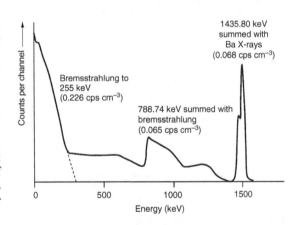

Figure 10.5 Schematic diagram of the background spectrum of a 38 mm × 38 mm BriLanCe 380 detector. The count rates quoted are integral counts beneath the feature labelled

10.6.8 Other new scintillators

There are a number of new scintillator materials under development to add to the list of conventional materials. The properties of some of these, GSO, LSA and YAP, for example, are listed in Table 10.1. Many of these materials have specific applications in specific fields, such as space research or medical imaging and are not suitable for gamma-ray spectrometry.

10.7 PHOTOMULTIPLIER TUBES

The output from the scintillation crystal is a quantity of light which must be measured and converted into an electrical signal. Conventionally this is done by a **photomultiplier tube** (PMT). The general arrangement is shown in Figure 10.6. The processes which take place within the PMT are:

- The light photon strikes a light sensitive layer, the **photocathode**, causing it to emit a photoelectron.
- The photoelectrons are focused electrostatically onto the first of a series of electron multiplier stages, called **dynodes**. These emit more electrons than they receive, thus amplifying the signal.
- The electrons from the first dynode are multiplied at the second dynode, and again at the third, all the way down the chain.
- The amplified signal is then collected at the anode and passed out to the measurement circuits.

The design of photomultipliers varies depending upon what they are to measure. For scintillation gamma-ray measurements, they are always of an end-window design with the photocathode deposited on the inside of the face of the tube. The whole structure is enclosed within an evacuated glass envelope and connections made via a multipin plug at the anode end.

10.8 THE PHOTOCATHODE

This, the light detecting element of the PMT, is made of a material which has a low energy barrier to the release of photoelectrons from its surface (in other words, it has a low *work function*). Typical photocathode materials are Na_2KSb activated with caesium, the so-called **multialkali** coating, and K_2CsSb activated with oxygen and caesium which is referred to as the **bialkali** coating. Sometimes, photocathode materials are referred to by code numbers to indicate their spectral response, for example, **S-11**, **S-13**, etc. The photocathode in scintillation systems is often of the bialkali type.

Not every photon received will produce a photoelectron. The energy of a typical scintillation photon in the blue region of the spectrum is about 3 eV. The electrons excited by such photons must migrate to the surface of the photocathode and still have sufficient energy to overcome the work function which might be 1.5 to 2 eV in the materials described. This places a limit on the thickness of the photocathode to a few tens of nanometres. At such thicknesses, the photocathode will only be semi-transparent, even to the scintillation light, by which this process further reduces the overall yield of photoelectrons.

The **quantum efficiency** of a photocathode material, the number of photoelectrons emitted per incident photon, might be 20–30 % at the optimum photon energy. Now, in NaI(Tl), it takes about 26 eV of energy absorbed by the detector to produce one photon. Taking the quantum efficiency into account as well, this means that it takes

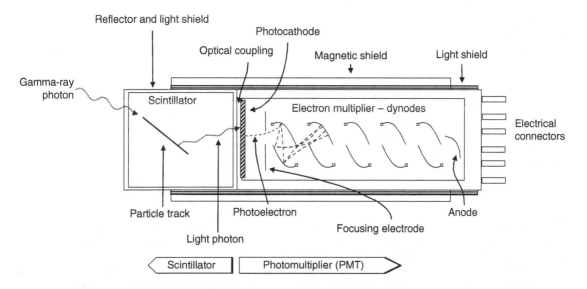

Figure 10.6   Schematic diagram of a scintillation detector comprising a scintillation crystal optically coupled to a photomultiplier tube

about 100 eV of gamma-ray energy to create the primary electrical signal at the photocathode. Compare this with the equivalent value for a germanium semiconductor of 2.96 eV. As we shall see later, this has implications for detector resolution.

10.9 THE DYNODE ELECTRON MULTIPLIER CHAIN

These electrodes are coated with a material which emits a number of secondary electrons in response to the absorption of a single electron. Dynode coatings might be of beryllium, magnesium oxides or Cs_3Sb. Modern alternatives are materials, such as gallium phosphide, that are described as having **negative electron affinity**. Because of the nature of their band structure, these materials essentially have no surface energy barrier. The consequence is that more secondary electrons are produced per incident electron than would otherwise be the case.

The multiplication factor, the ratio of the numbers of secondary and primary electrons, depends upon the potential difference between each consecutive pair of dynodes. Typical inter-dynode voltages are 80–120V and the number of stages 10–12. The incident electron must be able to create electrons within the dynode with energy at least equal to the band gap. Since this is 2–3 eV for typical dynode coatings, we might expect of the order of 30 electrons for every 100 V of potential difference. In fact, the multiplication factor is much less than this. Not every excited electron will move in the direction of the surface and, of those that do, not all will have sufficient energy when they get there to overcome the surface potential barrier and escape. In practice, multiplication factors are of the order of 4–6 at typical inter-dynode voltages.

Having escaped from the dynode, not all of the secondary electrons will reach the next and, of those that do, not all will produce further secondary electrons. If we take the gain factor for a single stage to be $k \times m$, where m is the multiplication factor and k takes into account the losses referred to, then the overall gain of a photomultiplier with N stages would be:

$$\text{Gain} = (k \times m)^N \qquad (10.1)$$

If we take a typical value for m of 5 and assume k is near to 1, this gives an overall multiplication of about 10^7 for ten stages. A much greater multiplication would be obtained with a negative electron affinity dynode coating where the single stage multiplication factor might be 55 or so. The actual physical arrangement of the dynode chain

varies. The early 'venetian blind' arrangement has been superseded by linear focused systems (as in Figure 10.6) and what is called the 'box grid design'. Photomultipliers for gamma spectrometry are likely to incorporate one of these latter types.

There are alternative forms of electron multiplier available, such as the continuous channel electron multiplier and the micro-channel plate. These have potential but are unlikely to be found in use for gamma spectrometry at the present time.

10.10 PHOTODIODE SCINTILLATION DETECTORS

The photomultiplier is limited by the spectral sensitivity of its photocathode at low energy (long wavelength) and by optical absorption of the scintillation light at higher energy. Another option is the semiconductor diode referred to as a **photodiode**. Silicon photodiodes have several advantages over the photomultiplier. They have a wider sensitivity range (see Figure 10.7) and higher quantum efficiency (up to 70 % compared to 25 % for photocathode materials).

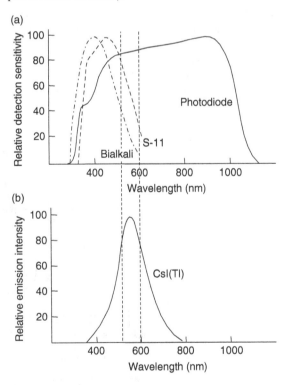

Figure 10.7 (a) The spectral response of a photodiode compared to typical photomultiplier tubes. (b) The emission spectrum of the CsI(Tl) scintillator

Another important advantage is that they are insensitive to magnetic fields. They are more rugged and smaller than the equivalent PMT. Because the charge carriers within them travel much shorter distances, this gives them an advantage in timing applications. There is, however, a disadvantage in that the signal size is small which means that electronic noise is much more important than it is in photomultiplier systems. The noise increases with the capacitance of the photodiode. As we saw in Chapter 3, Section 3.5 in connection with planar semiconductor detectors, the larger the detector, the larger the capacitance. For this reason, the size of photodiodes is limited to a few square centimetres at the present time.

Current commercially available detectors incorporating a photodiode (usually with a CsI(Tl) scintillator) are specified for use above 70 keV or so (limited below that by noise) and claim an improved resolution over PMT combinations above 500–600 keV.

10.11 CONSTRUCTION OF THE COMPLETE DETECTOR

The typical complete scintillation gamma-ray detector is shown in Figure 10.8, with the scintillator crystal optically coupled (of which more later) to the photomultiplier tube.

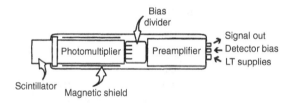

Figure 10.8 General arrangement of a complete scintillation detector

Of the light not emitted in the direction of the photocathode, a proportion will be lost from the surface of the crystal and not contribute to the detector signal. This is minimized by surrounding the detector on all sides by a reflector. A diffuse, rather than specular (shiny), reflector has been found to be best and the scintillation crystal is coated on all but the exit face with either magnesium or aluminium oxide. The crystal is then mounted within a thin aluminium can.

If the scintillator is hygroscopic, it is essential that it be hermetically sealed to avoid deterioration. In a permanently mounted arrangement, such a scintillator might be sealed directly onto the face of the PMT. A crystal intended to be demountable would be provided with a thin glass or quartz window sealed to the aluminium can.

In such an arrangement, the photomultiplier would be mounted in an enclosure which would allow a light tight seal with the scintillator.

The transfer of electrons from dynode to dynode is likely to be affected by magnetic fields in the vicinity of the detector. For this reason, photomultipliers are magnetically shielded, typically by incorporating a cylinder of a high-permeability material, such as μ-metal (a nickel–iron alloy), around the photomultiplier and within the outer enclosure.

The photomultiplier tube communicates with the outside world via the pins at its end. Through these are supplied the bias supply for each of the dynodes and for the anode. The tube will be plugged into a tube base which incorporates divider resistors to split the anode potential appropriately. It is not uncommon for the base to also house the preamplifier.

Photomultipliers are not particularly resistant to mechanical shock and vibration and should be protected from this. Special ruggedized arrangements may be needed if vibration is liable to be a problem and perhaps for detectors intended for field applications.

10.11.1 Detector shapes

The standard shape for a scintillation detector is a simple cylinder with its height equal to its diameter. The ease with which such detectors can be made with precisely reproducible dimensions and properties made the concept of a 'standard detector' a reality. For most routine purposes, the detector of choice is the NaI(Tl). Typical 'off-the-shelf' sizes are (in inches) $1 \times 1, 2 \times 2$ and 3×3, often still quoted in archaic units. End well detectors of similar sizes are also easily available.

For low-energy X-ray measurements, the detector crystal would be provided with a thin light-tight window, usually of beryllium. The high atomic number of NaI (and indeed most gamma-sensitive scintillators) means that X-rays will be completely absorbed within a very thin layer of scintillator. Accordingly, detectors designed for X-ray work will only be one or two millimetres thick.

In Chapter 13, I will describe special coincidence and active shielding systems which utilize scintillators. These will often be of scintillators other than NaI(Tl), particularly BGO, and the shape will be specific to the system.

10.11.2 Optical coupling of the scintillator to the photomultiplier

When light passes through an interface, there is a potential for losses by reflection at the interface. Light which is incident on the surface at more than the critical angle will be totally internally reflected (TIR in Figure 10.9). If the angle of incidence (θ) is less than the critical angle, almost all of the light will be transmitted (TR). A small amount will be reflected back – the so-called *Fresnel reflection* (FR). The critical angle (θ_C) depends upon the ratio of the refractive indices of the media on either side of the interface:

$$\theta_C = \sin^{-1}(n_1/n_0) \tag{10.2}$$

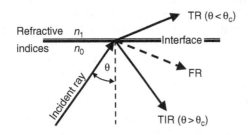

Figure 10.9   Reflection and transmission of light at an interface

To reduce transmission losses to a minimum, the refractive indices should be as near as possible equal. The refractive index of glasses is about 1.5 and a look at Table 10.1 reveals that most scintillator materials have a refractive index rather greater than that. (For example, the refractive index of NaI(Tl) is 1.85, from which the critical angle can be calculated to be 54. This means that, assuming the light approaches the interface from all directions, 60 % will be reflected back into the scintillator.) To minimize the light losses due to this mismatch, the scintillator is optically coupled to the photomultiplier by using high-viscosity silicone oil which facilitates the transmission of the light through the interface. Ruggedized systems may use epoxy cements instead. Because the quality of this interface is so important, uncoupling the scintillator crystal from the PMT should not be undertaken lightly. On the other hand, the performance of an ageing detector might be improved by renewing this coupling.

Some scintillator materials are brittle and the crystal is easily fractured if it suffers a blow. This may not prevent the detector from working altogether, unless the PMT is also damaged, but will almost certainly worsen the resolution of the detector. This is caused by interfaces within the crystal at the fracture planes which impair the passage of light from certain parts of the crystal to the photocathode. The result of this loss of light is a worsening of resolution. A detector crystal damaged in this way is beyond repair.

The collection of light cannot be relied upon to be uniform over the whole volume of a detector. For large detectors, particularly those with complicated shapes, light collection is improved by using more than one photomultiplier disposed around the crystal. In situations where space is limited and there is no room for a photomultiplier or where the environment is not suitable, the detector crystal may be interfaced to its photomultiplier via a light pipe, a length of transparent material within which internal reflection will permit the light to be transmitted over a distance, even around corners. Light pipes may also be used with very thin detectors to improve their resolutions. The effect of the light pipe is to spread the light over the whole of the photocathode, thus evening out any variations in response across it. The design of efficient light pipes is outside the scope of this present book.

10.12   THE RESOLUTION OF SCINTILLATION SYSTEMS

The resolution of scintillation detectors is considerably greater (i.e. worse) than that of semiconductor detectors for reasons which will become plain. Semiconductor resolution is expressed as the full width half maximum (FWHM) of a spectrum peak in energy, usually keV. It is conventional to express the scintillation detector resolution, W, as FWHM, calculated as a percentage of the peak energy, E, namely:

$$W = \text{FWHM} \times 100/E \tag{10.3}$$

To compare detectors, one with another, the standard peak energy used for scintillation systems is the 661.6 keV peak of ^{137}Cs. For a standard 3×3 NaI(Tl) detector, one can expect a resolution of about 7.5 %, equivalent to an FWHM of 50 keV. Expressed as a percentage, the resolution of a scintillation detector apparently improves as energy increases. However, even though numerically the resolution decreases, the peaks still become broader at higher energy.

As with semiconductor detector systems, the width of the peak in a spectrum depends upon the uncertainties associated with the production of the signal in the detector, collection of the signal, and the electronic noise added during transmission from detector to measuring instrument. Compared to the semiconductor detector, the sources of uncertainty in scintillation spectrometry are many but we can group them in the same way as we did in Chapter 8, Section 8.5.1 (here using widths relative to energy, W, rather than FWHM):

$$W^2 = W_S^2 + W_i^2 + W_D^2 \tag{10.4}$$

The first of these terms, W_S, accounts for statistical factors all related to the number of information carriers – electrons, photons or charge pairs, as appropriate. We can include in this uncertainties due to:

- production of electron–hole pairs;
- the number of photoelectrons emitted by the photocathode for each photon;
- the multiplication factor at the first dynode;
- the multiplication factors at each subsequent dynode and collection at the anode.

W_i, the **intrinsic effective line width**, takes account of non-linearity in the response of a scintillator to gamma-ray energy which comes about because of non-linearity in the conversion of the pair energy to light (see Section 10.12.2). Because the response is non-linear, the magnitude of these uncertainties will also depend upon energy in a complex manner. The maximum contribution to the overall width is about 5% at 400 keV.

The remaining term, W_D, includes a host of factors that we can attribute to various characteristics of the detector system as a whole. These factors, which are all independent of energy, are:

- the uniformity and purity of the detector material;
- delivery of the light to the photocathode;
- the uniformity of the photocathode and dynodes;
- electronic noise in the measurement electronics;
- electronic drift during measurements.

Unlike the semiconductor detector system where electronic noise can be a major cause of poor resolution, this is not a significant factor in scintillation systems, but we include it here for completeness.

10.12.1 Statistical uncertainties in the detection process

The statistical factors referred to above are by far the most significant sources of scintillation line broadening. Below 100 keV, the resolution is mainly determined by these uncertainties. (Compare this with semiconductor detectors where at low energy, electronic noise is usually the dominant factor – see Chapter 11.) The uncertainty is at its greatest at the point at which the number of information carriers is at a minimum, being mainly determined by the number of photoelectrons emitted by the photocathode.

A numerical example might put this into context. Figure 10.10 shows the processes involved in conversion of gamma-ray energy into photoelectron energy. Let us assume that a single gamma-ray of 661.6 keV (^{137}Cs) is completely absorbed in the scintillator. If this is NaI(TI), we can expect 11.3 % of this (74.76 keV) to re-appear

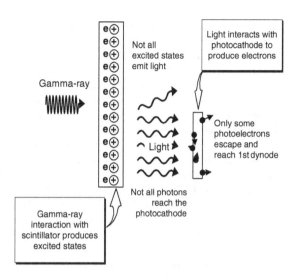

Figure 10.10 The processes within the scintillation detector which convert gamma-ray energy to photoelectron energy

as light photons of average energy 3 eV, i.e. on average, 24 920 photons.

If, say, 75 % of these are collected at the photocathode and assuming the quantum efficiency is 20 %, only 3738 photoelectrons will be produced ($24\,920 \times 0.75 \times 0.2$). If we further assume that Poisson statistics would be valid, we can estimate the relative uncertainty at this point in the process as 1.64 % ($100/\sqrt{3738}$). Adding in the uncertainty on the number of photons produced originally (0.630 % added in quadrature) and expressing as an FWHM, this is 4.1 %. We have not taken into account other sources of uncertainty, such as the intrinsic line width or the photocathode inhomogeneity, but nevertheless we can compare this with the resolution of an unremarkable germanium semiconductor detector of 1.62 keV at the same energy, equivalent to 0.24 %, which figure takes all sources of uncertainty into account. The same calculation for a gamma-ray of 100 keV gives 10.6 % scintillation resolution.

From this point in the signal chain, the number of electrons increases from stage to stage down the multiplier. At each stage, there will be a decreasing uncertainty on the number of extra electrons produced, each of which will contribute to the total uncertainty.

10.12.2 Factors associated with the scintillator crystal

One would hope that wherever a gamma-ray interacts within a crystal, the amount of light emitted would be the same. By and large, this is so, but there will inevitably be impurity sites and lattice imperfections within the crystal which can interfere with the transfer of energy. If the scintillator is activated, then an extra dimension is introduced because of the potential for inhomogeneous distribution of the activator atoms. It is important, therefore, that the scintillator be pure, homogeneous and of good optical quality. One can expect uncertainties due to these sources to contribute less that 2 % to the overall uncertainty.

A more subtle source of uncertainty is the inherent non-linearity of the response of the scintillator to the energetic electrons produced by the gamma-ray interaction (Figure 10.11). If every gamma-ray were absorbed completely on every occasion, there would be no problem. There would be a non-linearity of energy response but this would have no consequence as far as resolution was concerned. As it is, most gamma-rays are absorbed by a number of interactions, each producing electrons of different energy. From Figure 10.11 we can deduce that if a 661.6 keV gamma-ray were absorbed by a single photoelectric event, a smaller amount of light would be produced than if it were absorbed by several Compton scatters. This is because, in this case, the average primary electron energy would be lower. The scintillator response is higher at lower energies and therefore more light would be produced. There is then a variable spread of electron energies, even for the detection of identical gamma-rays. This extra uncertainty, the intrinsic effective line width referred to above, causes a further spread in the gamma-ray peak. Lanthanum bromide has a much lower intrinsic effective line width – just one of the factors contributing to the considerably better resolution of $LaBr_3(Ce)$ detectors compared to NaI(Tl) (see Menge *et al.* (2007) where the data in Figure 10.11 are compared with $LaBr_3(Ce)$).

Once the scintillation light has been emitted, it must find its way to the photocathode. Studies with narrow beams of gamma-rays have confirmed, not surprisingly, that the light from gamma-rays detected in parts of the scintillator far from the photocathode may be collected less efficiently than from detection events nearby. Again, this uncertainty on the amount of light collected increases the peak width. This may be a particular problem with large detectors and for this reason multiple photomultipliers become a necessity. Detectors with re-entrant and irregular shapes also have problems of light collection. A 3×3 well detector of NaI(Tl) might have a resolution of 9 % compared to the 7.5 % of the normal cylindrical detector.

10.12.3 The variation of resolution with gamma-ray energy

On considering the various terms of Equation (10.4), we find that the change in width of a scintillation detector peak is mainly dependent upon the statistical term, W_i, which is a function of the square root of the gamma-ray energy. The variation of the W_i term with energy is slight in comparison and the W_D term is independent of energy. Thus, the energy relationship might be approximated by:

$$W = a + b/\sqrt{E} \qquad (10.5)$$

where a and b are empirical constants. It is important to appreciate that the resolution of a scintillation detector, expressed conventionally as a percentage relative width, decreases with increasing energy. Knoll (1989) suggests an alternative empirical equation:

$$W = \sqrt{(a + bE)/E} \qquad (10.6)$$

10.13 ELECTRONICS FOR SCINTILLATION SYSTEMS

In comparison with the other sources of uncertainty discussed above, *electronic noise* in scintillation detector systems is a minor problem. More important, as we shall see, is *gain drift* caused by instability in the high-voltage supply. The priorities when selecting electronic modules for scintillation counting are somewhat different from those which determine a system for high-resolution (semiconductor) spectrometry.

10.13.1 High-voltage supply

The high-voltage units described in Chapter 4, Section 4.2 would not be satisfactory for scintillation detector

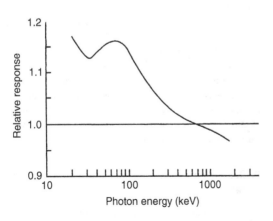

Figure 10.11 The response of the NaI(Tl) scintillator to photons as a function of energy

systems. In the first place, photomultipliers draw a greater current than a semiconductor bias supply unit can provide. The unit must be able to supply currents of the order of milliamperes. Also important is the stability of the supply.

In Section 10.9, it was explained that the dynode electron multiplication factor depended upon the potential difference between successive dynodes. This implies that the overall multiplication, G, of the complete photomultiplier chain should be related to the overall anode to cathode voltage, V, in the following manner:

$$G \propto V^N \tag{10.7}$$

where N is the number of dynodes in the chain. Thus, for a ten-stage photomultiplier we might expect that the gain would vary according to the tenth power of V. In practice, the dependence on voltage is less severe than this and dependencies of V^6 to V^9 may be more realistic. Regardless of the actual exponent, it is obvious that slight changes in voltage will cause a greater proportional change in photomultiplier gain. Such slight changes in gain would add an extra unnecessary uncertainty and degrade the system resolution. For this reason, the stability specification for a high voltage supply for use with photomultipliers is much tighter than that acceptable for semiconductor systems.

For example, the voltage regulation might be specified as within 0.001 % and the temperature stability as + 0.005 % per °C compared to 0.1 % and 0.08 %, respectively, for units intended for semiconductor detector systems. The long-term stability must also be good and figures of 0.01 % over a 1 h period and 0.03 % over 24 h would be typical.

It is desirable that the high-voltage unit should be able to supply either positive or negative polarity to suit different photomultiplier systems. Often, high-voltage supplies for use with photomultipliers have switched voltage increments of 500 or even 1000 V. Such units would be unsuitable for semiconductor detector systems where large bias voltage changes are best avoided.

10.13.2 Preamplifiers

As with all detectors, the pulse of current at the output, in this case the PMT anode, must be integrated to provide the signal. Because electronic noise is usually not a problem, preamplifiers for scintillation systems need not have a particularly low noise specification. All three types of preamplifier – voltage, current and charge-sensitive – are in common use. Charge-sensitive preamplifiers are often offered for routine use but low cost voltage-sensitive types are also common. For normal gamma spectrometry

purposes, there appears to be little reason to prefer one type rather than another. Current-sensitive preamplifiers are provided specifically for fast timing applications and would not be used for normal energy spectrometry.

Preamplifiers built into a photomultiplier base are available, making the complete scintillator/photomultiplier/ preamplifier combination a convenient package. As an alternative, the preamplifier might be built into a low-cost amplifier. In some combined units, a voltage-sensitive preamplifier is used, in others charge-sensitive.

I have assumed so far that the output signal is taken from the PMT anode but there are advantages in taking a signal from the last dynode before the anode. Where a PMT base provides sockets for connection to either of these, it is intended that the dynode output be used for energy spectrometry and the anode signal for timing purposes.

10.13.3 Amplifiers

As with the preamplifier, the scintillation amplifier need not be of such a demanding low noise specification as would be needed for semiconductor systems. In the manufacturers' catalogues, a distinction is commonly made between 'amplifier', suitable for low-resolution spectrometry, and 'spectroscopy amplifier' intended for high-resolution spectrometry using semiconductor detectors. Typical simple amplifier modules provide pole-zero cancellation and automatic base line restoration. The pulse shaping time options provided are often limited on such instruments and may need to be selected internally. Because of the faster rise time of scintillation pulses, the time constants provided are usually within the range 0.2 to 2 or 3 μs.

Developments in detector systems and in computers mean that it is now possible to purchase simple scintillations systems which plug directly into a computer, providing complete functionality in a portable system. The ORTEC *digiBASE*^TM system packs the preamplifier, high-voltage supply and digital signal processor into the base of the photomultiplier with just a single cable to connect to the USB port of the computer on which appropriate software has been installed.

10.13.4 Multi-channel analysers and spectrum analysis

Scintillation spectrometry does not place any special demands on the MCA system. In general, because of the poor resolution compared to semiconductor detectors, spectrum sizes can be much smaller. For example, a general rule of thumb suggests that it is sufficient to have four channels for each FWHM (see also Chapter 11,

Section 11.3.3). Applying that to a NaI(Tl) detector with a resolution of 7 % at 661.6 keV, as few as 128 channels could cover up to 1480 keV or so ($661.6 \times 0.07 \times 128/4$). In practice, larger spectrum sizes are used, perhaps only because MCAs have at least 1024, and often more, channels. Bearing in mind that larger spectrum sizes mean fewer counts in each channel, there is no point at all in using very large spectrum sizes.

The fact that small spectrum sizes can be used means that, with the addition of a multiplexer or mixer/router, several scintillation detectors can be served by the same MCA system. While this would be grossly inconvenient if each detector were in use by a separate person, it does make for a low cost multi-detector system.

Scintillation spectrometry, of necessity, is restricted to simple measurements of few nuclides. Very often, it will be sufficient to use the limited peak area measurement facilities of the MCA system (be it hardwired or software emulator) based upon manually set regions-of-interest. One should be cautious about attempting to use spectrum analysis programs for the analysis of scintillation spectra. Most programs will have been developed with semiconductor spectra in mind and may not be suitable, although ORTEC do provide *ScintiVision-32*, specifically aimed at low resolution spectrometry. If in doubt, the software supplier will be able to advise.

Because the width of peaks in NaI(Tl) scintillation spectra is a strong function of energy, the peaks at low energy are narrow and at high energy very broad. This causes difficulties when attempting to work over a large energy range. If a large spectrum size is chosen, the peaks at low energy will have sufficient channels in them for good spectrum analysis, but those at high energy will be spread over so many channels that peak search may fail. With a spectrum to suit high-energy peaks, low-energy peaks will tend to merge. PGT provide instrumentation with Quadratic Compression Conversion (QCC), specifically for scintillation detection. This uses an ADC that, in effect, has channel width proportional to peak width. The peaks in QCC spectra all have the same FWHM in terms of channel number. High-energy peaks that would span many channels in a conventional spectrum are squashed into fewer channels with larger numbers of counts in them, making spectrum analysis easier. The idea is attractive, but is not widely used.

10.14 COMPARISON OF SODIUM IODIDE AND GERMANIUM DETECTORS

Table 10.2 summarizes the differences between NaI(Tl) and HPGe detectors. In general terms, if spectrometry is of paramount importance then the HPGe must be the

Table 10.2 Comparison of NaI(Tl) and HPGe detectors

Sodium Iodide	HPGe
Cheaper ($\times 10$)	—
More efficient ($\times 10$)	—
Larger volumes available	—
Room-temperature operation	Low temperature operation (77 K)
Sensitive to temperature	Insensitive to temperature
Sensitive to anode voltage (V^7)	Insensitive to bias voltage
Poor energy resolution (6 %, 80 keV for 3×3 at 1332 keV)	Good energy resolution (0.15 %, 2 keV typical at 1332 keV)

detector of choice. In other situations, for example, if only one or two nuclides are to be measured and therefore energy resolution is not of particular concern, scintillation detector systems are a reliable low cost option.

In the early days of high resolution germanium detectors, there was a certain degree of controversy about whether it was better to use low resolution sodium iodide or high resolution germanium for low count rate measurements. The argument for the former was that the more efficient sodium iodide detector provided more counts within a peak from a given amount of radioactivity and therefore better statistics. (The argument has largely ceased to be relevant now that germanium detectors of similar efficiency to sodium iodide detectors are available, albeit at considerable expense.) In fact, as soon as germanium detectors became available with reasonable resolution they have always provided more precise peak areas and a lower limit of detection than sodium iodide detectors.

The reason lies in the fact that the counts in a germanium spectrum, even though fewer in number, are concentrated within a few channels, whereas the counts in the sodium iodide spectrum are spread over many channels. This means that peaks are easier to detect in the germanium spectrum and ultimately the limit of measurement is lower. The huge difference in resolution between the two detectors cannot be compensated by the increase in the count rate from the sodium iodide detector. That being said, there are low count rate situations where the need for high-resolution spectrometry is not paramount and the higher cost of a semiconductor detector is not justified.

Other situations can be envisaged where the detector environment is not suitable for germanium detectors. As long as the spectrometric demands are not too complex the relatively small size of the complete scintillation detector head may make it the detector of choice. In Chapter 13,

I shall explain the use of scintillation detectors in anti-Compton and background suppression systems.

PRACTICAL POINTS

- Until very recently, for all routine purposes the thallium-activated sodium iodide (NaI(TI)) detector has been the most suitable. The standard 3×3 detector has advantages in that there is a great deal of information in the literature which is directly relevant to it.
- $LaCl_3(Ce)$ and $LaBr_3(Ce)$ detectors have superior properties and are now commercially available. Except for low-background applications, they are worth considering.
- If space is limited or a higher absorption coefficient detector is needed, CsI(TI) or BGO right be an alternative.
- For timing applications, consider undoped CsI, BaF_2 or CsF.
- With the exception of the high-voltage unit, low-specification electronic modules are usually satisfactory for scintillation spectrometry. Because the photomultiplier gain is very dependent on voltage, the high-voltage unit must be very stable.
- Ambient temperature will affect the detector to some extent. You should seek to minimize room-temperature changes.
- Scintillation detectors and their photomultipliers are fragile. For field applications, ruggedized systems should be purchased.

FURTHER READING

- For general background information, the manufacturers' literature is worth consulting, especially the Harshaw/QS catalogue Scintillation Detectors (1992), available from agents for Harshaw/QS.

- Knoll, G.F. (1989). *Radiation Detection and Measurement*, 2nd Edn, John Wiley & Sons, Inc, New York, NY, USA.

- Birks, J.B. (1964). *The Theory and Practice of Scintillation Counting*, Pergamon Press, Oxford, UK.

- A very good, if brief, perspective view of scintillation detectors over the last 50 years with a discussion of newer scintillators, with a comprehensive bibliography:

Moszynski, M. (2003). Inorganic scintillation detectors in γ-ray spectrometry, *Nucl. Instr. Meth. Phys. Res., A*, **505**, 101–110.

- Details of many scintillators can be found on the Saint-Gobain website: *(http://www.detectors.saint-gobain.com)*, where there is a link to the 'BrilLanCe Products Whitepaper'.

- For a look at $LaCl_3(Ce)$, in particular:

Shah, K.S., Glodo, J., Klugerman, M., Cirignano, L., Moses, W.W., Derenzo, S.E. and Weber, M.J. (2003). $LaCl_3$: Ce scintillator for γ-ray detection, *Nucl. Instr. Meth. Phys. Res., A*, **505**, 76–81.

Menge, P.R., Gautier, G., Iltis, A., Rozsa, C. and Solovyev, V. (2007). Performance of large lanthanum bromide scintillators, *Nucl. Instr. Meth. Phys. Res., A*, **579**, 6–10.

Choosing and Setting up a Detector, and Checking its Specifications

11.1 INTRODUCTION

In Chapter 3, I argued that germanium is the best material for high-resolution gamma-ray detectors. The detection mechanism was discussed and the energy ranges, shapes and sizes of detector that are commercially available were referred to. Here, I look at what factors should be taken into consideration when selecting a detector system for a particular application. I readily acknowledge a heavy dependence on the manufacturers' literature in this section. The main factors to consider, remembering that a better specification inevitably means higher cost, would be as follows:

- **Energy range**. The upper limit of the energy range for most applications would be 3000 keV. There are few decay gamma-rays with energies above that. On the other hand, prompt gamma-ray energies extend to 10–15 MeV. Extending the energy range upwards makes no difference to the type of detector chosen – a coaxial detector would be appropriate.
- **Detector type**. If measurements at low energy (< 100 keV, say) are not required, then a p-type detector would be fine; n-type detectors are more expensive than p-types. If low-energy photons are relevant, then an n-type detector with a thin window is essential. Remember such windows are fragile. For ultra-low energies, special windows are available. If neutron damage to the detector is likely, an n-type detector should be purchased. Neutron damage in n-type detectors is repairable.
- **Resolution**. The typical detector resolution at this time is 1.9 keV at 1332 keV. Some detectors will have rather better resolution and some of the larger detectors will have poorer resolution because of charge collection

difficulties. In general, one would buy as good a resolution as possible within the constraints of cost, bearing in mind the other aspects of detector choice.

- **Size of detector**. The general observation 'bigger is better' is not necessarily true (see Chapter 13). Certainly, a larger detector will give more counts per Becquerel than a small one; very desirable when measuring very low activity samples. However, for a particular laboratory, it might be more sensible to purchase a larger number of smaller detectors, rather than a few larger ones.

  If sample activities are high, when counts per second are not limited, then a smaller detector may be more satisfactory, with shorter charge collection times, greater throughput and possibly better resolution. However, it can be argued that a large detector with a collimator might be preferable. The reasoning being that, because a larger detector has a larger peak-to-Compton ratio, smaller peaks will be more easily measured.
- **Shape of detector**. The standard 'square cylinder' shape is not necessarily optimum for all situations. In particular, if predominantly low energies are to be measured, the additional germanium in a deep detector, necessary to absorb high-energy gamma-rays, would give an unnecessarily high Compton continuum. A thinner detector, tailored to low energies, would also have better resolution. It is also possible to buy detectors tailored to the width of the samples expected to be measured, for example, ORTEC's Profile Detectors.
- **Count Rate**. Resistive feedback preamplifiers are fitted as standard. If high count rates are likely to be encountered, it is necessary to consider which type of preamplifier to order. Low to medium count rates are readily coped with by the standard resistive feedback

type. For high count rates, specify a transistor reset preamplifier. However, even if count rates are expected to be low, it may be necessary to have a system that will not 'lock out' if unexpected high count rates are encountered, for example, in emergency situations. A transistor feedback preamplifier should then be chosen.

Table 11.1 summarizes these choices for a few typical applications.

11.2 SETTING UP A GERMANIUM DETECTOR SYSTEM

The manufacturer will provide a manual covering the installation and, possibly, commissioning of the detector. Instructions in there should be followed. What follows here is an expansion of that advice. It is my belief that all significant events in the life of a detector should be entered in a log book located close to the detector itself. The time to open the book is when a detector is delivered, starting by pasting in the detector specification sheet, or a copy of it.

11.2.1 Installation – the detector environment

The counting room

Like gamma spectrometrists, gamma-ray detectors and their electronics work better when housed in a good environment. While a 'clean room' environment is not essential, a clean dustless room is an advantage. The ideal room would be cool, temperature controlled, with good lighting and no windows to avoid large temperature variations, and its own stabilized power supply. None of these are essential. Temperature control is certainly worthwhile. If

Table 11.1   Selecting a germanium detector

Application	Activity/count rate level	Energy range (keV)	Detector type	Comments[a]
General use	Low–high	50–3000	Large p-type coaxial	Possible TRP; gated integrator; LFC if high count rates expected
General use, extended to low energy	Low–high	10–3000	Large n-type coaxial	NB: good X-ray efficiency will enhance true coincidence summing. Special wide range detector designs are available
Environmental (plenty of sample)	Low	40–3000	Large p-type coaxial	Low background option
Environmental (small samples)	Low	30–3000	Well	Near 4π efficiency but large summing; not good for complex unknowns
Neutron activation	Medium–high	40–3000	p-type or n-type coaxial	Possibly TRP and LFC
Prompt gamma measurements	Medium–high	Up to 10 000	n-type repairable	Possible absorber for X-rays, risk of neutron damage
Post-accident monitoring	Low–high	40–3000	p-type coaxial	TRP for high throughput; LFC for transient high activity
Fissile material (safeguards)	Low (–high)	3–1000	n-type (short) or planar	Lung monitor needs are similar; large diameter or cluster of smaller detectors
Whole body monitor	Low	40–3000	large p-type or n-type	Unusual large 'sample'; clusters of detectors; shielded room
Portable survey (land/sea/air)	Low	40–3000	p-type coaxial	'Ruggedized' detector, portable cryostat or electrocooling, if handheld
Low energy X-rays	—	1–30	Si(Li)	Optical reset preamplifier for best resolution
	—	0.3–300	Ge ultra-low energy	Special HPGe detector designs available

[a] TRP, transistor reset preamplifier; LFC, loss-free counting.

the room has windows, to avoid ingress of dust, these are better sealed so that they cannot be opened. Bear in mind that dust and other airborne matter may contain natural radioactive material. Dust on sites where radioactive materials are handled will always carry the potential to be contaminated. If the room is on the sunny side of a building, it will probably be necessary to provide blinds to avoid high temperatures on sunny days.

The electrical supply

A clean electrical supply to the counters is recommended. It is better to avoid having electrical machinery on the same supply as the detector electronics. Even if that is not the case, it might be advisable to arrange for a stabilized supply and maybe local anti-spike, anti-surge devices. An **uninterruptible power supply** (UPS) provides peace of mind.

Placement of the detector

We know that all our building materials contain small amounts of uranium, thorium and potassium – all are radioactive to a small degree. Although we will have to provide shielding for our detector in any case, if we are intent on measuring low activity samples it will make sense to install the detector(s) in the centre of the counting room, rather than the more convenient arrangement around the walls.

11.2.2 Liquid nitrogen supply

Before taking delivery of a detector system, unless you intend to use electrical cooling you will need to have arranged a reliable source of liquid nitrogen and the mechanical means of filling the detector; pumps, tubing, funnels – whatever your scheme needs. Remember that all of the liquid nitrogen used to fill the cryostat evaporates into the counting room. Nitrogen is an asphyxiating gas. You might consider installing an oxygen monitor to check the laboratory atmosphere, particularly during detector filling, and if there are many detectors in one space. Liquid nitrogen can also cause cold burns on skin. You will need to provide personal protective equipment and to define safe procedures for transfer of liquid nitrogen from the bulk store to the counting room and into the cryostat(s) for approval by the Health and Safety Officer.

It is advisable to arrange a regular procedure for topping up the liquid nitrogen that leaves an ample safety margin, taking into account the size of the Dewar. For example, the holding time for 30 L Dewars may be quoted as 14 days. A weekly schedule for topping

up has the advantage of simplicity and will allow a good margin for emergencies, such as non-delivery of the liquid nitrogen. A log book to record the date of topping up is recommended. Detector Dewars can be filled with liquid nitrogen in a number of ways, as follows:

- Simply by pouring it into a funnel connected to the filling tube.
- By using a mechanical pump designed for the purpose.
- By pressurization of a reservoir connected to the filling tube. Pressurization can be achieved by closing the vent to the reservoir and waiting (slow) or by using nitrogen gas from a cylinder. In the latter case, it is important to use oxygen-free nitrogen; otherwise liquid oxygen will collect in the dewar.

All of the equipment used to transfer the nitrogen must be able to withstand the rapid cooling by the extremely cold liquid. Metal, nylon and some other polymeric materials will be satisfactory. One can only be sure that the Dewar is full by waiting for liquid nitrogen to overflow from the vent tube. It is unsatisfactory to have liquid nitrogen splashing around the Dewar and the preamplifier, or indeed around the counting room itself. A length of tubing able to withstand liquid nitrogen should be connected to the vent tube, positioned so as to direct the overflow somewhere safe but visible. Note that liquid nitrogen will crack many floor coverings. (It is possible, in normal use of the detector, to direct the nitrogen gas from this vent pipe into the counting chamber to flush out radon gas and so reduce the detector background count rate. This will be referred to again in Chapter 13.)

One manufacturer strongly recommends that all power be removed from the detector electronics while topping up the Dewar. I have not found this necessary for routine topping up and most users would agree. However, it is essential to have all power to the detector switched off when cooling the detector down from room temperature. Whether one tops up the liquid nitrogen during an actual spectrum measurement is debatable. Many detectors will be affected by microphony due to the mechanical noise of the liquid nitrogen passing into the Dewar and by boiling of the nitrogen. This may or may not cause unwanted electronic noise. Caution would suggest pausing a count during filling and for a few minutes afterwards to allow the nitrogen to settle down.

There are commercial units, some in the NIM format, for monitoring the level of nitrogen in the Dewar. These can give audible or visual warnings of a low level and can be arranged to trip out the HV and/or initiate an automatic nitrogen fill cycle.

11.2.3 Shielding

Gamma-ray detectors are very sensitive. Wherever the counting room is, it will have radioactive material around and inside it because of the uranium, thorium and potassium in the building materials, laboratory furniture and even people. In some circumstances, there may be radioactive sources present in the counting room, other than the one being measured. Shielding is, therefore, essential, especially if low activity measurements are to be made:

- A typical purchased shield will be a cylinder of lead, 100 mm thick, with a lid of the same thickness. The bottom of the shield, where the detector pokes through into the counting space, will again be a lead sheet, but perhaps only 30 mm thick, and split with semicircular cut-outs to fit around the detector cold finger.
- For many purposes, interlocking lead bricks of 50 mm thickness will be adequate. If using these, pay attention to access. You will need some sort of closable doorway and you must be able to load your maximum size of sample; Marinelli beakers are bulky. You will also need access for cleaning the interior and retrieving any dropped samples. Do not forget to shield the base of the shield enclosure – the floor will also be radioactive.
- Lead potentially contains ^{210}Pb. Whether purchasing a complete shield or lead bricks, you need to be assured that the lead is of suitable low activity quality. If buying bricks, it may be wise to reserve a set of bricks from a supplier's batch, and run a test for radioactive contaminants on a few samples of that batch. Severe cleaning of the surface of lead bricks has been shown to reduce background levels.
- Steel potentially contains ^{60}Co. Bear that in mind if steel plays any part in the construction – framework, top plate, base plate and sample holders. Pre-1950s steel is best, if obtainable.
- Aluminium contains traces of uranium and thorium. It is best avoided close to the detector.
- In Chapter ??, Section ??, I discussed backscatter, which can cause unwanted features in the spectrum. This will be minimized by having a larger shielded enclosure. If possible, have the lead-to-detector endcap distance greater than 100 mm.
- For an internal 'graded shield' (Chapter ??, Sections ?? and ??), consisting of sheets of cadmium (1–2 mm) and then copper (0.5–1.5 mm) linings, the lead is almost essential for work involving low-energy gammas or X-rays. This will minimize the presence of fluorescent lead X-rays in the spectra. The grading should also be applied to the lid and base of the shield. Recently there

has been a move to use tin instead of cadmium because of the toxicity of the latter. (Within Europe, attempts are currently being made to remove cadmium entirely from the workplace.)
- If you need an interior light, use a filament bulb, not a fluorescent tube. The former is less likely to transmit noise to the preamplifier.

It is likely that you will need some sort of shelf or jig within the shield to hold the sample. This must be constructed of low Z materials to minimize fluorescence and should be as light as possible, consistent with secure holding of samples, to minimize bremsstrahlung and Compton scattering. Rigid plastics are fine. Aluminium, for the reasons stated above, should be avoided.

11.2.4 Cabling

The detector will have a number of cables for connection to the pulse processing system. The following four cables are the most important:

- A multi-way connector to provide supply voltages ($\pm$ 12 and $\pm$ 24 volts) to the preamplifier. This will be supplied with the detector and should be connected to the socket on the rear of the amplifier.
- HV input – this will need a cable able to carry 5000 V and have SHV connectors; this will usually be stamped RG 59/U.
- The signal output – this will need a cable with BNC connectors and usually of 93 Ω impedance. It would be stamped RG 62/U. Systems matched to 50 Ω or 75 Ω would require different cables (see Table 11.2). It is important that cable of the correct impedance is used, particularly if long cable runs are necessary.
- The HV cut-out cable – this is a logic level and the specification of the cable is not critical. It will need BNC connectors.

Table 11.2 Common coaxial cables for spectrometry systems

Reference	Impedance, Ω	Connector	Use
RG 58/U	50	BNC	Signal
RG 59/U	75	SHV	High voltage
RG 62/U	93	BNC	Signal

There may also be other cables – a linear pulse output labelled 'Timing' or 'Output 2' (this will be identical to 'Output 1'), a test pulse input and, in the case of a transistor reset preamplifier, a gating pulse output.

Many more modern systems have much simpler connections. The ORTEC *DSpec*, for example, has a single multifunction cable terminating in a unit into which all of the cables from the detector are plugged. Connection to the computer is via a single USB cable. The *DSpec* unit has 'smart' features, such as the ability to establish the correct bias polarity automatically.

The following general observations are worth making:

- It is preferable to take mains supply for the various components of the system from the same mains power supply socket. This will minimize the chance of ground loops.
- View old cables with extreme suspicion. If in doubt, destroy rather than set aside.
- RG 58 and RG 62 signal cables must not be used for the bias supply. The SHV high-voltage connectors are incompatible with BNC signal sockets.

11.2.5 Installing the detector

Having arranged a home for the detector, we can now consider installing it. Let us assume you have just taken delivery of a new detector:

- Follow the unpacking instructions that accompanied it. Don't discard the packing materials – you may, at some time in the future, need to return the detector to the supplier. The Dewar and detector will usually be delivered in separate containers. Check both items for physical damage. Do not be tempted to fill the Dewar with liquid nitrogen while it is still in its packing case. Spilled nitrogen could be trapped between the Dewar and foam, which might cause damage to the Dewar's vacuum seal.
- If the detector has a thin beryllium window, make sure that the plastic cap provided with it is in place and securely fixed – with adhesive tape, if necessary. That window is very fragile and, if the cap falls off, could be broken while the detector is being handled.
- Many detectors manage without it, but it is not a bad idea to mount the Dewar on an anti-vibration support. If there is heavy machinery nearby in the building, this may be a necessity. A sheet of expanded polystyrene or polyurethane would be fine. Thin plywood on thick polyurethane is sometimes recommended, but note that wood is not a good material for low level counting systems because of the ^{40}K it contains.
- Place the detector in its shielding. It is easier to do this when the Dewar is empty, but you may wish to cool the detector, connect it up and test it before committing it to the shielding. If you are installing a new detector,

or one just returned from repair, you may wish to delay installation in the shielding until you have checked the specification (see Section 11.4).

- Fit the shielding around the stem of the detector. To avoid unwanted electronic noise, the shield should be electrically isolated from the detector and the cryostat. If physical contact between them is unavoidable, insulating material should be interposed.
- Fill the Dewar with liquid nitrogen until overflow from the vent tube is observed. The detector itself has a high thermal capacity. It will take at least 6 h to come to thermal equilibrium. The process of cooling the system from room temperature is discussed in depth in Chapter 12 under 'thermal cycling'.
- Check that the amplifier system is disconnected from the mains and connect the cables from the detector to the amplifier system. Keep the cable-run from preamplifier to amplifier as short as convenient, i.e. install the amplifier system close to the shield. Untidy cabling is a trip hazard and, if several detectors are installed in one place, liable to lead to confusion. Bind together all cables from the NIM-bin (or equivalent) to the preamplifier. This is easier if they are all the same length. When experience has shown that a particular route for cable-runs works satisfactorily, in that expected system specifications are achieved, then do not move the cables. Tape or cable-tie them in place.
- If there is a risk that a detector will become contaminated, and in particular if the detector is liable to be used to measure radioactive solutions, consider covering the endcap with thin plastic film, such as 'Cling-film'. Renew this regularly.

Unless there is a particular need to measure particularly low energies, it would be wise to keep the plastic cap protecting a beryllium window in place. If working with it removed, it would be wise to post a large notice warning users of the delicacy of the window, the disastrous effects of rupture and the terrible retribution that will be visited on any miserable wretch who has the misfortune to break it.

11.2.6 Preparation for powering-up

The detector is installed, the cabling connected and it has been cooling for at least 6 h. I assume in this section that a NIM system is being used and that it is not connected to the mains supply. Figure 11.1, which is based on the Canberra Model 2022 preamplifier, shows the connectors and LEDs available at a typical preamplifier. This is not

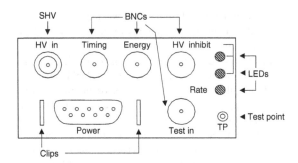

Figure 11.1 LEDs and connectors on a preamplifier, based on Canberra Nuclear Model 2002 (see text for further details)

how one would normally see the preamplifier; nevertheless, the figure is a useful representation of the ins-and-outs of the preamplifier. In practice, for most detectors the preamplifier will be within the detector housing and the signals may be provided by flying leads. The LEDs will be mounted on the bottom or side of the housing. Unfortunately, unless one is a limbo dancer with a very small head, these are very difficult to see when the detector is mounted within its shielding:

- Check that the NIM-bin, or other electronic system, is switched off.
- Check that the HV unit is the type that has continuously variable voltage rather than switchable blocks.
- Look at the detector specification sheet and check that the polarity of the high-voltage unit matches that required by the detector. If the output polarity needs altering, it will usually be necessary to remove the side panel of the module to make the adjustment. Make sure the HV unit is set at zero volts and switched off.
- Ensure all NIM units are securely screwed into the NIM-bin.
- The preamplifier requires a low-voltage power supply ($+12$ and $+24$ V). That is generally provided by the main amplifier via a multi-way cable. Check that that it is in place. Further check that the connector is well secured by clips or screws at both ends (it may be a captive cable at the preamplifier end). It is most important that the preamplifier circuits are powered up and functioning before any high voltage is switched on.
- Connect the HV inhibit from the HV bias supply to either the liquid nitrogen monitor (if installed) or to the HV INHIBIT on the preamplifier. With Tennelec systems, this temperature switch is internal to the preamplifier and no external connector is required.
- Connect the preamplifier output labelled 'ENERGY' or 'OUTPUT 1' to the amplifier input. It will be neater to

use the amplifier input socket on the back panel. Check that the input polarity switch on the amplifier matches the output polarity of the preamplifier pulse.

- Connect the amplifier UNIpolar output to the MCA input. Again, it will be neater to use the socket on the back panel of the amplifier. It is definitely recommended that the output be taken from the rear panel if impedance matching is necessary as that socket can be matched to the cable (usually 93 Ω).
- Make any other cabling connections. The system will be operated in DC-coupled mode. This will usually be preset in each component, but if an option is available ensure that DC-coupling is chosen.
- For the initial setting up, connect the amplifier output signals to an oscilloscope via a T-piece. The oscilloscope should be powered from the same circuit as the NIM-bin.
- Switch amplifier settings to those recommended on the detector specification sheet, or alternatively, put coarse gain to 50, time constant to 4 μs.

11.2.7 Powering-up and initial checks

Resistive feedback preamplifier systems

The detector is installed and ready to power-up. I assume initially that we are dealing with the more common RF or resistive feedback preamplifier and that the electronic system is NIM based. The instructions below are very detailed, intended for when installing a new detector, checking out a detector after repair, or for those wishing to understand more about their detector. On a routine basis, most people would not be as pernickety – 'switch on and wind it up' being more likely:

(1) Switch on the mains to the NIM-bin and to the MCA system. At this point, the ± 12 V and ± 24 V will be supplied to the preamplifier.
(2) The HV unit will have external LEDs showing the bias polarity setting; these operate even when there is no voltage output from the module. Make sure that the LED illuminated corresponds to the polarity required by the detector.
(3) Check that the HV inhibit LED on the preamplifier is green, confirming that the germanium is cool enough (red would indicate too warm).
(4) Check that the 'rate' or 'high count rate' LED is not red, or is only transiently red for a few seconds; this monitors both count rate and leakage current in the detector and neither should be triggering this LED at this stage. If leakage current is indicated, then

improper cooling may have taken place and reme-dial action, as described in Chapter 12, Section 12.3, should be taken.

(5) If everything appears satisfactory, put a radioactive source with a simple spectrum (such as ^{60}Co) near the detector.

(6) Start acquisition on the MCA. Because the HV is not switched on, we can expect a large amount of noise in the lower part of the spectrum. The oscilloscope will show a wide band of noise on the oscilloscope, perhaps 300–400 mV wide.

(7) Switch the detector bias to ON. Even with the HV unit output set to a nominal zero, there may be a change visible. In the example of Figure 11.2, the original noise width on the oscilloscope halved. This reduction is due to a reduction in the detector capacitance.

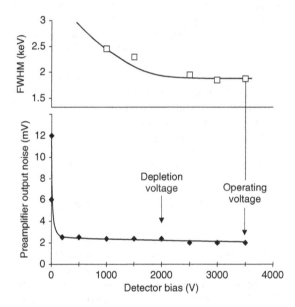

Figure 11.2 An example of noise width measured at the amplifier output as a function of detector bias. The upper plot shows how resolution improves up to the point at which full depletion is achieved

(8) A voltmeter put onto the test point of the preamplifier should read between -1 and $-2V$ and will not be very different from the value given on the specification sheet.

(9) Increase the bias to 200V. Bearing in mind the damage that could be caused to a preamplifier by large voltage fluctuations in the past, it has always been my practice to raise the bias slowly, say 100 volts per second. However, modern systems have

well protected electronics and this does not appear to be strictly necessary. Certainly, the manufacturers' service engineers are much more cavalier when altering the bias.

(10) Do not be alarmed if the system apparently 'dies' on each voltage change and all signals disappear. This should only last a few seconds.

(11) Clear the MCA and start it counting. Some sort of spectrum should be apparent – although maybe not a pretty sight.

(12) Continue a steady smooth increase of the bias, pausing every 500 V to check the noise width (which should decrease at first) and the test point voltage (which should stay constant); continue up to the recommended operating voltage. In practice, it is not necessary to measure the FWHM at this stage, but this was done for the purpose of illustration in Figure 11.2. Note that as the detector capacitance decreases and the noise level decreases, the FWHM improves accordingly.

(13) Once the depletion voltage (noted on the specification sheet) is reached, the noise level will stay constant. The 45 % n-type detector used to derive the data in Figure 11.2 had a depletion voltage of 2000 V and an operating voltage of 3500 V.

(14) When the operating voltage has been reached, allow the system to stabilize for 30 min before continuing with optimization of the electronics.

Transistor reset (TRP) preamplifier systems

The TRP delivers a stepped output, rather than a normal pulse output. Within it are inhibit pulse width and reset delay adjustments. These are tuned at the factory to the particular detector (see the manual) and do not need attention from the user. This type of preamplifier does not have a high count rate LED, because it cannot lock-out. There will be no preamplifier test point and instead of measuring noise, the time between resets (TBR – Section 12.2.2) is monitored. Powering-up does not differ too much from the check-list above, except that:

• Steps (5) to (12). Before any radioactive source is put on and before any bias is applied, connect the oscilloscope to the preamplifier output. A substantial voltage level will be seen; taking the Canberra 2101 as an example, the level will be -12 V for n-type detectors, and $+12$ V for p-type detectors.

• Bias should again be applied gently when it should be seen that the output becomes a step ramp, as seen earlier in Chapter 4, Figure 4.9. On an n-type detector this will move from -2 V to $+2$ V before resetting.

For a p-type detector, the voltage will move from $+2\,V$ to $-2\,V$.

- When the bias is at the recommended operating voltage, then the reset frequency should be firmly established at or near the TBR shown on the specification sheet and the dynamic range of the reset procedure should be (in this case) still about 4 V.
- Now the ^{60}Co source can be introduced. That will cause the TBR to shorten. When this occurs, connect to the MCA and check the presence of the characteristic two-peaked spectrum.
- Steps (13) and (14) apply as before.

11.2.8 Switching off the system

When switching off (or powering down in modern jargon), the detector bias should first be reduced to zero and then switched to OFF. Allow a short time for the bias voltage to decay away before switching off the NIM bin. If the detector is not to be used for some time, it would make sense to switch the system off, minimizing the risk to the preamplifier should the mains power supply fail and later, in an uncontrolled manner, reconnect. On a routine basis there is no value, other than concern for the environment, in switching the system off unless necessary because of the time taken to reach a working equilibrium after switching on again.

11.3 OPTIMIZING THE ELECTRONIC SYSTEM

11.3.1 General considerations

There is a variety of electronic units – some have been around for years and some are recent designs. It would not make sense to try and cover all of the detailed procedures for all modules. While, from time to time, dialogue between the manufacturers and the users may not be sweetness and light, setting up the electronic system is the time to depend upon their advice. In short, *read the manual*! I will here only mention general points with the object of explaining why the detector electronics are set up in a particular manner. If there is a conflict between the suggestions here and the manual, ignore this and follow the manual! Chapter 4 covered the purpose and mechanism of the circuits within the amplifier. In this section, I will cover implementation. I will refer to the detector specification. Table 11.3 is a generic example. Not every manufacturer's sheet will contain all of these items, and most will contain additional or alternative items. Nevertheless, it will give the user an idea of what to look for when he/she receives their new detector specification sheet.

11.3.2 DC level adjustment and baseline noise

All units in the counting chain are DC-coupled. On some amplifiers there is a screw potentiometer to adjust the DC level, the practical baseline level for unipolar pulses. Do not be alarmed if your amplifier does not have this adjustment; if this is the case pass directly to 11.3.3. Otherwise:

- Disconnect the input to the amplifier from the preamplifier.
- Connect the amplifier unipolar output to the oscilloscope, making sure that the DC coupling is selected. There should be no pulses.
- Switch the oscilloscope to the 'ground' setting. This will display the oscilloscope's zero level. You may wish to adjust the position of the oscilloscope trace so that it falls along one of the display graticules.
- Adjust the DC level screwdriver control on the front panel of the amplifier until there is no movement of the oscilloscope baseline on switching between the amplifier and the oscilloscope ground. The DC level is now correctly set.
- Reconnect the preamplifier input to the amplifier.
- Look at the baseline on the oscilloscope. The width of the noise in a good system should be about 5mV and not more than about 10 mV.
- There should be no regular oscillations. Change the time-base (horizontal time axis) over a wide range to check this. See also 'Troubleshooting' in Chapter 12.

11.3.3 Setting the conversion gain and energy range

It is first necessary to decide on the energy range required and from that the appropriate conversion gain (i.e. spectrum size). In Chapter 4, Section 4.6.8, it was explained that the optimum spectrum size is a balance between sufficient channels within the peaks of interest to aid computer deconvolution of peaks, but not so many that the statistical uncertainty on each channel suffers. The relationship in Equation (11.1) was suggested:

$$\text{Conversion gain} = \text{energy range}$$
$$\times 4/\text{FWHM of representative peak}$$
$$(11.1)$$

This will give a number that must then be rounded, up or down as common sense suggests, to an available spectrum size – 4096, 8192 or 16384.

For example, the energy range 0–2000 keV is suitable for many purposes. Suppose we intend using the detector described in Table 11.3. If we wish to arrange matters so

Table 11.3   Generic detector specification and performance data sheet[a]

DETECTOR SPECIFICATIONS AND PERFORMANCE DATA

Specifications
The purchase specifications and therefore the warranted performance of this
detector are as follows:

Model: #####   Serial Number: #####
Cryostat description: Vertical dipstick, type 2000B slimline
See below for warranted performance specifications.

Physical Characteristics

Geometry:	Coaxial, one open end, closed end facing window
Diameter:	80.5 mm
Active volume:	____ cc
Length:	107.3 mm
Distance from window:	4 mm
Well depth:	N/A mm
Well diameter:	N/A mm
Inactive Ge thickness:	700 μm

Electrical Characteristics

Depletion voltage:	+ 1500 V DC
Recommended bias voltage:	+ 3000 V DC
Leakage current at recommended bias:	0.01 nA
Preamplifier test point voltage at recommended voltage:	−1.6 V DC
Capacitance at recommended bias:	_____ pF

Measured Performance

Parameter	Warranted	Measured	Time constant
Resolution at 1.33 MeV	2.25 keV	2.15 keV	6 μs
Peak-to-Compton Ratio	78	83.4	6 μs
Relative efficiency at 1.33 MeV	100 %	114.2 %	6 μs
Peak shape factor (FWTM/FWHM)	2.00	1.93	6 μs
Peak shape factor (FWFM/FWHM)	3.00	2.74	6 μs
Resolution at 1.33 MeV	1.400 keV	0.912 keV	6 μs

Tested by: _____   Date: _____
Approved by: _____   Date: _____

[a] The data in this sheet are actual data, but from more than one detector. Identification data
have been removed.

that there are sufficient channels within the 1332.5 keV
peak (2.15 keV FWHM) then Equation (11.1) suggests
3721 channels (2000 × 4/2.15) or, rounded up, a spectrum size of 4096 channels. If we are more interested in
peaks at the lower end of the spectrum – the 122 keV
peak (FWHM 0.912 keV), for example – then a spectrum
size of 8772 channels, rounded down to 8192, would be
required. At intermediate energies, either 4096 or 8192
would be acceptable. Having set up the MCA system
appropriately, the amplifier gain needs to be adjusted. In
the list below, assume that we require a range of about

2000 keV over a 4096 channel spectrum. The energy scale
will be approximately 0.5 keV per channel:

- Check that there is no digital off-set on the ADC.
- Place an appropriate source on the detector and start a
  count. 'Appropriate' here means a source with an easily
  identifiable peak, or peaks, within the upper half of the
  spectrum. For example, with a range of 2000 keV, ^{60}Co
  with its 1173 and 1332 keV peaks would be ideal.
- Now adjust the amplifier gain, coarse and fine, so as to
  place a target peak where expected. For the 1332 keV,

the gain would be adjusted to place the peak centroid close to channel 2660. There is no need to make exact adjustment at this stage because other optimizations will affect the gain of the amplifier to a small degree.

- Wait until after the time constant and pole-zero adjustments have been made before doing a proper energy calibration.

11.3.4 Pole-zero (PZ) cancellation

Correct PZ cancellation is essential for good resolution at any count rate other than a very low one (Figure 11.3 – this is part of Figure 4.20). It is also an essential preliminary to the baseline restoration (BLR) described in Section 11.3.6. The correction is not required with the transistor reset preamplifier. It is a dynamic on-line process. Chapter 4, Section 4.6 contains background information on pulse overshoot or undershoot.

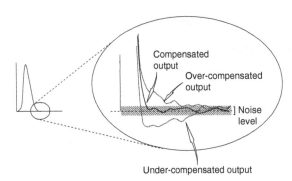

Figure 11.3   Pole-zero cancellation

There are three ways in which this adjustment can be made.

Oscilloscope and source

This is the most common method:

- Place a radioactive source near to the detector so as to give a few thousand counts per second in the spectrum. Unless the MCA system can be persuaded to display the overall count rate, it is helpful to have a rate meter in the system, to give an instant reading of the input count rate.
- Check the amplifier manual to see whether particular switches must be set; for example, a 'BLR mode' switch may have to be set to a PZ position.
- Choose and note the amplifier time constant.

- Connect the UNIpolar pulse output to the oscilloscope. Check that the oscilloscope is DC coupled.
- Observe the trailing edge of the pulse on a sensitive vertical scale of 100 or 50 or $20\,\text{mV}\,\text{cm}^{-1}$. (There will sometimes be too much noise on the latter.) It should look like Figure 11.3.[1]
- Now adjust the PZ screw control or controls so that the trailing edge returns to the baseline with no undershoot or overshoot. It is easier to do this from the overshoot situation, that is, the pulse tail dipping below the baseline; the adjustment from this position is made by turning the screw clockwise. Some amplifiers have separate coarse and fine controls of PZ; some, such as the Tennelec 245, have LEDs indicating the direction of the adjustment.
- Some older amplifiers may show the behaviour shown in Figure 11.4, where correcting for the initial overshoot increases the longer term undershoot. All I can recommend is a process of trial and error; moving the PZ control by units of 1/8 of a turn and measuring the resolution on each occasion. Eventually, an effective minimum deviation from the baseline will be found at that particular count rate, resulting in a minimum FWHM. Note that the pole-zero must be readjusted whenever the amplifier time constant is altered or there are substantial changes to the gain.

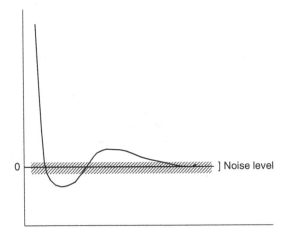

Figure 11.4   Pole-zero cancellation – an awkward situation

---

[1] Note: Some oscilloscopes (e.g. Tektronix 465 and 475) will overload when these sensitive ranges are used with input pulses of 5 to 10 V. A limiting device may be built into the amplifier (as in the Tennelec TC244) or can be used externally (e.g. Canberra's Schottky Clamp LB1502).

After successful pole-zero cancellation, peaks in the spectrum should show neither low-energy nor high-energy tails. Examining the spectrum on a logarithmic scale will emphasize tails. Practical experience suggests that peak width is less sensitive to over-compensation than under compensation. If in doubt, choose to do the former. Chapter 14, Section 14.3.2 suggests an iterative fine-tuning procedure that could be useful at high count rates.

Oscilloscope and square wave

This method, similar to the method above but using a square wave instead of detector pulses, is said to be more sensitive. Consult the amplifier manual for details of how to apply this method.

Automatic correction

A few more recently introduced amplifiers allow an automatic PZ cancellation simply by pressing a button on the front panel. No oscilloscope is needed but, if one is available, examining the amplifier output will allow the user to watch the correction as it happens. The author, notoriously sceptical of automatic procedures, has been pleasantly surprised by the accuracy of such corrections!

11.3.5 Incorporating a pulse generator

Chapter 4, Section 4.7.3 discussed the use of a pulser for making allowance for losses due to dead time and pile-up (random summing) in a system. This section provides practical advice. The pulse generator must provide pulses that adequately simulate the preamplifier output pulses. The type of pulser will depend on the type of preamplifier. The pulse amplitude must be variable over the whole MCA conversion range and very stable, so that a narrow peak can be placed in a convenient place in the spectrum clear of sample peaks. The pulse repetition rate must be adjustable and stable, triggered by a calibrated internal clock. A more complicated arrangement, which would have the virtue of providing a stream of pulses randomly distributed in time, would be to have the pulser triggered by the decay of a radioactive source. This would involve a subsidiary source and detector and a scaler to count the number of pulses produced.

The output impedance of the pulser should be $93\,\Omega$. It should be connected to the 'Test Pulse Input' on the preamplifier.

With a resistive feedback preamplifier

We saw in Chapter 4, Figure 4.7 that this type of preamplifier provides an output with a very fast rise-time and a long fall-time. To emulate this, a high quality pulse generator is required; unfortunately, this will be expensive. It must provide pulses with a rise-time, variable within the range 50 to 500 ns, and be able to set the fall-time over a wide range up to $500\,\mu s$. To set up the pulser:

- With no pulses from the pulse generator, carry out a good pole-zero correction on the detector pulses.
- Switch on the pulse generator, and observing the amplifier output with the oscilloscope scales set as though the PZ were to be corrected, adjust instead the decay-time of the pulse generator until good return of the pulser pulses to the baseline is seen with no overshoot or undershoot.
- The pulse amplitude can now be adjusted so as to place the 'pulser peak' in a region of the spectrum clear of peaks and with a flat baseline; this will often be at high energy.
- The pulse repetition rate should be set so that it is less than 10 % of the detector pulse rate.

With a transistor reset preamplifier

With the transistor reset preamplifier, the procedure is much simpler. The preamplifier output is a step pulse. A square wave generator can be used instead of the pulse generator described above. Correction of pole-zero is not necessary. It is worth checking with the oscilloscope that within the pulse train the detector pulses and pulser pulses behave in a similar fashion.

11.3.6 Baseline restoration (BLR)

This was described in Chapter 4, Section 4.7. The process is done automatically within the amplifier as long as it is enabled (the AUTO setting is usually satisfactory) and the appropriate count rate setting is selected. Unless the AUTO setting is selected, it will be necessary to set the BLR threshold manually. Consult the manual.

11.3.7 Optimum time constant

The manufacturer will recommend the shaping time constant to be used. The optimum time constant (Chapter 4, Section 4.4.4) depends upon the size, shape, charge collection characteristics and, maybe, the count rate. A general guideline suggests that, in order to achieve good charge collection, the shaping time constant should not be less than $10 \times$ the longest preamplifier pulse rise-time. A time constant of $3\,\mu s$ might be suitable for a small, say 20 %, HPGe detector, $8\,\mu s$ for a 150 % HPGe. Si(Li) detectors tend to need longer shaping times, reflecting the lower charge carrier mobilities in silicon, and NaI(Tl) detectors much shorter times.

With a new, or repaired, detector it may be worthwhile to check whether you agree with the manufacturer's suggestion. You may wish, in any case, to select a shorter time constant in order to cope with particularly high count rates, accepting a small resolution penalty in return.

The optimization procedure is as follows:

(1) Present a source of ^{60}Co to the detector so that the count rate is, say, 1000 cps. ^{60}Co is suggested as a suitable source so that you can compare the measured resolutions with that on the specification sheet.
(2) Set the time constant to a low value, say 1 μs.
(3) Make the PZ cancellation adjustment.
(4) Measure a spectrum for a long enough time for the 1332.5 keV to be well defined.
(5) Estimate the FWHM of the 1332.5 keV. This will often be done for you by your spectrum acquisition software. Because altering the shaping time constant is likely to alter the energy calibration, it may be necessary to correct that FWHM. A useful approximate estimate of FWHM to compare with the specification can be obtained by using the equation:

$$\text{FWHM (keV)} = \text{FWHM (channels)}$$
$$\times \frac{\text{apparent peak energy (keV)}}{\text{peak position (channels)}} \quad (11.2)$$

(6) Go back to step (2) and increase the time constant to the next setting. Repeat steps (3) to (5).
(7) When all amplifier settings have been covered, you should be able to plot FWHM against shaping time constant. The optimum is that which gives the lowest FWHM. You would expect that to be similar to the value on the specification sheet, with a shaping time constant similar to that recommended by the manufacturer.

If, in fact, you intend to use the detector at high count rate, the information you now have would allow you to estimate the resolution penalty if you were to use, say, a 2 μs shaping time instead of 4 μs. A small resolution penalty might be worth it to double the throughput. You may also wish to repeat the exercise above at a much higher count rate. It is likely that you would create two curves similar to those in Figure 11.5, which demonstrate that the optimum shaping time shifts to lower shaping times at very high count rate.

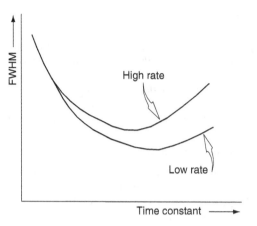

Figure 11.5   System resolution at high and low count rates as a function of amplifier time constant

11.4 CHECKING THE MANUFACTURER'S SPECIFICATION

11.4.1 The Manufacturer's Specification Sheet

With each new detector, or after a detector repair, the user will receive a specification sheet which should be retained. A typical sheet (see Table 11.3) will contain, in addition to model number, serial number, type of Dewar, etc., the following information:

- **The warranted parameters** – resolution at particular energies, peak width parameters, peak-to-Compton ratio and relative efficiency. This is the specification on which you purchased the detector and expect to be achieved.
- **The measured parameters**, as above. Having sold the detector to you, the manufacturer will do predelivery checks, providing the information to you in the 'measured' section of the specification sheet. It is usual for the measured values to be rather better than those warranted. If measured values are worse, you have cause for complaint!
- **Physical dimensions** of the detector, including such things as detector-to-cap distance. (These are needed when assessing summing, variation of count rate with sample-to-detector distance and for setting up MCNP and similar calibration models.)
- **Electrical parameters** of the detector – bias polarity, depletion voltage, operating voltage, suggested amplifier shaping time, and maybe preamplifier test point voltage.
- The window thickness and inactive germanium layer thickness.

Such a specification would refer to a standard coaxial detector. Specification sheets for other types of detector will contain other information and not all the parameters noted in Table 11.3 will be warranted. For example, for a well detector geometric details of the well and the active volume will be quoted. It is likely that only the detector resolution will be warranted. Peak-to-Compton ratio may be measured, but not warranted. For low-energy detectors, only the resolution will be warranted.

The analyst's first task when receiving a new detector, or after a detector repair, is to check the information in the specification by practical measurements. The industry standard procedures for measuring these parameters are given in ANSI/IEEE Std 325-1986 (see *Further Reading*). It is important that, before checking the specification, the settings of the detector system, particularly the time-constant, have been optimized. Any deficiency in setting up will inevitably be reflected in a poor peak shape. (The converse is, of course, also true; a poor peak shape means something is wrong!)

11.4.2 Detector resolution and peak shape

The first parameter to check is the resolution of the detector. In addition, checking this will also immediately confirm that the electronic system has been set up correctly. Unless that is the case, there is little point in pressing on with efficiency and peak-to-Compton measurements.

Resolution is a measure of the width of a full energy peak. It varies with energy and, when used as a statement of performance is measured at a particular energy. For wide range HPGe detectors, the resolution is always quoted at 1332.5 keV (^{60}Co). Often, a measurement will also be quoted for 122.1 keV (^{57}Co). Table 11.4 lists a number of gamma-ray energies traditionally used for specification purposes for different types of detector. The primary measure of width is Full-Width-at-Half-Maximum.

Table 11.4 Energies recommended for resolution measurements

Energy (keV)	Nuclide	HPGe		NaI(Tl)
		Large	Small	
1332.5	^{60}Co	*	—	—
661.7	^{137}Cs	—	—	*
122.1	^{57}Co	*	*	—
59.5	^{241}Am	—	—	*
22.1	^{109}Cd	—	—	—
5.9	^{55}Fe	—	*	—

The procedure is simple. Place a ^{60}Co source near to the detector so that the count rate is about 1000 cps and measure a spectrum for sufficient time that the uncertainty on the measured peak area is 1 % or less. Then, using the interpolative method illustrated in Figure 7.4 (Chapter 7), estimate the width of the peak at half its maximum height. Unless the peak is centred on a channel, estimating the peak height manually by interpolating between two channels is difficult. It is worthwhile making a slight adjustment to the amplifier gain to centre the peak on a channel. Plotting out channel data and graphically interpolating the channel contents is useful and results in a permanent reminder of the peak shape for comparison in the future when resolution problems beset one. Of course, if the spectrum analysis program provides an immediate measure of FWHM, most users would depend upon that.

Apart from the peak width being satisfactory, we would also like to be assured that the shape of the peak is good – it does not have a tail at low or high energy. This can be assessed by also measuring the width of the peak again at a lower height. Typically, the manufacturers will quote Full-Width-at-one-Tenth-Maximum (FWTM or FW0.1M), and sometimes at one fiftieth (FWFM or FW0.02M). The point is that if a peak does have a tail, which might not be evident at half peak height, it is more likely to affect the width lower down the peak. Figure 11.6 is an idealized

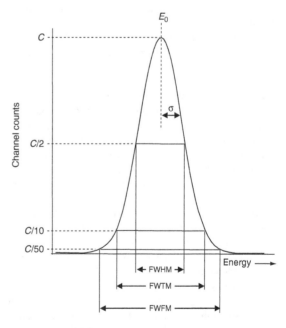

Figure 11.6 Detector resolution. Representation of a Gaussian peak with full-width-at-half-maximum, full-width-at-one-tenth-maximum and full-width-at-one-fiftieth-maximum

representation of a peak in the spectrum. In a well set-up system, this will be an approximation to a Gaussian, bearing in mind that a spectrum peak is a histogram, not a smooth mathematical curve. The figure shows the relationship between the three width measurements.

A Gaussian can be defined by three parameters: its position, its area and its width, or standard deviation. Equation (11.3) expresses this mathematically. The height of the Gaussian at a particular channel, C, a distance from the centroid of the peak, $E - E_0$, is calculated from A, the area (number of counts in the peak) and σ, the standard deviation. C is, in effect, the number of counts in the channel:

$$C = \frac{A}{\sigma\sqrt{2\pi}} \exp\left[\frac{-(E - E_0)^2}{2\sigma^2}\right] \qquad (11.3)$$

This equation allows us to deduce a number of things:

(1) FWHM $= 2.355 \times \sigma$.
(2) FWTM $= 4.292 \times \sigma$.
(3) FWFM $= 5.594 \times \sigma$.
(4) FWHM $= 0.939 \times A/C_0$, where C_0 is the height of the peak at the centroid. This is the basis of the quick manual method of estimating FWHM described in Chapter 7, Section 7.5.2.

From relationships (1) and (2) above, we can deduce that if a peak is indeed Gaussian in shape, the expected ratio of FWTM to FWHM will be 1.823. Therefore, if the ratio of our measured widths is substantially greater than that, we can deduce that the peak shape is not near enough Gaussian. How substantial is substantial? The manufacturers have traditionally used a limit of 2; anything less than that being regarded as a good peak shape. In fact, it is very easy to show that peaks can have a very noticeable tail and still have a FWTM/FWHM ratio less than 2. (Tails can be created simply by 'unadjusting' the pole-zero cancellation.) In the first edition of this book, John and I suggested that a limit of 1.9 is more realistic.

Similar considerations apply to the FWFM/FWHM ratio. A pure Gaussian would give a ratio of 2.376; the manufacturers tend to warrant a ratio of better than 3; I would prefer a ratio better than 2.5. (PGT currently use peak width ratios of 1.9 and 2.5 in their warranties.) This information is summarized in Table 11.5.

If these ratios are significantly less than the expected values, something is wrong. Most probably, insufficient counts will have been accumulated, and the result is a statistical accident. Alternatively, it might be that the height of the peak has been inaccurately measured. Unless

Table 11.5   Ideal and acceptable resolution ratios

Ratio	Value for Gaussian	Acceptable in practice
FWTM/FWHM	1.82	< 1.9
FWFM/FWHM	2.38	< 2.5

the peak is centred on a channel, it is difficult to estimate the *true* height.

From a specification check point of view, then, it is necessary to estimate both FWHM and FWTM. FWFM may be a little difficult to determine because, unless the peak contains a very large number of counts, the channels with counts close to 1/50 of full height will be close to the peak background continuum. Taking the specification in Table 11.3 as an example, the detector would be judged satisfactory relative to the warranted ratios, but not on the ratios given in Table 11.5.

In addition to such mathematical checks, there is no substitute for close visual examination of the spectrum. Set the MCA display scale to logarithmic, which emphasizes the lower parts of the peak, place the cursor at the peak centroid and compare the peak shapes on either side. The peak should be symmetrical. As it happens, the detector to which the peak width data in Table 11.3 referred did produce peaks with a visible low-energy tail – probably caused by incomplete charge collection in such a large detector.

The user should be aware that the value of the detector resolution is a major selling point for the supplier, and it is only natural and reasonable that the measurements reported on the specification sheet will have been obtained under near perfect conditions. Having said that, it is the experience of the author that resolution values close to the manufacturer's, and certainly better than the warranted values, will usually be obtained. The manufacturers will, of course, be using their highest specification equipment, setup expertly. They will also have performed the measurement at low count rate and by arranging the electronics so that there are a large number of channels in the peak. Figure 11.7 shows the measured FWHM at 1332.5 keV for a particular detector system. Each measurement was made with a different spectrum size. The resolution is apparently better the larger the spectrum size. This is an artefact due to the discrete nature of the data. With more channels, interpolation at the measurement width can be performed more accurately. The recommendation is that for this special case of FWHM measurement, the MCA should be set up so that there are at least 10 channels in the peak.

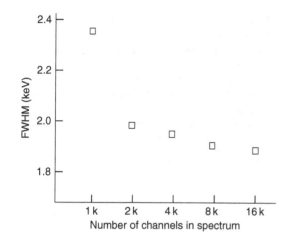

Figure 11.7 Variation of measured FWHM with number of channels in the spectrum. The manufacturer's resolutions were: warranted, 1.9 keV; measured, 1.82 keV

A suitable procedure would be:

- Place a ^{60}Co source near to the detector so that the count rate is no more than 1000 cps. The activity of the source does not need to be known. The details here are given for the 1332.5 keV peak. For measurement at other energies, a similar procedure should be followed, ensuring that one FWHM is spread over at least ten channels.
- Switch the amplifier time constant to that reported in the detector specification. Ensure that the electronic system is set up properly. Pay particular attention to baseline restoration and pole-zero correction.
- Arrange the amplifier gain and ADC/MCA so that the energy calibration scale is about 0.15 keV/channel and the 1332.5 peak is visible in the spectrum. It may be necessary to use the ADC digital offset to shift the spectrum to lower energies.
- Both the 1173.2 and 1332.5 keV peaks should be within the spectrum so that they can be used to provide a two point energy calibration.
- Check on an oscilloscope that pulses are not over-loading (that is, they are not flat-topped).
- Collect a spectrum until the centroid of the chosen peak can be readily calculated or assessed. Then, adjust the fine gain of the amplifier until the centroid sits squarely on an integral channel (to within 0.1 channel). This will make subsequent reading of the peak maximum much easier.
- Collect a spectrum until there are 10 000 counts in the 1332.5 keV peak.

- Perform an energy calibration and set up a region of interest (ROI) about the peak, running the ROI into three or four continuum channels on either side. This is best done on a logarithmic display scale. Use the MCA facilities to calculate the FWHM in energy units.
- If you are suspicious of the software FWHM calculation, then a physical plot of channel content as a function of channel number onto graph paper is not difficult. The FWHM will be in 'number of channels' in the first place. You may well find that the graph points on the sides of the peak fall on a straight line, easing the task of interpolation and measurement. If the peak you are measuring is sitting on a background continuum which is more than 1 % of the peak maximum, then that background must be subtracted when finding half the maximum (Figure 11.8).

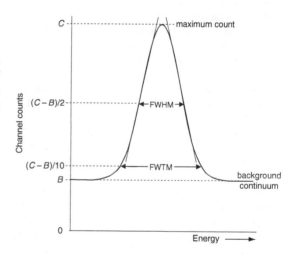

Figure 11.8 Demonstrating that the background *B* must be subtracted in determining half the maximum, and that at the FWHM level the data points are approximately linear on either side

Table 11.6 lists typical resolutions for detectors currently available. The range of resolution varies with the quality (and price) of the detector, and the size. In general, the larger the detector, the worse the resolution.

11.4.3 Detector efficiency

Coaxial detectors

When discussing detector efficiency informally, figures such as 15 and 45 % are bandied about. It is when

Table 11.6 Detector resolutions (in keV) available commercially[a]

Detector type	Photon energy (keV)		
	5.9	122	1332.5
Coaxial p-type	—	0.8 to 1.4	1.7 to 2.5
Coaxial n-type	0.66 to 1.2	0.8 to 1.4	1.7 to 2.7
n-type Short	0.3 to 0.5	0.6 to 0.72	—
n-type Well	—	1.1 to 1.4	1.9 to 2.3
Planar	0.14 to 0.4	0.48 to 0.72	—
Ultra-low energy	0.14 to 0.16	0.48 to 0.72	—
Si(Li)	0.16 to 0.25	—	—

[a] Data from suppliers' catalogues (1993).

the salesperson starts to talk about detectors of 150 % efficiency that unease sets in. As my late co-author wrote in the first edition *there is a whiff of the perpetual motion machine, of the philosophers' stone.* Of course, we are not challenging the laws of physics; these figures are relative, not absolute, efficiencies. The historical basis for this is that in the beginning was sodium iodide. When the newfangled germanium (actually Ge(Li)) detectors appeared, what everyone wanted to know was *how do they compare with my sodium iodide detector?* So, it became common to relate the germanium detector efficiency to that of a standard 3×3 NaI(Tl) detector (i.e. 3 in. diameter by 3 in. high – in these metric times, 76 mm × 76 mm). That particular detector was the most readily available, could be made very reproducibly (unlike HPGe detectors) and was present in most gamma spectrometry laboratories.

This figure-of-merit relative efficiency is defined as: the ratio of the counts per second in the 1332.5 keV peak of ^{60}Co when the source is counted at 250 mm source-to-detector distance on the axis of the detector, to the counts per second in the 1332.5 keV peak when the same source is measured at the same distance by a 3×3 NaI(Tl) detector. It is strictly only relevant to wide range coaxial detectors. Planar and other low-energy detectors are not designed to cover the energy range up to 1332.5 keV.

Fortunately, it is not necessary to possess a 3×3 NaI(Tl) detector in order to check this item of the specification. Because 3×3 NaI(Tl) detectors can be manufactured reproducibly, it is known that the count rate at 1332.5 keV produced by any such detector, with a ^{60}Co source at 250 mm, is 0.0012 cps per Becquerel. The procedure is as follows:

- Acquire a point source of ^{60}Co of known disintegration rate (about 10^5 Bq, known to within 1 % standard uncertainty).

- Place this at precisely 250 mm from the centre of the top of the endcap on the detector axis.
- Collect a spectrum in a properly set up system (with BLR in play, PZ corrected, etc.), using the amplifier time constant recommended by the manufacturer. Continue until at least 20 000 counts have been accumulated in the 1332.5 keV peak.
- Measure the number of counts in the 1332.5 keV peak and divide by the live time of the count.
- Use the following equation to calculate the relative efficiency as a percentage:

$$\text{Relative efficiency} = \frac{(\text{net cps in 1332.5 keV peak}) \times 100}{(\text{Bq of } ^{60}\text{Co at count time}) \times 0.0012} \, (\%)$$

$$(11.4)$$

Note that ^{60}Co is a common component of the system background and in principle could introduce a high bias to the measurement. Normally, this will be negligibly small in comparison to the source activity. Nevertheless, this should be borne in mind and an appropriate correction for background made if necessary.

In principle, it should be possible to calculate the relative efficiency from the detector dimensions. According to ORTEC, the relationship is not simple, but approximately 4.3 cm^3, or 23 g of germanium, are required for each 1 % of relative efficiency. If the active volume of the detector is quoted in your specification, it will be possible to relate your measured relative efficiency to that. Note that using the quoted detector diameter and height to calculate the overall detector volume would give a value greater than the actual active volume because of the presence of the dead layer, the contact hole drilled into the detector and the bulletization.

From the point of view of checking the detector specification, this single spectrum can be used to check all warranted parameters – resolution, peak shape parameters, relative efficiency and peak-to-Compton ratio.

Well detectors

It is not standard practice to warrant relative efficiency for well detectors, but if it were the procedure described about could be used unaltered. The ANSI/IEEE standard includes a procedure for determining the absolute efficiency of well detectors but this has not been adopted generally by the manufacturers and unless specially requested, it would not normally be on the specification sheet. Normally 'active volume' of germanium is quoted as a measure of efficiency. Unfortunately, this latter is a parameter that the user is unable to check. The ANSI/IEEE procedure for measuring a standard absolute efficiency is:

- Place a standardized ^{60}Co point source of known disintegration rate at the time of the count (about 10^4 Bq – one tenth the activity of that used at 250 mm) 10 mm above the bottom of the detector well.
- In order to minimize random summing, make sure that pile-up rejection is enabled and that the count rate is low.
- Collect a spectrum for a known live time.
- In the spectrum there will be full energy peaks at 1173.2 and 1322.5 keV and in addition a sum peak at 2505.7 keV. The sum peak represents counts lost from the full energy peak. Note the number of counts within the 1332.5 and 2505.7 keV peaks.
- Calculate the in-well efficiency at 1332.5 keV as:

$$\text{Efficiency} = \frac{(1332.5\,\text{keV counts} + 2505.7\,\text{counts})}{(\text{seconds live time} \times \text{Bq }^{60}\text{Co})}$$

$$(11.5)$$

11.4.4 Peak-to-Compton (P/C) ratio

This single factor has elements of both detector resolution and detector full energy peak efficiency – the bigger the detector, the greater the likelihood of complete gamma-ray absorption and therefore more counts in the full energy peak rather than the Compton continuum, and the better the resolution, the higher the full energy peak because the narrower it is. However, the detector environment will also have an effect. All material near the active volume of the detector will be responsible for Compton-scattered gammas entering the detector and thus increasing the level of the Compton continuum. This will include the detector mounting, the dead layer/contact thickness of the detector, the material used in the endcap and cryostat, as well as its geometrical arrangement and thickness.

The lead of the shielding also has an effect. Because of that, the maker's measurement will have been made without any lead shielding; one recommendation is to have the detector *at least three feet (one metre) away from other objects*. The author's practical experience is that installation in a typical shield makes little difference to the ratio. As described in the ANSI/IEEE standard, it is the ratio of the number of counts in the highest channel of the 1332.5keV ^{60}Co peak to the average channel count in the Compton continuum between 1040 and 1096 keV in that same spectrum (see Figure 11.9). The ratio is analogous to a signal-to-noise ratio. The procedure is very straightforward:

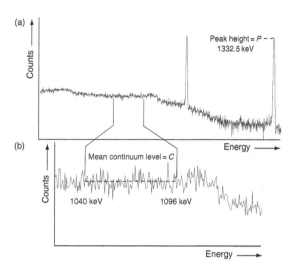

Figure 11.9 Peak-to-Compton ratio – P/C (^{60}Co source). (a) The general location of P and C (counts on a logarithmic scale), and (b) the constant nature of the continuum in the B region (counts on a linear scale)

- Measure a ^{60}Co spectrum with the source in a normal on-axis position. The source need not be calibrated as the spectrum acquired for the relative efficiency measurement would be ideal for this measurement.
- Note the count in the highest channel of the 1332.5 keV peak (P).
- Set up a region-of-interest from the channel nearest in energy to 1040 keV up to that nearest in energy to the 1096 keV peak.
- Calculate or read off the gross counts in that ROI and divide by the number of channels within it to get the average count per channel (C).
- Peak-to-Compton ratio is simply the ratio P/C.

Values of the peak-to-Compton ratio for 1332.5 keV range from about 40:1 for a small 10 % relative efficiency detector, to over 90:1 for some very large detectors (see Chapter 13, Figure 13.1). In practice, a larger P/C means better counting statistics in complex spectra with improved ability to measure low-energy peaks in the presence of a Compton continuum from higher-energy gammas.

The ANSI/IEEE standard also refers to measurement using the ^{137}Cs 661.7 keV peak, where the appropriate Compton continuum is from 358 to 382 keV. This particular ratio is rarely quoted. It could be useful when counting in the 300 to 400 keV region with reasonably large detectors; it will be of little value for specifically low-energy detectors designed to operate at less than, say, 150 keV.

11.4.5 Window thickness index

The ANSI/IEEE standard recommends a measure of the total window thickness in low-energy detectors; this will clearly affect the low-energy counting efficiency. The absorption in the 'window' is not just due to the thickness of beryllium or other material used in the endcap, but includes the detector dead layer and any cryostat mount. Two nuclides are suggested for this purpose: ^{133}Ba with energies of 31 (30.6 + 31.0), 53.16, 81.00 and 160.61 keV, and ^{109}Cd with energies of 22.0 (21.99 + 22.16) and 88.03 keV. The reported parameter is the ratio of peak areas. Not all manufacturers provide this information, but as an example the range of n-type detectors with Be windows supplied by ORTEC have a warranted value for the ratio of the 22 to 88 keV areas from ^{109}Cd of 17. The greater absorption of the 22 keV gamma-ray in an aluminium endcap could drop this ratio to as low as 9. To check this index:

- Acquire a ^{109}Cd source that will exhibit negligible self-absorption.
- Place this at least 100 mm from the endcap.
- Collect a spectrum containing the two peaks, checking that the 22 keV Ag Kα X-ray (in fact, 21.99 + 22.16 keV for Kα$_2$ and Kα$_1$, respectively) is distinct and resolvable from the 25 keV Kβ X-rays. Collect so as to obtain a few thousand counts in each peak.
- Record the net areas of the 22 and 88 keV peaks and divide to obtain the index.

11.4.6 Physical parameters

The manufacture specifies a number of physical parameters on the specification sheet. The user has no option but to accept these figures. However, they will be critical when setting up mathematical models such as MCNP programs and ISOCS (see Chapter 7, Section 7.7). The measurements will have been made with the detector at room temperature and un-mounted. Therefore, some laboratories have found it necessary to X-ray their detector **while cooled** in order to measure the physical parameters at operating temperature. There have been surprises; detectors mounted off-centre, for example! A useful trick when making such measurements is to stick a small piece of carefully measured platinum wire to the side of the detector cap. The image of this will be very clear on the X-ray and provide a reference distance.

Parameters, such as dead layer thickness, cannot be measured. Often, mathematical models have to adjust them empirically in such a way as to make the model fit practical measurements.

PRACTICAL POINTS

- Choosing a detector:
  Do you need a general purpose detector, and have no particular interest in lower-energy X-rays? If so, choose a p-type coaxial detector.
  Or will it be used for a special task? Consult Table 11.1 and the suppliers' catalogues.
  For measurement of low-energy photons, in particular, compare specifications from different suppliers, as there are several alternative types of detector available.
- Setting up the system:
  Prepare a clean location before it arrives.
  Then read the manual; implement its recommendations. Sections 11.2 and 11.3 should prove helpful.
- Check the manufacturer's specification as far as possible. Remember you have paid a lot of money for a particular efficiency and energy resolution; send the detector back if you can't approach the numbers on the specification sheet.
- Before ordering the detector, make sure you will have access to an assured supply of liquid nitrogen. If not, electrical cooling will be necessary.

FURTHER READING

- Currently accepted standard procedures for testing germanium detectors are given in the following report, available from The Institute of Electrical and Electronic Engineers, Service Center, 445 Hoes Lane, PO Box 1331, Piscataway, NJ 08855-1331, USA:

IEEE Standard Test Procedures for Germanium Gamma-ray Detectors, ANSI/IEEE Std 325-1986, IEEE, New York, NY, USA.

- An excellent document, with only minor eccentricities, that covers many of the setting up procedures of this chapter is available from the American National Standards Institute, 11 West 42nd Street, 13th Floor, New York, NY 10036, USA, as:

ANSI (1991). *Calibration and Use of Germanium Spectrometers for the Measurement of Gamma-ray Emission Rates of Radionuclides*, ANSI/N42.14-1991, IEEE, New York, NY, USA (a revision of ANSI N42.14-1978).

- As I mention repeatedly, do not ignore the suppliers' catalogues and application notes. Catalogues are very helpful when choosing a detector, and are likely to be more up-to-date than other sources of information. Detector and system manuals are essential when setting up.

12

Troubleshooting

This chapter is intended to help the gamma spectrometrist identify and, hopefully find a cure for, common instrumental gamma spectrometer problems. One particular manufacturer suggests that the most common problem is failure of the cryostat vacuum. (Those people who are cavalier about connecting up their bias-shutdown circuit should, perhaps, take note of that!) While I have certainly experienced that, I have personally endured more preamplifier problems – and with rather greater frequency in recent times. Although, it has to be said, that repair is more often achievable on-site, rather than 'return to base', than used to be the case.

This chapter starts with the fault finding chart. It is as wide ranging as possible but cannot hope to identify all of the quirks and idiosyncrasies of particular systems. Nevertheless, I hope it will provide a useful structure to the process of fault identification. The manufacturer referred to above also added that 'the number of things that can contribute to loss of resolution is almost limitless'. I don't think we need to be as despondent as that; most problems are easily identified and quickly resolved. The chart is followed by detailed descriptions and advice on particular problems identified within the chart itself.

12.1 FAULT-FINDING

12.1.1 Equipment required

It is my belief that all radiation measuring laboratories should have the following items available and relevant staff trained to use them:

- A simple voltmeter or 'multimeter'

  - For checking low-voltage power supplies. NIM power supplies have test points for $\pm 24V$, $\pm 12V$ and possibly $\pm 6V$ (use appropriate DC voltage range).

  - For checking electrical fuses (use appropriate low-resistance Ω range).
  - For checking the mains supply (use the 250 V, or greater, AC voltage range WITH CARE).
  - For checking that the bias voltage is present (use a high-DC voltage range WITH CARE; for safety, test the bias unit at the lower end of its voltage range).

- An oscilloscope for looking at pulses. Unless your amplifier has automatic correction built in to it, this item is essential for pole-zero cancellation. The specification does not need to be high, but must have a 50 MHz band-width, the ability to measure rise times down to a few nanoseconds and have a voltage sensitivity of 5mV per division. An oscilloscope with a storage option is a bonus, allowing a more considered examination of pulses.
- A pulse generator that can simulate detector pulses is perhaps not essential but is certainly desirable. It should provide pulses with variable rise time (perhaps 10 to 500 ns) and variable fall-time (perhaps 10 to 500 μs). The output from the pulser is put into the TEST INPUT of the preamplifier. It allows a distinction to be made between detector problems and pulse processing problems. Some laboratories routinely use a pulser for dead time and random summing correction purposes. For systems with TRP preamplifiers, a simpler pulser providing square pulses would be adequate.

Of course, before having to diagnose problems when they arise, it is essential that the spectrometrist understands, and knows what to expect from, a working system. Time spent using the oscilloscope to examine outputs from preamplifier and amplifier will be of great use in familiarization with the instrument and providing experience against which to judge a faulty system.

12.1.2 Fault-finding guide

Table 12.1 is a compilation of information from the manufacturers with some input from my own experience. Before seeking a complicated diagnosis, check two simple factors:

- Are the power supplies, especially NIM test points, all present?
- Is the cabling sound, are plugs on cables secure and are cables firmly connected to the various units?

The cause of the problem may appear very quickly, but if this is not the case and the difficulty is one of poor resolution, then:

- check the system with a pulse generator and oscilloscope at an early stage.

Within the chart, numbers separated by slashes, such as 4/1/1, refer internally to other symptoms/actions within the chart; references with numbers separated by dots,

Table 12.1   Fault-finding guide

Symptom	Possible cause	Action
1. No spectrum in MCA, but some counts present	1. Low activity; wrong display scaling	1. Alter MCA display to logarithmic
	2. Wrong MCA segment displayed	1. Display correct segment
	3. Improperly set ADC	1. Check digital offset, LLD and ULD on ADC and conversion gain
	4. Improperly set amplifier; peaks outside range	1. Check amplifier output with oscilloscope; adjust gain to give signal of proper amplitude
2. No counts at all in MCA	1. No signal input to MCA	1. Check all cables; are the plugs OK?
		2. Check amplifier output with oscilloscope to confirm
	2. Source activity too high; resistive feedback preamplifier blocked	1. Check high rate LED on preamplifier; move source away from detector (not needed with TRP)
	3. Pulses wrong polarity	1. Check amplifier output pulses are positive
	4. Pulses gated out by coincidence line	1. Disconnect coincidence cable or check coincidence/anticoincidence switch
3. No amplifier output	1. NIM power supply faulty	1. Check low voltages at NIM-bin test points; repair or replace
	2. No input to amplifier	1. Check preamplifier output with oscilloscope to confirm ($50\,mV\,cm^{-1}$)
		2. Check 3/1/1, power supply cable to preamplifier and then go to 4/5/6.
	3. Amplifier faulty	1. Check settings
		2. Check connections
		3. Consult manual; check that any differential input is being used properly
		4. Put pulse generator pulse through preamplifier test input
		5. Replace if necessary
4. No preamplifier output at working bias; high count rate LED flashes	1. Source activity too high	1. See 2/1/1
	2. Excessive detector leakage current	1. Decrease bias in 100 V steps until output pulses appear
		2. Check test point voltage versus bias (see Section 12.2)
		3. Try warm-up cycle (see Section 12.3)
5. No preamplifier output at any bias; high count rate LED flashes; flat baseline	1. Power supply failure	1. See 3/1/1 and 3/2
	2. Preamplifier failure (FET blown; bias voltage short circuit)	1. Contact manufacturer

6. As 5, but LED not flashing	1. Power supply failure	1. See 3/1/1, 3/2/1 and 3/2/2
	2. Cryostat failure	1. Contact manufacturer
7. All peaks have low-energy tail	1. Pole-zero correction wrong	1. Check amplifier output pulse on oscilloscope (Section 10.3.4); turn PZ clockwise to bring trailing edge up to baseline
	2. Poor charge collection (i) HV too low	1. Check bias unit output with voltmeter (care!) – replace if necessary 2. Check bias supply cabling/connectors
	3. Poor charge collection (ii) shaping time too short	1. Increase time constant, or 2. Use gated integrator
8. High-energy (> 1000 keV) peaks have low-energy tail; low-energy peaks (< 150 keV) not so much affected; pulser peak width OK	1. Poor charge collection due to low-energy neutron damage	1. Use annealing kit, or 2. Contact manufacturer for repair: DO NOT ALLOW DETECTOR TO WARM UP; this reduces chance of successful repair
9. All peaks have high energy tail to baseline	1. Pole-zero correction wrong	1. Check as in Section 10.3.4. Turn PZ counter-clockwise to correct overcompensation
	2. Pile-up (random summing)	1. Increase source-to-detector distance 2. Switch in PUR (if available)
	3. True coincidence summing (see Chapter 9)	1. Increase source-to-detector distance
10. Only a few peaks show tail (high or low)	1. Peaks are multiplets; two or more closely spaced gamma lines	1. No instrumental fix available; deconvolute after spectrum collection using software.
11. Poor resolution at high count rates	1. PZ wrong 2. Inappropriate time constant	1. Check PZ, as in Chapter 10, Section 10.3.4 1. Try shorter time constants or gated integrator
	3. Preamplifier PZ wrong	1. Usually factory preset. First set main-amplifier PZ, then set preamplifier PZ to give stable baseline. Do these tests at high count rate.
12. Poor resolution in gamma peaks and pulser; wobbling amplifier output baseline (oscilloscope)	1. Baseline restorer not working	1. Check BLR; optimize threshold if manual option selected (Section 10.3.5)
13. Poor resolution in gamma peaks and pulser. Sinusoidal oscillation on amplifier output (oscilloscope)	1. Noise on baseline 2. Radiofrequency pick-up (see Section 12.4.2)	1. Eliminate by filtering or isolating. 1. Use common mode rejection 2. Use extra EMI shielding 3. Re-route cabling 4. Move VDU away 5. Move/replace HV bias supply
	3. Ground loop; frequency 50–60 Hz (see Section 12.4.1)	1. Ensure one effective common ground 2. Check/replace cabling 3. Un-couple oscilloscope 4. Add ground-loop suppressor
14. Poor resolution as 13, but oscillation slowly damped. Sensitive to tapping on cryostat.	1. Microphonics (see Section 12.4.3)	1. Dewar on foam pad (Section 10.2.1) 2. Consider sound-proof enclosure 3. Try shorter time constant 4. Try symmetrical BLR mode 5. Clean and dry Dewar; check that cold finger is not touching Dewar bottom 6. Consider microphonically ruggedized detector 7. Send detector plus preamplifier for repair

Table 12.1   (Continued)

Symptom	Possible cause	Action
15. Poor resolution: spikes on preamplifier output of opposite polarity to signal pulses (oscilloscope)	1. HV breakdown in bias supply or preamplifier or cabling	1. Check bias voltage (see 4/2/2 and 4/2/3) 2. Try drying preamplifier components, if accessible, with warm air or methanol; SWITCH BIAS OFF FIRST! 3. Contact manufacturer
16. Poor resolution. Ragged baseline on oscilloscope with one or more of the following: +ve spikes, −ve spikes, square +ve and −ve pulses	1. Breakdown across surface of detector or an insulator	1. Check value of bias supply 2. Reduce bias in 100 V stages until baseline returns to normal (temporary measure) 3. Try warm-up cycle (see Section 12.3)
17. Poor resolution. No unusual pulses but wide baseline noise	1. Excess detector leakage current	1. Try actions 16/1/1 and 16/1/2 2. Warm-up cycle likely to succeed
18. Wandering peaks	1. Amplifier gain drift	1. Repair/replace amplifier
	2. Other faulty electronics	2. Check cables/connectors for intermittent connection of signals or ground
	3. Detector-to-preamplifier connection loose	1. Tap gently on detector to induce shift. Send for repair 2. Clean preamplifier contacts if possible
19. High liquid nitrogen loss (1): cold Dewar with water condensation, normal cryostat temperature	1. Loss of Dewar vacuum NB: It is not safe to leave the bias supply switched on if this is the case	1. Check nitrogen loss rate from whole system. Normal loss from cryostat plus Dewar is 1.2 to $1.8\,L\,day^{-1}$ (1/3 dewar, 2/3 cryostat). Density of liquid N_2: $0.807\,kg\,L^{-1}$ 2. Replace Dewar
20. High liquid nitrogen loss (2): cold cryostat with condensation, normal Dewar temperature NOTE: A cold cryostat is expected during thermal cycling and does not indicate a fault (see Section 12.3)	1. Loss of cryostat vacuum NB: It is not safe to leave the bias supply switched on if this is the case.	1. As 19/1/1 2. Contact manufacturer. 3. If the detector has a Be window, take great care. If the leaking detector is allowed to warm up, high internal pressure could be generated which could cause the Be window to explode. Cover the endcap; if practical, move the detector to a safe place; DO NOT WARM UP; contact the manufacturer immediately
21. Erratic baseline on amplifier output with TRP	1. PZ wrongly set	1. Turn PZ counter clockwise as far as it will go
22. Intermittent output on TRP system	1. High voltage breakdown in cables 2. Unstable bias 3. Moisture on bias network	1. Check cables and bias supply 1. Replace bias unit 1. Clean and dry as 15/1/2

refer, as elsewhere in the text, to sections elsewhere, so that 10.2.1 is a section of Chapter 10.

12.2 PREAMPLIFIER TEST POINT AND LEAKAGE CURRENT

Monitoring the preamplifier test point can give useful information about the state of the detector and preamplifier. Not all manufacturers provide this facility. Much of the information in this section was supplied by Canberra.

12.2.1 Resistive feedback (RF) preamplifiers

Figure 12.1 shows how the test point monitors the output of the charge integrator (Figure 4.9 (Chapter 4) is a diagram of the whole preamplifier). With no source on the detector, it can be used to estimate the current I_i leaking past the detector element, i.e. the **leakage current**; in practice, leakage round the surface of a germanium crystal is more significant than leakage through the bulk. The effect that I_i has on the voltage measured at the test point depends mainly on R_f, the feedback resistor, as described in Chapter 4, Section 4.3.1. R_f is a large value resistor – its actual size is a compromise; a large value is required for low noise, a small one for good performance at high count rates. R_f is typically $2 \times 10^9 \, \Omega$ for a coaxial detector, $5 \times 10^{10} \, \Omega$ for a planar detector.

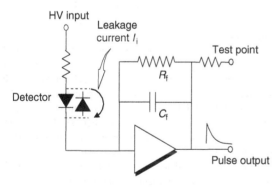

Figure 12.1 Simplified diagram showing test point and leakage current for a resistive feedback preamplifier

Leaving aside leakage, the voltage at the test point is determined by the preamplifier's FET. This will be constant from zero-bias up to the operating bias. This voltage will be quoted on the specification sheet. It is nominally $-1 \, V$ but can range from $-0.5 \, V$ to $-2.0 \, V$. Any current leakage will add to that constant value, making the test point voltage more negative if the bias voltage is positive and more positive if negative. The

leakage current is calculated by using the following equation:

$$I_i = (V_{TP} - V_0)/R_f \qquad (12.1)$$

where V_0 is the specification test point voltage and V_{TP} is the measured voltage. An example would be instructive. The detector used to gather the data in Figure 12.2 was a 45% coaxial HPGe detector with the following characteristics:

- depletion voltage, $+1500 \, V$;
- operating voltage, $+3000 \, V$;
- specification test point voltage (V_0), $-1.5 \, V$;
- leakage current (I_i) at recommended bias, $0.01 \, nA$;
- feedback resistor (R_f), $2 \times 10^9 \, \Omega$.

Under normal circumstances, with no source on the detector, the preamplifier test point voltage when operated at the recommended bias voltage was $-1.6 \, V$. Over time, the detector started to exhibit poor detector resolution to which no obvious solution was found (4/2 in Table 12.1).

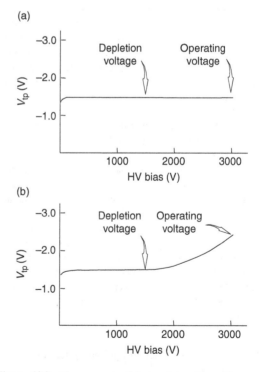

Figure 12.2 Measurement of test point voltage, V_{tp}, as a function of bias for (a) a good detector, and (b) the same detector with a significant leakage current

The test point voltage was measured at different voltages to generate the plot in Figure 12.2(b). The curve looks fine until the depletion voltage is exceeded when an increase (in this case, more negative) in test point voltage occurs. At the normal operating voltage, the test point voltage was −2.4 V. By using Equation (12.1), we can calculate that the leakage current has increased from the nominal 0.01 nA to 0.45 nA. Clearly, there is a problem. The problem was compounded when a ^{60}Co source was presented to the detector when, at only a few thousands counts per second, the test point voltage 'rose' to −2.7 V.

It would help if, in times of misery, one had available a printed copy of a test point voltage versus bias curve measured when the detector was working satisfactorily for comparison. This is, perhaps, another item to place in the detector log book when commissioning a new detector (see Chapter 11, Section 11.2). In this particular case, the assumption was made that the enhanced leakage current was due to condensation of gas traces onto the detector within its vacuum enclosure. A thermal cycle was undertaken (see below, Section 12.3) and good resolution was restored. Presumably, the thermal cycle successfully removed the surface contamination from the germanium crystal.

It is usual when determining the test point voltage/bias curve to find that on increasing the bias, the test point voltage flicks high transiently and then returns to the steady reading. This is particularly marked at low voltages and is due to the detector capacitance being charged. If indeed this does not occur, there could be a broken contact in the detector–bias line.

If the test point voltage reads 22–23 V, i.e. the supply voltage, then it is likely that the FET has been destroyed; any high value unaffected by bias voltage points to a defective FET.

In older detectors, there may be a slight, but acceptable, increase in leakage current beyond the depletion voltage, and that this becomes unacceptable at a breakdown voltage perhaps 2 kV higher. The lower end of the operating range is determined by the depletion voltage while the upper end is governed by the start of a fast rise in the leakage current. With the improved germanium material available now, the curve is essentially flat well past the depletion voltage and, in most cases, the leakage current does not increase at all up to 8000 V.

12.2.2 Transistor reset and pulsed optical reset preamplifiers

Auto-reset preamplifiers will not have an external test point. The quiescent preamplifier reset rate (that is, the

rate with no sources near the detector) is determined by the leakage current. The time between resets (TBR) can be used to assess the leakage current in an analogous way to measuring the test point voltage on a resistive feedback preamplifier. The reset rate can be measured by examining the preamplifier output on an oscilloscope. The rate will increase transiently following change of bias, due to the same capacitance effect as with the transients described in Section 12.2.1.

Under normal circumstances, a TRP will reset at intervals of from 0.1 to 1.0 s, depending on detector size, temperature, etc., and TBR will be constant as the applied bias is increased (compare Figure 12.2(a)). In a faulty system, the TBR will go smaller as bias is increased because the increased leakage current causes more frequent resetting.

12.3 THERMAL CYCLING OF THE DETECTOR

If the resolution of a detector has degraded with no apparent reason and all the usual checks and adjustments, such as pole-zero cancellation, fail to cure the problem it may be that a thermal cycle is necessary.[1]

12.3.1 The origin of the problem

Unidentified resolution problems may be a consequence of contamination of the detector itself, causing surface leakage currents. This contamination arises because of desorption of gases from the absorber within the vacuum enclosure.

Figure 12.3 shows the germanium crystal attached to a cold finger, all within a vacuum enclosure. At the bottom of the cold finger is a pack of very high surface area absorbent, sometimes referred to as the cryopump; this might be molecular sieve material or, in a low background cryostat, activated charcoal. Figure 3.16 (Chapter 3) shows the arrangement in a demountable detector where the absorber is mounted behind the detector. After the detector is mounted in its cryostat by the manufacturer, the vacuum enclosure will be evacuated. There will, however, still be traces of atmospheric gases within it. When the detector at room temperature is cooled (Chapter 11, Section 11.2.5), either by pouring liquid nitrogen into the Dewar or putting the detector assembly into an already full Dewar, the first thing to cool

---

[1] NB: The procedure described below is only applicable to modern HPGe detectors. If there are any lithium-drifted (Ge(Li)) detectors in use at this time, they must *not* be subjected to thermal cycling. If those detectors are warmed to room temperature, they will be damaged.

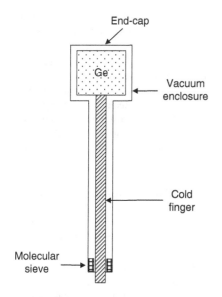

End-cap

Ge

Vacuum
enclosure

Cold
finger

Molecular
sieve

Figure 12.3 Representation of detector cryostat, molecular sieve and Dewar (not to scale)

will be the absorber pack, which will absorb most of the remaining gas traces, leaving a more perfect vacuum. In addition, gases absorbed on warmer parts of the cryostat and the detector itself will desorb and be trapped by the absorber. This process is aided by the fact that the detector itself, a large mass of germanium, has a high thermal capacity and will remain warmer than the absorber for some time. In the demountable detector arrangement, it is this differential cooling rate that allows the absorber to function. We are left with a clean detector under a very high vacuum. This situation will prevail as long as the absorber is kept at liquid nitrogen temperature. In a non-demountable system, this is aided by the fact that, due to the thermal gradient along the cold finger, the detector is a few degrees above the boiling point of liquid nitrogen.

However, during normal operation it is possible for small amounts of those absorbed gases to diffuse away from the absorber and find themselves on the detector. A substantial amount of such contamination will lead to a leakage current around the detector surface, increased electronic noise and degraded resolution. Note that if a detector is allowed to warm for any length of time, either by allowing the liquid nitrogen reservoir to run dry or by removing it from the Dewar, desorption of the gaseous impurities from the absorber will be invited – the low thermal capacity absorber will warm up quickly and the still cold detector will be available to condense the contamination on its surface.

12.3.2 The thermal cycling procedure

The cure for those unexplained resolution problems may be a thermal cycle. Indeed, even the need for a thermal cycle is not indicated by an increased leakage current – a thermal cycle may be useful even if only to satisfy a diagnostic question.

A thermal cycle cannot be rushed. It is essential that the detector is warmed completely to room temperature. Otherwise, if the detector is re-cooled before that, it will have more contamination on its surface than it had before cycling. An improper cycle could make things worse. This is worth remembering. If a detector is inadvertently left to warm partially, it will probably be necessary to complete the warming to room temperature before cooling again.

I must emphasize that before a detector is warmed, the HV must be switched off. If, as should be the case, the preamplifier HV cutout is connected, that will automatically switch off the HV as the temperature rises. However, when the detector cools down again, it would be preferable that the HV did not switch on again automatically without the user being satisfied that thermal equilibrium is established. (There is a hazard that if a pressure rise occurs, due to desorption of gas, and this happens with HV applied, there could be damaging electrical discharges.)

A recommended procedure is as follows:

- Ensure the high voltage is turned down and is switched off. Unplug the cable to be sure.
- If possible, slide the detector cryostat assembly out of the Dewar. This should not be difficult for standard dip-stick arrangements; you may need to free a number of bolts holding the assembly onto a flange on the Dewar vessel.
- If icing makes removal difficult, gently warm the Dewar with a hair-dryer to help free the assembly.
- Portable detectors have an integral Dewar and a different strategy is required. The manufacturer's manual will advise, but often one of the fill or vent ports should be blocked and the resultant boil-off pressure will increase and force liquid nitrogen out of the other port. The Dewar should then be flushed with room-temperature dry nitrogen gas; commercial 'oxygen-free nitrogen' is suitable. Canberra recommend an overnight purge at 3 to 5 litres per minute.
- If the detector is designed to operate vertically, then keep it vertical when out of the Dewar, to remove any possibility of the molecular sieve falling into the detector section.
- Allow the detector cryostat to warm up in a room at normal temperature and low humidity. This will take at least 16 h, although some would advise allowing 36 to 48 h. Do not attempt to speed the process with artificial

heat. Warming must be slow. Improper heating could cause damage to detector, insulation and preamplifier. During warm-up, the detector endcap may feel colder than usual. If the environment is humid, there could also be condensation on the cap. This is not a problem. It is because, as the absorber within the vacuum chamber releases its absorbed gases, the thermal conductivity of the gas space increases, allowing the still cold germanium to chill the endcap. Dry off any such condensation.

- While the detector is warming, the opportunity should be taken to dean and dry the Dewar. Empty it of liquid nitrogen. Set it on its side to allow the cold gases to flow out to speed the warm-up. Then invert to remove any water or debris that may be present.
- Once the detector is fully at room temperature, it can be cooled again by re-remounting the cryostat in the Dewar partly filled with liquid nitrogen. Fill the Dewar completely in the usual manner and wait 6 h for thermal equilibrium to be thoroughly established.
- Finally, top-up the liquid nitrogen again, apply the detector bias and test the system.

12.3.3 Frosted detector enclosure

If a detector cap is found to be seriously frosted, as opposed to being cool, there is a serious problem. It is likely that the vacuum chamber containing the detector crystal has sprung a leak. This has allowed air to enter the chamber, removing the thermal insulation afforded by the vacuum, as described above.

This is a potentially dangerous situation that should be handled with care. There are documented incidents where the endcap of a detector has exploded with the potential for serious bodily harm to occupants of the laboratory. In one such incident, the cause was thought to be a consequence of the fact that the boiling point of liquid nitrogen (77 K) is less than that of liquid oxygen (90 K). It is thought that the temperature of the cold finger and absorber were still below 90 K and oxygen from the air entering (what had been) the vacuum chamber condensed as liquid. Once the liquid nitrogen had fully evaporated, the rate of warming was such that the small hole that triggered the incident was not able to release the evaporating liquid oxygen fast enough so that the pressure within the chamber increased until the cap exploded. The risk of such an event happening is much greater if the cryostat has a thin window. My advice if one does find a seriously frosted cap is:

- Remove the sample and/or sample holder from the detector cap.

- Turn off the HV and shut down the electronics.
- If the detector is not mounted within a shield, move it carefully to an out-of-the-way corner of the laboratory. Post notices to keep people at a distance, so that if there is a catastrophic event nobody is likely to be hurt. Some sort of physical barrier might be contemplated.
- If the detector is mounted in its shielding, leave it there. Close the lid or door. Post a notice forbidding opening.
- Wait until it is clear that all the liquid nitrogen in the Dewar has evaporated and the detector cap is back to room temperature.

Note that removing the remaining liquid nitrogen from the Dewar or removing the detector from the Dewar would not be a good idea because it would increase the rate of warming and might precipitate the sequence of events above. By way of reassurance, it should be said that failure of the vacuum cap does not normally result in such extreme consequences. Nevertheless, the fact that they have been know to happen should make one err on the side of caution.

12.4 GROUND LOOPS, PICK-UP AND MICROPHONICS

12.4.1 Ground loops

Ground loops are one of those mysterious electronic happenings that seem inexplicable for those not versed in the electronic arts. Figure 12.4 suggests a way in which a ground loop might occur. Unit A is connected to the mains power and is grounded at (a). Unit B is similarly grounded at point (b). Unit A supplies pulses to unit B via a normal coaxial cable. We can trace a continuous path (a loop, in fact) from unit A, though the signal cable braiding, through unit B's mains cable ground to the mains ground and back to unit A through its mains cable.

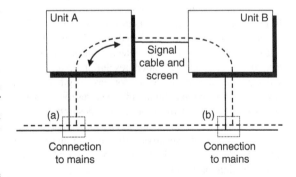

Figure 12.4 Representation of a ground loop

It is possible for a small variable current to flow around the loop. That variable current will give rise to a variable voltage which will cause a distortion of the signal voltage.

The symptoms can be degraded resolution and a very noticeable sinusoidal waveform at mains-supply frequency (50 or 60 Hz) on the signal – which, although often accompanying a ground loop, should really be regarded as pick-up from the AC power line.

Ground loops are most often found when different parts of a system are grounded at different points on the supply bus. They will be exacerbated by faulty cables, poorly attached coaxial plugs and by intermittent discontinuities in the grounding of signals caused by loose connectors. In general, the user should try for one effective common ground for all parts of the system; in practice, a considerable amount of trial and error could be required.

The following hints and tips may be useful. Some of these are sometimes incompatible:

- Check the tightness of all cables and connectors.
- If possible, put the amplifier, HV bias supply and ADC in the same NIM-bin, with preamplifier power coming from the amplifier.
- However, *do not* have the amplifier next to the ADC or HV supply (to minimize pick-up from these units).
- For systems with a non-NIM ADC, it may be possible to 'float' the amplifier and HV (that is, disconnect the earth connection from the power supply), providing a ground through the ADC input.
- Alternatively, earth the MCA chassis to the NIM-bin ground.
- If the HV supply is AC powered, ground this through the preamplifier.
- Commercial ground loop eliminators can be purchased (e.g. from Canberra Nuclear) which go onto cables; these interrupt or isolate the ground links in signal cables, HV bias lines and preamplifier power supplies.

Once you have a set-up free of ground loops, it is worthwhile fixing the cabling in place with cable ties to prevent casual uninformed 'tidying-up' inadvertently creating one.

12.4.2 Electromagnetic pick-up

Pick-up is the receipt of unwanted electromagnetic radiation by the pulse processing system which manifests itself as additional noise and degraded resolution. Some sources of such electromagnetic interference (EMI) might be designed to emit electromagnetic radiation, such as microwave ovens, radio and air traffic control radar, microwave telecommunications, TV and radio stations, and mobile phones (cell phones), in which case the phenomenon might be referred to as RFI – radiofrequency interference. (Radiofrequency covers a wide electromagnetic range from 3 kHz to 300 GHz.) Other sources, often involving rapidly changing magnetic fields, are less obvious: domestic appliances, power lines, lightning, VDUs, heavy machinery and switches – even the HV unit within the gamma spectrometry system.

Normal coaxial cables, with their grounded braiding surrounding the central conductor, do provide a measure of screening but if the EMI is of sufficient intensity that may be insufficient. More efficient shielding can be achieved by using doubly- or preferably triply-screened cable, which is readily available commercially. Such cables have longer, more rigid, BNC connectors.

When seeking to eliminate EMI as a source of resolution degradation you might consider the following:

- With the oscilloscope (time base set to $10\,\mathrm{ms\,cm^{-1}}$) look for 50 or 60 Hz mains frequency oscillations due to AC power line pick-up. If present, good earthing connections should eliminate it.
- Again, with the oscilloscope (time base set to $50\,\mathrm{\mu s\,cm^{-1}}$) look for 20 kHz oscillations. If present, these could be coming from the HV power supply. Keep the amplifier distant from the HV unit within the NIM-bin. If the problem persists, change the HV unit.
- If there are several cables running from the preamplifier to the amplifier, the lie of the cables could form a form of 'loop-antenna', which can be a particularly efficient way to pick up RFI. This can be minimized by bundling all the cables together using tape or cable ties to keep them in place. Tennelec ran these signal and power supply lines within one shielded bundle as a matter of course, as do modern instruments, such as the *ORTEC DSpec*.
- Grounding the cryostat may help suppress RFI, but could make the ground loop problem worse.

CRT monitors

Older CRT monitors can cause particular problems because their internal electronics generate rapidly changing magnetic fields and considerable radiofrequency emissions. These emissions are highly directional. It is possible that moving the monitor a few tens of centimetres could render the resolution of a hitherto working spectrometer unusable. Checking whether the source of the problem is the CRT is easy: collect a spectrum of a suitable source (the usual ^{60}Co would do) with the monitor on and measure the FWHM of a peak; clear the spectrum, switch the monitor off and start a count using the appropriate keystroke; after a suitable time, stop the count using a keystroke; switch the monitor on and measure the FWHM again. If the latter measure is satisfactory and the former not, the reason must be EMI from the monitor. Solutions:

- Move the monitor to a position where it does not interfere.
- Replace the monitor with a low-EMI-rated model or an LCD monitor.
- Try and arrange the cable run so that the preamplifier cable does not run past the monitor. (If you suspect that this is the problem, a simple check can be made by connecting the INPUT of the amplifier to the TEST IN of the preamplifier. As there should be no output from the preamplifier's TEST IN, any signal coming through the amplifier will be due to pick-up.)
- Some recent amplifiers have a transformer built into their input, which is designed to reduce the high-frequency noise associated with the raster on a CRT.

If the problem cannot be traced to one's own CRT, consider what might be happening at the other side of the counting room wall. Has a CRT on the other side been moved to within range of your electronics?

Common mode rejection

Figure 12.5 demonstrates a situation in which the cable transmitting the preamplifier pulses is picking up transient EMI signals – these might be from high power switching circuits, for example. The lower cable, labelled 'Normal' at the end, is shown with a preamplifier pulse plus an unpleasant looking transient. The way to remove such a transient is called **common mode rejection**. To do this, you will need an amplifier with built in rejection circuitry. The key feature to look out for is a **differential input**. To use this, an identical cable to that used for the detector signal is laid alongside it and connected to the amplifier's differential input, 'Diff'. At the preamplifier end this would not be connected to the pulse output but would be terminated in the same way as the signal cable. It is assumed that this cable will pickup the transient in exactly the same way as the signal cable. The common mode rejection circuitry in the amplifier subtracts the differential input from the signal input, leaving a transient-free signal. Amplifiers with this facility include the *ORTEC 972*, the *Canberra 2025* and the *Tennelec TC244*.

A variation on this principle would be to connect the differential cable to a reversed polarity output, so that the subtraction would, in addition to removing the transients, double the signal amplitude. Common mode rejection is likely to come into its own when the preamplifier-to-amplifier distance is large and/or the environment is very noisy.

Mains supply problems

Mains power lines are very seldom screened and are liable to pick-up from the whole panoply of sources of EMI. Figure 12.6(a) depicts spikes on a mains supply: short-duration high-amplitude voltage transients that could be due to lightning hitting power lines, utilities switching lines, operation of elevators, air-conditioners, etc. Figure 12.6(b) shows RFI pick-up on power lines, perhaps originating in transformers, fluorescent lights' auto-ignition systems, electric motors, etc.

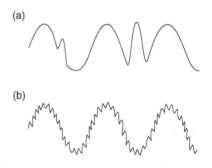

Figure 12.6   Problems on mains power supplies: (a) transients; (b) RFI/EMI

A good clean AC mains/line power supply is required. Use a supply that is 'conditioned', i.e. smoothed, filtered and non-spiky. Local protection against spikes and EMI is relatively cheap and easy to implement.

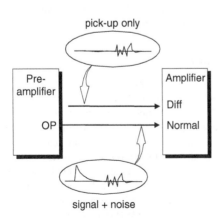

Figure 12.5   Common mode rejection device using the differential input of an amplifier

AC power supplies can also be subject to other problems, not related to EMI, e.g. Sags and surges, where the AC supply voltage drops or rises for periods of more than one cycle, and brownouts, which are longduration sags which could last several days. Moreover, there can also be blackouts or a complete power failure. If long-count periods are anticipated, some measure of protection against complete power loss would be valuable, and in some circumstances essential. Such protection must be able to operate before the mains voltage drops sufficiently to switch equipment off. Standby power systems, in which rechargeable batteries maintain the voltage for a short period, perhaps as little as 30 min, may allow a user to go through an orderly shut-down procedure but are of little value if the interruption happens overnight. Ideally, installation of a proper uninterruptible power supply (UPS) would keep the system running almost indefinitely, but at significant expense.

12.4.3 Microphonics

Mechanisms and checks

Microphonics in germanium detectors are not well understood. It is the case, though, that all detectors are sensitive to vibration and audio-noise, some more so than others. Defective detectors can be excessively sensitive, making them impossible to use. The effects on the spectrum can be a serious loss of resolution and, in some cases doubling of peaks (Figure 12.7). Sensitivity of a detector can be assessed by clapping hands near to the cryostat, or by *gently* tapping the endcap while watching the amplifier output on an oscilloscope using a voltage scale that will allow the baseline to be examined. (When tapping, remember that materials at liquid nitrogen temperature can be brittle.) Tapping most detectors will produce some

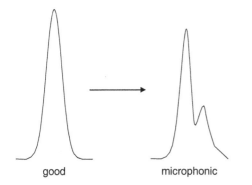

Figure 12.7 One possible type of peak degradation due to microphonics

sort of disturbance to the signal baseline that will die away immediately. Such behaviour does not necessarily indicate a microphonic problem. On the other hand, a detector with a faulty detector–FET connection might be so sensitive as to behave as a microphone, with a profound oscilloscope response when talking to the detector at normal conversational volume. (It is not polite to shout at one's detector!) In the fault-finder in Table 12.1 (14/1), I suggest that slowly damped baseline oscillations can be attributed to external vibrations.

The vibration is thought to move components within the cryostat, in particular, the electrical connection between the detector and the FET. Although the movements might be very small, of the order of 10^{-10} m, the very small change in capacitance can lead to a significant output signal. If the resolution of the detector is to be immune to vibrations, the mechanical stability of the mounting, connections and components of the cryostat must be high. Systems with a cooled FET are said to have better (i.e. lower effect) microphonic performance because of the shorter 'gate-lead' to the FET. Even with perfectly immovable detector coupling, there might still be a problem of resonance, particular vibration frequencies tuned to the natural frequency of the detector end-cap and other components. Users of portable detectors should be aware that microphonic sensitivity can vary markedly with the orientation of the detector.

An alternative source of microphonic noise is said to be ice inside the Dewar. Two mechanisms have been suggested:

- Ice crystals coalescing and then cracking apart in the liquid nitrogen, giving a continuous background of audio-noise.
- Ice bridges being formed between the cryostat and the Dewar casing, so that vibrations of the Dewar are transmitted to the cryostat.

These problems can be avoided by cleaning and drying the Dewar and are minimized by using *dry* liquid nitrogen. If necessary, the humidity in the counting room may need to be controlled. There are reports of microphonic problems caused by the bubbling of the liquid nitrogen in the Dewar as it boils. Since we cannot prevent this happening, a detector that is sensitive to such small vibrations would have to be rebuilt.

Solutions

These are listed in the fault-finding guide (Table 12.1):

- Chapter 11, Section 11.2.1 mentioned installing the detector on an anti-vibration support. This simple action has been shown to be very effective in situations where pumps and other vibrating machinery is installed on the same concrete raft as the counting room.

- If the detector is exposed to excessive audio noise, it may be worth constructing a sound-proofed enclosure around the detector.
- In normal situations, minimizing external noise with double glazing and, regrettably, discouraging staff from whistling, singing or impromptu Morris dance demonstrations, may be beneficial.
- Some microphonic effects may be filtered out at the amplifier. Moving to a shorter time constant could reduce the microphonic effect at the expense of increased system noise. If the amplifier has optional settings for the baseline restorer, try changing these to 'symmetrical' or 'high' or 'auto'.
- The importance of keeping your nitrogen dry is covered above in Section 12.4.3.
- Replacing an analogue pulse processing system with a digital system that incorporates low frequency rejection (LFR) may remove microphonic sensitivity completely. LFR incorporates additional filtering to digitally remove the unwanted vibration signal. It is this feature of digital pulse processing that has made possible complete hand held electrically cooled detector systems by removing the influence of vibration from the refrigerator.

Detector manufacturers are very conscious of the effects of microphonics and most will subject detectors to suitable testing before release for sale; some will report these tests to customers. PGT produce a range of detectors they call 'Quiet Ones', where special attention has been paid to anti-microphonic design. A parameter has been devised, the *microphonic rating*, which measures the relative increase in amplifier noise at a frequency where this increase is a maximum. Frequencies from 100 Hz to 10 kHz are scanned.

PRACTICAL POINTS

This entire chapter is 'practical'. To summarize:

(1) Acquire a multimeter, an oscilloscope and a pulse generator.
(2) Check power supplies on the NIM-bin.
(3) Check cabling is secure.
(4) Use the pulse generator to check whether the problem is in the detector or the electronics. That is, does the pulser peak in the spectrum have the same resolution problem as a gamma-ray peak?
(5) Go to the fault-finding guide (Table 12.1) and follow the diagnostics and remedies in a systematic fashion.

FURTHER READING

Such little information as is available on troubleshooting in high-resolution gamma spectrometry systems is scattered through the manuals of equipment suppliers: I would particularly commend those of Canberra, PGT and ORTEC.

13

Low Count Rate Systems

13.1 INTRODUCTION

Many gamma spectrometry laboratories will be concerned with measuring low levels of radioactivity in, what might be termed 'environmental samples' – these might be foodstuffs or wastes destined (hopefully) for consignment to normal civil waste disposal streams, or perhaps in whole body measurements. Some fundamental research projects involve extremely low intensity gamma-ray measurements, as we shall see in Section 13.4. From a regulatory point of view, there is constant pressure to measure lower and lower activity levels of a wide range of nuclides, sometimes with little theoretical justification.

Marine environmental measurements are a particular field where considerable efforts have been put into developing low activity measurements to such an extent that the gamma spectrometers are retreating underground, where cosmic ray backgrounds are much lower (Section 13.4.7). One particular reason for such a desire for ultimate MDA is that other techniques, such as ICPMS and AMS, are able to use smaller samples to achieve their ends, and radiometric techniques are beginning to be asked to make do with similar sized samples. Smaller samples facilitate collection, storage and preparation of samples and minimize expensive time on station for the survey ship.

In Section 5.6, I discussed the principles underlying the concepts of Critical Limit, Limit of Detection and its activity equivalent, the dubious Minimum Detectable Activity (MDA), which you will recall is *not* the minimum activity detectable. In that section, I pointed out that the common use of MDA as an indication of the upper limit of activity within a measured sample is not justifiable. However, it is the correct parameter to consider when discussing 'what if' in the context of evaluation of

a method. For a gamma spectrometry measurement, the MDA, in Becquerels, leaving aside corrections for decay and suchlike, can be expressed as:

$$\text{MDA} = \frac{L_D}{\varepsilon \times P_\gamma \times t_C} \qquad (13.1)$$

where L_D is the limit of detection in counts, ε is the efficiency of the detector at the energy of the measured gamma-ray, P_γ the gamma-ray emission probability and t_C the live time of the count. In Chapter 5, Section 5.6.4, we saw that the limit of detection will depend upon the degree of confidence we care to place on the certainty of detection. If we are to be 95 % confident of detecting a peak, the number of counts in it, the limit of detection, would be:

$$L_D = 2.71 + 3.29\sigma_0 \qquad (13.2)$$

where σ_0 is the uncertainty of the background. If, for the moment, we assume that we are measuring the number of counts in a region-of-interest, σ_0 can be shown to be $\sqrt{(2B)}$, where B is the number of background counts. Remember, that for gamma spectrometry, peak background is partly due to external radiation and partly counts due to scattering within the shield and the sample itself. If that total background count rate is B_R counts per channel per second, the measured background over a ROI n channels wide must be:

$$B = B_R \times t_C \times n \qquad (13.3)$$

The ROI width will be determined by the width of the gamma-ray peak – that, in turn, being related to the FWHM of the peak. Examples of the peak width necessary to measure particular fractions of the total peak area are given in Table 13.1. So, taking into account FWHM, we can calculate B as:

$$B = B_R \times t_C \times \text{FWHM} \times F/\text{ECAL} \qquad (13.4)$$

Table 13.1 Peak width necessary for particular peak area coverage

Proportion of peak area covered %	Coverage factor	Peak width, FWHM
95.45	2.000	1.699
98.76	2.500	2.123
99.00	2.576	2.188
99.68	2.944	2.500
99.73	3.000	2.548
99.96	3.533	3.000
99.99	4.000	3.397

$$\text{FWHM} = 2.355\sigma$$

Here, F is the factor necessary for the desired coverage, and ECAL is the energy calibration factor to convert the measured FWHM to channels.

Using Equation (13.4) to calculate σ_0, putting that into Equation (13.2) and that in turn into Equation (13.1), we end up with:

$$\text{MDA} = \frac{2.71 + 3.29\sqrt{2B_R \times t_C \times \text{FWHM} \times F/\text{ECAL}}}{\varepsilon \times P_\gamma \times t_C} \qquad (13.5)$$

Under most circumstances in gamma spectrometry, the factor 2.71 will be insignificant compared to the rest of the numerator and can be ignored. F, ECAL and P_γ will all be constant for a particular measurement, and ε must be proportional to relative efficiency, RE, and so Equation (13.5) can be reduced to:

$$\text{MDA} \propto \frac{\sqrt{B_R \times \text{FWHM}}}{\text{RE} \times \sqrt{t_C}} \qquad (13.6)$$

The inverse of this equation is sometimes used as a 'figure-of-merit' (FOM) for a counting system and is, in effect, the square root of the traditional S^2/B criterion. Equation (13.6) tells us that to lower our MDA, i.e. achieve a more sensitive analysis, we might:

- Decrease the background continuum by improving the shielding or using a detector with lower impurity concentrations within its constructional materials.

- Reduce the peak width (FWHM), perhaps by using a detector with better resolution.
- Increase the efficiency, perhaps by using a larger detector or perhaps by optimizing the shape of the sample relative to the detector.
- Increase the count time. Because t_C is a square root term, a halving of MDA will need four times the count time. If count times are long already, this might not be a viable option. Count time will often be constrained by the need to count a number of samples within a fixed time period (optimization of count time was discussed in Chapter 5, Section 5.5.3).

It is worth remembering that many nuclides emit a number of gamma-rays at different energies. It might be advantageous to use a higher energy gamma-ray, which would have a lower underlying background continuum, even if it had a lower emission probability. On the other hand, a lower energy gamma-ray would have a smaller peak width, but would be on a higher continuum – a complicated case of swings and roundabouts. I will discuss some of the issues surrounding improvement in MDA shortly, but first I must come back to a problem with Equation (13.6). This equation is, in effect, that used in ORTEC's 'Bigger is MUCH better' paper (Keyser *et al.*, 1990) derived from a much deeper mathematical treatment of the matter. Unfortunately, in forming that equation we assumed that we were simply measuring an ROI, for which $\sigma_0 = \sqrt{(2B)}$, rather than the peak area equation for L_D, in which case $\sigma_0 = \sqrt{[B(1+n/2m)]}$ (See Chapter 5, section 5.6.4 for the derivation; m is the number of channels used to estimate the continuum level on each side of the peak.) If we use that in Equation (13.3), we arrive at the following:

$$B = B_R \times t_C \times n(1 + n/2m) = B_R \times t_C \times (n + n^2/2m)$$
$$\approx B_R \times t_C \times n^2/2m \qquad (13.7)$$

The background region width, m, is selected within the spectrum analysis program and, for the measurement of any particular gamma-ray, would normally be constant. Certainly, for the purposes of comparing different counting schemes a fixed m would be fine. In Equation (13.7), the assumption that n is insignificant compared to $n^2/2m$ barely justifiable, but using it instead of Equation (13.3) and following the reasoning above results in an expression for MDA in which *FWHM* is *not* square rooted:

$$\text{MDA} \propto \frac{\sqrt{B_R} \times \text{FWHM}}{\text{RE} \times \sqrt{t_C}} \qquad (13.8)$$

I will use Equation (13.6) within this chapter, but the reader may like to see whether using the correct equation would alter the conclusions made. A spreadsheet to do this can be found on the website relating to this book.

In the remainder of this chapter, I deal with achieving high efficiency measurements and later with creating a low background environment for the detector to operate in (i.e. reducing B_R). Detector efficiency is determined by its size and its shape. As Equation (13.6) suggests, our MDA also depends upon the resolution. I shall discuss all of these factors. The paper by Keyser *et al.* referred to above contains a number of useful experimental data.

13.2 COUNTING WITH HIGH EFFICIENCY

13.2.1 MDA: efficiency and resolution

Consider purchasing a new detector. You wish to achieve the lowest possible MDA within a fixed budget. A chat with the manufacturers will reveal that there are a number of detectors available, some with excellent resolution, some with a large relative efficiency. The prices differ; higher resolution and higher efficiency both mean higher cost. In general, the larger the detector, the poorer the resolution. If funds are limited, is it better to compromise on efficiency or on detector size? Equation (13.6) allows a rational decision to be made. We need a measure of efficiency, ε. In principle, the efficiency referred to is the full-energy peak efficiency. Before purchasing the detector, we cannot know that. However, it will be proportional to the relative efficiency, which *is* quoted by the manufacturer. We must also take into account the fact that as the detector size increases, the background count rate, B_R, will increase in proportion. We can, therefore take B_R to be proportional to ε, or the relative efficiency. If we substitute that in Equation (13.6), we find that MDA is proportional to FWHM/$\sqrt{\varepsilon}$. Consider two detectors – one with a resolution of 1.7 keV and a relative efficiency of 30 %, the other a 45 % detector with 1.9 keV resolution. The MDA ratio will be:

$$\frac{\text{MDA}_2}{\text{MDA}_1} = \frac{30\%/\sqrt{1.7}}{45\%/\sqrt{1.9}} = 0.70 \qquad (13.9)$$

So, we can expect that the larger detector would provide a slightly lower MDA. In order to improve the MDA significantly, we would need a much bigger detector. For example, the 114 % detector referred to in Table 11.3 (Chapter 11) would provide an improvement of 0.30, albeit at considerable extra cost.

13.2.2 MDA: efficiency, background and counting period

There are two reasons for the continuum underlying peaks in a spectrum:

- Incomplete absorption of gamma-rays due to Compton scattering; the gamma-rays may be originating from the sample itself or from sources within the detector environment. If the sample contains significant radioactivity, it is likely that this continuum source will dominate. I will refer to this as B_S.
- Cosmic-ray interactions resulting in a continuum of photons (see Section 13.4.6). In addition to this, there will be a proportion of gamma-rays from sources outside the shielding that are Compton scattered while travelling through it. From those, the detector will 'see' a continuum of photons. In a low background system, these sources may dominate the peak-background level. This will be referred to as, B_E.

Thus, $B_R = B_S + B_E$.

Variation of B_S, the Compton continuum due to the sample, with detector size

We would expect a more efficient, larger detector to have a larger background count rate. The proportion of gamma-rays striking the detector that end up on the Compton continuum is related to the peak-to-Compton ratio (P/C). Figure 13.1 shows how this varies with relative efficiency. The data in this figure were derived

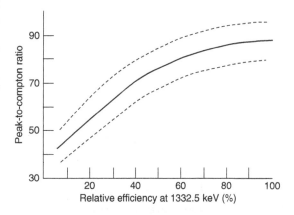

Figure 13.1 Peak-to-Compton ratio as a function of relative efficiency for coaxial p-type HPGe detectors. The full curve drawn is a mean of published data from ORTEC and Canberra while the dashed envelope indicates the range of peak-to-Compton ratios seen in different detectors

from measurements made by ORTEC and Canberra on a large number of p-type coaxial detectors and collated by John Hemingway. The central line is a best-guess fit to the data. The outer dashed lines represent the considerable scatter of the points, due to the vagaries of the manufacturing process. It does appear that for small detectors ($< 50\%$ relative efficiency) P/C is proportional to relative efficiency. For larger detectors there is a fall-off, which could be attributed, in part, to their poorer resolution making the height of the full-energy peak lower, although there may be other factors involved.

Figure 13.2 shows the measured background continuum level in a spectrum at four points clear of peaks, plotted against relative efficiency in a situation where B_S is very much greater than B_E. In this series of experiments by ORTEC (Keyser *et al.*, 1990), the background continuum was created by a measuring a mixture of ^{152}Eu, ^{154}Eu and ^{125}Sb. These nuclides emit gamma-rays throughout the spectrum range; the background was measured between peaks. Two things are evident. First, the continuum level is greater at lower energy. This is common experience and is simply due to the Compton continuum from each gamma-ray piling up on those from higher energy gamma-rays. Secondly, even though the original data points are somewhat scattered, it is clear that the B_S is not proportional to detector efficiency. The increase in background continuum level is more gradual as the detector size increases. So, taking into account Equation (13.8), and assuming that the count period and the resolution remain constant, as detector size increases, B_S (related to the

Compton level) increases at a lower rate than the full-energy peak efficiency (which is proportional to the size). The logical conclusion is that the MDA will decrease rather more than would be expected from the increase in detector size. To quantify that, we can resort to Equation (13.6) again. In this case, FWHM and t_C are constant and we can take B_S as equivalent to B_R and relative efficiency as a measure of ε. Then, taking the data from Figure 13.2, comparing a 30% detector and a 90% detector at 325 keV we get:

$$\frac{\text{MDA }(90\%)}{\text{MDA }(30\%)} = \frac{\sqrt{1.7}/90\%}{\sqrt{2.95}/30\%} = 0.44$$

Thus, we have an improvement in MDA of a factor of 2.28 for a three times the efficiency, at a cost ratio of 2.6 (using cost data given in the Keyser *et al.* (1990) paper). Calculations based on the curves for other energies give similar figures. Taking data from the original paper, Figure 13.3 shows how the relative MDA alters with detector size relative to that for a 30% detector. The data in this figure represent the means of calculations at all of the four energies considered in Figure 13.2. While this degree of improvement seems modest for a considerable cost, the proponents of 'super-large' detectors suggest that there are hidden benefits in terms of sample throughput. Assuming that for a particular count time, the 30% detector meets a required MDA. Equation (13.6) allows us to fix the MDA and FWHM and see how the count period necessary to achieve that same MDA alters with detector size. In fact, it is straightforward to show that the ratio of count periods for different

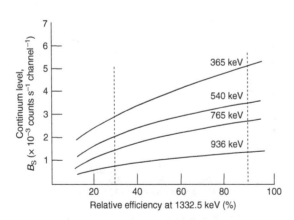

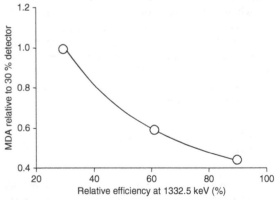

Figure 13.2 Examples of count rates in four continuum regions of a mixed nuclide spectrum (^{152}Eu, ^{154}Eu and ^{125}Sb) as a function of detector efficiency (data adapted from Keyser *et al.* (1990))

Figure 13.3 Relative MDA as a function of relative efficiency for coaxial p-type detectors and a fixed counting time. Case where Compton background, is dominant; $B_S > B_E$ (recalculated from data in Keyser *et al.* (1990))

detectors to give the same MDA is the square of the MDA ratio as calculated above:

$$\frac{t_{C1}}{t_{C2}} = \left(\frac{MDA_1}{MDA_2}\right)^2 \qquad (13.10)$$

Figure 13.4 plots the mean of data for all four energies of Figure 13.2. From that curve, we deduce that for the same MDA we would only have to count on a 90 % detector for about 20 % of the time needed on the 30 % detector. We would be able to measure five times as many samples on the 90 % detector within the same period of time and still achieve the same MDA. Furthermore, because the cost per percent relative efficiency decreases with detector size, a 90 % detector would be 14 % cheaper than three 30 % detectors (using data cited by Keyser *et al.*, 1990) and considerably cheaper than five!

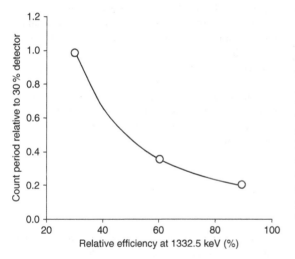

Figure 13.4 Relative counting time as a function of relative efficiency for coaxial p-type coaxial detectors, counting to a fixed MDA. Case where Compton background is dominant; $B_S > B_E$ (recalculated from data in Keyser *et al.* (1990))

While this seems a very worthwhile strategy, the author, being notoriously cynical, would not care to place all his eggs in this particular basket. The following counter-indications should be considered:

- There is the question of detector redundancy; detectors do fail from time to time. Can your laboratory afford to place complete faith in a single detector?
- The larger the detector, the greater is the degree of true coincidence summing. This was discussed at length in Chapter 8.

- For low energy gamma-rays, say less than 300 keV, a larger detector volume is unnecessary. It is more important how the germanium is disposed relative to the sample (see below). Greater width of a germanium crystal may improve efficiency and lower the MDA, but increasing the length of the crystal may just provide more germanium to interact with the background and, by increasing B_E, *increase* the MDA.
- The analysis above is only relevant to the case where the Compton continuum dominates the peak backgrounds. As we will see below, if the environmental background is dominant the relative count-time advantage disappears.
- The analysis above does not take into account the variation of FWHM and peak width.

Variation of B_E, the environmental background, with detector size

The spectra of very low activity samples, measured in a low background shield, will be dominated by the background due to the environment of the detector; B_E will be greater than B_S. Again, we can turn to experimental evidence for this. Figure 13.5 plots the background counts per second per channel for a large number of detectors, with relative efficiencies ranging from 17 % to over 100 %, at the particular non-peak energy of 464 keV (data from Keyser *et al.*, 1990). The line, if not linear, is certainly more nearly linear than the relationship between the Compton-continuum level and detector size (Figure 13.2). Data at 1443 and 2335 keV also show a proportional relationship. If B_E is proportional to ε (and relative efficiency), Equation (13.8) tells us that the MDA will decrease as the square root of the relative efficiency.

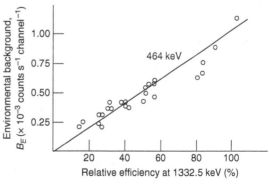

Figure 13.5 'Environmental' background count, $B_E > B_S$, at 464 keV for a variety of detectors in a low background shield as a function of detector efficiency (adapted from data in Keyser *et al.* (1990))

Figure 13.6 plots MDA relative to that of a 30 % detector and includes, for comparison, the data of Figure 13.3, together with other data, the dashed lines, which I will discuss presently. Clearly, the improvement in MDA is much less for a super-large detector if the background is primarily external to the detector/sample system, rather than internal.

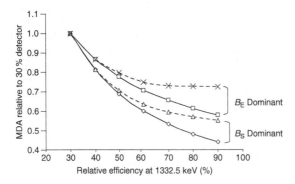

Figure 13.6 Relative MDA versus relative efficiency for a fixed counting time, calculated from Equation (13.8), for cases where B_E is dominant, where B_S is dominant and where FWHM alters with size of detector (continuous lines, FWHM constant; dashed lines, FWHM variable)

The relative count period for constant MDA (Figure 13.7) is also disappointing. In fact, throughput of one 90 % detector is exactly the same as for three 30 % detectors, but the cost benefit remains, at the expense of peace-of-mind.

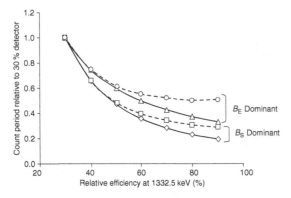

Figure 13.7 Relative counting time versus relative efficiency for a fixed MDA, calculated from Equation (13.10), for cases where B_E is dominant, where B_S is dominant and where FWHM alters with size of detector (continuous lines, FWHM constant; dashed lines, FWHM variable)

The effect of resolution on relative MDA and count period

In the two sections above, it was assumed that the FWHM of the detector is unchanging. In fact, it is likely that as detector size increases the FWHM will also rise. The effect of that will be to spread the peak over more channels and so increase the background beneath the peak, to the detriment of the MDA. Equation (13.8) allows us to estimate the effect of deterioration in FWHM by multiplying the relative MDAs calculated as described in the above two sections by the ratio of FWHM.

In Figure 13.6, the two dashed curves represent the MDA ratios, taking into account FWHM variation, for the cases where Compton and external background are dominant. In this exercise, the FWHM was held constant at 1.75 keV (1332.5 keV) up to 40 % relative efficiency and then gradually raised to 2.2 keV for the 90 % detector. The relative time-plot, Figure 13.7, also contains data for the variable FWHM cases. It can be seen that if the FWHM of the 90 % were indeed 2.2 keV there would be no throughput improvement relative to three 30 % detectors at all.

In deriving Equation (13.8), I explained that the background under the peak is $B_R \times t_C \times n(1 + n/2m)$ where n is the peak width and $2m$ is the number of channels used to estimate the background. For a strictly accurate comparison, FWHM (keV) in Equation (13.8) should be replaced by:

$$\sqrt{\mathrm{FWHM}\frac{F}{\mathrm{ECAL}}(1 + \mathrm{FWHM}\frac{F}{2m \times \mathrm{ECAL}})} \qquad (13.11)$$

where F is the coverage factor and ECAL the energy calibration. In fact, for reasonable values of F, ECAL, and m, the relative MDA and relative time values will be very similar, but slightly more favourable, values to those in Figures 13.5 and 13.6. However, using Equation (13.6) in these deliberations will give very different values.

Table 13.2 summarizes the calculation of relative MDAs and relative count times for the two cases where the Compton continuum from the sample dominates and when the sample activity is very low and the environmental background predominates.

Is bigger MUCH better?

The analysis used by the proponents of super-large detectors, and explained above, assumes that most of the background beneath the peaks in the spectrum is due to

Table 13.2 Equations used to estimate relative MDA and count period[a]

Parameter	B_S dominant		B_E dominant	
	Fixed FWHM	Variable FWHM	Fixed FWHM	Variable FWHM
Relative MDA	$\dfrac{\sqrt{B_{S1}}\mathrm{RE}_2}{\mathrm{RE}_1\sqrt{B_{S2}}} = \mathrm{RMDA}_S$	$\mathrm{RMDA}_S\dfrac{F_1}{F_2}$	$\dfrac{\mathrm{RE}_2}{\mathrm{RE}_1} = \mathrm{RMDA}_E$	$\mathrm{RMDA}_E\dfrac{F_1}{F_2}$
Relative t_C	$(\mathrm{RMDA}_S)^2$	$\left(\mathrm{RMDA}_S\dfrac{F_1}{F_2}\right)^2$	$(\mathrm{RMDA}_E)^2$	$\left(\mathrm{RMDA}_E\dfrac{F_1}{F_2}\right)^2$

[a] B_{S1} and B_{S2} continuum-background levels; RE, is relative efficiency; F_1 and F_2, are FWHMs.

Compton scattering within the detector shield – mainly from the sample itself. It also ignores the effect of poorer resolution of larger detectors. Under the conditions implied, bigger does appear to be better. However, the deductions made above would be completely different for detectors with much better, or worse, resolution than is typical. The deductions are also different if the continuum beneath the peaks is due primarily to sources of radiation external to the shielding.

In practice, if a low MDA is required, improve the shielding of the detector and buy the best resolution you can afford. Whether you go for a super-large detector or just a large one will depend upon whether you are prepared to depend upon a single detector, rather than two or three. Table 13.3 shows calculated sample-throughput data for a number of situations.

This table shows that the only guaranteed benefit of one 90 % detector rather than three 30 % detectors is, on year-2000 relative prices, a 14 % cost saving. Interestingly, two 60 % detectors would give better sample throughput than one 90 % detector, albeit at a rather greater cost. The spreadsheet used to calculate the data for Table 13.3 and for the preceding figures is present on the website relating to this book.

13.3 THE EFFECT OF DETECTOR SHAPE

So far, in this chapter, the emphasis has been on MDA and its relationship to size of the detector, culminating in a discussion of whether bigger is better, or not. There are situations, though, where bigger is clearly not better, where the shape of the detector is more important.

13.3.1 Low energy measurements

Figure 13.8 shows the thickness of germanium needed to absorb 99.5 % of gamma-rays, normally incident on the detector, as a function of energy (see also Figure 3.7). Clearly, if we are primarily interested in low energy gamma-rays a large thickness of germanium is not needed. According to Figure 13.8, 99.5 % of 100 keV gamma-rays will be absorbed within 18 mm of germanium; it could be argued that additional germanium is waste if gamma-rays above that energy are of no interest. Indeed, the situation is not improved by the fact that partial absorption in the remaining germanium will simply serve to increase the height of the Compton continuum. For low energy detection, thin detectors are preferred; planar detectors and Canberra's *LEGe* detectors are designed for

Table 13.3 Estimated throughput for two large detectors relative to a 30 % detector[a]

Relative efficiency (%)	Number of detectors	Resolution (keV)	Throughput to achieve same MDA				Relative cost
			B_S Dominant		B_E Dominant		
			FWHM = 1.75 keV	FWHM as given	FWHM = 1.75 keV	FWHM as given	
30	3	1.75	3	3	3	3	1.00
60	2	1.85	6	5	4	4	1.24
90	1	2.20	5	3	3	2	0.86

[a] For variable FWHM calculations, the coverage factor was 3, the energy calibration was 0.25 keV channel^{-1} and the background region width 3 channels.

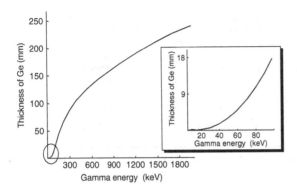

Figure 13.8 The thickness of germanium required to absorb 99.5 % of a normally incident beam of photons (calculation based on data in Debertin and Helmer (1988))

this purpose, and at slightly higher energies, the ORTEC *LO-AX* detectors. In Chapter 3, Section 3.5, I discussed detector shape in relation to detector capacitance and its effect on resolution. A great deal of technical expertise has gone into designing detectors with low capacitance and good resolution.

At low energy counting efficiency is most effectively increased by having a relatively thin, but large area, source on a thin, large diameter detector. Such detectors will range from 5 to 10 mm thickness for a planar detector and 20 to 30 mm for *LO-AX detectors* (see, for example, Figure 13.9). Detectors up to 70 mm diameter are currently available. Comparison of a 70 mm *LO-AX* detector with a 51 mm diameter planar detector, under the broad-beam conditions of whole body monitoring for plutonium and uranium isotopes in lungs, revealed a reduction in MDA by a factor of 0.75 over the energy range 30 to 400 keV (Twomey and Keyser, 1994).

The optimum relationship between sample size, detector size and measured energy range is now being paid more attention. The manufacturers now refer to the 'detector profile'. Indeed, current ORTEC catalogues include the 'PROFILE series of GEM detectors' to help buyers select the right detector for their anticipated sample size and shape.

13.3.2 Well detectors

The well detector was mentioned briefly in Chapter 3, Section 3.4.4 and its notional efficiency curve illustrated in Figure 3.6. This is a most obvious example of detector shape affecting efficiency and MDA. If the sample is, in effect, inside the detector we can expect higher efficiency compared to a standard detector, whose efficiency must always be less than 50 %. Well detector efficiency can be in excess of 90 % over an energy range 50 to 200 keV, depending upon the detector size, the thickness of its dead layer and the thickness of the sample well in the endcap. Figure 13.10 shows the construction of a typical well detector, together with the full-energy peak

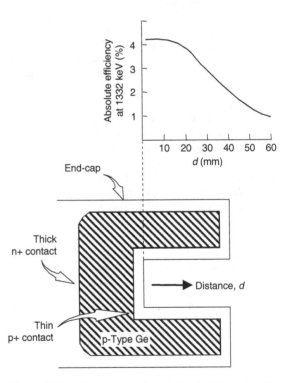

Figure 13.10 Cross-section of a typical well detector, together with the efficiency at 1332 keV from a point source as a function of distance from the bottom of the well (adapted from figures originally provided by Canberra Nuclear)

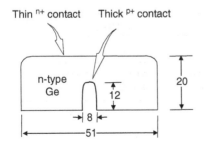

Figure 13.9 The construction of a typical *LO-AX* detector: all dimensions are in mm (adapted from figure in ORTEC catalog)

efficiency for the 1332 keV peak of ^{60}Co, for a particular detector, as a function of point source position within the well. Maximum efficiency can only be achieved with small sources at the bottom of the well. However, at that position, there is some latitude on positioning; moving the source up to 10 mm away from the bottom of the well caused less than 2 % change in efficiency. Such movement of a source away from a standard coaxial detector would cause a considerable change in efficiency.

Canberra has reported measurements where the efficiency of a well detector is five times better than an 80 % coaxial detector. However, the capacitance of a well detector will be higher than for an equivalent coaxial detector, resulting in worse resolution – perhaps 2.3 keV FWHM rather than 2.0 keV. As we saw above in Section 13.2.2, worse resolution means worse (i.e. higher) MDA, meaning that the improvement in MDA expected based on efficiency ratio would not quite be achieved.

A limitation of well detectors is, of course, the size of the well; that referred to in Figure 13.10 would only accommodate 4 or 5 cm of sample within its most sensitive region. For many applications, particularly the measurement of environmental samples, the desirable sample size would be at least an order of magnitude greater. The manufacturers PGT (Princeton Gamma-Tech Inc.) use a 'through hole' well configuration. The advantage of this is a larger usable sample volume in the highest efficiency part of the detector, which more than compensates for the marginal loss of solid angle, giving improved sensitivity per unit mass. PGT also cites technical reasons for claiming improved resolution compared to conventional well-detectors.

In Chapter 8, true coincidence summing (TCS) was discussed at length. I pointed out that the degree of TCS is related to the solid angle subtended by the sample at the detector. For a well detector, that solid angle is close to 4π; the probability of summing of coincident gamma-rays becomes almost one. This is summing at its worst! Generation of a meaningful efficiency calibration curve is, to say the least, problematic. This is a situation where measurements by direct comparison with reference sources should be a first choice option (see Chapter 8, Section 8.8.3). Alternatively, bearing in mind that for electron capture decay nuclides the X-rays emitted play a major role in summing, fitting the well with an absorbing liner may screen out those X-rays and reduce summing to manageable proportions.

To summarize, well detectors offer significant advantages in terms of counting efficiency, but these advantages are offset by the limitation on sample size and much-increased TCS. Before contemplating buying a well detector, the analyst would be well advised to consider carefully the decay schemes of all nuclides to be measured, looking for gamma-rays in cascade. If the nuclides to be measured do not sum, there would be no particular reason to reject a well-detector, but even then efficiency calibration is difficult and it may be better to rely on comparative measurements.

Nevertheless, well detectors do have important uses. Reyss *et al.* (1995) report their experience of a large low-background well detector installed in the underground facility, LSCE, at Mondane in France. This particular detector has a larger-than-average well, able to hold 14 cm^3 of sample. The high efficiency of the detector combined with the ultra-low background in LSC, gives MDAs for ^{226}Ra and ^{228}Ra in sea waters comparable to alpha spectrometry but with a simple one-stage separation (co-precipitation with barium sulfate) instead of the multi-stage separation of the alpha method. It is, apparently, also possible to reduce the sample size from 500 to 30 L.

13.3.3 Sample quantity and geometry

Quite clearly, if our MDA is calculated as a number of Becquerels, the limit in terms of Bq per unit mass (or volume) will be improved (i.e. lowered) by increasing the sample size. Bear in mind that the improvement will not be linear; as sample size increases, the average distance of sample from the detector will increase and the overall counting efficiency will decrease. Figure 13.11 shows the

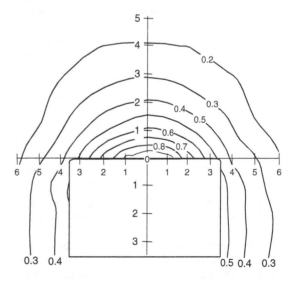

Figure 13.11 Response contours around a germanium p-type detector. Contours are for the 1332.5 keV full-energy peak from a point ^{60}Co source, relative to the centre of the endcap: all distances are in cm (adapted from drawing by R. Mercer)

result of measurements around a detector using a ^{60}Co point source. The contours are full-energy peak count rates for the 1332.5 keV gamma-ray relative to the count rate from the source at the centre of the endcap. It is clear that for this particular detector:

- Counting efficiency is greater if the sample is placed on top of the detector, rather than at the side. This is because the detector is mounted close to the face of the encapsulation and there is likely to be a greater distance between the side of the germanium and the sidewall of the endcap. In addition, the detector crystal is mounted within an aluminium cup. That will absorb a proportion of the gamma-rays before they reach the detector itself.
- There is a rapid fall-off in efficiency with distance from the endcap. Moving the detector 1 cm away from this particular detector would reduce efficiency by 40 %.
- There is a slight asymmetry in the response; the right-hand side as drawn (Figure 13.11) is more sensitive. This is likely to be due to the germanium not sitting centrally within the endcap. This is not an uncommon situation.

The contours in the figure support the idea of a hemisphere-plus-cylinder arrangement of material about the detector, and indeed John Hemingway showed that there is a small increase in sensitivity, in terms of counts per mass of sample, if a spherical disposition of sample is used. However, the small improvement would not be worth the inconvenience of such an awkward geometry. The conventional and more amenable geometry is the simple cylindrical re-entrant beaker called a **Marinelli**, or **Marinelli beaker**, after its originator – this is shown in Figure 13.12.

There is a specification by the IEC (see *Further Reading*) which describes two standard sizes. The dimensions are given in Table 13.4. Each has an annular layer of about 15 mm of material around the sides of the detector endcap, and calculation gives much the same thickness on top of the detector for each size $h = 14$ and 17.5 mm when filled with 450 and 1000 ml, respectively). If the contours of Figure 13.11 are typical, then this equal top-layer thickness may slightly under-exploit the efficiency potential of these standard Marinelli beakers.

A significant practical difficulty is that these dimensions do not necessarily accord with the actual dimensions of detector endcaps. As larger detectors have become available, the diameter of the endcap has had to be increased. Standard endcap sizes are listed in Table 13.5. From that, it is evident that detector sizes of more than 40 % cannot accommodate the smaller, 450 ml, IEC

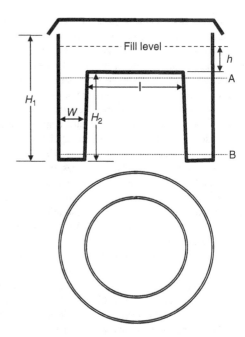

Figure 13.12   The IEC standard Marinelli re-entrant beaker. The beakers are designed to be filled to the level shown, where $h \approx W$

Table 13.4   Dimensions of standard IEC Marinelli beakers[a]

Volume (ml)	450 ± 2	1000 ± 2
H_1 (mm)	104	150
H_2 (mm)	68.3	100
I (mm)		
at A	76.9	119.2
at B	77.4	120.0
W (mm)		
at A	15.3	15.6
at B	14.8	14.8

[a] Wall thickness is 2 mm.

beaker and the 1 L size would have to be used. This larger beaker, with its 119 mm well diameter, would have a considerable gap between the well sides and the detector; 18 mm all round for a 40 % detector. Such a sloppy fit would demand some sort of arrangement for physical location of the beaker, and represents a significant loss of efficiency compared to one tailored to the detector.

Such considerations mean that a number of non-IEC standard Marinelli beakers are in use. If the analyst were to commission the manufacture of a beaker tailored to

Table 13.5 Germanium detector endcap diameters (in mm)[a]

Relative efficiency (%)	Supplier	
	X	Y
10–15	—	70 (standard)
< 40	76	—
28–65	—	82.6 (oversize)
40–50	83	—
50–70	89	—
60–120	—	95.3 (ultra-wide)
70–100	95	—
> 100	102	—

[a] Data from suppliers' catalogues.

their own detector system, the following considerations would have to be taken into account:

- The dimensions must take into account the required volume and the detector cap dimensions,
- The dimensions must also be determined in such a way that as much sample as possible is disposed within the regions of highest counting efficiency,
- The plastic used for construction must be suitably rigid; large floppy beakers are a hazard,
- If a sample changer is to be used, a secure lifting arrangement must be allowed for; there may be a limitation on the mass the changer can handle.
- If liquids are to be measured, there must be a leak-proof seal to the body.
- Certain foodstuffs (for example, potatoes) will tend to ferment when measured wet. In such cases, a vented beaker may be needed,

The sample size chosen (i.e. the volume of the beaker) will depend upon a number of considerations:

- How much sample is available?
- How much sample can be conveniently prepared for counting? This may depend upon economics as much as laboratory resources. A sample may need to be dried, ashed and ground to a small particle size. Can such extra labour intensive activity be justified? In a commercial situation, would the client be prepared to pay for it?
- What limit of detection is being aimed for? This will be related to sample mass and count period, as we know from Equation (13.1).
- Much of the sample within a Marinelli beaker is close to the detector. That means true coincidence summing (TCS) errors. With a larger beaker, more of the sample will be further away from the detector and, consequently, TCS errors will be lessened.

- However, if a larger sample is chosen, can appropriate self-absorption corrections be made?

Ideally, a Marinelli beaker would have a snug fit to the endcap – but not too snug! It must be possible to place and remove it easily, especially if a sample changer is involved. The re-entrant cavity of most Marinelli beakers is tapered slightly to facilitate this. One could suggest that the optimum minimum diameter of the cavity should be the endcap diameter plus 1 or 2 ml. If samples likely to putrefy have to be measured, the resulting increase in internal pressure during counting could make a tightly fitting Marinelli clamp itself onto the detector cap. For such samples, a vented beaker is advisable.

The depth of the beaker should take into account the shape and position of the detector within its cap. Figure 13.13 shows the top face of the detector crystal to be 5 mm inside the endcap; that is commonly but not universally true. The depth of the re-entrant cavity should bear some relationship to the length of the detector crystal, L. In terms of design, it would be reasonable to suggest

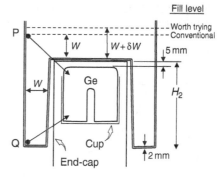

Figure 13.13 An indication of how the dimensions of a Marinelli beaker might (optimally) relate to germanium crystal dimensions and position in the cryostat (see text for further details)

that the longest distance from the top surface of the sample, P, to the top corner of the detector should be similar to the largest distance from the bottom of the sample, Q, to the bottom edge of the detector. If absorption in the paths to the sensitive volume of the detector is the same in both cases, then the optimum cavity depth is approximately:

$$H_2 = L + W + 14 \tag{13.12}$$

where all dimensions are in ml, W is the depth of the sample above the endcap and a wall thickness of 2 mm is assumed.

ANSI/IEEE N42.14:1999 makes the surprising general recommendation that Marinelli beakers of 4 L be used, specifically to minimize TCS. It also states that corrections for summing 'should either be estimated or applied'. I can only reiterate the advice of Chapter 8, that it is much more satisfactory to calibrate with a known quantity of the nuclide to be measured; then all cascade, geometry and density problems cancel. Remember that TCS is independent of count-rate; the effect is not smaller at 'environmental' levels.

With larger samples, there is always a decrease in efficiency, which is not compensated for by the increase in mass of sample. Figure 13.14 demonstrates the decreasing returns, in terms of MDA, with increasing volume. The measurements were of an aqueous source of ^{137}Cs in non-optimized beakers (in the sense of Equations 13.12) of 0.5, 0.9 and 2.5 L. Clearly, the largest beaker is the most effective, in the sense that it provides the lowest MDA. However, in this instance, using five times the volume, 0.5 to 2.5 L, improves the MDA only by a factor of 2.6. Different numbers would be obtained with a different detector, different gamma energies and different densities of material. Debertin and Helmer (1988) suggest generally that not much would be gained for gamma energies greater than 150 keV if the material thickness (W above) exceeds 50 mm. This would be equivalent to a volume of about 3.5 L.

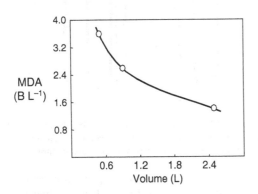

Figure 13.14 An experimentally determined example of how the MDA improves as the volume of a Marinelli beaker increases, but with decreasing returns. Data refer to ^{137}Cs in aqueous solution and a 18.5 % detector

Finally, it should be recognized that when using Marinelli beakers at count-rates that are close to background, the presence of the sample itself represents an extra layer of absorber to the gamma-rays. The background count rate (the 'environmental' background B_{E}) will be lower with the filled Marinelli beaker than without

it. Especially when low energy gamma-rays are to be measured, backgrounds should be determined with a Marinelli in place, filled with strictly inactive material of the same density as the sample. This is easier to recommend than to implement.

13.4 LOW BACKGROUND SYSTEMS

Because background has such an impact on MDA, it obviously makes sense to create detectors with an inherently low background and then install them in a low background environment. In order to create a low background system, it is important to examine where the detector background originates and then seek to reduce the sources. Sometimes that is not as straightforward as it sounds and, over the years, people have spent a large part of their professional lives seeking the ultimate low background. The impetus for seeking low background comes from:

• The desire to measure samples to lower and lower MDA.
• The need, in some cases, to measure small environmental samples to an acceptable MDA.
• The search for subtle physics phenomena such as double-beta decay.

Background radiation comes from the following sources:

• Radionuclides within the materials of the detector assembly.
• Radionuclides within the surroundings of the detector, including the air.
• Cosmic-ray interactions with both the detector itself and its surroundings.
• Radionuclides generated by cosmic ray interactions with the detector itself and its surroundings.

The relative importance of these sources depends upon the construction of the detector, of its shielding and the location of the detector; ultra-low background systems will usually be underground. A typical detector, with no extraordinary precautions taken might have 10 % of its background originating within itself, 40 % from its immediate environment, 10 % from radon in the air and the remaining 40 % from cosmic ray interactions. Reduction of these sources starts with selecting materials for the detector and shielding with particularly low amounts of the unwanted nuclides. Appendix D lists the gamma-rays most often detected in background spectra and their origin.

13.4.1 The background spectrum

An interesting, and useful, paper to consult in relation to this section is that by Bossew (2005), in which he discusses the gamma-rays detected in a spectrum representing 3.3 years of counting time. In fact, this spectrum is the sum of 333 separate background spectra, accumulated over 16 years. The paper also contains some discussion of the temporal variability of background, and in the case of the ^{222}Rn daughters, the monthly variability.

Assuming that there is no local contamination of the environment, within a background spectrum of a detector at, or near, ground level, there could be peaks originating from the following sources:

- Primordial nuclides: ^{235}U, ^{238}U, ^{232}Th and their daughters, and ^{40}K, within the detector and its environment (measurement of these NORM nuclides is covered in Chapter 16, Section 16.1).
- Anthropogenic nuclides: mainly ^{137}Cs, from nuclear weapon and Chernobyl fall-out, and ^{60}Co from steel manufacture. In particular circumstances, there could be other fission product nuclides.
- Activation products: nuclides created by natural sources of neutrons activating the detector itself and its surroundings by (n, γ), (n, α), (n, p) and (n, 2n) reactions: ^{27}Mg, ^{56}Fe, ^{60}Co, ^{63}Cu, ^{65}Cu, ^{71m}Ge, ^{73m}Ge, ^{75m}Ge, ^{77m}Ge, ^{115}Cd, ^{115m}Cd, ^{116m}In and possible others.
- Prompt gamma radiation from neutron capture of detector and surroundings: ^{64*}Cu, ^{114*}Cd.
- Excitation of stable nuclides within detector and surroundings: ^{63*}Cu, ^{65*}Cu, ^{72*}Ge, ^{73*}Ge, ^{74*}Ge, ^{76*}Ge, ^{206*}Pb and ^{207*}Pb.
- Cosmic-ray generated nuclide: ^{7}Be.
- The annihilation peak at 511 keV generated by pair-production events within the detector environment by high energy gamma-rays from the nuclides above and cosmic ray events.

All of the peaks due to the nuclides mentioned above will be sat on a continuum due to Compton scattering of the gamma-rays and backscattering and bremsstrahlung due to direct interaction of cosmic particles. This will be discussed further in Section 13.4.6.

Background reduction by removal of the sources of radionuclides within the materials of the detector and surroundings and siting the detector in a location of low background are referred to as **passive** methods. Methods that deduce which detector counts can be identified as originating from background and prevent them being recorded by the MCA system are termed **active** methods. These will be discussed in Section 13.5.

13.4.2 Low background detectors

In recent years, detector manufacturers, in response to a demand for lower and lower MDA, have addressed the problem of radioactivity in the detector assembly and most now will provide low background detectors constructed from special materials. Fortunately, the major component of a detector, the germanium crystal, is of extremely high purity in order for it to function effectively. Impurity concentrations of less than one part in 10^{12} make HPGe detector material one of the purest substances known. There are also no primordial naturally occurring radioactive germanium isotopes. However, as we shall see, the detector itself is a target for cosmic ray interactions that can give rise to background peaks due to activation and excitation.

Other materials routinely used in standard detectors contain small amounts of contaminants. Table 13.6 shows three common radionuclides and their concentrations in a range of materials used for the construction of detectors and their shielding.

The ^{208}Tl, in the ^{232}Th decay series, and ^{214}Bi in the ^{238}U decay series are used as indicators of the amount of their parents in the materials. Other members of each chain will also be present. Apart from these **primordial** nuclides (those surviving from the formation of the solar system), traces of **anthropogenic** (human-made) species are seen, such as ^{60}Co in steels and ^{137}Cs in molecular sieves. There are obvious reasons for some of these radioactive contents; aluminium always has traces of uranium and thorium within it, and it is not unreasonable to expect ^{40}K in molecular sieves. However, relatively large amounts (compared to other materials in the list) of uranium and thorium daughter nuclides in epoxies and printed circuit boards are unexpected. Clearly, for some materials, there is ample scope for reducing the activity the detector 'sees' by selecting a material with a lower activity. In a similar table to that above, Dassie (in a private communication) reported 4.3 Bq of ^{137}Cs kg^{-1} of CsI and 33 Bq of ^{40}K kg^{-1} of NaI, considerably more than in Table 13.6; this latter would be a problem were it to be used for low level Compton suppression systems.

When the manufacturers of detectors design a low-background detector, each constructional material will be considered; samples will be tested for radioactive contamination and sources of low activity materials sought. (Figures 3.15 and 3.16 in Chapter 3 are a useful reminder of the construction of the detector and its cryostat.) Aluminium is a significant problem. In most detector systems, it is used for the endcap and for other supporting purposes within the detector housing. For a low background detector, magnesium (ORTEC) or extra-high-purity aluminium, 99.999 % (Canberra) would be

Table 13.6 Primordial radionuclide concentrations in materials used in germanium and NaI detector systems[a]

Material	Radionuclide concentration (Bq kg^{-1})		
	^{208}Tl	^{214}Bi	^{40}K
Concentration equivalent to 1 Bq kg^{-1}	0.25 ppm Th	77 ppb U	33 ppm K
Aluminium	0.12–3.3	< 0.07–33	< 0.4–20
Beryllium	0.17	12	< 20
Copper	< 0.005	< 0.02–0.05	< 0.2
Copper (101)	< 0.0005	< 0.0009	< 0.009
Copper (OFHC)	< 0.005	< 0.02–0.17	< 0.2
Epoxy	0.8–67	1.3–880	< 20–2000
Grease, high-vacuum	< 0.02	< 0.2	< 0.2
Indium	< 0.02	< 0.05	< 0.4
Lead	< 0.0004	< 0.0007	< 0.0002
Molecular sieve	6.7–8.3	17–50	130–200
Mylar, aluminized	1.7	3.3	< 40
Oil, cutting	< 0.007	< 0.05	< 0.4
Plastic tubing	< 0.07	< 0.07	< 20
Printed circuit board	33	67	67
Quartz	0.1–1	< 0.4–17	< 4
Reflector materials	< 0.002–1.7	< 0.02–3.3	<0.09–5
Rubber, sponge	0.83–3.3	1.3–20	< 7–40
Silica, fused	< 0.4	< 0.2	< 2
Silicone, foam	0.33	0.83	< 4
Sodium iodide (Tl)	< 0.05	< 0.07	< 0.5
Solder	< 0.005	< 0.02	< 0.2
Steel, stainless	< 0.04	< 0.1	< 4
Steel, pre-1940	< 0.009	< 0.02	< 0.2
Teflon	< 0.005	< 0.02–0.2	< 0.4
Wire, Teflon-coated	< 0.07	< 0.02	< 0.4

[a] Data from Brodzinski *et al.* (1985) scaled to Bq; upper limits are rounded to 1 significant figure with other figures to 2 significant figures.

used. Even beryllium has its problems. The best available beryllium will contain some primordial radionuclides. Fortunately, carbon composite detector caps have become available recently, and are expected to have a lower radioactive content than the metal alternatives. One would expect them to be the cap of choice where low background is paramount, although Laubenstein *et al.* (2004) complain of 'high intrinsic contamination by ^{40}K and sometimes ^{226}Ra, which can limit their use in deep underground setups'.

The detector holder, clamps and cooling rod in a standard detector would normally be of OFHC copper, but selected copper of much higher purity would be used for a low background system. A beryllium window will be of specially selected material; stainless-steel screws replaced by brass or low-cobalt stainless-steel and rubber O-rings might be replaced by indium metal. Even such careful selection may have unexpected consequences; indium has a high capture cross-section for thermal neutrons.

In exceptional circumstances, a reduction in impurity peaks could result in the appearance of indium capture gamma-ray peaks in the background spectrum. Replacing stainless-steel items with high-purity copper also has a down-side; copper can be activated (n, α) by cosmic neutrons to ^{60}Co.

The vacuum-getter within the evacuated detector chamber is often, in normal systems, a molecular sieve – an aluminium/silicon material – that will unavoidably contain very significant amounts of uranium, thorium and potassium (see Table 13.6). For low background systems, this is replaced by activated charcoal, which, while not necessarily being completely free of radioactive materials, will have a lower activity than a molecular sieve.

The preamplifier is usually mounted below the detector within the overall housing. It comprises a fibreglass printed circuit board, aluminium structural items and a number of electrical items. Especially low activity resistors, capacitors and the FET are not available and, because the preamplifier is so close to the detector, it will be moved so as to avoid line-of-sight to the detector, or be shielded from the detector by suitable high Z materials, such as lead or tungsten. This is not an ideal solution because of fluorescence within that shield, giving rise to unwanted X-ray peaks in the spectrum. Canberra and PGT move the preamplifier away from the detector altogether, mounting it outside the shielding.

A prospective purchaser might consider it worthwhile to obtain evidence of the measured background of the specific detector before buying. It may not be possible to reproduce the performance of the manufacturer's low-background shield in which the test was done, but such data will give a benchmark against which the measurements after installation can be evaluated.

13.4.3 Detector shielding

In Chapter 2, Section 2.8, I discussed shielding in the context of interactions of gamma-rays. Some of the advice given in that section will be shown to be less than perfect in the pursuit of ultra-low background. To give some idea of context, I measured the background of a typical 45 % detector installed at ground level in a standard commercial cylindrical lead shield, graded with cadmium and copper, over 200 000 s. I found a total spectrum count rate of 5.5 cps and a count rate of 0.014 cps per keV at 55 keV – about 100 times less than that at 2300 keV. Perhaps nothing to get excited about, but these levels are orders of magnitude too high for people seriously seeking low background conditions. Calculation of the activity equivalent of the peaks in that background gave 60 Bq kg^{-1} of ^{40}K, 60 Bq kg^{-1} of ^{238}U and 10 Bq kg^{-1} of ^{232}Th in a 0.3 kg sample, representing a serious limitation on the practical MDA.

The first object of the shielding is to reduce the number of gamma-rays originating outside from reaching the detector. This means surrounding the detector with high Z material to absorb them, the most convenient being, of course, lead. Table 13.7 shows the calculated attenuation factors for a collimated beam of gamma-rays impinging normally on lead of thicknesses that are generally used for detector shields (see also Chapter 2, Table 2.1). (As it happens, 50 mm and 100 mm are the thicknesses of standard lead bricks sometimes used for DIY shielding.) Lower-energy radiation will be quite adequately stopped by 50 mm of lead, but it is necessary to remove high-energy gamma-rays as well because, even if we only intend to measure relatively low energies, Compton scattering of those high-energy gammas will contribute to the background continuum at low energy. More often than

not, the shield thickness will be 100 mm, although low-background shields will probably be 150 mm.

Table 13.7 Attenuation factors for a beam of gamma radiation on lead

Energy (keV)	Attenuation factor		
	5 cm Pb	10 cm Pb	15 cm Pb
200	$> 10^6$	$> 10^6$	$> 10^6$
500	9400	$> 10^6$	$> 10^6$
1000	56	3100	1.8×10^5
1500	19	370	7200
2000	14	190	2500
3000	11	120	1300

A difficulty with lead is its radioactive impurities. Table 13.6 give a misleading impression of the radioactivity of lead. While the chemical-refining processes for lead will remove most isotopes of impurity elements, they cannot remove the ^{210}Pb daughter of ^{238}U, which will be present in trace amounts in the lead ore. ^{210}Pb emits a 46.54 keV gamma-ray, not so much of a problem in itself, but its daughter ^{210}Bi (half-life 5.013 d) will be in secular equilibrium with it. ^{210}Bi emits a beta particle of maximum energy 1161 keV, which will result in a bremsstrahlung continuum extending down to low energy. Most modern lead will contain some ^{210}Pb. Fortunately its half-life is 'only' 22.3 years and there are (limited) sources of old lead (**aged** lead), more than a couple of hundred years old, in which the activity is much lower than modern lead. This material is obviously in short supply and it is more economical to use a layer (say 25 mm) of aged lead within a 100 or 150 mm modern lead shield. Aged, or 'low level lead', can be expected to have less than 25 Bq of ^{210}Pb per kg of lead, although Verplancke (1992) takes 'low-level lead' as having less than 10 Bq kg^{-1}. Laubenstein *et al.* (2004) are fortunate enough to have a shield of less than 5 Bq kg^{-1}.

13.4.4 The graded shield

In Chapter 2, Section 2.8, I described how a graded shield is valuable in reducing fluorescent Pb X-rays at 72 to 87 keV. As it happens, this would also absorb the ^{210}Pb gammas and reduce the ^{210}Bi bremsstrahlung. However, placing low Z materials close to the detector will cause a noticeable rise in backscatter. The graded shield lining thickness is a compromise between reducing the background by absorbing X-rays and ^{210}Pb on the one hand, and increasing the background by increasing backscatter.

For low background applications, the use of cadmium in the grading is not to be recommended. The cross-section of cadmium for the absorption of thermal neutrons is very high. The result of such absorption is the appearance of the 558.4 keV peak in the spectrum from the prompt gamma-ray emitted during the reaction. Cadmium is better replaced by tin, which being a neighbour in the Periodic Table, has similar X-ray absorption properties but a much lower neutron absorption cross-section.

Canberra in their *Application Note* (1995), in the light of practical measurements, suggest that if a detector is to be used at ground level, where cosmic interactions with the detector are dominant, the use of aged lead may not be justified. Their argument is that measurements show that the ^{210}Bi bremsstrahlung is not significant above 500 keV and that a suitable graded shield will absorb both ^{210}Pb gamma-rays and the bremsstrahlung. Nevertheless, Canberra's 'Ultra low background' shield does incorporate an inner aged lead shield.

In selecting an appropriate thickness of tin and copper, it should be remembered that, in addition to tending to raise the background continuum due to scattering, in principal both of these materials can be activated by cosmic-generated neutrons. In very long backgrounds, peaks due to activation, for example ^{63}Cu(n, α)^{60}Co, and to excitation, ^{63}Cu(n, n′)^{63*}Cu, will be evident (see Appendix D). In below-ground installations, where the cosmic flux is low, it is possible to use much greater thicknesses of copper for the graded shield – up to 15 cm (!) has been suggested for very deep locations.

13.4.5 Airborne activity

Radon isotopes, both radon, ^{222}Rn, and thoron, ^{220}Rn, are present in air as active gases emanating from traces of ^{238}U and ^{232}Th in building constructional materials and/or local soils and rock. Neither of these is particularly well endowed with gamma-ray emissions but their progeny are. These can be absorbed on dust particles and surfaces within the detector enclosure and give rise to characteristic peaks of ^{214}Pb, ^{214}Bi and ^{210}Pb, (from the ^{238}U series) and ^{212}Pb, ^{212}Bi and ^{208}Tl (in the ^{232}Th series) in the background. The difficulty with radon is that its concentration around the detector is likely to vary with time of day and season of the year, and with atmospheric pressure, wind speed, temperature, etc. A reliable and reproducible background from radon daughters is often difficult to achieve.

A more specific problem could be experienced when a detector is installed on a nuclear reactor site, particularly in a reactor hall. There, it may also detect ^{41}Ar (half-life 1.827 h, $E\gamma = 1293.6$ keV) produced by thermal neutron activation of argon in the air within the voids in the reactor structure, from where it can diffuse into the local atmosphere.

These problems can be minimized by the following alternatives:

- making the shield as air-tight as convenient, to prevent diffusion of radon into the shield;
- feeding the vent gas from the liquid nitrogen Dewar into the shielding as a slow continuous flush;
- flushing with cylinder nitrogen or radon-free air;
- filling the free space in the shield with sealed plastic bags containing nitrogen or filtered, radon-daughter-free air.

With all of these remedies, but in particular with the latter, there is the difficulty of arranging easy access for placing samples on the detector. Using the vent gas from the liquid nitrogen Dewar would seem to be the most attractive option. At least one commercial low-background shield incorporates a 'purge port' for that purpose. If self-installing such a port, it should be remembered that the nitrogen gas will be cold. If, as in a typical vertical detector/shield arrangement, there are gaps in the bottom of the shield, the cold nitrogen gas will tend to fall through instead of filling the whole shield. Maybe, a filling port at the top of the shield would be preferable, and certainly some attention to gaps where radon may diffuse in, or be carried in by the flow of nitrogen, would be useful. A paper by Hurtado *et al.* (2006) discusses, among other measures, the effect of radon flushing, using evaporated nitrogen from the cryostat, on a detector at ground level. Although the overall count rate is reduced, the magnitude is only slight and, although there appears to be a slight reduction in all of the ^{212,214}Bi and ^{212,214}Pb count rates, the significance is doubtful. The authors do comment that much more substantial flushing has been shown to provide a more significant reduction.

Notwithstanding these reservations, it is routine practice in ultra-low background counting rooms to purge, not only the detector space, but also the whole laboratory of radon with filtered clean air.

13.4.6 The effect of cosmic radiation

The earth's atmosphere receives a cosmic ray flux of about 70 % protons, 20 % alpha particles and 10 % other heavier ionized particles. Their energies are extremely high: 10^4 GeV to at least 10^{10} GeV. In the upper atmosphere, at about 25 km above the surface, these particles interact to produce secondary radiations of many sorts, largely pions (π-mesons). At surface levels, the pions interact further and about 70 % of the total flux generates muons, the remainder resulting in lower-energy photons, electrons and positrons. The muons are very penetrating and they are the prime source of fast neutrons. These fast neutrons cause activation and excitation within the surroundings of our detectors and with the detector itself.

Direct interaction of secondary cosmic radiation

Figure 13.15 displays a background spectrum measured over an energy range much greater than that necessary for normal gamma-ray spectrometry. The spectrum is thought to be a consequence of muon interactions:

- The steep decline up to 2000 keV, the 'normal' spectrometry range, is due to Compton events, backscatter and bremsstrahlung resulting from the decay of muons ($\mu^\pm$) into high-energy electrons and positrons. On this are superimposed the 511 keV annihilation radiation and all the gamma-ray peaks from the background nuclides and the peaks from activations described below.
- From 2 to 7 MeV, the continuum is attributed to cosmic particles passing through only part of the detector.
- The very broad peaked continuum centred at about 13 MeV is due to pions ($\pi^\pm$) and other particles, $\mu^\pm$, electrons, positrons and protons passing through the detector where they deposit about 6 or 7 MeV per cm of germanium traversed.

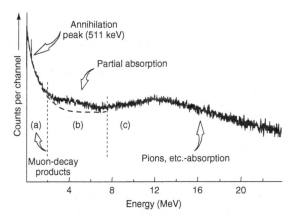

Figure 13.15 High-energy background spectrum mainly due to cosmic radiation, collected from a 2000 mm², 20 mm thick, LEGe detector in a low background cryostat mounted in a 100 mm thick low background lead shield; count period, 5 d (adapted from diagram of Canberra Semiconductors, BV)

The total cosmic ray flux giving events between 2 and 24 MeV is about 0.015 particles per second per square centimetre. The 'peak' at 13 MeV corresponds to about $0.08\,h^{-1}\,keV^{-1}$, which diminishes to about $0.01\,h^{-1}\,keV^{-1}$ at 24 MeV.

Cosmic neutron induced activity

The fast neutrons in the cosmic flux at sea level, sometimes called *tertiary cosmic radiation*, are created mainly by interactions of the muons with high Z material – the lead of the shielding being a crucial example. The magnitude of this flux in lead is of the order of 0.11 fast neutrons per minute per kilogram of lead. These neutrons stimulate the normal range of fast neutron reactions: (n, p), (n, α), (n, 2n) and (n, n'γ). (Heusser (1994) gives a more complete list). Appendix D lists the most commonly observed gamma-rays in background spectra, including those due to fast neutron reactions.

If there is a substantial amount of low Z material, including hydrogen and most polymers, within the neighbourhood of the detector they will tend to thermalize the fast neutrons. Thermal neutron fluxes of the order of $10^{-3}\,n\,cm^{-2}\,s^{-1}$ have been estimated. If thermal neutrons are present, a whole panoply of thermal neutron capture reactions is possible, potentially giving rise to prompt capture gamma-rays and decay gamma-rays from the capture products. In particular, materials such as Cd, F, B and Li which have high neutron cross-sections should also be avoided if prompt gamma-ray peaks are to be avoided.

A number of small peaks appear in long background spectra because of activation or excitation of the germanium of the detector. Table 13.8 shows the isotopic composition of germanium, together with some of the nuclear reactions that could be stimulated. There may also be evidence in the spectrum of fast neutron reactions with components of the shielding, lead, copper, cadmium, iron, for example. Table 13.9 lists some of the more prominent peaks that may be detected with an assignment to particular nuclear reactions. A number of these peaks are of particular interest:

- All of the peaks marked (b) in Table 13.9 are caused by excitation of a stable isotope of germanium to a higher energy level. When that energy level de-excites, a gamma-ray is emitted. However, these particular gamma-ray peaks are not Gaussian as we might expect, but are tailed to high energy. This is because the collision of the fast neutron with a germanium atom is inelastic – the germanium atom will recoil with an amount of energy depending upon the angle through

Table 13.8 Isotopes of germanium and the most likely neutron reactions

Isotope	Abundance %	Cross-section (barn)	Reactions
^{70}Ge	20.5	3.2	$(n, \gamma)^{71m}$Ge
^{72}Ge	27.4	0.98	$(n, \gamma)^{73m}$Ge, $(n, n')^{72*}$Ge
^{73}Ge	7.8	15	$(n, n')^{73*}$Ge
^{74}Ge	36.5	0.143, 0.24	$(n, \gamma)^{75m}$Ge, $(n, 2n)^{73m}$Ge, $(n, n')^{74*}$Ge
^{76}Ge	7.8	0.09, 0.05	$(n, \gamma)^{77m}$Ge, $(n, 2n)^{75m}$Ge, $(n, n')^{76*}$Ge

Table 13.9 Examples of gamma energies produced by cosmic neutrons in a shielded germanium detector

Energy (keV)[a]	Emission probability (%)	Nuclide[b]	Half-life	Reaction
53.4	10.5	^{73m}Ge	499 ms	^{72}Ge(n, γ), ^{74}Ge$(n, 2n)$
68.8 (b)	—	^{73*}Ge	Prompt	^{73}Ge(n, n')
139.7	38.8	^{75m}Ge	47.7 s	^{74}Ge(n, γ), ^{76}Ge$(n, 2n)$
159.7	11.3	^{77m}Ge	52.9 s	^{76}Ge(n, γ)
198.4	91	^{71m}Ge	20.4 ms	^{70}Ge(n, γ)
278.3	—	^{64*}Cu	Prompt	^{63}Cu(n, γ), ^{65}Cu$(n, 2n)$
558.4	—	^{114*}Cd	Prompt	^{113}Cd(n, γ)
569.7	—	^{207*}Pb, ^{76*}Ge	Prompt	^{207}Pb(n, n'), ^{76}Ge(n, n')
579.2	—	^{207*}Pb	Prompt	^{207}Pb(n, n')
595.8 (b)	—	^{74*}Ge	Prompt	^{74}Ge(n, n')
691.0 (b)	—	^{72*}Ge	Prompt	^{72}Ge(n, n')
803.3	—	^{206*}Pb	Prompt	^{206}Pb(n, n')
962.1	—	^{63*}Cu	Prompt	^{63}Cu(n, n')
1097.3	56.2	^{116m}In	54.1 min	^{115}In(n, γ)
1293.6	84.4	^{116m}In	54.1 min	^{115}In(n, γ)
2223	—	^{2*}H	Prompt	^{1}H(n, γ)

[a] Gamma-rays labelled (b) give broad, asymmetric peaks due to nuclear excitation (see text for further details).
[b] * indicates an excited state of the nuclide.

which the neutron is scattered. Both the gamma-ray emitted by the de-excitation gamma-ray and the recoiling atom will create electron–hole pairs, all of which will contribute to the detector signal. The total energy absorbed will be greater than the gamma-ray energy by a variable amount; hence, these peaks are tailed to higher energy (see Figure 13.16). Similar events taking place in the copper lining and the lead shield only contribute to the de-excitation gamma-ray.

- The 559.4 keV peak is caused by the prompt gamma-ray emitted when ^{113}Cd atoms in the graded shielding capture thermal neutrons. If the cadmium is replaced by tin, which has a much lower thermal neutron cross-section, this peak will disappear from the background spectrum.
- The peak at 198.4 keV, assigned to the very short-lived activation product, ^{71m}Ge, is at first examination difficult to understand. This nuclide emits two gamma-rays of 174.97 (91 %) and 23.44 (0.48 %) keV. Although one

might expect a small amount of summing, one would still expect to see a peak at 174.97 keV. There is none! In fact, the low emission probability of the 23.44 keV peak is due to a very high degree of internal conversion. This gives rise to a number of photoelectrons and Auger electrons *within the detector*, so that the 23.44 keV de-excitation energy is added to the 174.97 keV to give a spectrum peak at 198.4 keV.

The count rates of these induced activity peaks are not high. The largest peaks, seen from a 1 kg detector are at 595.8 and 691.0 keV with up to 500 counts per day, and at 139.7 and 198.3 keV with up to 200 counts per day.

Finally, as the cosmic ray fast neutron flux in a detector assembly derives largely from muon interactions in the lead shield and lead is a very poor neutron absorber, additional lead increases the neutron flux and the neutron-induced activities. The recommendation is that, unless the

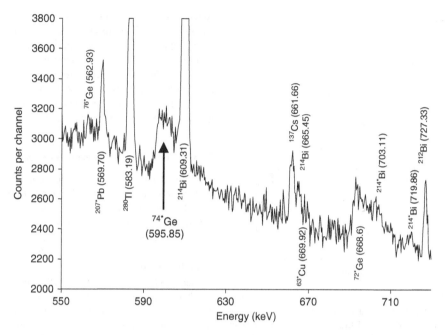

Figure 13.16 Region of a background spectrum in which asymmetric germanium excitation peaks can be observed (plotted from data provided by P. Bossew (Personal communication))

detector is installed underground, where the muon field is reduced, no more lead should be used than is necessary to screen out external gammas. In many instances, this would be a thickness of 100–150 mm.

13.4.7 Underground measurements

In terms of the continuum, detector background is dominated by the effects of the cosmic muon flux. This can be reduced by installing the detector underground, using the earth itself as shielding. Curve A in Figure 13.17 shows how the muon intensity decreases with depth. (The conventional unit of depth used in this field is 'm w.e.' – metres water equivalent. The fluxes in Figure 13.17 are plotted relative to the muon flux at the surface.) Curves B, C and D show the relative fluxes of secondary neutrons, neutrons formed by muon interactions with lead and with rock. Line E represents the relative neutron flux created by spontaneous fission and α, n reactions due to the uranium and thorium in the rock surrounding a detector, calculated assuming average concentrations of uranium and thorium in the continental upper crust, to be 10.7 ppm and 2.8 ppm, respectively. Deeper than 500 m w.e., the cosmic ray and neutron effects become less than the effect of the host rock. Even relatively shallow depths can reduce the muon flux by a useful amount (only 10 m w.e. could reduce it by 60 %) and the secondary neutron flux by even more. The

IAEA-MEL (Marine Environment Laboratory) (Povinec *et al.* (2004)) is sited in an underground car park only 35 m w.e below ground level.

The relative count rates, expressed in counts per day per kilogram of germanium, of various detectors of the CELLAR network are also plotted on Figure 13.17 for comparison. Count rates are plotted relative to an estimated count rate at ground level without shielding. (CELLAR is a pseudo-acronym for a collaboration of European Low-Level underground laboratories.) The deepest laboratory, LSCE, at Mondane in France, is at 4800 m w.e.

With the exception of the IAEA-MEL, all of the detectors have only passive shielding designed on the principles discussed above. The impressively low background count rates, between 30 and 300 cpd kg^{-1} Ge, are achieved because the earth shields the detector from the cosmic-ray background and the local environmental background due to the host rock is removed by lead shielding. A good part of the residual count rate could be due to radioactive impurities in the detector materials themselves.

It should also be noted that, at ground level, materials destined for shielding purposes, such as copper and iron, will be exposed to the muon related fast neutron flux and will be gaining long-lived nuclides, such as ^{60}Co (in copper) and ^{54}Mn (in iron), by activation. For that

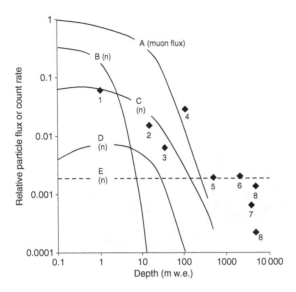

Figure 13.17 Muon and neutron fluxes at and below ground/sea level relative to that at ground/sea level, together with relative count rates of CELLAR underground detector systems. Flux rates are calculated relative to the muon flux rate at ground/sea level. Detector count rates (cps per kg of germanium) are calculated relative to an estimated ground level count rate: A, muon flux; B, secondary neutrons; C, muon induced neutrons in lead; D, muon induced neutrons on rock; E, neutrons from fission and α, n in rock (data adapted from Heusser (1996): 1, ARC Seibersdorf (Austria); 2, Max-Planck Institute (Germany); 3, IAEA-MEL (France); 4, VKTA (Germany); 5, IRMM (Belgium); 6, PTB (Germany); 7, LNGS (Italy); 8, LSCE (France) (data adapted from Laubenstein *et al.* (2004)

reason, laboratories seriously aiming for ultra-low background will purchase and take delivery of materials as soon as possible after manufacture and take them underground to avoid unnecessary activation. Gurriarian *et al.* (2004), reporting on the calibration of a large ultra-low background well detector in use in the LSCE facility, note the detection of a number of nuclides from ground level activation: ^{125}Sn (solder), ^{60}Co (copper), ^{57}Co, ^{58}Co, ^{54}Mn, and ^{65}Zn (germanium). The last group of nuclides can be clearly identified as being within the detector itself because their peaks have associated peaks due to summing with the electron capture X-rays, which are normally absorbed before reaching the active germanium.

13.5 ACTIVE BACKGROUND REDUCTION

Whether dealing with low background detectors or normal laboratory systems, the MDA depends upon the continuum level beneath the measured gamma-ray peaks.

Active background reduction is a means of reducing that continuum. I will discuss two aspects of this; reduction in the Compton continuum within the spectrum and reduction of the cosmic ray continuum discussed above. Both involve additional detectors to detect gamma-rays that are either leaving the detector, in the case of Compton suppression, or might be about to enter the detector, in the case of background suppression.

13.5.1 Compton suppression systems

Compton scattered gamma-rays leaving the detector represent incomplete absorption of a gamma-ray. That means that the detection event will result in a count, not in the full-energy peak, as hoped, but on the Compton continuum. If we can find a way of reducing that, all peaks standing on it will be measured with a lower uncertainty and a better MDA.

The simple trick is to surround the HPGe detector with another, high efficiency detector, which may be referred to as the **veto detector, guard detector** or **shield detector**, which will inform us whenever a gamma-ray is scattered out of the HPGe. That can then allow us to stop the ADC/MCA from recording the event. Traditionally, NaI(Tl) scintillators were used for this purpose, but if space is limited the more expensive, but more effective, bismuth germanate (BGO) could be substituted (there is information on scintillation systems in Chapter 10). The poor energy resolution of scintillators is not an issue here – the most important thing is to make sure that every scattered gamma-ray is detected.

Figure 13.18 shows a schematic diagram of a simple Compton suppression system. The coincidence unit detects when pulses in the HPGe and the guard detector appear together. The width of pulses from scintillation detectors is much shorter than HPGe pulses. The guard detector output pulses need to be delayed by a precise amount until the correct time for rejecting the HPGe pulse within the ADC/MCA. In fact, it is possible to dispense with the coincidence unit and depend on the fact that, if the guard pulse is converted to a logic pulse and delayed by the correct amount, it will automatically reject any HPGe pulse that happens to be there at the time. The paper by Hurtado *et al.* (2006), referred to in Section 13.4.5, compares experience with coincidence gating and simple delayed gating.

The fact that a high Z material surrounds the HPGe spectroscopy detector is useful in that it will provide extra shielding for it – both passive and active. An external gamma-ray will have to pass through the guard detector to reach the HPGe. If it interacts with both detectors, the event will be rejected. As a side effect, escape peaks will

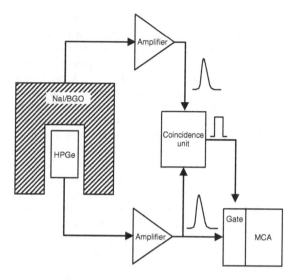

Figure 13.18 An anticoincidence circuit for Compton suppression

be removed from the spectrum as the annihilation photons escaping from the HPGe are detected by the veto detector.

A well-designed system should be able to achieve reductions in the Compton continuum of a factor of 8 to 10, improving the peak-to-Compton ratio of a detector with, from say 50 to 1 to 500 to 1. ORTEC claim peak-to-Compton ratios of up to 940 to 1. Equation (13.8) told us that the MDA is proportional to the square root of the background continuum level. Reducing that by a factor of 9 improves the MDA by a factor of 3.

Design considerations – spectrometry (HPGe) detector

The efficiency of the Compton suppression depends upon the guard detector being able to detect the scattered gamma-rays. It has to be remembered that, while the gamma-rays from the source may enter the detector through an entrance window, the scattered gamma-rays will escape in all directions. Any feature of the HPGe detector that would hinder them from reaching the guard detector will reduce the effectiveness of the suppression:

- A p-type detector should not be used, because its thick dead layer will tend to absorb scattered gamma-rays. For detectors mounted in a conventional aluminium endcap, an n-type detector may have twice the Compton suppression factor of a p-type detector of the same efficiency.

- A normal aluminium endcap will absorb escaping radiation. Use a low-density endcap, such as carbon composite.
- The cup holding the germanium crystal in place should be of low density. Aluminium is preferred to copper.

Note that these requirements could conflict with the demands for low background, in that copper could be chosen for its lower radioactivity content. However, the higher density of copper, whether used for endcap or detector cup, will absorb more scattered gamma-rays.

Design considerations – guard detector

The thickness of the scintillator must be sufficient to have a good probability of absorbing the scattered gamma-ray. In principle, it need not absorb the gamma-ray completely; all that is necessary is for it to interact with the guard detector significantly. However, if absorption is incomplete, scattered gamma-rays from the guard could find their way back to the HPGe and complicate matters. To get some idea of the required thickness, we can consider a 1500 keV gamma-ray scattered through 45°. The HPGe detector will absorb 690 keV and 810 keV will be lost (Chapter 2, Equation (2.5)). There is a 90 % probability of the scattered gamma being absorbed in 93 mm of NaI(Tl), and a 99 % probability of absorption in 185 mm. If BGO is used, these thicknesses can be reduced by a factor of 0.4. A typical guard detector might be 350 mm in diameter and 350 mm in length.

As far as possible, the guard detector should surround the spectrometry detector. In order to achieve this, it may be necessary to use more than one guard detector. Figure 13.19 shows one such arrangement. This ORTEC

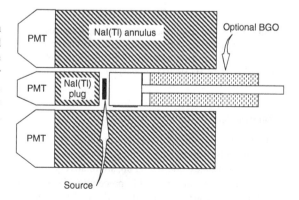

Figure 13.19 An arrangement using an annulus and a plug detector: PMT, photomultiplier tube; BGO is optional (see text for further details)

system has a 230×230 mm axial annulus of sodium iodide, with a 76×76 mm plug detector. The presence of the plug detector is said to eliminate the Compton edge from the spectrum. In this design, the scintillator is mounted around the cryostat cooling rod; this is similar to ORTEC's 'Duet' design, the which increases the probability of interception of the forward scattered photons. Note that the HPGE detector is not placed centrally within the annular guard detector. This is because most scattering is in the forward direction and therefore the HPGe detector is positioned to allow more scintillator at its back than at its face. Furthermore, the energy of backscattered photons is low, compared to forward scattered photons (Chapter 2, Figure 2.15), so that a lesser thickness of guard detector is required behind the source. The design of Figure 13.20 allows for an external source. Again, allowance is made for the dominance of forward scattering. *Canberra Application Note AN-D-8901* shows several other configurations.

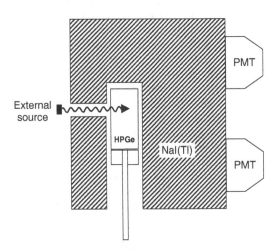

Figure 13.20   An arrangement with an external source or beam of radiation, showing the more efficient asymmetric disposition of the NaI(Tl) guard detector: PMT, photo multiplier tube (adapted from the Harshaw/QS catalogue)

Disadvantages of Compton suppression

(1) Sample size is limited, whether mounted internally or externally with respect to the guard detector.
(2) The electronics, and setting up, are relatively complex (a summary of an NIST procedure, Stover and Lamaze (2005), exemplifies that). The ORTEC system mentioned above needs thirteen NIM units, which include four amplifiers and four delay units – all need to be set up.

(3) A single, very large HPGe will usually be a less complicated alternative, available at a similar cost, and possibly with lower MDA. Using Equation (13.8), we can compare a 20 % detector, suppression factor 9, with an unsuppressed 100 % detector. Assuming that the Compton continuum is dominated by the sample activity ($B_S > B_E$, referring to Section 13.2.2.1), by using Figure 13.2 we can estimate the continuum for the larger detector to be about twice that of the unsuppressed 20 % detector. Equation (13.8) gives us an MDA ratio of 0.85 in favour of the larger detector. However, that assumes both detectors have the same resolution. If, as is likely, the larger detector has a larger resolution, or if the Compton continuum is dominated by the environmental background (and is, therefore proportional to detector size) the smaller, suppressed system could still give a better MDA.

(4) The most serious disadvantage arises when nuclides are measured that have gamma-rays in cascade. If one of these gamma-rays is detected in the spectrometry detector and another of the same decay cascade in the guard detector, a count will be lost from the full-energy peak – in fact, all full-energy peaks related to the cascade will be suppressed. Even gamma-rays scattered back from the shielding into the guard detector and beta-particle bremsstrahlung entering the guard detector, all of which will be coincident with the gamma-rays, can result in full-energy peak suppression. In broad terms, the 'better' the system, that is, the greater the Compton suppression ratio, the lower the germanium count rate for these cascading peaks. The reader will appreciate that true coincidence summing (Chapter 8) is a similar problem but in that case, the difficulty is the accidental collection of cascading gammas; here, the equipment is actually designed to maximize the removal of gamma-rays that appear at the same time.

Obviously, this could have a dramatic effect on MDA; a 10 % loss in efficiency translates to a 10 fold increase in MDA (efficiency is in the denominator of Equation (13.8)), and could completely alter the judgement as to whether a particular Compton suppressed system is better than an unsuppressed, much larger detector. Of course, if one only wishes to measure nuclides without cascading decay schemes, for example, ^{137}Cs, such considerations are irrelevant. Each nuclide, and its decay scheme must be considered individually, exactly as one would when judging true coincidence summing with a conventional detector. It is worthwhile considering arrangements to measure an unsuppressed spectrum at the same time as the suppressed one and using the most appropriate one to estimate activities.

13.5.2 Veto guard detectors

If one's preoccupation is to reduce the spectrum continuum due to external radiation, rather than the Compton continuum, a secondary detector designed to intercept background before it reaches the spectrometry detector is required. This is usually achieved by mounting a large area detector, usually a plastic scintillator, sometimes a multiwire proportional counter, on top of the lead shield (Figure 13.21). Because a large fraction of the cosmic rays travel vertically, it is not usual to worry about the sides of the detector arrangement. The veto detector is most effective at suppressing the continuum due to high-energy interactions in the germanium (Figure 13.15) and will have some effect on prompt gamma radiations from fast neutron reactions. Ideally, the veto detector should have its own shielding to minimize low energy interactions from the local environment, which would cause unwanted, spurious coincidences.

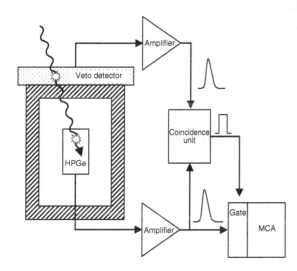

Figure 13.21 Anticoincidence arrangement for a cosmic ray veto detector

The surface system mentioned above in the context of optimization also incorporated a veto detector. Hurtado *et al.* (2006) examined four different gating scenarios to allow the plastic scintillator veto detector to prevent background counts from being recorded. Two of those, in effect, involved detecting coincidences between veto counts and HPGe counts while the other two used a

simpler system gating the detector for every veto count, on the basis that if a count is recorded in the veto detector any HPGe count is irrelevant. All four systems needed to be set up carefully with regard to how the electronic systems detect the moment in time when a veto pulse is accepted and the delay needed to intercept the HPGe pulse at the right moment in time. In all cases, it was demonstrated that, when set up correctly, the veto gating had no effect on the HPGe detection efficiency. With the exception of one particular peak, none of the systems made any significant reduction in activation peak areas, but did reduce to the overall spectrum count rate to about 73 % of its former level. Perhaps surprisingly, the simpler systems gave a slightly greater reduction than the others.

Active shields in the form of veto detectors are undoubtedly worthwhile above ground and down to a depth of about 100 m w.e. underground. The plastic scintillator veto detector shielding the IAEA-MEL detectors (Povinec *et al.* (2004)) installed at a depth of 35 m w.e. and referred to above, reduces the backgrounds to a level equivalent to 250 m w.e. depth. Reduction factors are between 4 and 11 for the various detectors. However, as depth increases the improvement is less. Above ground improvements of factors of 4 to 10 are achievable. At 500 m w.e., there might be only a 40 % reduction in background and only a few percent at 3000 m w.e.

13.6 ULTRA-LOW-LEVEL SYSTEMS

A small number of groups of workers have put a large effort into reducing backgrounds to many orders of magnitude below those available from a commercial low background system. Of the CELLAR collaboration, most laboratories are concerned with the measurement of environmental samples, including the measurement of constructional materials to support the design of other event lower background systems. A smaller number of laboratories are concerned with highly sensitive nuclear physics experiments, in particular the search for the rare double beta decay ($\beta\beta$) in ^{76}Ge.

The account of one particular laboratory's quest for ultimate low background, going to what John Hemingway, in the first edition of this book, termed 'awesome extremes', is fascinating. It illustrates many of the issues raised in previous sections and that can perhaps justify inclusion of a brief summary. Starting with a commercial low background detector, Brodzinski *et al.* (1985, 1990) followed this extraordinary path:

- The detector was set up with NaI(Tl) anticoincidence, heavy passive shielding of lead and old iron to absorb external gammas, and borated paraffin to absorb cosmic

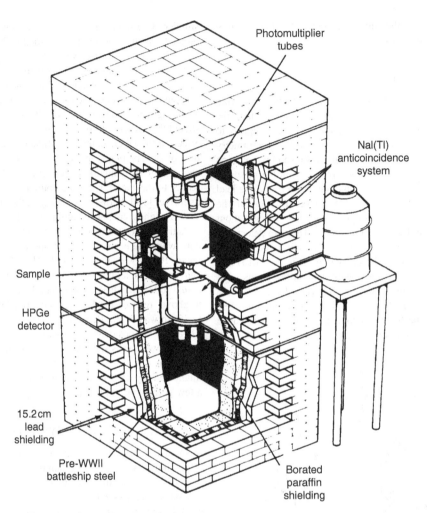

Photomultiplier tubes

NaI(Tl) anticoincidence system

Sample

HPGe detector

15.2 cm lead shielding

Pre-WWII battleship steel

Borated paraffin shielding

Figure 13.22 The initial system – standard low background germanium detector in an anticoincidence shield and a heavy multi-layered passive shield (reproduced by permission of R.L. Brodzinski)

neutrons, as shown in Figure 13.22. This assembly produced the upper spectrum in Figure 13.23, with count rates spanning the range 2×10^{-1} to 2×10^{-3} counts keV^{-1} min^{-1}.

- The detector cryostat was then rebuilt to incorporate low background components. The NaI(Tl) anticoincidence system, which contained ^{226}Ra, was replaced by an external plastic scintillator veto detector. A layer of cadmium and borated wax was added, to thermalize fast neutrons and absorb the thermal neutrons. This resulted in the second spectrum and count rates down to 1×10^{-2} to 3×10^{-5} counts keV^{-1} min^{-l}.

- The whole assembly was then transferred to a gold mine at 1428 m below ground level to minimize cosmic ray

interactions. The resulting third spectrum was dominated by bremsstrahlung from ^{210}Bi arising from ^{210}Pb in the shield. An inner shield of 73 mm of copper was added to absorb that, resulting in count rates of 1×10^{-3} to 1×10^{-5} counts keV^{-1} min^{-1}.

- Some residual peaks in the spectrum were identified as ^{54}Mn, ^{59}Fe, ^{56}Co, ^{57}Co, ^{58}Co and ^{60}Co, presumed to be largely reaction products of energetic cosmic rays with copper while it was being manufactured and then stored above ground. The inner copper was replaced by lead cast from ingots of 448 year old lead (some twenty half-lives of ^{210}Pb) recovered from a Spanish galleon.

- Peaks from ^{40}K, ^{54}Mn, ^{56}Co, ^{57}Co, ^{58}Co, ^{60}Co and ^{65}Zn were still detectable and bremsstrahlung suspected to be

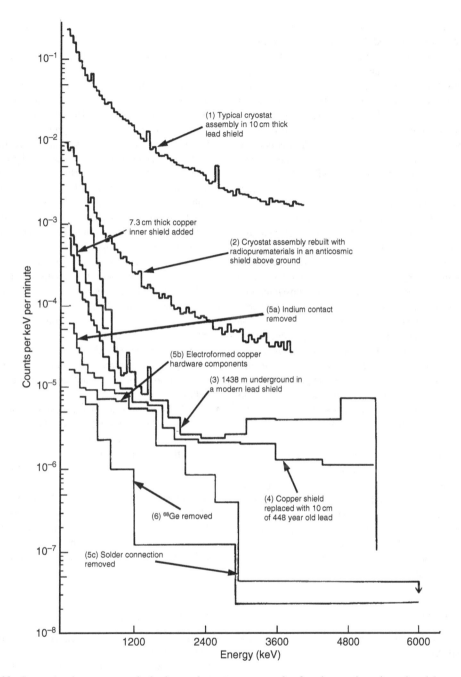

Figure 13.23 Successive improvements in background spectra as a result of various actions (reproduced by permission of R.L. Brodzinski)

from ^{68}Ge, ^{115}In, and ^{210}Po. The broad 'peak' at about 5000 keV was identified as a 5.2 MeV alpha in the ^{210}Pb decay chain, which came from surface ^{210}Po in a soldered connection directly on the germanium crystal.

This was confirmed by measurements over 20 months when the count-rate in this region fell at a rate consistent with the 138 day half-life of ^{210}Po. The count rates were now 6×10^{-4} to 1×10^{-6} counts keV^{-1} min^{-1}.

- The primordial ^{115}In (half-life 4.41×10^{14} year, a pure β emitter) was removed by remaking the soldered connection with the ancient lead. All OFHC copper components in the cryostat were re-fabricated by electroforming from highest purity copper sulfate solution, and thus the penultimate spectrum obtained, with count rates of 2×10^{-5} to 5×10^{-8} counts keV^{-1} min^{-1}.

- Residual activities were thought to be cosmogenic impurities in the germanium detector. It was decided to manufacture two new 1 kg detectors with the object of eliminating ^{68}Ge (half-life 270.8 d, EC), produced by energetic cosmic neutrons via the ^{70}Ge(n, 3n) reaction. The normal zone-refining process (Chapter 3, Section 3.4.2) cannot of course, separate this; it is necessary to prepare the detector from germanium that has not been exposed to cosmic radiation and to minimize exposure at all stages of manufacture. The germanium was specially taken from a mine 200 m deep. It was stored underground, where possible, and, because air freight is subject to a much higher neutron flux, moved by surface transport within a neutron moderating 'sugar castle'. It was rapidly processed in 'purged' systems to exclude the possibility of mixing with normal high purity germanium. When the finished detector was installed in the mine, the germanium had been only exposed to ground level cosmic ray neutrons for two weeks. The result was background count rates of 8×10^{-6} to 2×10^{-8} counts keV^{-1} min^{-1}.

This decrease of background by some five orders of magnitude from what was originally a very respectable low background performance is a remarkable tribute to years of meticulous painstaking work. The lessons learned are now, of course, applied to more recent quests for ultra-low background work. It is interesting that Laubenstein *et al.* (2004) suggest that further reduction in deep laboratory backgrounds might be achieved by even growing the detector crystal underground.

PRACTICAL POINTS

For increased sensitivity or smaller minimum detectable activity (MDA):

- Use a larger detector; efficiency is more important than resolution.
- Use a detector of appropriate shape for sample size and gamma energy.
- For gamma-ray energies < 300 keV, a short detector of large diameter will be most efficient.
- Use a suitably large quantity of sample.

- Use close geometry, yet beware of true coincidence summing.
- If the background continuum to the peak of interest is largely determined by the Compton scattering of gamma-rays from the sample and the energies of interest are not too low, then to improve throughput, consider buying one large detector rather than a number of smaller ones of the same total efficiency.
- If sample count rates are small compared to the environmental background rate, because background will be proportional to detector size there will be no advantage, in terms of throughput, in buying one large detector rather than an equivalent number of smaller ones.
- Small samples will be counted very efficiently in a well detector, but cascading gammas have a high probability of summing. The sample size constrains the MDA.
- Marinelli beakers enable large sample volumes to be counted; there are decreasing returns with increasing volume. For highest sensitivity, the Marinelli dimensions should be matched to the detector dimensions.
- At very low levels, detectors constructed of selected low background materials are essential. Look for internal gamma shields between the detector and preamplifier, or a remote preamplifier. Obtain as much information from the manufacturer as possible on its behaviour.
- Use 100 mm of lead in the passive shield, with, if possible, a layer of aged lead on the inside, especially if lower energies are of interest.
- Feed the nitrogen boiling off from the Dewar into the shielded space to minimize the effect of radon daughters.
- A Compton suppression shield should be chosen only after some thought. It is likely to be satisfactory only for the detection of gammas that are not in cascade. A veto guard detector will reduce cosmic ray events, being especially useful with high energies.
- If inspiration is required, read Section 13.6.

FURTHER READING

The manufacturers of gamma spectrometry equipment provide useful information on all of the aspects of this chapter. Much of that is available on the Internet. I found the following to be of particular use:

- Canberra's list of application notes which can be found at: *http://www.canberra.com/literature/967.asp.*

- ORTEC's list of application notes which can be found at: *http://www.ortec-online.com/application-notes/application-notes.htm.*

- On MDA, detector size and shape:

IEC (1981). *Germanium semiconductor detector gamma-ray efficiency determination using a standard reentrant beaker geometry*, Publication 697, International Electrotechnical Commission, Geneva, Switzerland.

PGT Inc., *Well-Detectors with the Through-Hole Advantage*: *http://www.pgt.com/Nuclear/Well.html.*

Keyser, R.M., Twomey, T.R. and Wagner, S.E. (1990). The benefits of using super-large germanium gamma-ray detectors for the quantitative determination of environmental radionuclides, *Radioact. Radiochem.*, **1**, 47–57 (also at *http://www.orteconline.com/pdf/biggerbetter.pdf*)

Twomey, T.R. and Keyser, R.M. (1994). Advances in large diameter, low energy HPGe detectors, *Nucl. Instr. Meth. Phys. Res., A*, **339**, 78–86.

ANSI/IEEE N42.14 (1999). *Calibration and Use of Germanium Spectrometers for the Measurement of Gamma-Ray Emission Rates of Radionuclides*, American National Standards Institute, New York, NY, USA (available at *http://webstore.ansi.org*).

- On the composition of a natural background:

Heusser, G. (1993). Cosmic ray induced background in Ge spectrometry, *Nucl. Instr. Meth. Phys. Res., B*, **83**, 223–228.

Heusser, G. (1996). Cosmic ray interaction study in low-level Ge spectrometry, *Nucl. Instr. Meth. Phys. Res., A*, **369**, 539–543.

Wordel., R., Mouchel, D., Altzitzoglou., T., Heusser, G., Quintana Arnes, B. and Meynendonckx, P. (1996). Study of neutron and muon background in low-level germanium gamma-ray spectrometry, *Nucl. Instr. Meth. Phys. Res., A*, **369**, 557–562.

Bossew, P. (2005) A very long-term HPGe-background gamma spectrum, *App. Radiat. Isotopes*, **62**, 635–644.

- On low background counting:

Canberra. *Considerations for Environmental Gamma Spectroscopy Systems*: *http://www.canberra.com/pdf/Literature/envapn.pdf.*

Canberra. *Ultra Low-Background Detector Systems*: *http://www.canberra.com/pdf/Literature/ultralowbg.pdf.*

Canberra. *Ultra Low-Background Shield – Model 777*: *http://www.canberra.com/pdf/Products/Detectors_pdf/2m777.pdf.*

Verplancke, V. (1992). Low level gamma spectroscopy; Low, Lower, *Lowest, Nucl. Instr. Meth. Phys. Res., A*, **312**, 174–182.

Reyss, J.-L., Schmidt, S., Legeleux, S. and Bonté, P. (1995). Large, low background well-type detectors for measurements of environmental radioactivity, *Nucl. Instr. Meth. Phys. Res., A*, **357**, 391–397.

Hurtado, S., García-León, M. and García-Tenorio, R. (2006). Optimized background reduction in low-level gamma-ray spectrometry at a surface laboratory, *Appl. Radiat. Isotopes*, **64**, 1006–1012.

- On underground counting systems:

Gurriaran, R., Barker, E., Bouisset, P., Cagnat, X. and Ferguson, C. (2004). Calibration of a very large ultra-low background well-type Ge detector for environmental sample measurements in an underground laboratory, *Nucl. Instr. Meth. Phys. Res., A*, **524**, 264–272.

Laubenstein, M., Hult, M., Gasparro, J., Arnold, D., Neumaier, S., Heusser, G., Köhler, M., Povinec, P., Reyss, J.-L., Schwaiger, M. and Theodórsson, P. (2004). Underground measurements of radioactivity, *App. Radiat. Isotopes*, **61**, 167–172.

- On active background reduction:

Canberra. *Compton Suppression... Made Easy*, Application Note AN-D-8901 (available at *http://www.canberra.com/pdf/Literature/comptonsupp.pdf*).

Povinec, P.P., Comanducci, J.-F. and Levy-Palomo., I. (2004). IAEA-MEL's underground counting laboratory in Monaco – background characteristics of HPGe detectors with anti-cosmic shielding, *Appl. Radiat. Isotopes*, **61**, 85–93.

Stover, T. and Lamaze, G. (2005). Compton suppression for neutron activation analysis applications at the National Institute of Standards and Technology (NIST), *Nucl. Instr. Meth. Phys. Res., B*, **241**, 223–227.

- The ultra-low-level work of Brodzinski *et al.* – outlined in Section 13.6 – was taken from the following publications:

Brodzinski, R.L., Brown, D.P., Evans, J.C., Hensley, W.K., Reeves, J.H., Wogman, N.A., Avignone, F.T. and Miley, H.S. (1985). An ultralow background germanium gamma-ray spectrometer, *Nucl. Instr. Meth. Phys. Res., A*, **239**, 207–213.

Brodzinski, R.L., Miley, H.S., Reeves, J.H. Avignone, F.T. (1990). Further reductions of radioactive backgrounds in ultra-sensitive germanium detectors, *Nucl. Instr. Meth. Phys. Res., A*, **292**, 337–342.

- A useful book on achieving the satisfactory conditions for low-level counting, given a favourable review by Knoll in 1998, is:

Theodorsson, P. (1996). *Measurement of Weak Radioactivity*, World Scientific Publishing Company, Singapore.

14

High Count Rate Systems

14.1 INTRODUCTION

Measurements at high count rate have been referred to many times in earlier chapters. In this chapter, I deal with detector systems designed specifically for counting at high count rates. I must record my indebtedness to the manufacturers' literature, an excellent source of advice on such technical matters; they, after all, designed their equipment with such applications in mind.

The main thrust of this chapter is achieving high **throughput** – maximizing the number of pulses being recorded in the spectrum, when the rate of pulses entering the system is high. High count rates usually mean loss of resolution, peak shifts and errors in dead time measurement. It will become apparent that when setting up the system there will have to be a compromise between throughput and resolution, and perhaps other factors.

The generic gamma spectrometry electronic system was illustrated in Chapter 4, Figure 4.1. From the detector, which generates the pulses, to the ADC, which measures their height, each individual part of the system has its own count rate limitation. In some cases, this will be the rate of pulses passing through the system; in others, the energy imposed on the unit per second. Table 14.1 lists the throughput limitations of the various parts of the detector system in general terms. Note that this table does not suggest that those ultimate throughputs would be usable. Resolution or dead time measurement accuracy might be the limitation in practice.

Perhaps I should define the term 'high count rate'. To some extent, this depends upon the context; for an analyst measuring environmental samples, in comparison, almost all measurements made by an activation analyst would be at high count rate. For the purpose of this chapter, high count rate will be taken to mean anything above 100 000 counts per second – that is, *input* count rate, not throughput into the gamma spectrum. That is not to suggest that count rates below that can

be taken as 'low'. There may be significant count rate issues, such as random summing (pile-up), at, say, 40 000 cps, which we might choose to regard as a 'moderate' count rate.

How we assess the input count rate in the first place is not immediately obvious. We see in Table 14.1 that the amplifier and the ADC are limiting components in the pulse throughput. The pulse rate neither at the output of the amplifier nor at the ADC output creating the gamma spectrum will represent the true input count rate. Fortunately, amplifiers for high count rate gamma spectrometry, as part of their pile-up rejection circuitry, incorporate a fast discriminator, which provides a logic pulse for almost every input pulse to the amplifier (see Chapter 4, Figure 4.25). These pulses are available at the rear of the amplifier, labelled ICR (for Input Count rate) or CRM (indicating it is intended for a Count rate Meter) and enable a more realistic estimate of input count rate to be made.

Why do we need to count at high rates? In principle, the problems of high count rate might be ameliorated by simply taking a smaller sample, by moving the sample further from the detector and letting the inverse square law work for us, or by restricting the view that the detector has of the sample using a collimator. (There are examples in the literature of automatic systems designed to alter the source-to-detector distance or degree of collimation to bring the count rate into a tolerable range.) There are, however, situations where such obvious strategies are not necessarily available; activation analysis is a particular example where one will not necessarily know the count rate in advance and would have to count under standard conditions regardless. There are applications where, although count rates may normally be low, a *capability* for high count rate measurement is needed. Examples are reactor site and stack monitoring, where possible high count rates after an accidental release of

Table 14.1 Component throughput limitations

Component	Limit due to:	Limiting mechanism	Dead time/pulse (μs)	Rate limit (cps)[a]
Germanium detector	Count rate	Charge collection time	0.2 to 0.5	7×10^5 to 2×10^6
Preamplifier:				
Resistive-feedback	Energy rate	Dynamic range	—	4×10^4 to 7×10^{5b}
Transistor reset	Energy rate	Dynamic range plus resetting time	~ 0.2	2×10^6
Pulse processor (amplifier)	Count rate	Pulse width; pile-up; overload recovery	2 to 100	4×10^3 to 2×10^5
ADC:				
Wilkinson	Energy rate	Conversion time	5 to 80	4×10^3 to 7×10^4
Successive Approximation	Count rate	Conversion time	1 to 25	1.5×10^4 to 4×10^5
MCA Memory	Count rate	Data-storage time	1 to 3	1×10^5 to 4×10^5

[a] These rates are rounded illustrative maximums. Pulse-rate limit = $10^6 \times$ dead-time (μs)/2.718.
[b] Preamplifier limits are based upon energy rates in MeV s^{-1}. The limits quoted are for 1 MeV gamma-rays.

radioactive material have to be taken into account when designing the monitoring system.

Measurements of samples containing short half-life nuclides may involve very high count rates at the beginning of the count, decreasing during the count period. Sometimes, when measuring low intensity gamma-rays in the presence of a high matrix activity, it may be necessary to accept a high count rate in order to be able to accumulate sufficient target counts to provide adequate statistical confidence in the measurement. Another, perhaps more mundane, reason for choosing to work at high count rate might be sample throughput; higher count rate, achieved by increasing the sample size or reducing source-to-detector distance, providing more counts within a shorter time without sacrificing statistical confidence. However, even disregarding the possible economic limitations of more specialized (and therefore more expensive) equipment, the effect of higher continuum levels beneath target peaks may have undesirable consequences in terms of MDA.

14.2 DETECTOR THROUGHPUT

The output pulse rate from the detector is related to the time it takes to collect all of the charge from a detection event. In Chapter 3, Section 3.6, I showed how this time varies with the collecting voltage applied, and Figure 3.10 shows that for field strengths of greater than about 3×10^5 V m^{-1}, the velocity of an electron in germanium is about 10^5 m s^{-1}, but the velocity of a hole is only about 8×10^4 m s^{-1}. The time needed to collect all of the charge depends upon the distances travelled by the

electrons and the holes, and in turn on the size of the detector and the location of the electrons and holes, relative to the collection electrodes, when they are created. To simplify matters, consider hypothetical coaxial detectors with lengths equal to their diameter (Figure 14.1) and assume that the holes created near to the anode of a detector will take the longest time to be collected at the cathode. The radius of the detector, and distance that the holes will travel, can be calculated from the volume of the detector, which we can estimate from the empirical relationship:

$$\text{Detector volume (cm}^3) = 4.3 \times \text{Relative efficiency}$$
$$(\% \text{ at } 1332.5 \text{ keV}) \qquad (14.1)$$

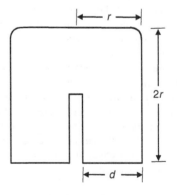

Figure 14.1 The dimensions of a 'bulletized' square cylindrical coaxial detector

We can now calculate an approximate maximum charge-collection time for detectors of different sizes. Figure 14.2 plots that charge collection time against relative efficiency. It demonstrates that, even with the largest detectors, more than 10^6 pulses could be generated within a detector per second. Even with a charge collection time of 600 ns, 1.6 million regularly spaced pulses could be output by the detector every second. To take account of randomly spaced pulses that figure must be divided by e (2.7182, the exponential factor), giving an estimate of about 600 000 randomly arriving pulses per second throughput.

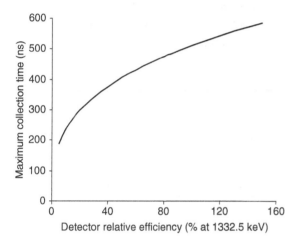

Figure 14.2 Calculated maximum charge collection times for coaxial detectors as a function of relative efficiency at 1332.5 keV

Thus, the detector is not a limitation on detector throughput at the count rates we are considering. However, as I noted in Chapter 4, conventional wisdom suggests that the shaping time of the amplifier should be set somewhat greater than the notional charge collection time and that will, as we shall see, significantly limit our count rates.

Figure 14.2 reminds us that a smaller detector will have a shorter collection time and may be more appropriate in some circumstances. Maybe there is merit in using a small rather than a large detector as, with a highly radioactive source, it is unlikely that there will be a general shortage of counts. The smaller number of pulses, with their faster rise times and collection times will be processed more reliably and with less pile-up by the rest of the system. However, a larger detector may be necessary to get good statistics on high energy but relatively low intensity gamma-rays. Larger detectors have better peak-to-Compton ratios. That means that the spectrum peaks we particularly wish to measure will be on a lower background continuum relative to peak height. In addition, if these peaks are

at higher energy, the superior full-energy peak efficiency of the larger detector helps achieve satisfactory statistical precision on the peak areas. Twomey *et al.* (1991) argue that collimation on a large detector so as to give the same total count rate as an uncollimated small detector will produce a superior result at all energies. They quote data comparing a 12 % detector and a collimated 120 % detector; the larger detector has a somewhat lower continuum below 500 keV (higher above 500 keV), but it has the same full-energy peak efficiency in this region as the smaller one, thus improving the low energy signal-to-noise ratio. Whether such considerations justify the extra cost of a very large detector would depend upon circumstances.

14.3 PREAMPLIFIERS FOR HIGH COUNT RATE

The mechanisms of the resistive feedback (RF) preamplifiers and transistor reset preamplifiers (TRPs) were discussed in Chapter 4, Section 4.3. Some of their properties are compared in Table 14.2, and some of the differences between them are explored in the following sections.

14.3.1 Energy rate saturation

Resistive feedback preamplifiers have a fundamental limitation that severely restricts their use at high count rate. Within the preamplifier, the feedback circuit acts to remove charge from the feedback capacitor, which accumulates the detector signal, and to restore the pre-pulse voltage level. At low count rates that works well, but at higher count rates the feedback circuit cannot remove charge fast enough. The result is considerable pulse pile-up, and the average voltage at the output from the preamplifier rises. As count rate continues to rise, the voltage rises to such a level that the operation of the transistors within the preamplifier is affected. Initially, that results in non-linearity and ultimately the transistors cease to function and there is no output from the preamplifier at all. We then say that the preamplifier has 'saturated' or 'locked-out' or 'paralyzed', according to personal preference – it no longer works! This is not a serious blow to the preamplifier's health; normal operation resumes as soon as the input count rate is reduced sufficiently, but it is not amusing for the analyst whose detector has apparently ceased to function.

For emergency monitoring systems, such behaviour is clearly not acceptable. It is when there is an emergency that count rates are expected to be high. A detector system that shuts down as soon as it is required to do some real work will not do. Throughput can be improved by reducing the value of the feedback resistor

Table 14.2   A comparison of preamplifier properties

Resistive feedback (RF)	Transistor reset (TRP)
Tail pulse output	Step pulse output
Pole zero collection essential (not easy at high rates)	Pole zero not required
Saturates due to energy rate limit, typically $\leq 2 \times 10^5\,\mathrm{MeV\,s^{-1}}$. When saturated, no output	Does not saturate; usable to energy rate of at least $10 \times RF$ preamplifier ($> 2 \times 10^6\,\mathrm{MeV\,s^{-1}}$)
Dynamic range about 20V, then distorts/shuts down	Dynamic range about 4V, then resets
Can use to higher energy rate by choosing smaller R_f, at expense of worse FWHM	Better resolution than an RF preamplifier with small R_f
Serous peak broadening at high count rates	Less peak broadening than RF at high rates
No dead time added to system	Additional dead time due to reset and overshoot
Test-point voltage can be monitored to check malfunction	No test point available
Cheaper	More expensive

Table 14.3   Effect of changing the feedback resistor in resistive-feedback preamplifiers

R_f (GΩ)	Energy rate limit (MeV s^{-1})	FWHM (keV) at	
		122.1 keV	1332.5 keV
2	2×10^5	1.00	1.81
1	4×10^5	1.02	1.85
0.5	8×10^5	1.08	1.93
0.2	2×10^6	1.25	2.13

(R_f), but at the expense of resolution. Some data given by Canberra are reproduced in Table 14.3. An order of magnitude improvement in throughput is only achieved at the expense of an 18 % increase in peak width at 1332.5 keV and a much greater loss of resolution at low energy. Changing the feedback resistor is not a do-it-yourself option. A low value R_f would normally be specified before purchase.

A more sensible solution would be to use instead a transistor-reset preamplifier (Chapter 4, Section 4.3.2). Such preamplifiers cannot saturate. Figure 14.3 shows a comparison between two closely matched detectors with different preamplifiers feeding the same optimized high-throughput pulse-processing chain (Canberra *2101 TRP* and *2002 RF* preamplifiers, both used with the *2024* gated integrator, set to 0.25 μs shaping time, and the *582* ADC; the source nuclide was ^{60}Co). The RF preamplifier shuts down soon after 2×10^5 cps, but the TRP keeps going until at least 8×10^5 cps, albeit at a very low fractional throughput.

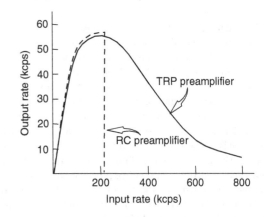

Figure 14.3   Throughput curves for two complete spectrometry systems using a transistor reset preamplifier (continuous line) and a resistive feedback preamplifier (dashed line), with each feeding the same high count rate amplifier and ADC (reproduced by permission of Canberra Nuclear)

14.3.2 Energy resolution

Table 14.3 tabulates the loss of resolution, due to increased noise, with smaller values of R_f. If one insisted on using an RF preamplifier, one would have to consider what degree of resolution loss would be acceptable compared to the increased throughput.

Figure 14.4 has data from the same detector/ preamplifier combination as Figure 14.3, but with a rear end

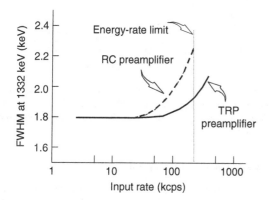

Figure 14.4 Resolution variation with count rate for systems with TRP (continuous line) and RF (dashed line) preamplifiers, with each feeding the same high resolution amplifier and ADC (reproduced by permission of Canberra Nuclear)

optimized for good resolution (Canberra *2025 AFI*, shaping time 2 μs, and the *8077* ADC). At low count rates there can be little to choose, in terms of peak resolution, between the two types of preamplifier, but the higher the input rate, the greater the difference in performance is. The advantage of the TRP is clear.

14.3.3 Dead time

There is essentially no dead time associated with the RF preamplifier itself. On the other hand, the TRP imposes a dead time period during each reset process. However, that might only be 2 μs per reset cycle. The rate of resets will obviously depend upon the count rate, but also upon the average pulse energy passing through the preamplifier – higher average energy means fewer pulses per reset cycle and hence more resets per second. More significant is the effect of the reset on the amplifier, which does not take kindly to the rapid fall in input voltage from, say, 4 V down to zero. This drives the amplifier into overload. Amplifier recovery takes an appreciable time, much longer than the 2 μs reset time that caused it (Figure 14.5).

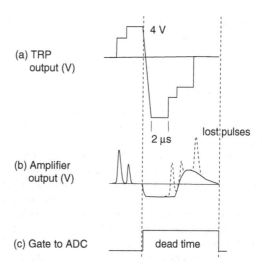

Figure 14.5 Dead time implications of the TRP, showing (a) the TRP output, with dead time during the reset, (b) the effect on the amplifier due to overload, and (c) the gating pulse to the ADC which must include the overload period

The dynamic range of TRPs is the voltage limit, typically 2 or 4 V, at which resetting takes place. In order to retain good linearity, this is much lower than the equivalent limit for RF preamplifiers (20 V). Clearly, a dynamic range of 4 V would be preferred to one of 2 V because there would be fewer resets for a given number of input pulses. In a test using the Canberra Nuclear *2101* preamplifier, with a 4 V dynamic range, receiving relatively high-energy pulses from ^{60}Co, the reset frequency was measured as one per 150 events – equivalent to 13 ns dead time per event. In comparison with the shaping time in the amplifier, which determines the overall width of a pulse, this is almost negligible. Even doubling the dead time by using a preamplifier with a 2 V dynamic range would make very little difference. There will be some variation in reset frequency with detector size; for a larger detector, with its higher peak-to-Compton ratio, the average energy per pulse will be higher.

14.4 AMPLIFIERS

The essential functions of an amplifier were discussed in Chapter 4, Section 4.4. I suggested there that 'pulse processor' would be a more appropriate name for this item than the historic 'amplifier'. This is particularly true when considering high count rate systems. The data in Table 14.1 showed us that the pulse processor is the critical restraint on pulse throughput, mainly due to pulse pile-up (random summing) within it. The high cost of these items reflects the technology within them to allow them to cope with high count rates.

14.4.1 Time constants and pile-up

Figure 14.14 (Chapter 4) showed the shape of the standard semi-Gaussian unipolar pulse, while Table 4.1 listed the various timing relationships of the pulse in terms of the amplifier shaping time. The peaking time of a pulse is about twice the shaping time and the overall width of the pulse 5 to 6 times the shaping time (Canberra suggest a factor of 6.2). A high specification amplifier might have shaping time constants available from 0.25 to 12 μs, giving output pulse widths between 1.2 and 70 μs. The longer the time constant, the longer the output pulse and the fewer pulses that can pass through the amplifier. However, that is not the whole story because, as the count rate increases, so does the probability that two pulses will overlap, i.e. pile-up. In fact, the number of piled-up pulses is proportional to the square of the shaping time constant. At 100 000 pulses per second, using a shaping time of 2 μs, there is a 67 % chance that any pulse will be piled-up. Because piled-up pulses are rejected by the PUR circuits within the amplifier (Chapter 4, Section 4.4.8) they contribute nothing to the spectrum and reduce throughput even further.

The obvious solution is to use short shaping times; at 0.25 μs, the probability of pile-up is only 13 % at 100 000 pps. However, that is not practicable with a normal amplifier. Figure 14.2 showed us that the charge-collection time can be as much as 0.6 μs, and, unless the shaping time is a least ten times that, there will be ballistic deficit (Chapter 4, Section 4.4.5), a consequence of inadequate charge collection that leads to a disastrous loss of resolution. Even for a small detector, with a charge collection time of 0.2 μs the shaping time would need to be at least 2 μs, resulting in output pulses 11 μs long (using the factor in Chapter 4, Table 4.1). The maximum possible throughput would then be about 33 000 pps (i.e. $10^6/11\,\mu s/e$). In practice, because of the extensive summing, that is reduced to fewer than 20 000 cps, as curve (b) in Figure 14.6 shows, when the input count rate is only 65 000 cps.

Nevertheless, at moderate count rates, reducing the shaping time constant can be a useful strategy. If the optimum were 4 μs, at which resolution were, say 1.8 keV, then using 1 μs would quadruple the throughput at the expense of increasing the resolution to perhaps 2.3 keV (see Figure 4.8 below). That may not be a serious limitation, but would still not achieve the extremely high count rates we are aiming for in this chapter.

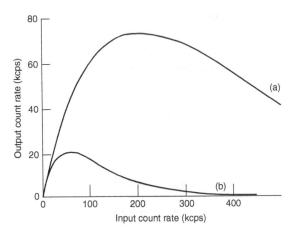

Figure 14.6 Typical throughput curves: (a) gated integrator using 0.25 μs shaping; (b) system using 2 μs semi-Gaussian shaping. In both cases, other system components are suitably 'fast'

14.4.2 The gated integrator

A small gain in throughput can be obtained by using triangular shaping within the amplifier (Chapter 4, Section 4.4.3), but for a dramatic improvement a gated integrator (Chapter 4, Section 4.4.5) is essential if one is determined to stay with analogue pulse processing. The gated integrator takes account of charge being collected beyond the peaking time of the conventional semi-Gaussian amplifier. Figure 14.6 compares the throughput of the conventional semi-Gaussian amplifier referred to above using 2 μs shaping, with a gated integrator running at 0.25 μs. The improvement is extraordinary.

Within the gated integrator amplifier itself, the integration is continued for about ten times the shaping time to ensure all the 'pre filter' pulse contributes (semi-Gaussian and gated integrator pulse shapes are compared in Figure 14.15 below). When discussing gated integrators, the '**integration time**' may be quoted rather than shaping time. The shaping actually performed by the integrator may not be the usual semi-Gaussian shaping. ORTEC claim an enhanced performance in their gated integrator amplifier model *973* by using a 'camel' or quasi-rectangular pulse before the integration stage, which results in less noise.

There is a downside to the gated integrator in that resolution will be poorer than when using the semi-Gaussian shaping optimized for good resolution (Figure 14.7 plus Table 14.5 below). However, that is a small price to pay for a many-fold increase in throughput. However, at low count rate, it is an unnecessary price, and the gated-integrator amplifier will also have an output socket for

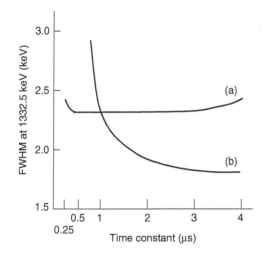

Figure 14.7 Typical resolutions as a function of shaping time for (a) gated integrator, and (b) semi-Gaussian amplifiers

semi-Gaussian pulses to be used at lower count rates. In Chapter 4, Section 4.5, I outlined an alternative to the gated integrator, where with a ballistic-deficit collector (Ortec *675* or internal to the Tennelec *245*) an empirical correction is made to each pulse based on the pulse rise-time. This 'resolution enhancement' process is not a complete correction for ballistic deficit. It is satisfactory for large detectors and moderately high count rates when used with triangular pulse shaping and where shaping times of $> 2\mu s$ are adequate. At higher rates where time constants of $< 1\mu s$ are needed, then the gated-integrator procedure gives better results. (At the time of this update, it would appear that such modules are no longer readily available.)

14.4.3 Pole zero correction

The function of pole zero correction is to match the shaping circuits of the amplifier to the fall time of the preamplifier pulse. It is an essential adjustment when RF preamplifiers are used but unnecessary on TRP preamplifiers because there is no fall time on its stepped output to match. Nevertheless, users should be aware that the PZ adjustment should be set to its limit, as indicated in its manual, for correct operation.

At high count rate, for resistive feedback preamplifiers, the correction can be more difficult to achieve than at lower rates. The automatic PZ button on some amplifiers is a welcome user friendly feature but, at high count rate where settings are more critical, the user may prefer the comfort of visual feedback while performing the correction manually. The operations of Chapter 10,

Section 10.3.4 using the oscilloscope display will give good adjustment, but fine-tuning following this could give the best result. With the oscilloscope disconnected, after each small adjustment of the PZ potentiometer (say 1/8 of a turn) measure a spectrum containing ^{60}Co and check the width of the 1332.5 keV peak, selecting that setting which gives the smallest width. The exercise should then be repeated at a high count rate to confirm that the optimum position has been achieved.

14.4.4 Amplifier stability – peak shift

It is inevitable that at very high count rate there will be some shift in the peak position. Ideally, energy calibration should be done at count rates similar to those expected in actual measurements. If this cannot be done, then it is important that peak shift should be small enough so that peak identification is not hampered at high count rate. The spectrum analysis software will have some sort of **'identification window'** (Twomey *et al.* (1991)), which might be $0.4 \times$ FWHM. If the peak shift is greater than this, peaks might fail to be identified. (It should not be overlooked that, if appropriate peaks can be found in the spectrum, an internal energy calibration can be performed on each spectrum. That may not, however, be a convenient strategy for routine use.)

The specification for the Canberra *2024* amplifier says that at $2\mu s$ semi-Gaussian shaping, a 9 V pulse height (from 1332.5 keV) will shift position by less than 0.024 % for a change of input rate from 2 to 100 kcps. For an FWHM $= 2$ keV, this is equivalent to less than $0.16 \times$ FWHM, and thus falls within the criterion. The Ortec *973U* gated integrator specification for much the same situation but with input rate change from 1 to 300 kcps produces a shift of less than 0.03 %; this is commendable and acceptable for that higher rate situation.

14.4.5 Amplifier stability – resolution

Resolution degradation at high rate is another inevitable factor that we would seek to minimize. Figure 14.8 shows that both gated integrator and semi-Gaussian amplifiers deteriorate. Again, a good performance is claimed by the *973U* gated integrator; the FWHM broadens by less than 10 % up to 300 kcps.

The issue of changing resolution should be looked at in the context of the spectrum analysis software. Does the software define the peak-integration window by measuring the width of each peak or does it rely on the FWHM calibration to specify the width? If the former, then the likelihood is that the peak area would be

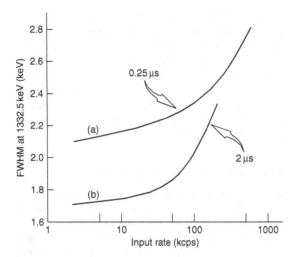

Figure 14.8 Examples of resolution as a function of input count rate for (a) gated integrator, and (b) semi-Gaussian (unipolar) shaping

measured accurately, albeit with a larger uncertainty. If the latter, the peak area would be incorrect. Even if the peak area is measured correctly, what does the software do if the peak is too wide? Does the software reject it as being too wide, does it attempt a futile deconvolution or does it provide a result?

14.4.6 Overload recovery

I have suggested that at highest count rates, the TRP is essential but that it does impose reset periods that drive the main amplifier into overload, and that the time needed to recover from that overload imposes a greater dead time loss than the reset itself.

For this reason, amplifiers intended to cope with high count rates are designed to have short recovery periods. Current specifications, irrespective of source, say that the amplifier recovers to within 2 % of the rated output within 2.5 non-overloaded pulse widths at maximum gain (amplification), where the degree of overload could be × 1000, for the Canberra *2024* or × 400 for the ORTEC *973*. By assuming a 2 µs shaping time and pulse width 5.6 times that, the resulting dead time could be as long as 28 µs at each reset. If, as suggested in Section 14.3.5, 150 events are processed before each reset, then the dead time burden per event due to the resetting, including overload recovery, is only some 0.21 µs. This is not quite negligible, but placed against a pulse width of 11 µs is only a minor contribution to the overall dead time.

The degree of overload will depend upon the dynamic range of the TRP; a 4 V reset will give greater overload than 2 V, although, of course, 4 V resets would happen only half as often as 2 V resets. The time for recovery from overload is an exponential function of the magnitude of the overload with larger overloads taking a disproportionate time. In addition to amplifier shaping time, the recovery time will also depend to some extent on amplifier gain. The recovery could also be affected by instabilities in the TRP resetting procedure, giving rise to what might be termed 'after-effects'.

14.5 DIGITAL PULSE PROCESSING

Digital signal processing (DSP from ORTEC), or digital signal analysis (DSA from Canberra), was introduced in Chapter 4, Section 4.11. Performance data from the manufacturers indicates that for all but the most demanding high count rate situations, DSP has considerable advantages over the analogue processing described in the preceding sections. The principles of DSP were covered in Chapter 4, Section 4.11, where it was pointed out that, once the input pulse has been digitized by the very fast flash ADC, there are no other analogue processes that can be affected by the count rate. The digital filtering takes place in isolation from the analogue pulse stream and is unaffected by many of the difficulties that beset the fully analogue systems. Nevertheless, the front-end circuits receiving the pulse stream, before the digitization, are unavoidably analogue and so there are still unwanted shifts in peak shape and position at high count rates.

There are many comparisons within the literature of the performance of DSP with analogue systems extolling, rightly, their improved higher count rate performance. Figure 14.9 compares measured throughput data for analogue and digital systems connected to the same detector system when optimized for good resolution. Figure 14.10 compares measured throughput of a gated-integrator system with that of DSP, both optimized for high count rate. In both situations, the improvement is dramatic. When throughput is optimized, even close to 400 kcps there is still considerable throughput. Figures 14.9 and 14.10 are partial data from different manufacturer' literature but when the full data are compared like-for-like, there is little to choose between them.

The increase in throughput is, however, only achieved at the expense of resolution. Figure 14.11 shows the change in resolution with count rate for the two systems. At low count rate, DSP has worse resolution than the gated integrator but the change in resolution, as count rate

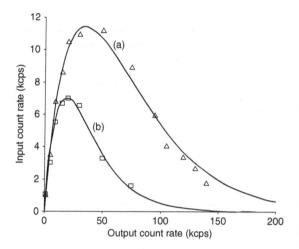

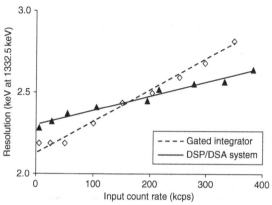

Figure 14.9 Comparison of throughput of (a) digital, and (b) analogue systems when optimized for resolution. Symbols represent measured data and the continuous lines the calculated throughputs according to Equation (14.5) for an 10 % n-type detector. (data taken from ORTEC literature)

Figure 14.11 Resolution change with input count rate for analogue (◇) and digital (▲) systems when optimized for throughput. Lines represent the best fits, for a (11 % n-type detector. (data taken from Canberra Application Note)

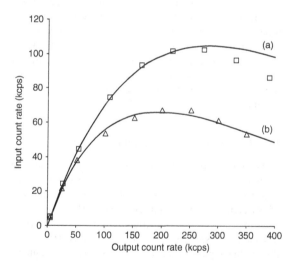

Figure 14.10 Comparison of throughput of (a) digital, and (b) analogue (gated integrator) systems when optimized for throughput. Symbols represent measured data and the continuous lines the calculated throughputs according to Equation (14.5) for an 11 % n-type detector (data taken from Canberra Application Note)

If such a resolution penalty were unacceptable, a compromise would have to be made. The DSP settings would be optimized in terms of resolution, rather than throughput, and the count rate would have to be severely limited. For the analogue system, the semi-Gaussian output might have to be used rather than the gated-integrator output and the shaping time adjusted to give proper charge collection. Under those circumstances, there might be little to choose between the two systems. This is illustrated by the data given in Table 14.4, where measurements of the pulse output of a 10 % relative efficiency n-type detector using analogue pulse processors are compared with those made using the first generation of the ORTEC *DSPec* digital processor. When optimized for resolution, both systems achieve the same resolution. When optimized for throughput, both systems loose out on resolution, the digital system being somewhat worse. In this particular case, the DSP throughput is not much greater than the analogue system, but the much lower degradation in resolution and smaller peak shift at very high count rate would make it the system of choice.

Because of differences in charge collection time, throughput will also depend on the size of the detector. The same source of data summarized in Table 14.4 also provided information about similar measurements made on a 140 % relative efficiency p-type GEM detector. In general, the differences in performance were similar to those shown in Table 14.4, but when optimized for resolution, the DSP suffers a much greater resolution loss at high count rate. Peak position stability for the analogue system was much worse than for the smaller detector, but the DSP

increases, is much lower. Some performance figures for more recent DSP systems than that used for the measurements of Figure 14.11 show rather smaller changes in resolution.

Table 14.4 Performance comparison of analogue and digital pulse processing systems[a]

Optimization system	Optimized for resolution		Optimized for throughput	
	Analogue	DSPec	Analogue	DSPec
System parameters	672 amp 6 μs 7 μs SA ADC	Rise time: 8.8 μs Flat top: 1.2 μs	973 GI 2.5 μs 1.5 μs SA ADC	Rise time: 0.8 μs Flat top: 1.2 μs
Throughput (Out @ In kcps)	7 @ 20	11.3 @ 50	57 @ 40	62 @ 140
Resolution at 1000 cps (keV at 1332.5 keV)	1.77	1.78	2.42	2.54
Resolution degradation	38 % at 75 kcps	9 % at 75 kcps	14 % at 140 kcps	2 % at 140 kcps
Peak shift at 140 kcps (keV)	8	0.22	0.13	0.11

[a] Data derived from ORTEC report on the first generation *DSPec*. Detector: 10 % n-type *GAMMA-X-PLUS* detector.

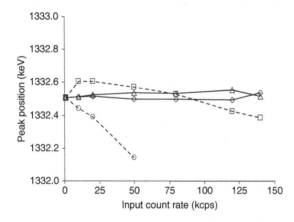

Figure 14.12 Peak shift with input count rate for analogue (dashed lines) and digital systems (continuous lines), optimized for resolution and throughput, for a 140 % p-type detector. (data taken from ORTEC literature)

system maintained its advantage, showing only a slight shift, even at 140 kcps input count rate (Figure 14.12).

14.6 THE ADC AND MCA

I explained in Chapter 4 that there are two types of ADC – the Wilkinson and the Successive Approximation. In this chapter, Section 4.6.6, I discussed the choice of ADC in a general manner. Clearly, for high count rate situations the choice is easy – the faster the better! Simplicity would suggest a Successive Approximation ADC, with its fixed conversion (dead) time, rather than the Wilkinson, with a conversion time dependent on pulse height. However, if the *average* gamma-ray energy were to fall in, or below, channel 360, even a 450 MHz Wilkinson ADC would

give a conversion time less than the fastest fixed conversion time ADCs (0.8 μs). That is not, however, a likely prospect.

Once the pulse has been digitized, the relevant channel content in the MCA memory has to be incremented. That process is unlikely to take much time, a couple of microseconds or so. In some cases, it might be performed in parallel with other processes leading to no extra dead time at all. With the fastest Successive Approximation ADCs, the overall dead time associated with the MCA need be no more than a few microseconds. Modern fast SA ADCs can often measure the pulse height, and store the count, within the overall pulse width, making no additional contribution to the total dead time.

14.7 DEAD TIMES AND THROUGHPUT

Correction for dead time is necessary at all count rates, but at high count rate, the accuracy of the correction is crucial to achieving reliable measurements. The various methods of correction – use of a pulser, Gedcke–Hale correction, loss-free counting and the virtual pulser – were described in Chapter 4, Section 4.7. In this present section, I will discuss the implications of high dead-times on throughput. In this section, 'throughput' will be taken to mean the number of pulses per second accumulated in the MCA memory. This will be looked at in comparison with the pulse rate introduced into the electronic system, referred to as the *input count rate*. I will assume that, as we are aiming for high count rates, the electronic system is of high specification.

As we have seen above, each component of the electronic system has its contribution to make to overall dead time: TRP reset, amplifier 'busy' signals, pile-up rejection gate, ADC 'busy' and MCA store-to-memory. It would appear that there is no industry standard for timing, polarity or duration of dead time signals that must be transmitted from module to module. For high count rate applications in particular, there is a strong incentive to purchase a complete system from a single manufacturer.

14.7.1 Extendable and non-extendable dead time

Each time a single pulse passes through the electronic system and is recorded in the MCA, a dead time period will be generated, say τ s long. If we have a succession of pulses, the overall dead time will be $R_m\tau$, where R_m is the count rate measured in the MCA. The time, in each second, during which the system is live, i.e. can accept pulses, is $1 - R_m\tau$. The true count rate, R_0, must be, therefore, the number of counts recorded divided by the length of time the spectrometer was live:

$$R_0 = R_m/(1 - R_m\tau) \tag{14.2}$$

Rearranged to a form more relevant for considering throughput, this becomes:

$$R_m = R_0/(1 + R_0\tau) \tag{14.3}$$

This type of dead time correction is described as **non-extending**. However, there will be situations, especially at high count rate, where a second pulse might arrive during the dead time period of the first (Figure 14.13). If

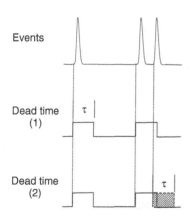

Figure 14.13 Representation of non-extending (1) and extending (2) dead time behaviour

the electronic system extends the dead time period to take account of that second pulse, the situation would be that of **extending dead time**. In this case, the mathematics,

based on the Poisson distribution, leads to the following equation for dead time correction:

$$R_m = R_0/\exp(R_0\tau) \tag{14.4}$$

At low count rates, where pile-up is negligible, the exponential term can be approximated by $(1 + R_0\tau)$. Equation (14.4) then becomes identical to Equation (14.3). Figure 14.14 shows the relationship between measured and true count rate as calculated using extending and non-extending dead times. (In this representation, the count rates are multiplied by the resolution time in order to provide graphs that are independent of amplifier shaping time.) A non-extending correction would suggest that the output count rate always increases with count rate to a saturation value. The graph that takes into account an extending dead time bears more than a passing resemblance to the measured throughput curves of Figures 14.9. This is, of course, no coincidence. In Section 4.4.8 (and Figure 4.23) in Chapter 4, the manner in which the amplifier pile-up rejection circuitry takes piled-up pulses into account by extending the dead time was discussed. It is generally assumed that the dead time losses within the amplifier system will be of the *extending* type.

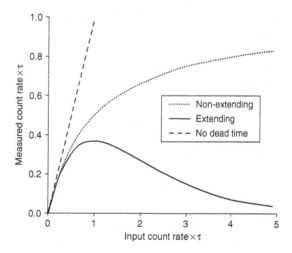

Figure 14.14 Extending and non-extending dead times, scaled to remove the influence of shaping time

Because of the shape of the throughput curve, it is evident that simple measurement of throughput cannot, by itself, give any idea of the input count rate. At anything less than maximum, for every output count rate there are two possible input count rates. In practice, of course, the MCA displayed dead time will immediately give an indication as to whether the input count rate is high or low. That, however, depends on whether the system is measuring dead time correctly. It cannot be stated too strongly that it is essential that all gating connections between preamplifier, amplifier and ADC are in place, and adjustments made according to the manufacturer's manual, if valid quantitative results are required.

Differentiation of Equation (14.4), with respect to R_0, leads us to the conclusion that R_m has a maximum where $R_0 \times \tau = 1$ (see Figure 14.12), at which point $R_0 = e \times R_m$ (where e is the exponential factor, 2.7182). Therefore, at maximum throughput:

$$R_m = 1/(2.718 \times \tau) \quad \text{and} \quad R_0 = 1/\tau \qquad (14.5)$$

The fractional throughput, R_m/R_0, must be 1/e, i.e. 36.8 %, regardless of amplifier shaping time. From a practical point of view, if the input count rate is higher than that for maximum throughput it would make sense to find some way of reducing the count rate closer to that maximum. Otherwise, there will be fewer counts in the spectrum for a given time period than one could achieve.

When it comes to calculating throughput, we have to assign a value to τ. At first sight, one might suggest that the pulse width, T_W, should be used. However, Jenkins *et al.* (1981) argued that it is more correct to use the sum of the pulse width and the peaking time, T_P, i.e. $\tau = T_W + T_P$ (Table 4.1, Chapter 4, lists empirical factors relating these widths to the amplifier shaping time). The peaking time is itself part of the whole pulse width and so has double weighting within τ. This allows for the fact that two pulses arriving within the peaking time would both be removed by the pile-up rejection circuit. The same reasoning is used within the Gedcke–Hale live time correction procedure (Chapter 4, Section 4.7.2). There are arguments in favour of using the linear gate time, T_{LG}, instead of T_P, because that is the time it takes the ADC to recognize a peak.

While the amplifier imposes an extending dead time on the pulse stream, that is not the case for the ADC, which gives a non-extending dead time, determined by the extra time needed to perform the pulse height measurement and memory storage time. The amplifier and MCA dead time operate in series and can be represented mathematically in a variation of Equation (14.4):

$$R_m = R_0/[\exp{(R_0\tau)} + R_0 T_A \times (T_A > 0)] \qquad (14.6)$$

where T_A is the additional time, beyond the pulse width, needed for MCA operations. The logical function ($T_A > 0$) ensures that this factor is only taken into account when T_A is greater than 0. T_A is shown in Figure 14.3 in relation to the overall MCA measurement time, T_M:

$T_A = T_M - T_W + T_P$. For semi-Gaussian pulse outputs, the factors listed in Table 4.6 (Chapter 4) can be used to calculate peaking time, T_P, and the overall peak processing time, T_W, from the shaping time. However, these are defined differently for different types of pulse processor. The various relevant waveforms are shown in Figure 14.15.

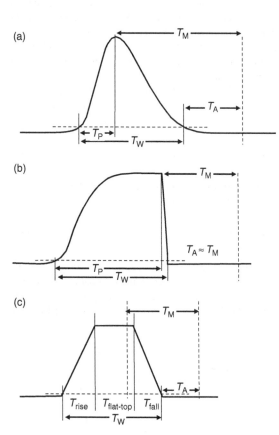

Figure 14.15 Representations of peaking time (T_P), width (T_W), ADC processing time (T_M) and additional processing time (T_A) for (a) semi-Gaussian pulses, (b) gated integrator pulses, and (c) DSP quasi-trapezoidal shaping

14.7.2 Gated integrators

For the gated integrator amplifier, the peaking time (usually called the **integration time**, T_I, in this context) is about ten times the shaping time – in gated integrator terms, the **pre-filter shaping time**. The overall peak width is only slightly more than the peaking time. Because T_P is almost equal to T_W, it is inevitable that the MCA processing time will be significant and the T_A term of

Equation (14.6) be non-zero. In this case, $\tau = 2 \times 10 \times$ shaping time, or $2 \times T_I$. The maximum throughput is then:

$$R_{max} = 1/(2.718 \times 2 \times T_I + T_A) \qquad (14.7)$$

For an ideal ADC, where T_A was negligible, this would result, for a gated integrator running with $0.25\,\mu s$ shaping time, in a maximum throughput of about 74 kcps, called the *gated integrator limit*, which appears to rule out any improvement using normal semi-Gaussian or triangular filtering. However, ORTEC have managed to reduce the integration time to $1.5\,\mu s$ in their *973U* gated integrator by using a 'camel' pre-filter instead of semi-Gaussian. This then allows a maximum throughput of 123 kcps. That must then be reduced by the ADC overheads. A $1.5\,\mu s$ fixed dead time ADC would only allow 104 kcps maximum throughput – however, that is comparable to some digital systems

14.7.3 DSP systems

For the digital processors, the peaking time can be taken as the sum of the rise time and the flat top width. The overall peak width is then that time plus the fall time, which is always equal to the rise time: $T_P = T_{RISE} + T_{FLAT\ TOP}$ and $T_W = 2 \times T_{RISE} + T_{FLAT\ TOP}$. In digital systems, the digitization is concomitant with the pulse reception and we would not expect it to make a separate contribution to Equation (14.6). T_A should, therefore, always be zero.

14.7.4 Theory versus practice

To give some degree of credulity to these theoretical fancies, we can apply Equation (14.6) to calculate the expected throughput for the practical situations represented by the data in Figures 14.9 and 14.10. Table 14.5 summarises this exercise.

Semi-Gaussian analogue versus digital processor

The data in Figure 14.9 were taken from performance figures in the first generation *DSPec* brochure, and were for a system optimized for resolution. The agreement is reasonable but certain adjustments to widths had to be made to achieve good agreement. T_P was replaced by T_{LG} and it was necessary to increase the T_W to 6.5 times the shaping time. The predicted and measured maximum throughputs were comfortably similar.

The digital processor data, resolution optimized, were taken from the same source. (This should not be compared with the Canberra data in Figure 14.10, which is throughput optimized. When throughput optimized, the ORTEC system gives similar performance.) Fitting Equation (14.6) to this data proved problematic. A reasonable fit could only be achieved by introducing an additional processing time of $9.5\,\mu s$, implying that there are sources of dead time unaccounted for. When fitted, the predicted maximum throughput was close to that measured, but at a much lower input rate than that measured.

Gated integrator analogue versus digital processor

Figure 14.10 compares the two types of processor when throughput-optimized. Data for this comparison were

Table 14.5 Application of the theoretical equation (Equation (14.6)) to measured throughput curves

Pulse processor[a]	Optimization	System parameter	Resolution (keV)	MCA time (μs)	Total time (μs)	Throughput (IP kcps @ OP kcps)	
						Calculated	Measured
(1) *672* analogue + *919* MCB	Resolution	$6\,\mu s$, triangular shaping	1.77	0	37	7 @ 19	7 @ 20
(2) *DSPec* (1st generation)	Resolution	Rise time, $8.8\,\mu s$; flat top, $1.2\,\mu s$	1.78	9.5	28	11 @ 35	11 @ 50
(3) *2024* gated integrator + *AccuSpec-B*	Throughput	$0.25\,\mu s$ shaping	2.15	1.2	3.8	66 @ 195	67 @ 250
(4) *2060* DSA	Throughput	Rise time, $0.72\,\mu s$; flat top, $0.68\,\mu s$	2.30	0	2.1	105 @ 284	102 @ 274

[a] Systems 1 and 2 (ORTEC) were supplied with pulses from a 10 % *GAMMA-X-PLUS* n-type detector, with Systems 3 and 4 (Canberra) with pulses from an 11 % n-type HPGe.

taken from a Canberra Application Note. Fitting the gated integrator data was straightforward. The overall peak width, T_W, was made slightly longer than the peaking time by making it 10.5 times, rather than 10 times, the shaping time. The allowance for processing time, T_A, was made 1.2 ms, slightly larger than the fixed ADC dead time of 0.8 ms. The fit was then reasonable and the predicted maximum throughput agreed very well with the measured value, but again at a much lower predicted input count rate.

Fitting the digital processor data was straightforward, again using a pulse width of 10.5 times the shaping time. Maximum throughput and input count rate at maximum agreed well with the measured data. However, in this case, the fit beyond the maximum throughput is not ideal. It would appear that there are extra sources of count rate dependent dead time not included in Equation (14.6). It may be that, in this particular system, TRP dead time effects are becoming significant at very high count rate.

14.8 SYSTEM CHECKS

Apart from the normal routine checks, one would have to assure oneself that the electronic system was working satisfactorily – at high count rate, there is the additional burden of confirming that counting losses are being adequately accounted for. There are procedures that have been widely used for many years. In 1990, Gehrke proposed the particular procedure below, which is now enshrined in the US standards ANSI N41.14 (revised). It is a test of the precision of automatic or semi-automatic dead time correction of whatever type – by a measured correction factor, by PUR and LTC circuits, by the pulser method, by the virtual pulser, or by any combination of these. The procedure is as follows:

(1) Set up the equipment according to the manufacturer's instructions. For example, if the system requires them, cables should be in place from a TRP (inhibit out) to amplifier (gate in), from amplifier (busy out) to MCA (busy in), from amplifier (GI inhibit out) to MCA (PUR in), etc.

(2) Determine the best time constant for the measurements intended, taking into account the required throughput and acceptable loss of resolution. For semi-Gaussian shaping at high count rates, this would not normally be greater than 2 μs. For RF preamplifier systems, pole zero cancellation would need to be checked.

(3) Tape a ^{137}Cs source securely in position near to the detector so as to give an input count rate, as measured at the ICR or CRM output of the amplifier, of 500 cps (see Section 14.1).

(4) Acquire a spectrum for a known time until there are at least 50 000 counts in the net peak area of the 661.7 keV peak. Record the net cps in the peak, the uncertainty on that count rate and the FWHM – all as measured by the usual method.

(5) Without moving the ^{137}Cs source, place a high activity ^{57}Co source at such a position as to double the count rate. The ^{57}Co source should be of such an activity that the maximum count rate required can be achieved by moving it closer to the detector. Acquire a second spectrum, again collecting at least 50 000 counts in the ^{137}Cs full-energy peak and again recording the net cps in the peak, the uncertainty on that count rate and the FWHM. If the counting losses are being adequately corrected for, the ^{137}Cs count rate should not be significantly different from the first measurement.

(6) A series of counts should now be performed after moving the ^{57}Co source progressively closer to the detector until the maximum count rate is achieved, or until the ^{137}Cs count rate becomes significantly different from the initial count rate. The acceptable magnitude of that difference, 1, 5, 10 % or whatever, would be a matter for the analyst to judge. Count rates higher than that final measurement would have to be regarded as beyond the usable range. Gehrke recommends that if differences in area exceed one third of the total acceptable uncertainty in normal measurements, then the useful counting-rate limit of the spectrometer has been exceeded. It is possible that the count rate limit might be determined by the FWHM becoming unacceptably large.

This particular test checks the counting loss correction when the gamma-ray energy is modest (661.7 keV) in the presence of a high count rate of lower energy gamma-rays (122.1 and 136.5 keV). Similar tests using other nuclides, providing higher energy gamma-rays for the background, or a 'pattern' of gamma-rays more representative of actual counting situations, could be devised by the individual analyst. Acceptable results should be obtainable at input pulse rates of up to 100 000 s^{-1}. With ultra-high count rate systems, this upper limit may go to 400 000 s^{-1}.

This test is similar to the procedure outlined earlier in Chapter 7, Section 7.6.8 to derive a random summing correction factor for conventional systems. It was pointed out that, regardless of the use of pile-up rejection circuitry, there will always be a small proportion of random coincidences that are within the resolution time of the fast differentiator and cannot be rejected. Whether the reason for lost counts is random summing or inadequacies in the live correction, the correction factor derived in using the procedure above will provide a better estimate of true count rate.

PRACTICAL POINTS

- Avoid high count rates, if possible, by increasing the sample-to-detector distance or by collimation.
- The detector is unlikely to be a problem in limiting count throughput. A small coaxial detector may be best in general, except that high-energy gammas will be better on a large coaxial detector, with low energy gammas better on a planar or *LEGe* or *LO-AX* of large area.
- Transistor reset preamplifiers do not shut down at high rates as do resistive feedback preamplifiers. They also show better resolution at high rates. However, the re-setting process introduces extra dead time, particularly due to overloading of the amplifier.
- The amplifier should be chosen carefully. This is likely to be the bottleneck at very high count rates.
- Short time constants are essential for high throughput, but may well bring in ballistic deficit effects. Overcome these by a ballistic deficit corrector or (at the highest rates) by using gated integrator pulses.
- The very fastest ADCs use the successive approximation technique, but a fast Wilkinson ADC could be superior to a mediocre Successive Approximation device, the choice depending on the effective gamma energy.
- Digital pulse processors are certainly worth considering.
- Qualitative identification of nuclides at high count rates is relatively reliable. Quantitative measurement requires large dead time correction factors. This is much more uncertain. In the high rate situation, it would be advisable to purchase all components from a single supplier in order to ensure that gating and timing pulses are completely compatible.
- Run the check of Section 14.6 to determine the upper count rate boundary of the system, where the limit is the acceptable uncertainty in the measured peak area counts per second.

FURTHER READING

- The manufacturer's websites are a useful, and convenient, source of information:

Twomey, T.R., Keyser, R.M., Simpson, M.L. and Wagner S.E. (1991). High-count-rate spectroscopy with Ge detectors: quantitative evaluation of the performance of high-rate systems (*http://www.ortec-online.com/pdf/hcrpaper.pdf*). Originally published: *Radioact. Radiochem.*, **2**(3), 28–48.

Canberra Application Note (1993). *A Practical Guide to High Count-rate Germanium Gamma Spectroscopy* (*http://www.canberra.com/pdf/Literature/nan0013.pdf*).

Canberra Application Note (1999). *Performance of Digital Signal Processors for Gamma Spectrometry* (*http://www.canberra.com/pdf/Literature/a0338.pdf*).

- Data for throughput comparisons above came from the Canberra Application Note referred to above and the following:

ORTEC Sales Brochure (1991). *DSPEC Digital Gamma-ray Spectrometer.* (This is useful for comparative data, but at the time of this update is out of print and can only be obtained by special request from Ametek, Spectrum House, 1 Millar Business Centre, Fishponds Close, Wokingham, RG41 2T2, UK).

- Two papers comparing DSP systems with analogue systems are:

Vo, D.T., Russo, P.A. and Sampson, T.E. (1998). *Comparisons Between Digital Gamma-ray Spectrometry (DSPec) and Standard Nuclear Instrumentation Methods* (*http://www.ortec-online.com/pdf/losalamospaper.pdf*). Originally published as Los Alamos Report, LA-13393-MS, Los Alamos, NM, USA.

Keyser, R.M. and Twomey, T.R. (2003). *Developments in High-Performance HPGe Detector Spectrometer Systems for Safeguards Applications* (*http://www.ortec-online.com/pdf/paperinmm.pdf*).

- The following is a valuable general account of high count-rate systems:

Hall, D. and Sengstock, G.E. (1991). Introduction to high count-rate germanium gamma-ray spectrometry, *Radioact. Radiochem.*, **2**, 22–46.

- A more detailed account with an especially useful discussion of one method of dead time correction (based on Chapter 4 of Jenkins *et al.* (1981)) is:

Twomey, T.R., Keyser, R.M., Simpson, M.L. and Wagner, S.E. (1991). High-count-rate spectrometry with Ge detectors: quantitative evaluation of the performance of high-rate systems (*http://www.ortec-online.com/pdf/hcrpaper.pdf*) (originally published: *Radioact. Radiochem.*, **2**(3), 28–48.

Jenkins R., Gould, R.W. and Gedcke, D. (1981). *Quantitative X-ray Spectrometry*, Marcel Dekker, New York, NY, USA.

- The system count-rate test is based on:

Gehrke, R.J. (1990). Tests to measure the performance of a Ge gamma-ray spectrometer and its analysis software, *Radioact. Radiochem.*, **1**, 19–31.

15

Ensuring Quality in Gamma-Ray Spectrometry

15.1 INTRODUCTION

There can be few analysts who would not desire high-quality results. Yet the assurance of quality is often way down the list of priorities when it comes to resources, be it time or money. High quality results can, of course, be achieved simply by employing an abundance of high-calibre analysts who have the understanding, the time and the inclination to assess every result as it is obtained. In practice, this is seldom possible. In many laboratories, the sheer weight of sample numbers prevents more than a cursory assessment of routine measurements. That being so, can we be certain that routine results are reliable and that anomalous results would be recognized?

Within the laboratory, we may be confident that the results we provide are consistent and accurate. Nevertheless, could we convince our customer of that? (Here, I will regard the customer as the recipient of our results, internal or external, whether or not we have a formal trade agreement with them.) What would happen if the senior analyst were not present? Would the results be as reliable if the main detector system became unavailable? To be able to assure a customer that, whatever the circumstances, the results we supply will be of consistent quality demands a more formal assessment of the whole process of measurement and organization.

Often, the initials QA, standing for **quality assurance**, engender dismay (or even despair!), with mental images of paperwork piled to the ceiling and rigid measurement disciplines stifling expert judgment. Indeed, if a quality system is set up there will inevitably be paperwork. However, in a laboratory that is already working effectively, in a quality aware manner, the necessary documentation should largely be a matter of codifying the status quo. (*Write down everything you do; then do everything you have written down.*) Often, even the first tentative steps towards writing the quality documentation expose weaknesses in the existing system and, from that point of view alone, are worth taking.

Increasingly, laboratories intending to improve their reputation will seek some sort of accreditation. In this chapter, I am not going to look formally at quality assurance or give advice on how to become accredited, although I will mention accreditation later. However, I will examine the ways in which a routine laboratory might organize itself so that it can assure its customers, even if only on an informal basis, that its results can be relied upon. In the long term, this might provide a basis for more formal accreditation.

Quality comes about from an attitude of mind, not simply by writing quality manuals or buying the latest equipment or QA software. Above all, it is important that everyone involved in the analysis process should be 'quality aware'. 'Everyone' can be taken to include everyone from management, who must supply the resources and leadership, through the analysts to those who perform the mundane but necessary operations such as washing up, who, unless they realize how critical their activities are in preventing cross-contamination, could completely invalidate whole batches of measurements.

If, when we quote a result, we are to be confident that it is an accurate estimate of the 'true' activity and the quoted uncertainty is a realistic estimate of the actual uncertainties inherent in the measurement, we must be sure that:

- our equipment is in working order;
- our nuclear data are valid;

Practical Gamma-ray Spectrometry – 2nd Edition   Gordon R. Gilmore
© 2008 John Wiley & Sons, Ltd

- our standards are fit for the purpose and traceable to external standards;
- our spectrum analysis was valid;
- we can justify our results to the customer and be able to go back to the original data to respond to queries, even some time after the measurement.

All of this is possible only if the measurement follows a properly validated standard procedure. In order to assure continuity and accountability, this must be properly documented. In addition to the more immediate preoccupations of method and equipment, there must be an overall administrative system in operation which ensures that equipment is checked at appropriate intervals, that information is archived properly, and that unexpected results are assessed properly. The system must allow for necessary change. If, at some point in time, it becomes necessary to alter the procedure, the changes must be fully validated and documented. Situations will inevitably arise for which the standard procedure is not appropriate. In such cases, the system must allow the actual procedure followed to be recorded and the information stored with the results in case of future query. In reality, this is probably easier to put into practice than would seem at first sight.

It might be argued by those whose measurements are only used in a semi-quantitative fashion that a QA system is not necessary in their case. Not true. Quality assurance means that, whatever the degree of quality of the result on an absolute scale, the expected quality can be relied upon. (A trite comparison might be the warranty on a used car that gives confidence that the car is fit for its purpose as a used car but does not imply that it will give the performance of a brand new one.)

It was rather disturbing to read in the recent report of an IAEA intercomparison (IAEA (2007)) that of the 327 laboratories reporting results, only 50 % had formally validated their method. Of all of the results reported, 18 % of the results from laboratories that were accredited were judged statistically not acceptable. For the 64 % of respondents who were not accredited, that figure rose to 31 %. Clearly, there is merit in working towards a situation where one could become accredited, even if one doesn't actually seek it.

15.2 NUCLEAR DATA

The need to use the best available nuclear data has been mentioned from time to time throughout this book. To reiterate briefly; as far as possible, use only data which have been critically evaluated. In general, use the most recent data available. I can only refer the reader to Appendix A again, where sources of nuclear data are compared. Appendix B contains data for the nuclides listed in the first edition of this book (the TECDOC-619 nuclides), updated to the most recent re-evaluation enshrined in the IAEA XGAMMA list. The nuclear-data libraries provided by the spectrum analysis program vendor should not be used unless the source of their data is known and is of satisfactory quality. (Their business is to provide you with software, not nuclear data.)

Once you have standardized on a particular set of data, it is then necessary to ensure that this is used throughout the laboratory, on all computers and by all personnel. If subsequently, it is felt that the data should be amended, perhaps when a more recent evaluation becomes available, then the changes, and the date on which the changes take place, must be documented. This is necessary so that any queries about a particular set of results may be addressed in the light of the actual nuclear data used at the time of measurement. It is important that a change in the data set should be made on all systems within the laboratory at the same time, not by a piecemeal 'creeping update'.

15.3 RADIONUCLIDE STANDARDS

It must be obvious that the accuracy of the whole measurement process depends ultimately on the quality of the standards used. Standards should be traceable, by which I mean that they should bear a known relationship to a national or even international standard. Of course, it is impractical to have every calibration standard compared directly with national standards held (in the UK) at the National Physical Laboratory (NPL). Instead, there is a hierarchy of standards, each of which is traceable upwards to through the national standard to international standards.

At the local level, we can achieve traceability by purchasing standards from laboratories that have been accredited for their preparation by the body responsible for the national standard. In the UK, the appropriate body is UKAS (The United Kingdom Accreditation Service, part of NPL) and in the USA, the National Institute of Standards and Technology (NIST). Each country has its own body which in turn participates in the international measurement system. (The overall scheme is shown in Figure 15.1.) By this means, traceability is maintained from local to international level.

Each calibrated standard should be accompanied by a certificate that should at least state the following:

- A reference time and date to which the certification refers.
- A description of the source – geometry, matrix composition and mass.

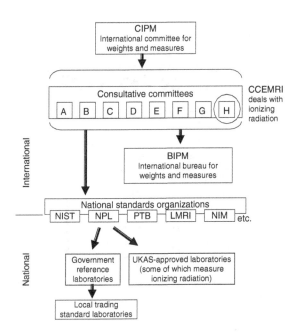

Figure 15.1 The hierarchy of standards organizations which provides a framework for traceability from the bench-top to internationally agreed standards (the national hierarchy for the UK is shown as an example)

- For each nuclide, the following should be specified:

  – activity at the reference time and date;
  – the Type A, Type B and overall uncertainties on the activity and the confidence level to which they refer;
  – the half-life, with its uncertainty, used to calculate the activity.

- A declaration that the source is traceable to a national standard – NIST, NPL, or whatever.
- A date and signature certifying the data.

Other useful information might be included. In particular, the source of the nuclear data used in the calibration is useful confirmation that the laboratory is 'on the ball'. Details of the method of calibration are of interest but not essential. If the supplier is accredited for source calibration by a national body, then one might infer traceability to the national standards without an explicit statement to that effect. However, one should be alert. UKAS accreditation is specific to a particular method or measurement. A laboratory accredited only for alpha source calibration would not necessarily be accredited for gamma source calibration.

Many calibration standards are produced as point sources but for most practical efficiency calibrations, a source in the same geometry as the samples is necessary. It is possible to purchase custom standards prepared in one's own containers. For example, one might supply a Marinelli beaker. In most cases, though, laboratories would prefer to purchase a calibrated standard solution and prepare their own sub standards. This is the point at which traceability can so easily be lost. Calibrated solutions could be orders of magnitude greater in activity than that needed for a *working standard*. It is essential that the method of preparation should be defined, documented and strictly adhered to. In principle, more than two dilutions from the original calibrated solution may destroy traceability. This means that for the preparation of standards at near environmental levels there is a potential problem. It is only relatively recently that very low level calibrated solutions have been available. If the calibrated solution is used over a long period of time, there must be checks in place to ensure that the solution has not evaporated significantly, perhaps by keeping an evaporation log, or otherwise degraded, by precipitation, for example. It is common experience that ^{113}Sn will readily come out of solution if the acidity of the QCYK reference material is not high enough.

15.4 MAINTAINING CONFIDENCE IN THE EQUIPMENT

It is essential that a record be kept of how the spectrometry equipment is set up and of any adjustments subsequently made. In my opinion, each system should have a logbook opened when it is installed. Since detectors might be moved from system to system, each detector should have its own logbook kept from the moment it is received into the laboratory. In this day of the 'paperless laboratory', the logbook could, of course, be held on computer. However, my personal preference would be for the old-fashioned paper logbook. It can be there at hand for instant use and never needs to wait while the network server re-boots – during which time the intention to record an event can so often evaporate. There is, however, the problem of how one backs-up the logbook.

15.4.1 Setting up and maintenance procedures

A most useful, and highly recommended, document that provides an excellent model of a standard scheme for the calibration and use of gamma spectrometry systems is published by the American National Standards Institute as standard ANSI N42.14-1999 (Committee N42 is the Accredited Standards Committee). Any laboratory following this standard is well on the way to having a universally acceptable measurement scheme;

the standard covers installation, calibration, measurement, performance tests of the equipment and of the analysis software and verification of the entire process. An appendix provides advice on a host of setting up procedures, including the preparation of working standards from calibrated solutions. I can recommend a more recent review by Gehrke and Davidson (2005) of setting up procedures with a view to eliminating or controlling artefacts within spectra. The physical setup of the spectrometer, electronic set up, fluorescent peaks and summing are among the topics discussed. This paper reproduces many spectra illustrating the points being made.

Once the spectrometer has been set up, a maintenance schedule should be defined so that various adjustments are checked on a regular basis at appropriate intervals. For example, the DC offset and pole-zero settings might be checked on a quarterly basis but the energy calibration checked on a daily basis. The ANSI document recommends efficiency and resolution checks on a daily or weekly basis. An efficiency check on a daily basis does seem rather enthusiastic, but in such matters, it is wise to err on the side of too frequent rather than seldom. If efficiency is monitored simply by measuring a check sample, there should be little problem.

Whenever any module in the instrumental chain is altered, the system should be checked completely. Even replacing the HT bias supply, which might not be expected to alter any of the calibrations, could have unforeseen effects on the spectrometer resolution because of electromagnetic interference with the amplifier. A complete system check is the only way to ensure complete confidence in the system.

15.4.2 Control charts

Apart from keeping a record of the state of the equipment, the routine calibration parameters should be monitored by means of control charts. One might monitor, for example, the resolution at two energies, the energy calibration factors, and the full-energy peak efficiency at two energies. A simple example is shown in Figure 15.2.

The principle of the control chart is to plot the measured values of the control parameter and compare them with a mean, or expected, value and with various control limits. These would be normally set above and below the mean at levels corresponding to 95 % and 99.8 % confidence limits. Assuming that the process or parameter is in control, there is only a one-in-20 chance of a measurement falling outside the 95 % levels, called the **warning levels** (UWL and LWL in Figure 15.2) and only a one-in-500 chance of it falling outside the 99.8 % **action levels** (UAL

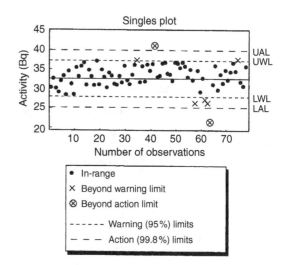

Figure 15.2   A control chart plot of individual measurements of activity of a test sample

and LAL, sometimes also called *control levels*). Results beyond a warning limit may be ignored, but results beyond the action limit cannot be. The detector system must be removed from use until it has been adjusted if necessary, checked out, by defined procedures, and proved to be providing justifiable results. In Figure 15.2, the 42nd control point is beyond the upper action limit but subsequent measurements suggest that this was merely chance. However, the fact that four of the seven points from 58 to 64 are beyond the warning limit and one of these is beyond the action limit indicated that action was necessary. In this case, control was regained by adjusting the system.

A word of warning: a simple laboratory check procedure might be to measure a standard or check sample before every batch of samples. What happens if this check is outside the designated limits? The laboratory procedure might decree that, if this check is beyond the action limit, the check measurement should be repeated and if the new value is acceptable then sample measurement can continue. It is important that the original check measurement should not be discarded. It must take its place on the control chart. If such out of range measurements are discarded, the recognition of a genuine problem will be much delayed. It should be remembered that, statistically, one in 20 measurements can be expected to be beyond the warning limit. If no measurements are ever beyond it, the warning limit is not set correctly and the usefulness of the chart is diminished.

The control chart has the advantage that changes over time can be seen clearly and, on occasion, significant departures from the acceptable range predicted. The predictive value of control charts should not be overlooked. However, unless they are assessed frequently by a competent analyst, trends suggesting future problems will be missed.

The chart in Figure 15.2 does not make best use of the information available. A much better approach, that which would be followed in most industrial control situations where control charts are routinely plotted, is to plot *grouped* data. This smoothes the data so that underlying features can be more easily perceived. (This is analogous to smoothing the data in a spectrum peak search as described in Chapter 9.) Figure 15.3 plots the data of Figure 15.2 in this way. Each point represents the mean of four individual measurements. What is now apparent is the upward trend of the earlier measurements. While it is possible to pick this up in Figure 15.2, it is by no means as obvious. Because such a chart, referred to as a **means control chart**, suppresses the statistical scatter a second chart, the **standard deviation control chart**, must also be plotted. (In process control, it would be normal to use ranges, rather than standard deviations, because they

are easier to plot manually. Since in a modern gamma-spectrometry laboratory the data are most likely to be calculated and plotted by computer, there is no reason not to use the basic statistical factor, standard deviation.) The standard deviation control chart should expose lack of control in the measurement that does not, in itself, alter the mean value. The chart derived from the Figure 15.2 data is shown in Figure 15.3 (a). One thing to notice about the standard deviation control chart is that the action and warning limits are not symmetrically disposed about the mean.

It is not essential to use a group size of four as used for Figure 15.3. (In statistical parlance, 'sample size' would be referred to. Here, I use the term 'group' to avoid confusion.) This is entirely at the discretion of the analyst. In practice, grouping four or five measurements is common. Table 15.1 gives the factors needed to plot the different control limits for different group sizes.

15.4.3 Setting up a control chart

Setting up a control chart, as with most aspects of gamma spectrometry, is by no means as obvious as it would appear at first sight. Although it might seem trite to say so, the first action must be to define what it is one wishes to control the performance of – the amplifier, the gamma spectrometer or the measurement method as a whole. For example, it is often suggested that a control chart monitoring the measured energy of one or more peaks should be plotted. Is this of any value in assessing the overall system performance? Not necessarily. If the spectrum analysis happens to involve a recalibration based upon the spectrum data, which would correct for it, small degrees of energy drift would have little or no effect on the result. In any case, if, as the ANSI N42 Standard suggests, the energy calibration is carried out on a daily basis, monitoring peak energies would have little value. On the other hand, monitoring the peak position (i.e. channel number rather than energy) would give a very clear idea of the amplifier/ADC stability. Incidentally, in this case it might also be sensible to monitor temperature, either ambient or that of the module itself, since gain and temperature are likely to be correlated.

One parameter that surely must be kept in control is the resolution of the system. I would suggest that this be monitored at high and low energy. As an early warning of impending resolution loss, monitoring the leakage current of the detector may also be useful.

Some might consider it more important to monitor the activity of a check sample. If this was analysed regularly in the same way as actual samples this would allow a control chart to be plotted which would monitor the

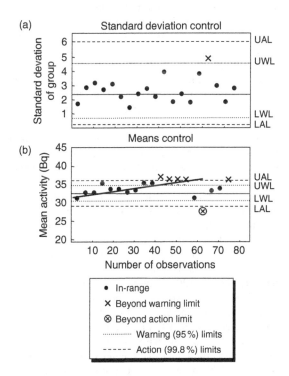

Figure 15.3 Mean and standard deviation control charts for grouped data from Figure 15.2 ($n = 4$)

Table 15.1 Factors for calculating control limits (data taken from Moroney (1990))

Limit	Upper action	Upper warning	Lower warning	Lower action
Confidence limit (%)	99.8	95.0	95.0	99.8
Means control limit[a]	3.09	1.96	−1.96	−3.09
Standard deviation control limit[b]	—	—	—	—
Group size				
2	4.12	2.87	0.04	0.00
3	2.98	2.17	0.18	0.04
4	2.57	1.93	0.28	0.10
5	2.34	1.81	0.37	0.16
6	2.21	1.72	0.42	0.21
7	2.11	1.66	0.46	0.26
8	2.04	1.62	0.50	0.29

[a] Mean + factor × mean standard deviation.
[b] Factor × mean standard deviation.

whole analysis, apart from sample preparation. While this is certainly useful, it will not necessarily provide as early a prediction of problems as monitoring the resolution. This is particularly so if the analysis is performed comparatively when errors common to sample and standard will tend to cancel out. A check sample should ideally be identical in every way to actual samples, with the exception that it contains sufficient of one or more of the nuclides to be measured to provide well-defined peaks within a short count period. If this check sample is to be used over months or years, the nuclides monitored must have long and well-evaluated half-lives.

In many radiometric procedures, it would be necessary to measure detector background. Is this of value in gamma spectrometry? In most cases, the background to a peak is the Compton continuum due to other activities in the sample. In this context, the *natural* detector background is of no consequence. There are two situations where background might be monitored, although in neither case would it be appropriate to construct control charts as a check on performance:

- Over a short count period as a check to confirm that there has been no gross contamination of the detector – useful if samples of a wide range of activities are measured on the same detector.
- Long background counts where peaked background corrections are needed. However, it should be recognized that some of the components of detector background can vary over time due to external cosmic events and, internally, due to variations in radon emission within a counting room. This variability must be accounted for within the uncertainty that must accompany the peaked background correction. The only way to do this is to combine a number of separate background measurements taken over a long period of time.

Having decided which parameters to plot, we need a mean value and a mean standard uncertainty for each, against which we can compare the check measurements. If the parameter is the activity of a reference source, we may be able to use the reference value or otherwise we must use the mean of measured values. The mean standard uncertainty must be calculated from the spread of measured values. If the parameter is activity, there is only limited merit in using the estimate of counting uncertainty provided by the spectrum analysis program. This will not include external sources of uncertainty that, if we wish to control the overall measurement process, must be considered. It may be, for example, that if the source must be placed at the centre of the detector cap, different operators have a different perception of where that is. That additional uncertainty is part of the overall measurement uncertainty and should be taken into account.

When a control chart is to be set up anew, it may be that there is no historical body of measurements that can be referred to in order to provide the mean and mean standard uncertainty estimates. In that situation, a large number of measurements, say 20 groups, must be taken within a short period, preferably involving all operators and at all times of the day, so that initial control values can be calculated. After the control chart has been running for some time, it may be necessary to revise the mean or mean standard deviation according to actual running conditions. For the purposes of this reassessment, out-of-range values might be omitted, but only if the excursion has been confirmed as a genuine extraordinary event.

It is now becoming standard practice for facilities for control charting to be included within the spectrum-analysis package. There are many software packages available for general control charting which would be equally valid for gamma spectrometry purposes, although the convenience of direct transfer of results from the analysis package would be lost. For those whose inclination is for a more personalized chart, any spreadsheet program that provides graphical-display would be suitable.

15.5 GAINING CONFIDENCE IN THE SPECTRUM ANALYSIS

In Chapter 9, I discussed at length the algorithms likely to form part of spectrum analysis programs. Whether they perform satisfactorily depends to some extent on how they are coded into the program. One cannot assume that the same algorithm implemented in different programs, or *even different versions of the same program*, will give precisely the same result. The reasons are connected with the detail of the program coding, the compiler used and the computer on which the program is run. Such matters need not concern us here but the implication is that the computer program should be evaluated to check that consistent and accurate results are obtained.

The ANSI N42 Standard referred to above recommends a number of tests for:

- automatic peak finding;
- independence of peak area from peak-height-to-baseline ratio;
- doublet peak finding and fitting.

It also suggests that '*Documentation of acceptable results by the software vendor shall be sufficient evidence of the capability of the software algorithms to justify not repeating the tests . . .*'. If users decide to accept such proof from the vendor, they must ensure that the tests have been carried out precisely according to the ANSI N42 Standards on the appropriate version of the program. Written proof would be needed for incorporation into the quality-assurance documentation. At the time of writing, I am unaware of *any* spectrum analysis software that has been rigorously validated to the satisfaction of accreditation bodies. However, pressure is being applied in that direction and the situation may change. Since validation applies only to a particular version, any upgrade of the software should be fully evaluated before replacing the current version.

Those who would doubt the need for software evaluation should read the report by Sanderson (1988,) where a single spectrum was submitted to seven well-respected

analysis programs and the results compared. This was not particularly reassuring. For example, the measurement of the ^{137}Cs peak, hardly problematical one would think, provided results ranging from 81 to 314 090 Bq. The conclusion of the author was that '*Completely automatic data analysis of complex gamma-ray spectra with IBM PC computers has not been achieved in this study.*' That study was some time ago and the programs have been improved. A re-evaluation in 1992 by the same author (Decker and Sanderson (1992)) using new versions of some of the original programs showed much improved performance. However, the results still demonstrate inconsistencies between programs. For example, what are we to make of the fact that three programs quote results for ^{212}Bi ranging from 23 to 44 Bq (probably statistically consistent if the uncertainties were taken into account) but two others quote less-than figures of 15.6 and 16.1? Why should some programs that cannot detect the nuclide quote a limit of measurement lower than the amount which, based on other analyses of the same spectrum, is actually present? Would it not be convenient if software vendors could provide us with standard software performance indicators, as they do for detectors, which would allow potential purchasers to compare different programs?

More recent software performance studies have been published and are discussed in more detail below. The report on the NPL standard spectra by Woods *et al.* (1997) considered nine spectrum analysis programs. Another independent comparison of twelve software packages has been published by the IAEA as TECDOC-1011 (IAEA (1998)) using the 1995 IAEA test spectra (see below) and another report following the 2002 software intercomparison using the 2002 IAEA test spectra has been published (Arnold *et al.* (2005)).

15.5.1 Test spectra

In most branches of analytical science, methods are checked and monitored using reference materials. When purchased, these are accompanied by a certificate that specifies the concentration, with an associated uncertainty, for a number of chemical entities. These will have been analysed by at least two different analytical methods and provide benchmarks against which methods and laboratories can be assessed. Indeed, gamma-ray spectrometrists plying their trade in the service of neutron activation analysis would utilize such reference materials routinely. The corollary for testing gamma-ray spectra would be to have available reference spectra that contained peaks at known positions containing known numbers of counts. It would be unrealistic, perhaps, to expect one spectrum to test all aspects of spectrum analysis but a small number of spectra should suffice. Surprisingly, although the concept is a hardly a difficult one, there are no test spectra generally available which are completely acceptable to users, software vendors and regulatory bodies as a whole.

The ideal test spectrum would have both large peaks and small peaks on both high and low continuum backgrounds to test the ability of the software to find and measure accurately peaks in a range of spectrum environments without reporting spurious peaks. There would need to be a test for doublet resolution, with a range of peak separations and peak ratios. In order to be assured that the analysis program would not be confused by spectral artefacts, such as backscatter peaks and Compton edges, these should also figure somewhere within the spectra.

Before looking at how test spectra can be prepared and those available, we must consider how they are to be used. In principle, all calculations of radioactivity reduce to a comparison between standard and sample peak areas. This might be directly, as in neutron activation analysis, or indirectly via an efficiency calibration curve. It follows, therefore, that an analysis program need not produce absolute peak area estimates as long as the program is consistent and the area estimate is indeed proportional to the *true* peak area. If an analysis program gives areas that are consistently 5 % low, then the efficiency calibration derived using the program will be 5 % low; the error will cancel out when the final calculations are performed. A set of test spectra should, therefore, include a standard or reference spectrum containing well-defined peaks to which peaks in all other test spectra can be related. This will best simulate the actual manner in which spectra are used.

The creation of test spectra which fulfil the criteria mentioned above is not a trivial matter. No matter how the spectra are created, there is scope for claims that they do not represent realistic detector spectra. In particular, the generation of small, poorly defined peaks with known areas in a realistic spectrum environment is problematical. It is certainly not acceptable to produce small peaks simply by counting onto an existing high background continuum. The statistical uncertainties of counting would make it impossible to know exactly how many counts had actually been accumulated. (Remember, in Chapter 5, Section 5.4.1, it was explained that we can never *measure* the background under a peak, only *estimate* it.) We might measure a particular source under standard conditions to very high precision. For example, let us say that there are 1 million counts in the peak when measured over 50 000 s of live time. Suppose then that, without moving the source or otherwise altering the system, a spectrum with a high Compton continuum is loaded and the source again counted for 5 s. The expected number of counts in

the peak would be 100 counts. However, in practice, all that can be said is that the actual number of counts will be within the range 80 to 120 counts in 95 % of cases. One may argue that channel-by-channel subtraction of the original Compton continuum would provide the number of counts added – so it would. However, then subtraction of the counted background underlying the peak would be needed. This background can only be estimated; once again there is an uncertainty in the actual number of counts within the peak.

One might consider preparing a small peak, stripping off the underlying background and digitally adding the remaining, precisely known counts, to the spectrum. Unfortunately, in doing this the statistical distribution of the counts is disturbed making the peak, at least in principle, invalid. This is best understood by a practical example. Let us suppose that we have a channel within a peak containing 2500 counts, including a continuum contribution of 1000 counts. The uncertainty on this total count is 50 ($\sqrt{2500}$). If we subtract 1000 counts from this channel, the uncertainty remains 50 because there is no uncertainty on the 1000. If we now add this corrected count of 1500 to a new background of 2500 counts, the total uncertainty on that channel content will be 71 counts $[\sqrt{(50^2 + 2500)}]$, rather than the 63 counts expected from the channel count $[\sqrt{(1500 + 2500)}]$.

In practice, if the baseline subtracted from the peak is small and the continuum to which it is added is large, the upset in the statistical distribution can be neglected. This would seem to be the only way in which test peaks can be generated with known position and area. On the other hand, if the subtracted baseline is much larger than the final baseline the statistical scatter will be severely underestimated by the channel contents alone.

At the time the first edition of this book was prepared, I stated that the only easily available test spectra were those produced by the IAEA and distributed in 1976. Fortunately, this is no longer the case and two further sets of spectra have been produced by the IAEA and a set by the NPL in the UK. These are all discussed below. The spectra used by Sanderson for his 1992 re-evaluation referred to above are also available, although they are of limited use.

15.5.2 Computer-generated test spectra

Since we know the shape of our detector peaks and the shape of the artefacts in our spectra, then there would seem to be no problem in generating spectra mathematically. In principle, one would simply define the shape of the background, superimpose mathematically valid statistical scatter and then add peak shape functions, each with scatter added. The mere suggestion of computer generated test spectra generates adverse reaction among potential users. Indeed, this is one of the oft-quoted objections to the IAEA G1 test spectra. I will discuss this further below and show that the objection is unfounded.

The IAEA G1 test spectra

This is a set of nine spectra distributed originally in 1976 as an intercomparison exercise (IAEA (1979)). For some time, the distribution of these spectra was limited by the medium on which they were available – punched tape, computer card or mainframe computer magnetic tape – but are now readily available on PC-compatible floppy disks and on the Internet. The spectra are in the public domain and can be copied freely. The spectra were prepared by measuring pure radionuclides with high precision – just below 10^6 counts in the highest channel of each spectrum – using a detector of, what was at the time, 'average' performance. This was a 60 cc Ge(Li) detector of resolution 2.8 keV FWHM at 1332.5 keV and a peak-to-Compton ratio of 40:1. It was noted that that peaks had a slight asymmetry. Spectra were recorded in 2000 channels with gain of about 0.5 keV/channel, giving a range of about 1000 keV. One FWHM is equivalent to 5.6 channels at 1332 keV.

The spectra were smoothed slightly to suppress residual statistical uncertainty and then combined in various ways after shifting by whole numbers of channels and dividing each channel count by a known factor. The reason for shifting peaks was originally to provide peaks at positions unknown to the participants in the original intercomparison exercise. In order to allow programs to be tested in the presence of high and low peak backgrounds, about 10 000 counts per channel was added to the bottom 1000 channels and about 200 counts per channel to the top 1000 channels. The 'join' between these two regions was adjusted to simulate a Compton edge. (It must be said that this is the one unsatisfactory aspect of these spectra in that this pseudo-Compton edge is much larger than that encountered in *real* spectra and most analysis programs identify spurious peaks at this position.) The spectra as constructed at this point had peaks in known positions and of known ratios to their reference peaks but were essentially 'noiseless'. The spectra were then subjected to a random number generator process to simulate the effects of Poisson-distributed counting statistics. While this process will alter individual channels, it will not alter the position of the peaks and should not alter the number of counts within the peaks. The spectra

in the set, shown in Figure 15.4, are described in the following:

- *G1100 – reference spectrum.* This contains 20 peaks to be regarded as independent nuclides. The spectrum is the sum of the complete spectra of all 20 pseudo-nuclides over the whole energy range. Each peak contains near to 65 000 counts and can be measured with an uncertainty of about 0.4 %.

- *G1200 – peak search test spectrum.* This contains 22 peaks all shifted and attenuated relative to G1100. Many of these are difficult to detect and measure. This spectrum also serves as a very good test of the ability of programs to measure the area of poorly defined peaks. Experience has shown that many programs can detect 17 peaks without reporting spurious peaks. The highest number of real peaks detectable is, perhaps, 19, at the expense of reporting spurious peaks. The pseudo-Compton edge in this spectrum seems to cause problems with many programs and, as this is an unreal feature, any peaks reported in the range 1020 to 1032 keV should be ignored for the purposes of assessment. (Nevertheless, it is instructive to examine the way in which the program handles this feature as an indicator of what could happen, perhaps in a less extreme fashion, to real Compton edges.) I am impressed by the claim of De Geer (2005) to have detected all 22 peaks in this spectrum with only 1 spurious peak detection using software developed for the CTBTO.

- *G1300 to G1305 – consistency test spectra.* These six spectra each contain 22 peaks and are derived from the same prototype spectrum but subjected to the 'noise generation' process separately. The idea of these spectra is to check the consistency of analysis of spectra that are identical except for the statistical scatter. Twenty peaks are in the same position as the G1100 reference peaks but attenuated. Another two peaks are attenuated and shifted. One notable feature of this set of spectra is that the peak at channel 1010.6 is near to the pseudo-Compton edge and, because of this, the area measurement may be affected by the downward slope at the high-energy side of the spectrum.

- *G1400 – deconvolution test spectrum.* This spectrum contains nine well-defined doublet peaks formed by shifting and attenuating peaks from G1100. The peak separations are 1, 3 and 6 channels and the peak ratios are 10:1, 3:1 and 1:1. These separations represent about 0.2, 0.5 and 1 FWHM, respectively. This spectrum is a severe test of deconvolution programs and most are found wanting.

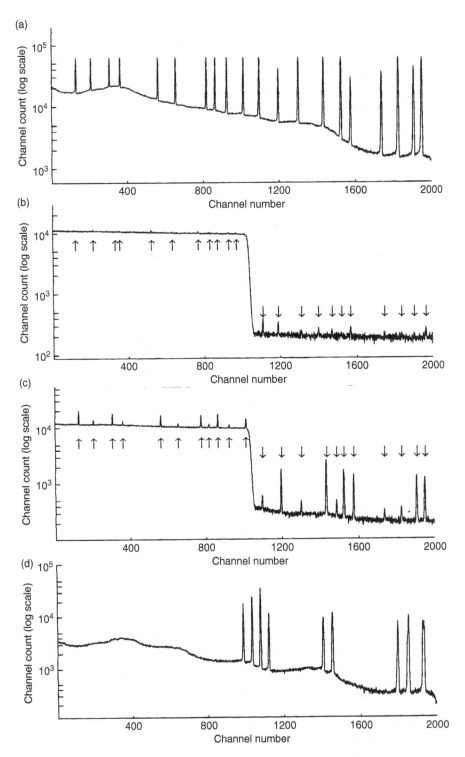

Figure 15.4 The IAEA G1 intercomparison/test spectra: (a) spectrum G1100; (b) spectrum G1200; (c) spectra G1300-G1305; (d) spectrum G1400 (*y*-axes, counts per channel; *x*-axes, channel number)

Objections to the IAEA G1 spectra

It should be pointed out that there are objections to the continued use of these spectra in some quarters. These objections revolve around the fact that the spectra are of only 2000 channels, are 'computer generated' and are measured on an old detector. In rebuttal of these objections, the following observations can be made.

The fact that only 2000 channels are available is irrelevant as the spectra can be regarded as a 2000 channel slice of a 4096-channel spectrum – or 8192 or 16384 channel spectra, for that matter. What matters is the number of channels within a peak. These spectra were recorded at about 0.5 keV/channel and I deduced in Chapter 5, Section 5.5.2 that this is, in many respects, optimal for best peak area measurement precision.

The argument that because the spectra have been 'computer generated' they are not relevant is not valid. The shapes of the peaks have not been altered by the mathematical manipulation. In fact, the correspondence of peak shape between standard and sample spectra may be better than in an actual measurement set where differences in count rate could, in principle at least, cause peak shapes to alter.

The only real problem may lie in the fact that the peaks in the IAEA spectra are slightly asymmetric. In practice, any detector system has the potential to produce spectra where the peaks are not ideal. For example, one would continue to use a detector after sustaining slight neutron damage and one would hope that the analysis software would cope. Indeed, peaks in spectra measured by modern HPGe detectors of 100–150 % relative efficiency also have low energy tails because of the limitations on charge collection in such large pieces of germanium. In fact, the peaks in the G1 spectra are not grossly asymmetric, as can be seen in Figure 15.5, comparing the actual shape of the peak at channel 1011 with a pure Gaussian with an underlying step function.

Notwithstanding the deficiencies of the G1 spectra, Nielsen and Pálsson (1998) used the G1300 spectra from that 20-odd year old set of spectra to compare ten spectrum analysis programs, many of them not widely known, but including *GammaVision*, *Genie-PC* and the author's own *CompAct*. Results for *GammaVision*, *Genie-PC* and *GammaTrac* were submitted by more than one laboratory, giving altogether results for 15 analyses. They were able to demonstrate that less than half the programs gave good agreement with expected peak areas and reasonable uncertainties. Only four programs gave results that could be deemed in statistical control. Apart from demonstrating that there are differences between peak area measurement programs, even with simple un-interfered peaks, the data show that different users of the same program can get

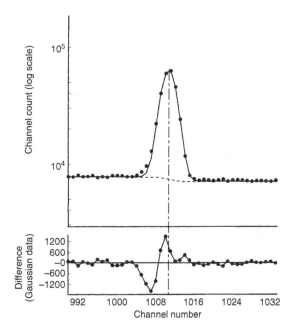

Figure 15.5 Comparison of the peak at about channel 1011 in the G1100 spectrum with a pure Gaussian (continuous line) plus step background function (short dashed line)

different results. Two users of *GammaVision* gave results assessed as 'low accuracy' while the third user gained 'high-accuracy' status.

The Sanderson test spectra

These test spectra, used for the re-evaluation of the spectrum analysis programs mentioned above, were prepared using a spectrum of a mixed gamma-ray standard filter paper measured on a 1.96 keV resolution n-type HPGe detector of 15.2 % relative efficiency. Peaks from this spectrum were added to a peak-free random background spectrum. This background was mathematically generated after assessing the actual background level in real spectra. Spectra were devised to test peak search, doublet resolution with equal and asymmetric peak ratios and an efficiency test. A spectrum of an environmental air filter sample was included as a test of a 'real life' situation. These spectra are available on the Internet.

The CTBTO spectrum

It is essential that the worldwide network of laboratories involved in monitoring high volume air filters for evidence of nuclear weapons tests should be highly proficient. To this end, the organization has developed test spectrum, created by MCNP but based on a historical spectrum measured after an actual weapon test. This has been internally validated and used for proficiency testing within the network of monitoring laboratories (Karhua *et al.* (2006)). Such a well-evaluated spectrum would be a valuable resource for other laboratories and I am hoping that, with a little encouragement, the CTBTO will release it into the public domain.

Programs for mathematically creating test spectra

It is a straightforward process to create a computer program to generate test spectra. It is only necessary to generate a continuum, superimpose Gaussian peaks containing a known number of counts at known positions within the spectrum, apply a randomization process to the numbers of counts in each channel to simulate counting uncertainty, and then put the whole spectrum into one of the standard spectrum formats. The difficulty is providing a distribution of counts which truly represents counting uncertainty. Nevertheless, leaving aside the more complicated Monte Carlo procedures that could generate complete spectra, many such programs must have been written, although few have been brought to public attention. I am aware of two in particular: *Peak-Maker*, described by De Geer (2004) where he uses the program to generate test peaks for his examination of Currie detection limits, and *SpecMaker* available from the author. The latter produces complete spectra of test peaks superimposed on either a mathematically generated background continuum or on an existing background spectrum.

While the spectra created by such tools may not have the advantage of, what we might describe as, public acceptability in the way that the published test spectra have, they are useful in gaining an understanding of the performance of one's spectrum analysis software in peak-measurement situations that one can create to order.

15.5.3 Test spectra created by counting

As I said above, none of the sets of test spectra available are ideal and, while I welcome their availability, it would be possible to criticize each of those below as being unrealistic or lacking in some particular aspect. Nevertheless, they are useful tools in understanding the performance of one's own software. If the 'correct' peak areas or activities are not achieved by one's program there must be a reason, and finding that reason will enhance one's appreciation of its limitations, perhaps leaving the way open for additional procedures outside the program itself to correct for them.

The 1995 IAEA test spectra

This set of eight spectra, created and made available by Menno Blaauw (1997) for the purpose of the 1995 IAEA spectrum analysis program intercomparison, consist of a calibration spectrum, a spectrum of ^{226}Ra in equilibrium with its daughters, a similar spectrum distorted by a high count rate (to assess the effect of random summing on the measurements) and results of additions performed to obtain doublets with known peak area ratios. Care has been taken to ensure that the statistical integrity of the spectra has been maintained. The spectra are 8192 channels with a fairly typical energy scale of 0.4 keV per channel. They can be downloaded from the Internet, together with necessary information and a computer program to compare statistically the analysis results with the reference results, including making allowance for missed peaks and false positives.

It is intended that the spectra will be used to assess the capability of the spectrum analysis program to measure the peak positions and areas accurately and with realistic uncertainty. The creators claim that the reference peak areas they quote are 'absolute and traceable' (Blaauw (1999)). At first sight, this claim appears difficult to understand. If a spectrum is generated by counting, then one can never know exactly how many counts will be in a particular peak within a spectrum because of the inescapable uncertainties of counting. One can, however, with additional counting and analysis, derive an expected value for each peak area with a small uncertainty. The creators of the spectra believe that they have determined expected values which are not dependent on the software used to measure them and, in that sense, are *absolute*. The provision of the program for assessment of the analysis results is a mixed blessing. The program runs under DOS and in these days of graphical user interfaces where drag-and-drop operation of programs is routine many people will find command line operation baffling.

The major problem with these spectra is that the calibration spectrum is limited to the energy range 122.1 keV to 1332.5 keV. The assessment program assumes that peaks as high as 3000 keV might be measured, thus inviting the analyst to extrapolate the energy and width calibrations. That would be bad practice and any comparison between expected and measured peak energy above 1332.5 keV would surely be suspect.

The 1997 NPL test spectra

These spectra were generated by the NPL as a means of allowing users to test the ability of their software in respect of peak and nuclide identification, activity measurement and the ability to cope with analytical problems, such as true coincidence summing, overlapping peaks (including the special case where the Doppler-broadened 511.0 keV peak is involved) and random summing. Three sets of spectra, representing certified known amounts of known nuclides, were created using n-type HPGe detectors as follows:

- Set 1: calibration spectrum and two standard spectra, A and B, measured at 550 mm source-to-detector distance. These spectra were measured on an 18 % relative efficiency detector within a graded shield.
- Set 2: background spectrum, calibration spectrum and two standard spectra, A and B, measured on the same detector system as Set 1 spectra but the sources were placed on the cap of the detector to enhance the true coincidence summing and random summing problems.
- Set 3 spectra: background, calibration and standard spectra, A and B, were measured 150 mm from an 11 % detector mounted inside an ungraded lead shield that was close to the detector. The intention was to provide spectra in which there is significant scattering of the gamma radiation and generation of fluorescent X-rays.

Set A spectra contained only ^{60}Co, ^{85}Sr and ^{137}Cs, while Set B spectra contained nuclides expected to present a more difficult analysis problem with true coincidence summing errors likely: ^{125}Sb, ^{137}Cs, ^{134}Cs, ^{154}Eu, ^{155}Eu and, as an un-certified impurity, ^{152}Eu. The intention is that the user should analyse the spectra using only the basic information provided with the spectra – essentially 'blind'. The results can then be checked against the reference values provided in a sealed envelope.

A comprehensive report describing the preparation of the spectra and their analysis using spectrum analysis programs from four suppliers has been published (Woods, 1997). The report includes analyses provided by the manufacturers themselves and by an expert user. There is also a useful appendix containing a personal appraisal of the various programs by the expert user which comments on such matters as usability of the programs. Unfortunately, the comparison is marred by the fact that when analysing the spectrum measured at high count rate, it would appear that neither the manufacturers nor the expert user took advantage of the information provided and corrected the spectra for random summing. The report incorrectly attributes all of the bias in the analyses of the high count-rate spectra to true coincidence summing alone.

It is clear from the reported results that true coincidence summing errors are present, as expected. The report comments on the fact that none of the software packages

makes an automatic true coincidence summing correction. This is hardly fair comment bearing in mind that information that would allow such a correction, in particular spectra to allow total efficiency calibration, are not supplied. Nevertheless, the spectrum set is a valuable resource for assessing the performance of one's own software and the way in which one applies it.

The 2002 IAEA test spectra

This set of spectra was created following a meeting of the IAEA Advisory Group on Metrology to allow assessment of the quality of spectrum analysis programs for low level radioactivity measurements. These spectra are intended to assess the ability of the spectrum analysis program to estimate activities rather than peak area. (The reference peak areas are provided for one particular spectrum. That will allow assessment of peak area measurement in the same way as in the 1995 IAEA spectra, albeit at much lower activities.)

The test spectra are accompanied by the spectra of calibration sources and background spectra, together with the appropriate documentation. Because measurement of low-activity sources will often be made close to the detector, errors due to true coincidence summing could be expected. Some programs do allow such corrections to be made and to facilitate that point source spectra of a number of nuclides needed to generate a total efficiency calibration are provided.

A comparison of the ability of seven spectrum analysis programs to analyse these spectra has been reported (Arnold (2005)). Apart from comparing the actual activities reported by the various programs, the authors comment on the, sometimes considerable, differences between the nuclear data provided in the nuclear data library supplied with the program. Quite so. It is my contention that such nuclear data libraries should be altered to accord with the sources discussed in Appendix A before analysis is even considered.

It is sobering that a table of analysis functions judged to be subject to 'partial or complete failure' highlights the fact that none of the programs could be relied upon to perform all necessary functions reliably. These are the very programs that on a day-to-day basis are providing results around the world for monitoring and regulatory purposes.

15.5.4 Assessing spectrum analysis performance

In order to compare programs one with another and with an 'ideal', performance indicators are needed. Every study of spectrum analysis uses a different measure of performance, often presented as tables of information. It is difficult to pick out from these tables just which program is better in particular situations. Ideally, we would have standard ways of assessing the performance that reduce to individual indexes. Ideally, these indexes should be easy to calculate and small in number. I would suggest the following as a reasonable attempt as defining standard indexes and I will illustrate their use in assessing the analysis of the G1 and Sanderson test spectra.

Peak search index. In a spectrum search, it is obviously important that as many peaks as possible should be detected but equally that as few spurious peaks as possible are reported. The index used by Keyser (1990) would seem to be useful:

$$\text{Peak detection quality factor} = Q_D = 1 - (S - L)/E$$
$$(15.1)$$

where S is the number of spurious peaks reported, L the number of peaks not detected (i.e. lost) and E the expected number of peaks. The maximum value for Q_D is 1 and negative values would arise when the total number of spurious peaks and non-detected peaks exceeds the number of peaks expected. Clearly, this index will depend upon the sensitivity setting of the search algorithm, and some effort would have to be put into optimizing this. If we take the number of true peaks measured to be M, then re-arranging Equation (15.1) we can also calculate this index as:

$$Q_D = (M - S)/E \qquad (15.2)$$

Peak location index. Having deduced the position of a peak, we need an index to indicate how close the estimated position is to the true position. Here, we need not only a measure of the magnitude of the difference but also some indication of positional bias. A simple way to do this is to consider only the differences, D, between the measured, C_M, and expected, C_E, positions expressed as either channels or energy, i.e. $D = C_M - C_E$.

This difference would be calculated for each peak in the test spectrum and the mean and standard deviation calculated. If the mean is significantly different from zero, a bias in the peak location is indicated. The standard deviation, σ_D, gives some measure of the uncertainty of the peak location algorithm; it should be as small as possible.

Peak area index. Again, we need some sort of indication of the magnitude of the deviation of the estimated area

from the true area and some indication of bias. Keyser's 'area estimation quality factor' Q_S is:

$$Q_S = 1 - N_S(x)/N \qquad (15.3)$$

where N is the number of peaks considered and $N_S(x)$ is the number of peaks having a difference from the expected area of greater than x %. This factor would be 1 if all peak area differences were less that x %. A problem with this is that the index has no meaning unless all peaks are measured to the same degree of uncertainty. A small peak will have a larger uncertainty and can legitimately have a greater difference. We could adjust the factor by taking $N_S(x)$ as being the number of peaks where the area was more than one standard deviation from the expected area. Even with this modification, the factor is not particularly useful when comparing results since 32 % of differences can, statistically, be expected to be greater than 1 standard deviation in any case. Hence, any value of Q_S of more than 0.68 would be acceptable. Only factors grossly below 0.68 would be significant.

While the index could be tinkered with, a more useful exercise is to observe the manner in which the errors in the area measurements are distributed around zero. (Here, of course, I really do mean 'error' rather than 'uncertainty'.) A suitable parameter is the normalized difference, D_A:

$$D_A = (A_M - A_E)/s \qquad (15.4)$$

where A_M and A_E are the observed and expected areas and s is the standard deviation of the measured area. So, for each peak area measurement in a test spectrum we calculate D_A and then the mean and standard deviation of the whole set. The normalized differences should be distributed symmetrically about zero with a standard deviation of 1. If the mean of these differences is significantly different from zero we can deduce that there is a bias in the peak area measurement, and a standard deviation, σ_A, much greater than 1 would suggest that the peak area algorithm was introducing sources of uncertainty (or even error!) in addition to counting uncertainty.

Summary. The five indexes, including standard deviations, are then:

- Peak search index: $Q_D = (M - S)/E$ for which the target is no more than 1.
- Peak location indexes: mean and standard deviation of $C_M - C_E$ for which the target is 0 ± 0 channels or keV.
- Peak area accuracy indexes: mean of $(A_M - A_E)/s$ for which the target is 0, and the standard deviation, σ_A, for which the target is no more than 1.

The same indexes are applicable to either singlet or doublet-test spectra. A full assessment does demand a fair amount of simple but time-consuming calculation and is best done by using a computer spreadsheet. Once set-up, this can be used to assess and compare different programs and would be a valuable aid to appreciating the performance of one's existing software, even if there is no intention to change.

Table 15.2 shows these indexes derived for a selection of programs when applied to the G1 and Sanderson test spectra. (The programs are not identified by name because the survey was incomplete. In some cases, the program version used has since been superseded and in others, the

analysis was not necessarily optimized. Nevertheless, a general picture of how the indexes can be used emerges.) The G1200 peak search indexes (PSI) are similar for the various programs when applied to singlet peaks but vary widely when applied to doublet-peak tests, reflecting considerable differences in performance. (Note that the PSI for the G1 spectra will not necessarily be the same as for the Sanderson spectra because of the different mix of easy and hard-to-measure peaks.)

None of the peak-location mean indexes suggests any bias in determining peak position. The variability index suggests that the effective uncertainty on peak position is of the order of 0.05 keV for well-defined peaks (G1300

Table 15.2 Examples of spectrum analysis performance indicators for the G1 and Sanderson test spectra. Different programs are indicated by A, B, C, D and E

Target	Peaks	PSI	Location indexes		Area indexes	
		1.00	0.000	0.000	0.00	1.0
G1200 peak-search test (22 peaks in 1 spectrum)						
A	17	0.68	0.056	0.395	0.08	0.89
B	13	0.45	0.019	0.282	−0.06	0.88
C	17	0.68	0.074	0.317	−0.01	0.62
D	15	0.68	0.049	0.693	0.77	0.71
G1300 measurement-consistency test (22 peaks in 1 of 6 spectra)						
A	22	1.00	0.005	0.098	0.16	1.02
C	22	1.00	0.030	0.099	0.20	0.94
D	22	0.98	−0.023	0.175	0.35	0.80
G1400 doublet-resolution test (9 doublets, 18 peaks in 1 spectrum)						
A	18	1.00	*	*	0.12	0.81
C	18	1.00	−0.288	0.612	−0.076	29.99
B	11	0.61	−0.126	0.374	0.14	1.09
D	17	0.83	*	*	0.58	14.79
Sanderson peak-search test (24 peaks, 3 in each of 8 spectra)						
A	14	0.58	−0.020	0.125	−0.00	0.64
C	14	0.58	−0.000	0.035	0.28	0.60
E	17	0.63	0.011	0.076	−0.03	0.87
Sanderson peak-separation test (60 peaks, 3 doublets in each of 10 spectra)						
A	60	1.00	*	*	−0.03	0.56
C	52	0.87	−0.001	0.032	0.3	0.96
E	44	0.73	−0.110	0.025	0.15	1.03
Sanderson low/high doublets (36 peaks, 3 doublets in each of 6 spectra)						
A	42	1.00	*	*	0.09	0.48
C	40	0.86	0.025	0.154	0.27	0.42
E	22	0.52	−0.042	0.336	1.19	3.67
Sanderson high/low doublets (42 peaks, 3 doublets in each of 7 doublets)						
A	36	1.00	*	*	−0.04	0.29
C	34	0.86	−0.004	0.052	−0.68	6.47
E	14	0.39	0.116	0.166	0.13	0.96

* indicates cases where a library directed fit using the expected energy was used.

set), becoming much worse for peaks near the limit of detection (G1200) and potentially very poor when doublets are deconvoluted.

The peak area bias index is, in most cases, small. A particular exception is the index for the analysis of G1200 using program C where, although not large, it is greater than for other programs. This slight positive bias could, perhaps, be attributed to the method used to determine peak limits in this particular program. The area scatter indexes are by and large less than 1 (as hoped), although there one or two notable exceptions. This index is particularly high for some of the doublet resolution results. Having said that, the G1 doublet resolution test is severe and demands more from the software than its specification would expect. Nevertheless, this index, as with the others, does give some idea of the relative performance of different software. Program A appears to perform particularly well in analysing the Sanderson spectra. The area-scatter indexes being particularly low (e.g. 0.29 for the low/high doublets). It may be that this is due to the details of the fitting process used by program A, which uses a digital model of the peaks rather than an analytical function. The obvious model to use is the relevant peak in the standard spectrum. Since this peak is also used to create the peaks in the test spectra (by simple attenuation), there is a closer agreement between sample and standard peak shapes than would be expected on statistical grounds. This reveals not so much a problem in the software, but a limitation of the test spectra themselves.

Table 15.2 does not include any of the other useful qualitative information that can be gleaned from the Sanderson test spectrum analyses, such as the minimum separation and minimum peak ratios for reliable measurement.

15.5.5 Intercomparison exercises

While a satisfactory software validation does provide a great deal of confidence, there is more to a radioactivity measurement than the spectrum analysis. It is desirable to have some idea of the performance of the analytical system as a whole. This can best be done by *external reference*. By this I mean the use of test samples and participation in intercomparison exercises. I would recommend that every laboratory should seek to test itself in this way on a routine basis. Intercomparisons involve a considerable amount of work for the organizers and consequently are relatively infrequent events. All the more reason to participate when the opportunity arises.

For the purposes of testing simple activity measurements, test samples, whether internal or external, would be

samples with a radioactive content known to the originator but not to the analyst. If possible, they should be identical in appearance to ordinary run-of-the-mill samples. Where significant sample preparation is involved, which might have a profound effect on the overall accuracy of measurement (and I could include activation analysis methods in this), intercomparison samples might be typical sample materials prepared in such a way that the many sub-samples distributed to analysts will be to all intents and purposes identical in composition. In such cases, the nuclide activity would not necessarily be known in advance.

The intercomparisons organized by the NPL on an annual basis are particularly valuable exercises for the UK. They continue to highlight the fact that so little account is being taken of true coincidence summing (see Chapter 8 and Figures 8.5 and 8.6). It cannot be emphasized too strongly that if your intercomparison results do not agree with expectation, the reasons for the difference must be sought and, if necessary, the analysis method altered appropriately.

The NPL intercomparisons have been limited in their scope in that all of the samples have been aqueous solutions. Participants regularly request that the NPL should provide intercomparison samples with the test nuclides dispersed in other matrices: soils, vegetable matter, etc. While the NPL have been sympathetic to the request, until 2007 (as this edition was in preparation) when a concrete sample was provided, they had not been able to fulfil it. We should not be surprised by this. The difficulties of producing such a material – homogeneous, stable, able to be sub-sampled for distribution and at an acceptable cost to the participants – are considerable.

The NPL intercomparisons are useful because the actual radioactive content of each sample is known. Intercomparison exercises for activation analysis have often involved samples for which the analyte concentration was not known. In such situations, one's own performance must then judged against that of other laboratories by comparison with the mean of all results reported. There are difficulties in this sort of intercomparison in that, in principle at least, it is possible for the majority of laboratories to report incorrect results. There have been cases where, for particular determinands, the organizing body felt unable to suggest an average content because of the wide range and statistical inconsistency of the results reported. (As it happened, had the true activities not been known in the 1989 NPL intercomparison study, the mean result deduced for [134]Cs, for example, would have been very significantly less than the true one.)

15.5.6 Assessment of intercomparison exercises

Traditionally, the u-score is used to determine whether, or not, the results from a laboratory are statistically consistent with the expected results. The u-score is calculated as follows:

$$u\text{-score} = |M - E| / \sqrt{(\sigma_M^2 + \sigma_E^2)} \qquad (15.5)$$

where M is a measured result, E the expected result and the σ's the corresponding standard uncertainties. If a result, together with its uncertainty, is consistent with the known value, the u-score will be less than 1.64. If the u-score is greater than 3.29, one can be sure that the measured result differs significantly from that expected. Values between these two values represent differing degrees of confidence in consistency or inconsistency. Consistency can be taken to mean that the differences between expected and observed result are consistent with the degree of uncertainty on the two values. (An alternative to the u-score is the z-score – a signed u-score. Recent NPL intercomparison assessments have used that, rather than the u-score.)

A weakness of relying solely on the u-score as a measure of performance is that there is ample scope for laboratories to over-estimate their uncertainties to ensure that the true result is within that larger uncertainty. The u-score of an individual measurement on its own does not take into account any bias. Even visual examination of the 2003 NPL intercomparison results reveals at least one laboratory with a consistent positive bias, taken over all nuclides measured, even though the u-scores for that laboratory are low because of the large uncertainties quoted. If laboratories are to judge themselves based on intercomparison exercises, there is need for an alternative scoring which takes into account both uncertainty and bias.

To address this problem, a number of visual display methods have been applied to intercomparison results to emphasize those that are within statistical reason and those not. One such method is the *Naji plot* (Figure 15.6). In this, the z-score is plotted against R_2, the ratio of the measured and expected standard uncertainty of the result – $(\sigma_M/\sigma_E)^2$. In Figure 15.6, the individual analytical results are shown plotted as diamonds. The three curves represent 1, 2 and 3 standard uncertainty limits. Within the inner curve, results can be deemed not significantly different from the expected values. Beyond the outer curve, results are definitely significantly different. Other positions on the plot correspond to other degrees of confidence. The NPL have used the Naji plot and variations on it in their intercomparison reports. Pommé (2206) has suggested an

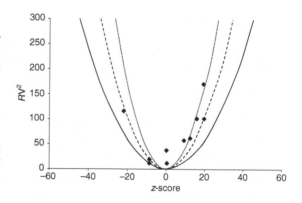

Figure 15.6 Naji plot of intercomparison data. Each marked point represents a radioactivity measurement compared to the expected value: ($\cdots\cdots$) $1\,\mathrm{s}/u = 1.65$; (- - - -) $2\,\mathrm{s}/u = 1.96$; (——) $3\,\mathrm{s}/u = 2.54$; user points

alternative form of plot in which the confidence limits appear as straight lines.

15.6 MAINTAINING RECORDS

I have mentioned already the need for analysis methods to be documented and followed faithfully, but more is needed to assure quality. In QA terms, *'if it is isn't written down, it hasn't been done.'* There is, therefore, a need for a strict procedure for recording the receipt of samples, the analytical process and the results. The reporting system must require the analyst to confirm that procedures have been followed. This is by no means as obvious a task as it might appear. For example, it is not acceptable to have information written into personal laboratory notebooks. Any information about an analysis must be entered into a laboratory notebook which forms part of the whole QA documentation. It may seem pernickety to have to specify the manner in which manual corrections are made to written records but this sort of detail is essential if results are to be interpreted unambiguously in the future. (For the record, the correct procedure is to strike out the old value with a single line so that the original can still be seen, and the new value written above, signed and dated.) I can do no more than draw the reader's attention to these matters and suggest that expert guidance is sought when setting up the minutiae of a QA system.

All of this information must be archived in such a way that the complete analysis process can be retraced and checked at any time in the future, up to the time at which records can be disposed of, which may be as much as 30 years from creation. Again, there are hidden problems here that must be addressed. If, as is likely, all of this information is stored on computer files, the means to access those files must be maintained with the archive. It is very unlikely that in 30 years' time we will be using the same programs, or even the same computer architecture. The archive must be accompanied by a copy of the software needed to access the data and, if the computer system becomes obsolete, a suitable computer.

The individual gamma-ray spectra may need to be archived, along with all of the other information. A 4096 channel spectrum needs a minimum of, say, 17 000 bytes of disk storage – that is, about 60 spectra per megabyte. A busy laboratory will soon fill its available hard disk space and some consideration must be given to the way in which spectra are archived – on floppy disk, on removable hard disk, optical floppy disk, CD-ROM, compressed or non-compressed (or even on old fashioned paper!). A file format, CAM (configuration access method), has been devised for the Canberra systems which contain every piece of information about the measurement configuration (including information about the amplifier, ADC and peripherals, such as sample-changers), the calibration data, sample parameters, geometry parameters, the spectrum data itself, the spectrum analysis configuration and the results of the analysis, and more. These files are considerably larger than the simple spectrum file and, while being valuable from a QA archive point of view, do compound the problems of storage. Again, we must remember the need to maintain the means of reading such files for the full archive period.

Fortunately, the daunting task of setting up a sample-documentation scheme can be eased by resorting to computer programs written for the purpose, so-called LIMS – Laboratory Information Management Systems. There are many of these available, and in use in many commercial analysis laboratories.

15.7 ACCREDITATION

The day is approaching when any laboratory seeking external work will be expected to have some sort of accreditation. Already, many major companies have made it policy not to use external services unless they have appropriate accreditation.

Each developed country has its own accreditation bodies and, in order to promote cross-border recognition of the individual national quality standards, there are mutual recognition agreements between the various bodies. In the UK, the appropriate body is UKAS, the United Kingdom Accreditation Service.

Accreditation of radioactivity measurements poses problems for the analyst, especially if short-lived nuclides are involved because of the need, for documentation

purposes, to specify precise measurement conditions. It is my experience in a neutron activation analysis laboratory, dealing with nuclides with half-lives as short as a few seconds, that it is not always possible to adhere to rigidly defined decay and count periods. In such circumstances, the optimum count conditions depend upon the ratio of the analyte activity to the matrix activity, which may depend upon the presence of many nuclides, the activity of which may not be predictable in advance. The analyst could find himself or herself in the position of having to use non-optimal counting conditions, and thus a lower quality result, in order to comply with the accreditation documentation. A skilled analyst would select suitable conditions on a sample-by-sample basis.

At one time, discussions with UKAS personnel suggested that there was little role for professional judgment in the operation of an accredited method. (There is, of course, a considerable role for the professional in setting up the system in the first place.) As time has passed, it would seem that there is now more understanding of the constraints of radioactivity measurement and UKAS are well used to accreditation of radiometric laboratories.

This is no place for a comprehensive guide to the accreditation process. The proper source for guidance is the accrediting body itself, which will readily provide information. I would urge anyone contemplating accreditation to talk to someone who has already accomplished it. With practical advice from such a source, one might avoid many of the pitfalls along the way. The general process leading to accreditation will be along the following lines:

- Approach the awarding body for initial guidance and literature.
- Basing it upon that advice, write the quality documentation. This could involve the assistance of external consultants. It would also be valuable to include staff consultation in the preparation of documentation. If methods are not already documented, it is possible that those performing the measurements know more about their detail than those charged with the task of documenting them.
- Return to the awarding body to agree the documentation. This is likely to be an iterative process.
- After agreement, train the staff and implement the system in the laboratory.
- Request assessment by the awarding body. If the assessment is satisfactory, a certificate and an accreditation number will be awarded. You will then be entitled to refer to your laboratory as 'XYZ accredited'.

You should be aware that accreditation can be revoked. All accredited laboratories will be re-assessed on a regular, probably yearly, basis, at which time adherence to agreed procedures will be examined and non-compliances reported for correction within a short time scale. These on-going costs should be budgeted for. Serious non-compliance with the agreed quality documentation could result in removal of accreditation.

Although I have not touched upon training, this is an integral part of a quality system. Just as the standards, hardware and software must be suitable for the purpose, so must the 'humanware' – the operators and supervisors. It is essential that that staff should be appropriately trained. This is, if anything, more important during the initial implementation of the quality-assurance programme. It will be necessary to impress upon staff the merits and logic of the new system and to get their support. Resentful staff, resistant to change, are unlikely to exhibit much in the way of quality awareness.

PRACTICAL POINTS

The following points suggest a general strategy for ensuring quality of measurement:

- Even if not seeking formal accreditation it is, in any case, worthwhile setting up a quality system. This might be based upon the ANSI-N42 document.
- Use only evaluated nuclear data.
- Use only traceable standards.
- Set up logbooks for the equipment and keep track of performance by plotting appropriate control charts. As a minimum, charts monitoring the high (say 1332.5 keV) and low (say 122.1 keV) energy resolution and the measured activity of a control sample should be maintained.
- Submit test spectra to the software. This will give some idea of its strengths and weaknesses. These may need to be taken account of in the overall analysis system. Software updates will need to be re-validated.
- In order to ensure accountability, set-up an administrative system that ensures that all samples are measured according to the documented methods. Set up an archive system for these records and the associated spectra.
- Don't be complacent! Test yourself, or your staff, against external criteria. Participate in intercomparisons. Make sure the lessons from such participation are identified and implemented.
- Make sure that staff understand why things are done the way they are, why their rôle is important to the final results, and train them properly.

Bear in mind that quality is not an automatic consequence of the documentation. Consistent, dependable quality comes about from what might be called a 'quality culture' within the laboratory.

FURTHER READING

- A useful introduction to quality assurance issues (even though written with chemical analysis in mind) is given in:

Huber, L. (1993). *Good Laboratory Practice – A Primer*, Publication 12-59091-6259E, Hewlett Packard, Germany.

Garfield, F.M. (1985). Laboratory quality assurance – A rationale for credibility, *Trends Anal. Chem.*, **4**, 162–166.

- The construction of control charts is described in the following:

Moroney, M.J. (1990). *Facts from Figures*, Penguin, London, UK.

Seymour, R., Beal, T., Sergent, F., Clark, W.H.C. and Gleason, G. (1993). Quality control and statistical process control for nuclear analytical measurements, *Radioact. Radiochem.*, **5**, 1–23.

Switsur, R. (1990). Statistical quality control graphs in radiocarbon dating, *Radiocarbon*, **32**, 347–354.

Oakland, J.S. and Followell, R.F. (1990). *Statistical Process Control, a Practical Guide*, Heinemann Newnes, Oxford, UK.

- On test spectra and evaluation of software:

Parr, R.M., Houtermans, H. and Schaerf, J. (1979). The IAEA intercomparison of methods of processing Ge(Li) gamma-ray spectra – preliminary report, in *Computers in Activation Analysis and Gamma-Ray Spectrometry*, IAEA CONF-780421, Carpenter, B.S., D'Agostine, M.D. and Yule, H.P. (Eds), International Atomic Energy Authority, Vienna, Austria.

Zagyvai, P., Parr, R.M. and Nagy, L.G. (1985). Additional results for the 'G-1' IAEA intercomparison of methods for processing Ge(Li) gamma-ray spectra, *J. Radioanal. Nucl. Chem.*, **89**, 589–607.

Sanderson, C.G. (1988). An evaluation of commercial IBM PC software for the analysis of low level environmental gamma-ray spectra, *Environ. Int.*, **14**, 379–384.

Decker, K.M. and Sanderson, C.G. (1992). A re-evaluation of commercial IBM PC software for the analysis of low-level environmental gamma-ray spectra, *Int. J. Radiat. Appl. Instrum.*, **43**, 323–337.

Keyser, R.M. (1990). Using standard spectra to develop and test gamma-ray analysis software, *Nucl. Instr. Meth. Phys. Res. A*, **286**, 409–414.

Koskelo, M.I. and Mercier, M.T. (1990). Verification of gamma spectroscopy programs: a standardized approach, *Nucl. Instr. Meth. Phys. Res., A*, **299**, 318–321.

Seymour, R.S. and Cox, J.E. (1991). HPGe gamma spectroscopy measurement of natural radionuclides in water with a focus on current hardware and software technologies, in *Monitoring Water in the 1990s: Meeting New Challenges*, ASTM STP 1102, J.R. Hall and G.D. Alyson (Eds), American

Society for Testing and Materials, Philadelphia, PA, USA, pp. 96–123.

Blaauw, M., Fernandez, V.O. and Westmeier, W. (1997). IAEA gamma-ray spectra for testing of spectrum analysis software, *Nucl. Instr. Meth. Phys. Res., A*, **387**, 410–415.

Blaauw, M., Fernandez, V.O., van Espen, P, Bernasconi, P.G., Noy, R.C., Dung, H.M. and Molla, N.I. (1997). The 1995 IAEA intercomparison of gamma-ray spectrum analysis software, *Nucl. Instr. Meth. Phys. Res., A*, **387**, 416–432.

Nielsen, S.P. and Pálsson, S.E. (1998). An intercomparison of software for processing Ge γ-ray spectra, *Nucl. Instr. Meth. Phys. Res., A*, **416**, 415–424.

Blaauw, M. (1999). The Reference Peak Areas of the 1995 Test Spectra for Gamma-Ray Spectum Analysis Programs are Absolute and Traceable, *Nucl. Instr. Meth. Phys. Res., A*, **432**, 74–76.

Los Arcos, J.M., Menno Blaauw, M., Fazinic, S. and Kolotov, V.P. (2005). The 2002 IAEA test spectra for low-level γ-ray spectrometry software, *Nucl. Instr. Meth. Phys. Res., A*, **536**, 189–195.

IAEA (1998). *Intercomparison of gamma-ray analysis software packages*, IAEA-TECDOC-1011, International Atomic Energy Authority, Vienna, Austria.

Woods, S.A., Hemingway, J.D., Bowles, N.E. and Makepeace, J.L. (1997). *Standard Gamma-Ray Spectra for the Comparison of Spectral Analysis Software*, NPL REPORT CIRM 2, National Physical Laboratory, Teddington, UK.

Arnold, D., Menno Blaauw, M., Fazinic, S. and Kolotov, V.P. (2005). The 2002 IAEA intercomparison of software for low-level γ-ray spectrometry, *Nucl. Instr. Meth. Phys. Res., A*, **536**, 196–210.

Karhua, P., De Geer, L.-E., McWilliams, E., Plenteda, R. and Werzi, R. (2006). Proficiency test for gamma spectroscopic analysis with a simulated fission product reference spectrum, *Appl. Radiat. Isotopes*, **64**, 1334–1339.

- Test spectrum programs:

De Geer, L.-E. (2005). Currie detection limits in gamma-ray spectroscopy, *Appl. Radiat. Isotopes*, **61**, 151–160.

See below for information on the *SpecMaker* program.

- Procedures for testing and setting up detector systems:

ANSI (1999). *American National Standard for calibration and use of germanium spectrometers for the measurement of gamma-ray emission rates of radionuclides*, ANSI N42. 14–1999, American National Standards Institute, New York, NY, USA (*available at http://webstore.ansi.org*).

Koskelo, M.J. and Schwenn, H.R. (2000). *Verification of gamma spectroscopy programs: N42.14 and beyond*. This paper on using the ANSI N42 tests as a basis for testing the quality of spectrum analysis programs is available at *http:// www.canberra.com/pdf/literature/g2k-paper.pdf*.

Gehrke, R.J. and Davidson, J.R. (2005). Acquisition of quality γ-ray spectra with HPGe spectrometers, *Appl. Radiat. Isotopes*, **62**, 479–499.

- A suggestion for visualization of intercomparison data:

Pommé, S. (2006). An intuitive visualisation of intercomparison results applied to the KCDB, *Appl. Radiat. Isotopes*, **64**, 1158–1162.

- Detailed information on accreditation is best obtained from the national body concerned.

- Intercomparison exercises:

IAEA (2007). *Report on IAEA-CU-2006-03 (spiked soil, water and grass)*, International Atomic Energy Authority, Vienna, Austria (available at *www.iaea.org/programmes/aqcs/ icpt/opt06.pdf*).

Information on recent and ongoing intercomparisons is available at: *http://www.iaea.org/programmes/aqcs/interlab_ studies.shtml*.

Reports on the various NPL Environmental Radioactivity Comparison Exercises can be obtained from the NPL, Teddington, UK and HM Stationery Office, London, UK.

Decker, K.A. (2001). *EML GAMMA SPECTROMETRY DATA EVALUATION PROGRAM*, EML-612, Environmental Measurements Laboratory, US Department of Energy, New York, NY, USA (available at *http://www.eml.st.dhs.gov/ publications/reports/eml602.pdf*).

INTERNET SOURCES OF INFORMATION

- The website of Nuclear Training Services Ltd carries the IAEA TECDOC-619 data and some test spectra:

  – IAEA TECDOC-619 data: *http://www.gammaspectro metry.co.uk/iaea*.
  – Test spectra, in general, and download of IAEA G1 and Sanderson spectra: *http://www.gammaspectrometry. co.uk/testspectra*.
  – The *SpecMaker* program can be obtained free of charge at *http://www.gammaspectrometry.co.uk/specmaker*.

- 1995 IAEA Test Spectra including download: *www.tnw. tudelft.nl* then use search box using 'iaea' as keyword.
- 2002 IAEA Test Spectra including download: *www.tnw. tudelft.nl* then use search box using 'iaea' as keyword.
- ANSI documents can be purchased online at: *http://www.ansi.org/* and at *http://ieeexplore.ieee.org/xpl/ standards.jsp*.

16

Gamma Spectrometry of Naturally Occurring Radioactive Materials (NORM)

16.1 INTRODUCTION

Materials of, what we might call, 'environmental' origin – soils, waters and suchlike – are measured either to determine background levels of radiation or to assess the level of contamination as a consequence of human activity. The nuclides usually measured by gamma spectrometry are the cosmogenic nuclides: ^{40}K, ^{235}U, ^{238}U and ^{232}Th. The uranium isotopes and ^{232}Th will be accompanied by their daughter nuclides. Natural materials containing these nuclides are often referred to by the acronym NORM – Naturally Occurring Radioactive Materials – although, in some contexts, NORM may refer to the nuclides themselves. There are, of course, other naturally occurring nuclides, such as ^{14}C, which are produced continuously by nuclear reactions between high-energy particles with oxygen and nitrogen in the earth's atmosphere. Of those, only ^{7}Be is measurable by gamma spectrometry and I include it below for completeness.

Gamma spectrometry of NORM is difficult for a number of reasons. First, the activity levels are low and, if statistically significant results are to be obtained, need long count periods, ideally on a gamma spectrometer whose construction and location are optimized for low activity measurements (see Chapter 13). The second difficulty is the matter of spectrometer background. Appendix C lists a large number of peaks that one might see in background spectra. Many of these are due to the NORM nuclides in the surroundings of the detector. In addition, there may be evidence of prompt neutron-capture gamma-rays and artefacts such as the annihilation peak and fluorescence X-rays. Depending upon the local environment, there may also be evidence of contamination from neutron-capture and fission-product nuclides, ^{60}Co and ^{137}Cs being fairly common examples. Any activity in the sample itself must be detected on top of all that background activity. In many cases, it will be necessary to make a peaked-background correction in addition to the normal peak background continuum subtraction. All of those difficulties are then compounded by the fact that there are a large number of mutual spectral interferences between the many nuclides in the decay series of uranium and thorium.

I will initially discuss only the measurement of samples that can reasonably be described as 'natural'. Measurement of materials in which the natural abundance or equilibrium of the decay series have been disturbed by chemical intervention, such as reprocessing, decommissioning, industrial or remediation wastes, will be discussed in Chapter 17, Section 17.2.

16.2 THE NORM DECAY SERIES

Uranium and thorium are not stable; they decay mainly by alpha-particle emission to nuclides that themselves are radioactive. Natural uranium is composed of three long-lived isotopes, ^{238}U, a smaller proportion of ^{235}U and an even smaller proportion of ^{234}U, the decay-series daughter of ^{238}U. Natural thorium has one single isotope, ^{232}Th. Each of these nuclides decays to an unstable daughter leading, in turn, to a whole series of nuclides that terminate in one or other of the stable isotopes of lead. Under normal circumstances, in a natural material, the ^{235}U/^{238}U ratio will be fixed and all nuclides in each of the series will be in equilibrium.

Gamma spectrometry of materials containing these nuclides can only be effectively done with a detailed understanding of the decay chains of the nuclides involved. Figures 16.1, 16.2 and 16.3 show the three decay series. These are incomplete in that they do not show a number of minor branches. Those, however, are of no practical significant to gamma spectrometry.

(1)	^{238}U 4.468 × 10^9 y
	↓α
(2)	^{234}Th 24.10 d
	↓β　　　↓β
(3)	^{234m}Pa 1.17 m
	↓IT
	^{234}Pa 6.70 h
	↓β　　　↓β
(4)	^{234}U 2.455 × 10^5 y
	↓α
(5)	^{230}Th 7.538 × 10^4 y
	↓α
(6)	^{226}Ra 1600 y
	↓α
(7)	^{222}Rn 3.8232 d
	↓α
(8)	^{218}Po 3.094 m
	↓α
(9)	^{214}Pb 26.8 m
	↓β
(10)	^{214}Bi 19.9 m
	↓β
(11)	^{214}Po 162.3 μs
	↓α
(12)	^{210}Pb 22.3 y
	↓β
(13)	^{210}Bi 5.013 d
	↓β
(14)	^{210}Po 138.4 d
	↓α
	^{206}Pb **STABLE**

14 decay stages
8 alpha particles
Nuclides underlined are measurable
by gamma spectrometry

Figure 16.1　The uranium decay series – ^{238}U

16.2.1 The uranium series – ^{238}U

The nuclides in this series are listed in Figure 16.1. ^{238}U comprises 99.25 % of natural uranium. That decays by alpha emission to ^{234}Th which in turn decays to ^{234m}Pa and so on until stable ^{206}Pb is reached. If we look at the half-lives of the various nuclides they are all much less than the half-life of ^{238}U. This means that, following the principles discussed in Chapter 1, Section 1.8, in a natural, undisturbed source of uranium, every daughter nuclide will be in secular equilibrium with the ^{238}U. The activity of each daughter nuclide will be equal to the ^{238}U activity. There are 14 radionuclides in the chain and so the total activity of such a source will be 14 times that of the parent, or of any individual nuclide.

There are, of course, cases where the half-life of a particular intermediate parent nuclide is less than that of the daughter (^{234m}Pa/^{234}U, for example). If we were dealing only with ^{234m}Pa, we would expect there to be no radioactive equilibrium. However, we must remember that, for sources older than 10 half-lives of the longest-lived progenitor, the half-life of each component, in this case the ^{234m}Pa, is effectively that of the ^{238}U. In practice, this means that the measurement of any of the nuclides in the decay chain can be taken as an estimate of the ^{238}U activity and of all other nuclides in the chain. In practice, we would measure more than one nuclide.

Not all of the nuclides in the series emit significant gamma radiation and, of those, only the six underlined in Figure 16.1 can be measured with ease. It is common to measure those nuclides and use the results to achieve a 'best estimate' of the parent activity. In doing so, agreement between the early members of the chain, ^{234}Th, ^{234m}Pa and ^{226}Ra, and the later members, ^{214}Pb, ^{214}Bi and ^{210}Pb (but see below), confirms that the series is in decay-equilibrium.

The qualification 'undisturbed' above is important. If the source is treated in any way so that some, or all, of the daughter nuclides are removed, the overall chain equilibrium will be broken.

16.2.2 The actinium series – ^{235}U

^{235}U comprises 0.72 % of natural uranium. Although only a small proportion of the element, its shorter half-life means that, in terms of radiations emitted, its spectro-metric significance is comparable to ^{238}U. The decay series, shown in Figure 16.2, involves 12 nuclides in 11 decay stages and the emission of 7 alpha particles (ignoring a number of minor decay branches).

(1)   <u>^{235}U 7.04 × 10^8 y</u>

          ↓α

(2) ^{231}Th 25.52 h

          ↓β

(3) ^{231}Pa 3.276 × 10^4 y

          ↓α

(4) ^{227}Ac 21.772 y

          ↓β

(5)   <u>^{227}Th 18.718 d</u>
        + α (1.38 %) to ^{223}Fr 22.00 m, then β

          ↓α

(6)   <u>^{223}Ra 11.43 d</u>

          ↓α

(7)   <u>^{219}Rn 3.96 s</u>

          ↓α

(8) ^{215}Po 1.781 ms

          ↓α

(9) ^{211}Pb 36.1 m

          ↓β

(10) ^{211}Bi 2.14 m

          ↓α

(11) ^{207}Tl 4.77 m
        + β (0.273%) ^{211}Po 516 ms then α

          ↓β

    ^{207}Pb STABLE

11 decay stages
7 alpha particles
Only ^{235}U is measurable by gamma
spectrometry

Figure 16.2 The actinium decay series – ^{235}U

Within this series, only ^{235}U itself can readily be measured, although ^{227}Th, ^{223}Ra and ^{219}Rn can be measured with more difficulty. Even though the uncertainties may be high, measurement of the daughter nuclides can provide useful support information confirming the direct ^{235}U measurement or giving insight into the disruption of the decay series.

Unfortunately, the major gamma-ray emitted by ^{235}U, at 185.72 keV, is almost at the same energy as that emitted by ^{226}Ra – 185.99 keV. Resolution of this mutual interference is difficult and will be discussed in full in Section 16.3.5. The total activity of the nuclides within the series is eleven times the ^{235}U activity.

16.2.3 The thorium series – ^{232}Th

Natural thorium is 100 % ^{232}Th. The decay series is shown in Figure 16.3. Six alpha particles are emitted during ten

(1) ^{232}Th 1.405 × 10^{10} y

          ↓α

(2) ^{228}Ra 5.75 y

          ↓β

(3)   <u>^{228}Ac 6.15 h</u>

          ↓β

(4) ^{228}Th 1.9127 y

          ↓α

(5)   <u>^{224}Ra 3.627 d</u>

          ↓α

(6) ^{220}Rn 55.8 s

          ↓α

(7) ^{216}Po 150 ms

          ↓α

(8)   <u>^{212}Pb 10.64 h</u>

          ↓β

(9)   <u>^{212}Bi 60.54 m</u>

      ↓β (64.06%)   ↓α (35.94%)

(10) ^{212}Po 0.300 μs   <u>^{208}Tl 3.060 m</u>

        ↓α   ↓β

    ^{208}Pb STABLE

10 decay stages
6 alpha particles
Nuclides underlined are measurable by gamma
spectrometry

Figure 16.3 The thorium decay series – ^{232}Th

decay stages. Four nuclides can be measured easily by gamma spectrometry: ^{228}Ac, ^{212}Pb, ^{212}Bi and ^{208}Tl. The decay of ^{212}Bi is branched – only 35.94 % of decays produce ^{208}Tl by alpha decay. The beta decay branch produces ^{212}Po that cannot be measured by gamma spectrometry. If a ^{208}Tl measurement is to be used to estimate the thorium activity, it must be divided by 0.3594 to correct for the branching.

16.2.4 Radon loss

All of the decay series above have within them a radon isotope. Radon is a gas. It will normally be trapped within a solid sample but if allowed to escape, for example, by grinding the sample, the equilibrium between the post-radon nuclides, many of which have short half-lives and decay rapidly, will be lost. In principle, this would alter the total activity of the sample and the dose rate from the sample. However, the half-lives of ^{219}Rn, in the actinium series, and ^{220}Rn in the thorium series, are very short and even if radon escapes, equilibrium will be re-established within minutes. That is not the case in the uranium series.

The seventh item in the ^{238}U decay chain is ^{222}Rn, with a half-life of 3.825 d. After loss of ^{222}Rn, there is ample time for the decay of the daughter nuclides preceding ^{210}Pb before re-growth of the ^{222}Rn. If, as is often the case, post-radon nuclides were measured to estimate ^{238}U activity, loss of radon would affect the whole activity measurement process. The solution is simple – encapsulate the sample and wait for about 10 half-lives of the ^{222}Rn to allow equilibrium to be re-established – say one month. Having said that, experience shows that it is, in fact, possible to grind some geological materials without apparent loss of radon. However, that cannot be relied upon. Different materials have different radon-emanating powers, which will depend upon the moisture content and other factors.

16.2.5 Natural disturbance of the decay series

In general, if a material of natural origin is examined the expected equilibrium within the decay series will be found. There are occasions, however, when that is not so. Groundwater passing through rocks can dissolve some of the elements and transport them elsewhere where they may be deposited. For most nuclides, that is not a problem. Those in the water will quickly decay and within the host rock will be quickly re-established. An exception is ^{210}Pb. Its 22.7 year half-life means that ^{210}Pb could be transported from one place to another, leaving a deficit in the host rock and an excess in the groundwater or at some other place where chemical conditions would cause the ^{210}Pb to be deposited or absorbed. For such reasons, it is not wise to rely on measurement of ^{210}Pb alone as an estimate of the ^{238}U activity.

16.3 GAMMA SPECTROMETRY OF THE NORM NUCLIDES

Table 16.1 List all major gamma-rays of the NORM nuclides. The nuclides ^{7}Be and ^{40}K are included for completeness. In the following sections, I have assumed that the activities being measured are low – environmental levels, we might say. The consequences are that the counting samples may be large and will almost always be measured close to the detector and, therefore, it will

be necessary to take account of the natural background to the detector and be aware of the possibility of true coincidence summing. If activities are high and the samples can be measured at some distance from the detector, all of the comments below relating to true coincidence summing can be disregarded, although if high enough, random summing may become a consideration.

I have assumed that my readers will be using a spectrometer with a normal energy range of 30 to, say, 2300 keV. There are many methods reported in the literature for the measurement of uranium isotopes using low-energy detectors. These methods are more applicable to high-activity samples and are beyond the scope I wish to cover in this section.

16.3.1 Measurement of ^{7}Be

^{7}Be is continuously generated in the atmosphere by spallation reactions of charged particles on oxygen and nitrogen. The nuclide is found in the gamma spectra of some natural waters and on environmental air filters. Measurement is simple – the only gamma-ray emitted, 477.60 keV, has no spectral interferences and there is usually no evidence of the nuclide in the background spectrum. The nuclide has a relatively short half-life and is not supported by decay of a parent; therefore, decay corrections will usually be required to the time of sampling.

16.3.2 Measurement of ^{40}K

^{40}K is very evident in background spectra. It is present as 0.17 % of natural potassium and is present in wood and building materials and even in the bodies of the gamma spectrometrists. The substantial presence of ^{40}K in the detector background and in many samples, with its long Compton continuum, severely restricts the limit of detection of the many nuclides emitting gamma-rays at lower energies. The gamma spectrometry of ^{40}K is straightforward but peaked-background correction is always necessary. There is a spectral interference from the 1459.91 keV peak of ^{228}Ac, which must be taken into account even when the activity of the ^{232}Th daughters is low and the peak shape is not noticeably affected.

16.3.3 Gamma spectrometry of the uranium/thorium series nuclides

The three decay series discussed in Section 16.2 contain many nuclides. Not all are measurable by gamma spectrometry but of those that can be, several of them have very complicated decay schemes. As a prime example, Browne and Firestone (see Appendix A) list some hundreds of gamma-rays emitted by [214]Bi, although most are of very low emission probability. Measurement of such nuclides is likely to be affected by true coincidence summing and mutual spectral interference between the nuclides can be expected. The choice of gamma-rays to be measured should be made with care.

For nuclides such as [214]Bi and [228]Ac, nuclide libraries will often contain many gamma-rays. There seems little point in attempting to measure all of these. Many will be of low emission probability and/or subject to summing. Even if those peaks are detected, the value of the results, unless allowance is made for errors, is likely to be low. More of a problem is spurious detection of such small peaks. In *GammaVision*, for example, if a low-intensity gamma-ray is spuriously detected, because the overall mean for the nuclide is not correctly weighted, it can affect the final result considerably. My advice would be to trim the nuclide library to include only, as far as possible, peaks with high emission probability, free from summing. It may be better to accept a lower-intensity peak free of summing, rather than a higher-intensity alternative that would have to be corrected for summing. Within the library for a particular nuclide, it may be necessary to include other peaks in order to help with deconvolution of peaks relating to other nuclides.

The decay scheme of [235]U is such that there are potential true summing possibilities. It has been suggested that the 143.76 and 163.33 keV peaks should not be used because of the possibility of summing with a 19.6 keV gamma-ray – the 143.76 keV summing out, the 163.33 keV summing in. However, it does appear that the emission probability of the 19.6 keV gamma-ray is very low and would, in any case, be absorbed to a large extent by the sample and sample container in many cases. If a p-type detector were used, it would be absorbed completely in the detector cap and dead layer. Summing effects on the gamma-rays normally used to measure [235]U are likely to be small.

Table 16.1 lists all of the nuclides in the uranium and thorium decay series which one might hope to measure. The comments column lists factors which should be taken into account when setting up the analysis and the nuclide library for the task. There are many potential spectral interferences and some of the more relevant ones are listed in Tables 16.2 and 16.3 and are discussed below in Section 16.3.5. The resolution of the major interference

between [226]Ra and [235]U at 186 keV is important enough to discuss separately in Section 16.3.5.

16.3.4 Allowance for natural background

Given reasonable uranium and thorium levels in a sample and a long enough count period, all of the nuclides highlighted in Figures 16.1, 16.2 and 16.3 will be visible in the gamma spectrum. However, it should not be forgotten that the detector environment also contains potassium, uranium and thorium in the constructional materials of the walls and floor of the building. A background spectrum will reveal that most of the gamma-rays you intend to measure will be evident. A peaked-background correction will be necessary. One should look carefully at the algorithms used by the spectrum analysis software. *Gamma Vision*, for example, provides facilities for deducting the background count rate on a peak-by-peak basis but does not take into account the uncertainty on the background-count rate. The result is an unnecessarily high false positive detection rate.

Peaked-background correction of the uranium and thorium daughter nuclide peaks is not as straightforward as might be imagined because of the variability of the background itself. Within the counting room there must be a background level of [222]Rn. The post radon nuclides, [214]Bi and [214]Pb, are daughters of [222]Rn and have short half-lives. If the concentration of radon in a room alters, for example, if the door is left open, we can expect that the contribution of those nuclides to the peaked-background corrections would also be variable. In my opinion, peaked-background corrections should be determined from a number of background measurements taken over a period of time, at different times of day perhaps, so that a more realistic mean correction and its actual, rather than counting, uncertainty can be established. Measures for reducing the radon-daughter background were discussed in Chapter 13, Section 13.4.5.

16.3.5 Resolution of the 186 keV peak

Of particular concern in the gamma spectrometry of NORM is the mutual interference between [235]U (185.72 keV) and [226]Ra (186.21 keV). These peaks are so close together that deconvolution in real environmental spectra is unlikely to give results that one can have confidence in. In principle, it would be possible to perform a peak stripping operation using other peaks in the [235]U spectrum to estimate its contribution to the 186 keV peak. Unfortunately, the emission probability of the next most intense peak at 143.76 keV is only 1/5 of that of the

Table 16.1   Most significant gamma-rays emitted by the NORM nuclides

Nuclide	Source of P_γ data[a]	Half-life[b]	Gamma-ray energy (keV)[c]	Emission probability, P_γ (%)[d]	Comments
[7]Be	LARA	53.22 d	**477.60**	10.44 (4)	—
[40]K	XGAMMA	4.563×10^{11} d	**1460.82**	10.66 (13)	Probable interference from [228]Ac
[235]U Series					
[235]U	LARA	2.571×10^{11} d	**185.72**	57.2 (8)	44.8 % of composite peak – [223]Ra, [226]Ra and [230]Th interference correction needed
	—	—	**143.76**	10.96 (8)	[230]Th interference – correction needed
	—	—	**163.33**	5.08 (7)	—
	—	—	205.31	5.01 (7)	Many interferences – not recommended
[227]Th	DDEP	18.718 d	**235.96**	12.6 (6)	—
	—	—	**256.23**	6.8 (4)	—
[223]Ra	LARA	11.43 d	**269.46**	13.7 (4)	Interference from [228]Ac
[219]Rn	LARA	3.96 s	**271.23**	10.8 (7)	Interference from [228]Ac and [223]Ra
	—	—	**401.81**	6.4 (5)	—
[238]U Series					
[238]U	DDEP	1.632×10^{12} d	49.55	0.0697 (26)	Unusable – serious [227]Th interference
[234]Th	LARA	24.10 d	**63.28**	4.8 (6)	—
	—	—	92.37	2.81 (26)	Measured together.
	—	—	92.79	2.77 (26)	Serious interference from [228]Ac when present. X-ray interferences.
[234m]Pa	GRG	1.17 m	**1001.03**	1.021 (15)	No interferences. Slight summing in possible
	—	—	766.37	0.391 (9)	Interferences from [214]Pb and [211]Pb
	—	—	258.19	0.075 (3)	Serious interference from [214]Pb
[226]Ra	XGAMMA	5.862×10^5 d (1600 y)	**186.21**	3.555 (19)	57.1 % of composite peak – [235]U and [230]Th interference correction needed
[214]Pb	DDEP	26.8 m	**351.93**	35.60 (7)	—
	—	—	**295.22**	18.414 (36)	Insignificant interference from [212]Bi
	—	—	242.00	7.268 (22)	Interference from [224]Ra and deconvolution with 238.63 keV of [212]Pb needed
[214]Bi	DDEP	19.9 m	609.31	45.49 (19)	Subject to TCS
	—	—	**1764.49**	15.31 (5)	—
	—	—	1120.29	14.91 (3)	Subject to TCS

Nuclide	Data source	Half-life	Gamma energy (keV)	Emission probability (%)	Comments
	—	—	1238.11	5.831(14)	Subject to TCS
	—	—	2204.21	4.913 (23)	—
^{210}Pb	LARA	8.14×10^3 d	**46.54**	4.25 (5)	Slight interference from ^{231}Pa
^{232}Th Series					
^{232}Th	LARA	5.13×10^{12} d	63.81	0.27 (2)	Unusable – serious ^{234}Th interference
^{228}Ac	LARA	6.15 h	**911.20**	25.8 (4)	Subject to TCS
	—	—	**968.97**	15.8 (3)	Subject to TCS
	—	—	338.32	11.27 (19)	Subject to TCS – slight interferences from ^{223}Ra and ^{214}Bi
	—	—	964.77	4.99 (9)	Subject to TCS – interference from ^{214}Bi
^{212}Pb	XGAMMA	10.64 h	**238.63**	43.6 (3)	Deconvolution with 242.00 keV of ^{214}Pb needed
	—	—	300.09	3.18 (13)	Subject to TCS – slight interference from ^{231}Pa
^{212}Bi	XGAMMA	60.54 m	727.33	6.74 (12)	Subject to TCS – serious interference from ^{228}Ac
	—	—	**1620.74**	1.51 (3)	—
^{208}Tl	XGAMMA (Corrected for ^{212}Bi branching)	3.060 m	2614.51	99.7 (2)	Subject to TCS
		—	**583.19**	85.0 (5)	Subject to TCS
		—	**860.56**	12.5 (2)	Subject to TCS
		—	510.7	22.6 (2)	Subject to TCS and difficulty resolving from 511 keV annihilation peak

[a] Data sources – see Appendix A for description; values marked GRG are empirical values – see Table 16.4.
[b] All half-lives are taken from LARA.
[c] Energies are rounded to two decimal places.
[d] Emission probabilities are quoted as in the source.

Table 16.2 Corrections for spectral interference in measurement of ^{226}Ra and ^{235}U

Nuclide	Decay series	Isotopic abundance in U$_{nat}$ (%)	Specific activity (Bq/g U$_{nat}$)	Gamma-ray energy (keV)	Gamma-emission probability (%)	Fraction of counts in peak	Standard uncertainty (%)
144 keV peak ratios/interference							
^{235}U	^{235}U	0.720	575.7	143.76	10.96	0.720	1.14
^{230}Th	^{238}U	99.275	12 346	143.87	0.049	0.069	10.24
^{223}Ra	^{235}U	0.720	575.7	144.23	3.22	0.211	2.93
		Correction to ^{235}U value using 144 keV peak:				0.720	1.14
186 keV peak ratios/interference							
^{235}U	^{235}U	0.720	575.7	185.72	57.2	0.428	1.55
^{226}Ra	^{238}U	99.275	12 346	186.21	3.555	0.571	0.86
^{230}Th	^{238}U	99.275	12 346	186.05	0.0088	0.001	10.25
		Factor applied to ^{226}Ra value to correct for interferences:				0.571	0.86
		Factor applied to ^{226}Ra value to give ^{235}U value:				0.02 660	2.16
	Factor applied to ^{226}Ra upper limit to give ^{235}U limit when 186 keV peak not detected:					0.06 215	1.50

185.72 keV peak and itself needs correction for spectral interference. Bearing in mind that the peaks in the spectra are of environmental samples and are often of high uncertainty, any attempt at peak stripping gives ^{226}Ra results of very poor quality.

However, if one can be sure that the ^{226}Ra is in radioactive equilibrium with its parent ^{238}U and that the ^{235}U/^{238}U isotopic ratio is the expected natural value, the counts in the 186 keV peak can be mathematically apportioned between ^{226}Ra and ^{235}U. From the specific activity of ^{235}U and ^{238}U (which can be taken from data tables or calculated from the isotope half-lives) and the gamma-ray emission probabilities, it is straightforward to calculate the proportion of counts in the 186 keV peak due to ^{226}Ra and to ^{235}U. At the same time, it is also possible to make a small correction for an interference due to ^{230}Th. Table 16.3 lists the information needed for this. Corrections for ^{230}Th and ^{223}Ra interferences, which one should apply to results for ^{235}U calculated from the 143.76 keV peak, are also included.

During the spectrum analysis, the 185.72 keV peak should be removed from the ^{235}U entry in the nuclide library. The analysis will then be performed assuming that the entire 186 keV peak is due to ^{226}Ra. We can then use the following correction factors to correct the ^{226}Ra value and derive an additional result for ^{235}U:

Corrected ^{226}Ra = 0.5709 × Apparent ^{226}Ra
Estimated ^{235}U = 0.02662 × Apparent ^{226}Ra

Table 16.3 also includes the relative standard uncertainties of the correction factors that must be taken into account when calculating the overall uncertainties of the results.

If, after peaked-background correction, the 186 keV peak area is not significant, an apparent ^{226}Ra upper limit should be calculated. This limit can then be multiplied by the factor 0.062 24 (the ratio of the gamma-ray emission probabilities of ^{235}U and ^{226}Ra) to give an estimated upper limit for ^{235}U, which will be lower than that achieved by using the minor peaks of ^{235}U.

Unless the ^{226}Ra is in equilibrium with ^{238}U and the ^{235}U/^{238}U ratio natural, this procedure will give misleading results. Measurements on samples from places where the chemical or isotopic composition has been altered need special consideration. This will be discussed in Chapter 17, Section 17.2.1.

16.3.6 Other spectral interferences and summing

Table 16.3 lists the significant spectral interferences in NORM spectra when using the gamma-rays suggested in Table 16.1. The relative count rates are provided to give some 'feel' for the relative magnitude of the interferences. These were calculated for equal masses of uranium (^{238}U + ^{235}U) and thorium and 100 times as much potassium. If there is a large excess of either of these nuclides, some of the interferences may become negligible but, equally likely, other interferences not mentioned here may become significant.

Some of the interferences listed may be resolved automatically by allowing the spectrum analysis program to deconvolute peaks; others may need correction by other means. For example, while deconvolution of the 242.00 keV peak of ^{214}Pb and the 238.63 keV of ^{212}Pb may give satisfactory results for both peaks, it is by no means certain that including the 240.99 keV peak of ^{224}Ra will provide acceptable values for all three peaks.

Just how one decides to make these corrections depends upon how much faith one has in the abilities of one's software. My advice, from a personal perspective, would be to rely on a few easily measured peaks that do not need deconvolution and, if possible, are not subject to true coincidence summing. For example, to measure ^{214}Bi a more accurate result will be obtained by using only the 1764.49 keV peak, even though the 609.31 keV gamma-ray has a much greater emission probability and none of the other major peaks suffer spectral interference, simply because this is the only usable peak not subject to significant summing. Similarly, the non-summed 238.63 keV peak of ^{212}Pb gives a more reliable result than the 300.09 keV peak, and it may be better to use only the non-summed, non-interfered 63.28 keV peak of ^{234}Th rather than risk deconvolution problems with the doublet 92.37 + 92.79 keV peak.

It is unfortunate that none of the peaks of ^{228}Ac or of ^{208}Tl are free of summing although there is scope for electing to measure only the one or two least-summed peaks. The major gamma-ray of ^{208}Tl, 2614.51 keV, is a consequence of a final transition to the ground state. All other nuclear de-excitations pass through that level and, inevitably, the summing errors on the measurement of that gamma-ray are very large. There would be little point in extending the energy scale of one's gamma spectrometer in order to measure the 2614.51 keV gamma-ray, in spite of its high emission probability. Another problem peak in the gamma spectrometry of ^{208}Tl is that at 510.7 keV. This would need deconvolution from the 511.0 keV annihilation peak present in the background. Because of the *Doppler broadening* of the annihilation peak, the spectrum analysis programs available are unlikely to be able to do that reliably and the peak is best ignored.

Table 16.3 Spectral Interferences in the gamma spectrometry of NORM nuclides[a]

Nuclide	Nuclide in peak	Gamma-ray energy (keV)		Probability of gamma emission (%)	Decay series	Proportion in peak (%)
^{235}U Series						
^{235}U	^{235}U	205.31		5.01	^{235}U	90.3
	^{227}Th	204.98		0.16	^{235}U	2.9
	^{228}Th	205.99		0.0185	^{232}Th	2.3
	^{227}Th	206.08		0.25	^{235}U	4.5
^{223}Ra	^{223}Ra	269.46		13.7	^{235}U	28.0
	^{228}Ac	270.24		3.46	^{232}Th	49.9
	^{219}Rn	271.23		10.8	^{232}Th	22.1
^{219}Rn	^{219}Rn	271.23	See ^{223}Ra above	—	—	—
^{238}U Series						
^{238}U	^{238}U	49.55		0.0697	^{238}U	15.4
	^{227}Th	50.13		8.2	^{235}U	84.6
^{234}Th	^{234}Th	63.28	See ^{232}Th below	—	—	—
^{234m}Pa	^{234m}Pa	766.37		0.391	^{238}U	82.7
	^{214}Pb	765.96		0.053	^{238}U	11.2
	^{211}Pb	766.51		0.617	^{235}U	6.1
^{234m}Pa	^{234m}Pa	258.19		0.0754	^{238}U	12.2
	^{228}Ac	257.52		0.030	^{232}Th	1.6
	^{214}Pb	258.87		0.5318	^{238}U	86.2
^{214}Pb	^{214}Pb	242.00		7.268	^{235}U	31.7
	^{212}Pb	238.63		43.6	^{232}Th	62.4
	^{224}Ra	240.99		4.12	^{238}U	5.9
^{210}Pb	^{210}Pb	46.54		4.25	^{238}U	99.8
	^{231}Pa	46.35		0.223	^{235}U	0.2
^{232}Th Series						
^{232}Th	^{232}Th	63.81		0.27	^{232}Th	1.8
	^{234}Th	63.28		4.8	^{235}U	98.2
^{228}Ac	^{228}Ac	338.32		11.27	^{232}Th	96.6
	^{223}Ra	338.28		2.79	^{238}U	3.4
^{228}Ac	^{228}Ac	964.77		4.99	^{232}Th	81.9
	^{214}Bi	964.08		0.363	^{238}U	18.1
^{212}Pb	^{212}Pb	238.63	See ^{214}Pb above	—	—	—
^{212}Pb	^{212}Pb	300.09		3.18	^{232}Th	82.8
	^{227}Th	300.50		0.014	^{235}U	0.1
	^{231}Pa	300.07		2.47	^{235}U	9.1
	^{227}Th	299.98		2.16	^{235}U	8.0
^{212}Bi	^{212}Bi	727.33		6.74	^{232}Th	91.6
	^{228}Ac	726.86		0.62	^{232}Th	8.4
^{208}Tl	^{208}Tl	510.7		22.6	^{232}Th	98.0
	^{228}Ac	508.96		0.45	^{232}Th	2.0
	Annihilation	511.00		—	—	Variable
^{40}K	^{40}K	1460.82		10.66	—	94.8
	^{228}Ac	1459.14		0.83	^{232}Th	5.2

[a] Relative peak areas are calculated assuming equal masses of uranium (^{235}U + ^{238}U) and thorium and 100 times as much potassium. Other interferences in the measurement of ^{226}Ra and ^{235}U are listed in Table 16.2.

16.4 NUCLEAR DATA OF THE NORM NUCLIDES

In spite of their ubiquity, the quality of the gamma-ray emission probability data for the NORM nuclides has been poorer than for most other commonly measured nuclides. For example, the *Radiochemical Manual*, taking data from the UKHEDD database, quotes uncertainties on the gamma-emission probabilities of the gamma-rays of ^{228}Ac that, with one exception, are 15 % of the emission probability itself (the exception is the 911.20 keV peak). Compare this to less than 2 % for most oft-measured nuclides. Recently Morel *et al.* (2004) have provided improved emission-probability data for ^{226}Ra and daughters, which are in accord, although not identical to, those in the recently evaluated data, XGAMMA, published by the IAEA (see Appendix A). I have found that using this data provides better agreement between ^{226}Ra and ^{214}Bi/Pb in an equilibrium situation and, when reference material are analysed, better agreement with expected activities than older data.

Of particular concern is the gamma-ray emission probability data for ^{234m}Pa. When using commonly accepted emission-probability data, it is not uncommon for gamma spectrometrists, including the author, to observe that measured values for ^{234m}Pa are noticeably higher than those for ^{234}Th – nuclides that are unlikely not to be in secular equilibrium. This implies that the ^{234m}Pa peak area in the spectrum is higher than expected. Spectral interference with other NORM nuclides does not appear to be responsible, nor does summing in. ^{234m}Pa has a complicated decay scheme and summing in to the 1001.03 keV peak normally used for analysis is possible, but not to the extent needed to explain the observed inconsistencies. Measurements of reference materials using the 'accepted' emission probability consistently give results for ^{234m}Pa higher than expected, even when ^{234}Th results measured in the same spectrum are as expected.

More than one analyst has suspected that the nuclear data are in error. Table 16.4 shows a number of selected values for the emission probability of the 1001.03 keV gamma-ray. There is a range of values from which to choose. The empirical value attributed to Gilmore, I derived from measurements of ^{234}Th and ^{234m}Pa in a large number of samples, many of them reference materials. The quoted value gave the best overall agreement between the nuclide activities. Unfortunately, the IAEA data evaluation, as it must, supports the literature values. This is an issue that deserves to be resolved once and for all.

Table 16.4 Emission-probability values for the 1001.03 keV gamma-ray of ^{234m}Pa

Probability (%)	Source of data
0.589	ICRP 38 – discredited value
0.832 ± 0.010	IAEA Evaluation XGAMMA 2007
0.835 ± 0.011	*Radiochemical Manual* (UKHEDD 2.2) and Adsley *et al.* (1998)
0.837 ± 0.010	ENSDF 2002 (Table of Isotopes)
0.839 ± 0.012	LARA (1999)
0.91 ± 0.05	Sutton *et al.* (1993)
0.92 ± 0.02	Anilkumar (1999)
1.021 ± 0.082	G.R. Gilmore (empirical, 2003)

16.5 MEASUREMENT OF CHEMICALLY MODIFIED NORM

Geological materials, which almost always contain small amounts of uranium and thorium, are encountered in various industrial situations. In some cases, the geological material is feedstock for a chemical industry; nuclear fuel production is an obvious example, phosphate extraction and manufacture of gypsum are others. In further cases, the material encountered is an inconvenient accompaniment to the product itself: coal ash contains a higher NORM concentration than the coal it came from; oil- and gas-field sludges contain radiologically significant NORM.

If the chemical processes involved result in separation of uranium or thorium from their daughter nuclides, equilibrium within their decay chains will be lost. Some of the daughter nuclides may end up in waste streams, others in the product. For example, commercial sources of cerium oxide can contain significant (and possibly radiologically embarrassing) thorium impurity levels that have followed the cerium from ore to product. The manufacture of laboratory-reagent uranium compounds separates the uranium isotopes from their daughters, profoundly affecting the expected nuclide activities. Even more difficult are materials originating from nuclear fuel production where the isotopic composition may also have been altered.

Even when an industrial site is abandoned, the problem may remain as contamination of the site itself. Remediation of such sites may lead to the measurement of soil samples, for example, to determine the appropriate method of disposal. Measurements of any materials from an industrial environment must be considered on a case-by-case basis to establish how the various decay-series equilibriums have been disrupted. One should not forget that even if samples are of natural origin, chemical fractionation of particular elements is possible – in particular, enhancement of ^{226}Ra and ^{210}Pb activities. This section will discuss a number of cases by way of example. Samples originating from nuclear sites will be discussed in Chapter 17.

16.5.1 Measurement of separated uranium

Consider the ^{238}U decay series (Figure 16.1). If uranium is chemically separated from a uranium containing feedstock, ^{235}U, ^{238}U and ^{234}U will all be selectively concentrated. Ideally, all of the daughter nuclides will be taken into other process streams. Most of the daughter nuclides have short half-lives and will decay rapidly. Some, ^{230}Th, ^{210}Pb and daughters have longer half-lives and can cause disposal problems in the plant. For example, in certain processes ^{210}Pb (and its alpha-active daughter, ^{210}Po) may be found absorbed in all sorts of unexpected locations, such as chimney dusts.

Within the separated uranium fraction, the ^{234}Th and $^{234(m)}$Pa nuclides will re-grow to their secular equilibrium values relatively quickly but, once separated, ^{230}Th, the long lived daughter of ^{234}U, is, to all intents and purposes, lost forever. That means that chemically separated samples of uranium will only contain nuclides within the first four steps of the chain. Nuclides such as ^{226}Ra, ^{214}Bi and ^{214}Pb, which are often measured to estimate ^{238}U activities, will be completely absent. For this reason, one cannot expect the gamma-ray spectrum of laboratory reagent uranium salts to be the same as natural uranium.

16.5.2 Measurement of separated thorium

The situation is particularly complicated for the ^{232}Th decay chain (Figure 16.3). Under a normal equilibrium situation one would find that the easily measurable nuclides, ^{228}Ac, ^{212}Pb, ^{212}Bi and ^{208}Tl, could all be measured and their activities taken as an estimate of the ^{232}Th activity (bearing in mind that the ^{208}Tl activity would have to be corrected for branching in the decay of ^{212}Bi). However, after chemical separation of thorium, none of these nuclides will necessarily have an activity equal to that of the ^{232}Th.

Imagine that we have chemically isolated ^{232}Th and ^{228}Th and that all of the daughter nuclides have been lost to waste streams. Looking at the daughters of ^{232}Th, we see that, because of its 5.75 year half-life, it will take many years for the ^{228}Ra to grow back to its pre-separation

equilibrium value. ^{228}Ac, with a much shorter half-life, will be in secular equilibrium with the ^{228}Ra and will also re-grow on a 5.75 year half-life time-scale.

Considering the ^{228}Th daughters, the ^{224}Ra activity will grow back to equilibrium within a month or so, but then will follow the 1.913 year half-life decay of the ^{228}Th. However, the situation is complicated by the fact that, after initially falling, the ^{228}Th activity will eventually start to increase again as the ^{228}Ra daughter of ^{232}Th, and the parent of ^{228}Th, grows in with a half-life of 5.75 years (see Figures 16.4 and 16.5). In fact, it takes some 40 years to re-establish equilibrium throughout the whole series.

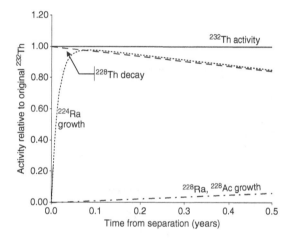

Figure 16.4 The decay of thorium isotopes and the growth of daughters in separated thorium over a short time-scale (0.5 years)

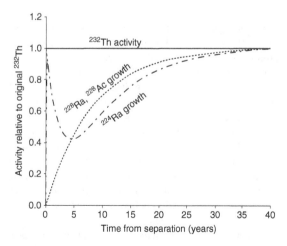

Figure 16.5 The growth of daughter nuclides in separated thorium over a 40 year time-scale

In principle, if the time of separation of the thorium isotopes from their daughters is known, the activity of all members of the series could be calculated from the measured activities of ^{228}Ac, ^{212}Pb, ^{212}Bi and ^{208}Tl. In practice, the time of separation is seldom known. Is it possible to glean any reliable information about the activity of the ^{232}Th itself and the total activity within a sample?

The ratio of pre-^{228}Th activities to post-^{228}Th activities will vary with time. This offers hope of using their ratio to estimate the time from separation. This is fine if the ratio is less than one. (Figure 16.6) If the ratio is one or more, there is ambiguity: for example, if the ratio is 1.05,

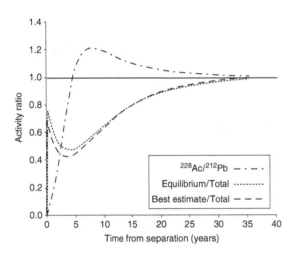

Figure 16.6 The ratio of early decay series nuclides (e.g. ^{228}Ac) relative to late-series nuclides (e.g. ^{212}Pb) in separated thorium as a function of time

the growth period could be close to 5 years or close to 30 years. However, modelling of the growth of the thorium isotope daughters reveals that we can make approximations. Even if only one post-^{228}Th nuclide is measured, say ^{212}Pb, and that activity is multiplied by 10 (the number of active nuclides in the thorium series) the result will always be within 25 % of the correct total activity as long as the time from separation is more than one month. Furthermore, after one year, such an estimate can be guaranteed to be no more than 15 % below the true value. While this seems unattractive from a scientific perspective, from the point of view of waste disposal, where large uncertainties on the activity of certain nuclides is acceptable, such an estimate might be satisfactory.

When the ^{228}Ac/^{212}Pb activity ratio is greater than one, a better estimate, identified as the 'Best Estimate' in Figures 16.6, can be calculated as:

Total series activity $= \left(^{228}\text{Ac} \times 7 + {}^{212}\text{Pb activity} \times 3\right)$

$$\times 1.052 \qquad (16.1)$$

where ^{228}Ac is taken to represent the first seven nuclides in the series and ^{212}Pb the last three. The empirical factor is included to minimise the errors. Using this equation the error on the total activity estimate lies between -10% and $+5\%$ (Figure 16.7)

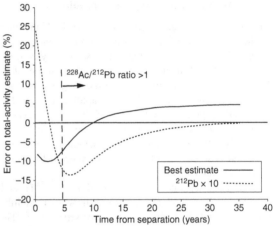

Figure 16.7 Relative error of total activity estimates based on nuclide ratio and the ^{212}Pb × 10 estimate as a function of time

Gamma spectrometry measurements cannot, however, give a direct estimate of the ^{232}Th activity unless equilibrium is assured or the sample history is known.

16.5.3 'Non-natural' thorium

It could be argued that no thorium is 'natural' – that is, 100 % ^{232}Th. All thorium must originate from thorium-containing geological material and all geological materials contain uranium to a greater or lesser extent. Within the ^{238}U decay series is ^{230}Th with a half-life of 7.538×10^4 y. When thorium is chemically separated from the host ore, the ^{230}Th will accompany the ^{232}Th. There must always be at least a small amount of ^{230}Th within the separated thorium. The ratio of half-lives is such that only 5 parts-per-million of uranium accompanying the thorium in the ore will result in an amount of ^{230}Th similar to the ^{232}Th activity. A common source of thorium is monazite ores, which have low uranium contents, but many other higher-grade ores contain significant concentrations of uranium.

Table 16.5   Gamma spectrometry analysis of a sample of gypsum

Nuclide	Container			
	Polystyrene/polyethene (33 days after sealing)		Glass (42 days after sealing)	
	Activity (Bq g^{-1})	±1 standard uncertainty	Activity (Bq g^{-1})	±1 standard uncertainty
^{238}U series nuclides				
^{234}Th	< 0.0097	—	< 0.075	—
^{234m}Pa	< 0.27	—	< 0.19	—
^{226}Ra	0.566	± 0.032	0.557	± 0.034
^{214}Pb	0.489	± 0.017	0.519	± 0.017
^{214}Bi	0.481	± 0.023	0.541	± 0.027
^{235}U series nuclides				
^{235}U	< 0.0076	—	< 0.011	—
^{227}Th	0.0101	± 0.0039	0.0143	± 0.0040
^{223}Ra	0.012	± 0.0036	0.0104	± 0.0050
^{219}Rn	0.0171	± 0.0040	0.0106	± 0.0060
^{232}Th series nuclides				
^{228}Ac	0.0291	± 0.0033	0.0266	± 0.0040
^{224}Ra	< 0.021	—	< 0.044	—
^{212}Pb	0.0054	± 0.0011	0.0084	± 0.0020
^{212}Bi	< 0.0090	—	< 0.017	—
^{208}Tl[a]	0.0052	± 0.0018	0.0045	± 0.0031

[a] The ^{208}Tl value was divided by the branching ratio (0.3594) to give a ^{232}Th equivalent.

Radiological-hazard assessments for thorium generally ignore ^{230}Th. Nevertheless, in some cases, it could pose a radiological problem greater than the thorium itself. ^{230}Th does emit gamma-rays but only with low probability, so that measurement by gamma spectrometry is difficult, but the alpha-particle emission is significant.

16.5.4 Measurement of gypsum – a cautionary tale

Building materials, bricks, concrete, plaster, etc. all contain trace amounts of uranium and thorium, together with their daughters – including the gaseous radon isotopes. Because of this, all buildings will contain a background level of radon within them. Internally, much use is made of plaster boarding. This is made of gypsum, a hydrated calcium sulfate. It is prepared by chemical treatment of limestone, which will inevitably contain some uranium and thorium impurities. During the chemical process, radium, which is chemically similar to calcium, will be incorporated into the gypsum and hence into the boarding. Within the boarding, radon isotopes, most importantly ^{222}Rn, will re-grow to equilibrium and will diffuse out into the building. Because of this, the levels of ^{226}Ra in gypsum are regulated; hence the need for gamma spectrometry.

What happens to those uranium and thorium impurities, and their daughters, during production? Gamma spectrometry of a typical commercial material, three months after manufacture and four weeks after preparation of the counting sample, gave the results shown in Table 16.5. The immediate observation is that the equilibriums of all of the decay series have been disturbed, but not in an obvious manner.

Because of their chemical similarity, we can safely assume that the radium isotopes in the raw material will follow the calcium, and be found in the finished gypsum. However, we cannot be sure of the fate of the uranium and thorium impurities. We can make the following observations:

- The absence of ^{234}Th (from ^{238}U) and ^{235}U shows that the uranium isotopes have not followed the radium into the gypsum to any significant extent.
- At first sight, the presence of ^{227}Th suggests that the thorium isotopes have been taken into the gypsum. We would not expect to detect ^{227}Th because of its relatively short half-life. However, if we could, we would also expect to detect ^{234}Th. Its absence suggests that the ^{227}Th is supported by its 21.77 year ^{227}Ac parent.

- The measured activities of ^{227}Th, ^{223}Ra and ^{219}Rn are similar, suggesting that the latter part of the ^{235}U decay series is in equilibrium.
- ^{228}Ac, within the ^{232}Th decay series, has been detected. Because of its short half-life, it must be in secular equilibrium with ^{228}Ra. Its activity is much greater than its daughters, as would be expected if we assume that on separation they all decayed and are now re-growing, as described in Section 16.5.2.
- Note that, because we cannot be sure that all, or indeed any, of the thorium isotopes have ended up in the gypsum, Equation (16.1) will not be relevant in this case.
- The activities of ^{214}Bi and ^{214}Pb were similar to each other but significantly lower than the parent ^{226}Ra. The counting sample had been sealed long enough for us to expect radioactive equilibrium between ^{226}Ra, ^{220}Rn and the daughters. The absence of a significant amount of ^{238}U means that the difference cannot be due to a contribution from ^{235}U to the 186 keV peak.
- Re-measurement of the material, 42 days after sealing in a glass container, gave consistent results between ^{226}Ra and the daughters. Apparently, radon gas was escaping from the plastic counting container.

The final point indicates a particular problem with gypsum. It would appear that, unlike many natural geological materials, radon is easily lost from the gypsum (it can be said to have a high emanating power) and, unfortunately, this can diffuse through plastic containers. Radon was found to escape from polystyrene containers with glued-on polyethene caps and from polypropylene containers. Without further investigation, I can only suggest that glass or metal counting containers would seem to be a safer option.

In the example shown in Table 16.5, the interpretation of the results was aided by the (apparent) absence of uranium. We could say quite reasonably that the 186 keV peak in the gamma spectrum was all due to ^{226}Ra. It is likely that in some cases at least some of the uranium from the raw material could be present in the gypsum. The interpretation then becomes more difficult and one must then correct the ^{226}Ra result for the presence of ^{235}U.

16.5.5 General observations

It is an unfortunate fact that, within the natural decay chains, the nuclides most easily measured by gamma-ray spectrometry are in the later decay chain stages – beyond the point at which equilibrium might have been disturbed.

A reasonable approach one might take is, wherever possible, to measure at least one nuclide before the potential break and at least one afterwards. If post- and pre-break activity-measurements agree, then one can deduce that equilibrium is established and a weighted mean of all of the measurements can be calculated and the result applied to the whole chain. If the results are not consistent (within the uncertainties of the measurements), then one must deduce that the source is not in radioactive equilibrium, and make what use one can of the data.

For ^{238}U, the situation is relatively simple. If ^{214}Bi/Pb cannot be detected, the sample must be separated uranium. If ^{214}Bi/Pb is detected, but the results disagree with, say, ^{234}Th or ^{226}Ra measurements, then equilibrium is not established and the sample can be left sealed for 30 days or so before re-measurement. As a matter of routine, samples to be measured for ^{238}U by gamma spectrometry should be kept sealed after preparation.

The measurement of ^{235}U is not easy because of the difficulties with interference from ^{226}Ra. Measurement of other nuclides in the actinium series will not provide particularly high quality measurements but may be of value in supporting the direct measurement.

In the case of the ^{232}Th series, ^{228}Ac can be determined before the chemical break and ^{212}Pb afterwards. If these nuclides provide inconsistent results, one can be sure that the source is not in radioactive equilibrium. Unfortunately, the converse is not true because of the complicated growth/decay situation that can result in ambiguous results. Unless the history of a chemically separated sample of thorium is known, gamma-ray spectrometry will not give a complete analysis of the complete decay chain. However, as noted in Section 16.5.2, some general information may be gleaned.

FURTHER READING

- On the ^{234m}Pa emission probability:

Adsley, I, Backhouse, J.S., Nichols, A.L. and Toole, J. (1998). U-238 Decay Chain: Resolution of Observed Anomalies in the Measured Secular Equilibrium Between Th-234 and Daughter Pa-234m, *Appl. Radiat. Isotopes*, **49**, 1337–1344.

Sutton, G.A., Napier, S.T., John, M. and Taylor, A. (1993). Uranium-238 decay chain data, *Sci. Total Environ.*, **130/141**, 393–401.

Anilkumar, S., Krishnan, N. and Abani, M.C. (1999). Application of fundamental parameter method for investigation of the branching intensity of 1001 keV gamma energy of ^{234m}Pa, *Appl. Radiat. Isotopes*, **51**, 725–728.

- The recently published evaluated IAEA data (2007) can be accessed at *http://www-nds.iaea.org/xgamma_standards/*.

17

Applications

There are very many applications of gamma-ray spectrometry. Having already discussed environmental measurements in Chapter 16, in this chapter I discuss a number which, to me, seem to be of particular interest. Each of them deserves a much broader treatment but time and space limit me to a general introduction. Each of them draws upon the principles developed in previous chapters, the idea being to illustrate how academic, and perhaps theoretical, ideas find their expression in practical uses. As it happens, the examples are related in the sense that in these applications gamma spectrometry could be said to help make ordinary life safer; measurements in support of the Comprehensive Test Ban Treaty (CTBT) help to prevent the proliferation of nuclear weapons, waste monitoring helps to ensure that radioactive waste is disposed of properly and safeguards measurements make sure that nuclear material is properly accounted for.

17.1 GAMMA SPECTROMETRY AND THE CTBT

17.1.1 Background

It could be argued that measurements in support of the Comprehensive Test Ban Treaty (CTBT) are the most important application of gamma spectrometry on the world stage. In September 1996, the General Assembly of the United Nations agreed by 158 votes to 3 to adopt the CTBT. This prohibits all nuclear-weapon tests of any yield in all places for all time. The treaty is the culmination of over forty years of effort, and after more than 2000 nuclear test explosions. It was signed within days by 71 states, including the five nuclear-weapon states: China, France, Russia, the United Kingdom and the United States. However, before the treaty comes into force, the rules are that all 44 'nuclear-capable' states – countries

that in 1996 possessed nuclear research or power reactors – have to both sign and ratify the treaty. At present, 41 of those have signed the treaty, the 'missing' three being India, Pakistan and North Korea, but not all of those who have signed have ratified it – Israel, Iran, China and, most regrettable, the first country to sign the treaty, the USA. Until such time as the treaty comes into effect, the UN have established a Preparatory Commission for the Comprehensive Nuclear Test Ban Treaty Organization (CTBTO); this operates from Vienna. The main task of the Preparatory Commission is to ensure that a global verification regime is operational by the time the CTBT comes into force. Once ratification is completed, CTBTO will report to the UN and member states if it believes an explosion has taken place. These notes are largely taken from CTBTO documents.

17.1.2 The global verification regime

The International Monitoring System (IMS) is a network of monitoring sensors, which search for, detect and provide evidence of, possible nuclear explosions for verification of non-compliance to the treaty. There are 321 monitoring stations, spread uniformly around the globe, many necessarily in remote areas. Detonation of a nuclear device releases huge amounts of energy, which interacts with the environment to propagate sound vibrations through solid earth, oceans and the atmosphere. Radioactive nuclear products are also created which may well leak into the atmosphere. The task of the monitoring stations is to detect these effects. The following types of monitor have been deployed:

- Seismographs, which can distinguish between an underground nuclear explosion and the many earthquakes that occur on a daily basis (on average, 100 per day) around the globe. There are 50 primary stations of this type and 120 auxiliary stations.

- Hydroacoustic stations, to measure acoustic waves in the oceans. Similar to seismic stations, examination of the data can separate signals from natural events, such as submarine volcanoes and earthquakes and man-made underwater explosions.
- Infrasound stations (60) detect the very low frequency sound that can be detected in the atmosphere using microbarometers (acoustic pressure sensors).
- For the measurement of radioactive releases, 80 stations monitor air particulates on a daily basis – of those, 40 also measure radioactive noble gases directly in the atmosphere.

Gamma spectrometry has the advantage of being uniquely able to identify and quantify the products of a nuclear weapon test without the possible ambiguities of interpreting vibration and sound.

17.1.3 Nuclides released in a nuclear explosion

The radioactive nuclides produced by detonation of a nuclear weapon are many and varied. In addition to the fission product nuclides, which we would expect, there are many nuclides created by activation of the components of the device itself and residues of the fuel. In principle, detect these nuclides and you have your 'smoking gun'. However, many of the nuclides created in a nuclear explosion could also be released by other, regular processes. The serious potential international repercussions of falsely claiming a test has occurred are such that in monitoring these nuclides care has to be taken that false positives are avoided and that only nuclides with real relevance are taken into account. Lars-Erik de Geer of the CTBTO (De Geer (1999)) examined the nuclides released by a nuclear explosion, according to the following sources:

- Residues of fuel materials – ^{3}H, ^{6}Li, ^{7}Li, ^{232}U, ^{233}U, ^{234}U, ^{235}U, ^{238}U, ^{239}Pu, ^{240}Pu, ^{241}Pu and ^{241}Am.
- Non-fission product reactions of fuel materials – transuranics caused by neutron capture and subsequent decay within the fuel.
- Fission products – very many!
- Activation products:
  - of non-fuel materials, i.e. of the constructional materials;
  - in the rocks surrounding an underground explosion;
  - in the ground below a near-surface explosion;
  - in sea water around or near a sea or sea–surface explosion;

- in the air around an atmospheric explosion;
- of neutron fluence detectors incorporated into the device – yttrium, gold and iridium;
- of tracers added to the device for research purposes.

In principle, measurement of the pattern of nuclides released can provide information about the type of nuclear device and whether its detonation was above or below ground. The CTBTO (1999) has produced a list of 47 relevant fission products, of which 28 have been detected in past monitoring exercises, and 45 relevant non-fission product nuclides, of which 17 have been previously detected. In this context, 'relevant' means that the nuclide would be produced in sufficient amount and its half-life is long enough to give a reasonable chance of detection some days after the test.

The task of the IMS radionuclide stations is to monitor these nuclides and deliver a justifiable report to the IDC (the International Data Centre) in Vienna on a daily basis as to whether a nuclear test has occurred or not. The aim is to detect any explosion greater than 1 kt (kiloton, non-SI, but the conventional unit used in this context), which is taken to be a 1.4×10^{23} fissions per event (1 kt is 1/10 of the size of the bomb dropped on Hiroshima in 1945). Figure 17.1 shows the activities of 23 of the 47 relevant fission products released by such an explosion after 5, 10 and 30 days.

The activities are very large. Most illicit nuclear tests are likely to be underground and only a fraction of the radioactive products will find their way to the atmosphere and a very much smaller fraction to measurement stations. Models used to design the measurement systems assume that only 10 % of the 1.5×10^{16} atoms of ^{133}Xe created by a 1 kt ^{235}U based explosion, and 50 % of the 5.5×10^{15} atoms of ^{140}Ba escape to the atmosphere. Dilution factors of about 10^{18} and 10^{20}, respectively, mean that the activities expected at a monitoring station are very small.

A couple of features stand out in Figure 17.1. First, is the fact that some of the nuclides considered have relatively short half-lives, and unless detection is achieved promptly, their usefulness declines. Secondly, the low activity yield of ^{134}Cs compared to ^{136}Cs and ^{137}Cs is worthy of note. This is because ^{134}Cs is shielded from the $A = 134$ isobaric precursors by the stable ^{134}Xe. The yield shown is the direct fission yield, a factor of 8000 lower than the chain yield leading to ^{137}Cs. There is a similar situation with regard to ^{136}Cs, which is shielded by ^{136}Xe. However, the half-life of ^{136}Cs is so short (13.16 d) that the activity created by fission is only slightly lower than that of ^{137}Cs.

Anybody active in measuring post-Chernobyl air filters, and suchlike, will remember that ^{134}Cs was very

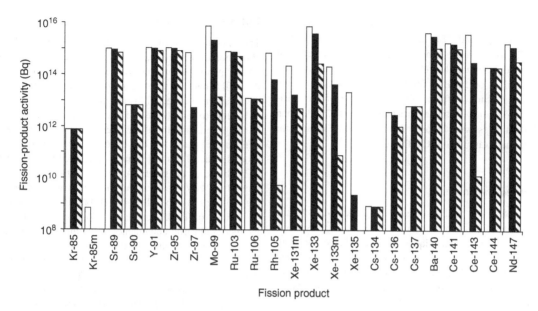

Figure 17.1   Activity of 24 fission product nuclides, created by a 1 kt nuclear explosion, after 5 (□) 10 (■) and 30 (▨) days. Data is taken from Comley (2001)

obvious in their spectra. That was because of activation of stable ^{133}Cs. This is at the end of the high yield $A = 133$ isobar and, as long as fission continues, increases in amount. It can then accept thermal neutrons to build up ^{134}Cs on a time-scale related to its half-life of 2.065 years. Fuel in a nuclear reactor has time to build up its ^{134}Cs – nuclear weapon fuel does not. Significant ^{134}Cs on an air filter would allow the IMS to assign a particular release to a reactor accident (declared or not), rather than a weapon test. Because reactor fission is largely thermal and bomb fission involves fast neutrons, the relative amounts of fission products at different points on the fission-yield distribution (see Chapter 1, Figure 1.29) can also distinguish between different types of release.

High demands are placed on reliability of the monitoring systems. They are, as far as possible, automatic in operation. They must have no more than seven days in a row down time and no more than 15 days in an entire year and must be providing data 95 % of the time. Certified laboratories must commit themselves to regular proficiency tests and the analysis of 'round-robin' exercises.

17.1.4 Measuring the radionuclides

The IMS network has 80 radionuclide stations that collect radioactive particulates on air filters. In addition, 40 of these also measure radioactive noble gases directly. There are also 16 laboratories around the world that are certified

to perform detailed examination of samples collected by the monitoring station. In the event of a Level 5 alert, the highest, in which more than one abnormal anthropogenic treaty relevant nuclides are detected, the analysis would be reviewed by one or other of these stations for confirmation.

Radioactive particulate monitoring

Taking into account dilution of the volatilized fission products before reaching it, the monitoring station has to have the capability of measuring down to $10 \, \mu Bq \, m^{-3}$ of ^{140}Ba. The measurement system is called RASA – Radionuclide Aerosol Sampler/Analyser. Detectors of at least 40 % relative efficiency are specified to allow the required limit to be achieved in a reasonable time. Direct calibration over the range 88 to 1836 keV using spiked filter standards is used. Monte Carlo and other mathematical methods are not used. The measurement scheme is:

- Collect sample on air filter, collection efficiency 80 % for 0.2 μm particles, for 24 h at a flow rate of at least $500 \, m^3 \, h^{-1}$.
- Allow a decay of no more that 24 h to allow short-lived natural nuclides (radon daughters – see Appendix D) to decay.
- Measure the gamma spectrum for at least 20 h.
- Analyse the spectrum and report results within 72 h.

99.9 % of spectra contain nothing but ^{212}Pb (within the ^{232}Th series) and ^{7}Be, a naturally generated nuclide usually found in air sample spectra. Any occurrence of anthropogenic nuclides relevant to treaty infringement is regarded as potentially significant and results in action being taken to verify the analysis by one of the certified laboratories. Within a particular period in 2003, when 1280 samples were measured, only 1 gave reason for additional examination (CTBTO (2003)).

Noble gas monitoring.

The noble gas analysis system is called ARSA, the Automated Radioxenon Sampler/Analyser. This uses two different detection techniques, β–γ coincidence or high-resolution gamma spectrometry. The nuclides sought are ^{131m}Xr, ^{133}Xe, ^{133m}Xe and ^{135}Xe, with a limit of measurement of 1 mBq m^{-3} for ^{133}Xe. The measurement scheme is as follows:

- Collect a sample of about 10 m^3 of air at a flow rate of 0.4 m^3 h^{-1} over a period of 24 h.
- After removal of water and CO_2, xenon is absorbed on cold charcoal, purified and concentrated.
- Radon is removed by gas chromatography.
- Measurement by gamma spectrometry or β–γ coincidence over 24 h.
- Report results within 48 h.

There are spectrometric difficulties, but the key nuclide, ^{133}Xe, is relatively easily determined. By itself, the detection of ^{133}Xe is not indicative of a treaty violation, because the isotope is used in nuclear medicine. Discrimination can be achieved by taking into account the overall isotopic composition of the xenon.

Measurement of the whole range of Xe isotopes is difficult because of the low energy of some of the gamma-rays and the fact that ^{131m}Xe and ^{133m}Xe only have low probability gamma emissions. All of the isotopes emit X-rays; the metastable states emit Xe X-rays and the others Cs X-rays. In a paper by Stocki *et al.* (2004), introducing an improved method of xenon isotope analysis (called SPALAX) using a gas permeation membrane and a noble-gas specific absorbent, the authors discuss the difficulties of deconvolution of the X-ray peaks in their spectra measured with a low energy detector. Good spectrum fits could only be obtained by using software that recognized the fact that X-ray peaks are not Gaussian but are a Voight function (see Chapter 1, Section 1.7.4).

17.1.5 Current status

The importance of the task allotted to the IMS radionuclide monitoring sites has led to a deep reappraisal of the whole measurement process, resulting in improvements that will benefit the whole gamma spectrometry fraternity. Two particular aspects illustrate this.

In order to ease the task of searching for target nuclides in the gamma spectra of air filters, the first process is to eliminate all peaks that can be identified as natural background peaks. The most significant nuclide in this respect is ^{212}Pb and its daughters. A particular triplet of peaks at around 150 keV caused some concern. It was established that this was due to ^{212}Pb or ^{212}Bi but the decay scheme of neither of these nuclides could account for them. Neither could true coincidence summing. The source of the triplet was eventually identified as a little known X–X summing, rather than the normal γ–γ summing. Two particular transitions in the ^{212}Pb decay are particularly highly converted. Since the two transitions are in cascade, the internal conversion electrons are emitted simultaneously and can sum. Two Bi K_{α_2} X-rays sum to 149.63 keV, two Bi K_{α_1} X-rays sum to 154.22 keV and the sum of one of each is 151.92 keV. Other, similar X–X coincidences have also been identified.

The positive identification of peaks in the spectrum, however small, and the suppression of false positive identifications is of prime importance in this field. Mistakes in identification could, in principle, have international consequences. De Geer (2005) at the CTBTO, has re-examined the Currie decision limits with respect to peak detection and found them wanting (decision limits were discussed in Chapter 5). From the IMS perspective, a 5 % risk (the usual factor chosen) of mis-identification is not good enough. De Geer deduces that the optimum peak measurement width will vary with the underlying background continuum level and will depend upon the degree of risk required. For example, for a background level of 300 counts per keV, the optimum peak width is only 1.25 times the FWHM of the peak, encompassing 85.9 % of the total peak area – somewhat narrower than most people normally would be comfortable with. Implementation of this new way of looking at critical limits has, apparently, reduced the staff time involved in manual judgement of spectra by a factor of 5–10.

Returning to the current status of the CTBT monitoring network, the verification network is in place and samples are being collected on a routine basis and the IMS continues to monitor and report to the IDC. It appears that the world is acting as if the CTBT is already in force and hopes that the rogue non-signatories and non-verifiers will, in time, join in. The USA supports the continuation of the voluntary ban and is happy for the IMS to continue its work – as well it might; it makes no contribution to any of the activities that would support the treaty when it comes into force. One can only hope that the enthusiasm of scientists, including gamma spectrometrists, watching out for illicit tests in support of a safer world will eventually be matched by those politicians with more parochial concerns.

17.2 GAMMA SPECTROMETRY OF NUCLEAR INDUSTRY WASTES

In this section, I will discuss the measurement of materials originating from the nuclear industry destined for low activity disposal sites. These might include waste material resulting from normal operation of the site or from decommissioning of a nuclear site. The range of materials submitted for measurement is large: soils, magnox swarf, leachings from contaminated items, concrete (sometimes including reinforcing bars as well!), floor sweepings, sludges from tanks and drains, etc. In many cases, the chemical composition is unknown and the sample is unlikely to be homogeneous. Achieving accurate, representative results is impossible within the constraints of time and money. A compromise is always necessary. There must be agreement between client and analyst that a particular procedure for homogenization and sample measurement will be deemed to provide results 'fit for purpose'. Fortunately, from a disposal point of view, the acceptable uncertainty on disposal activity is generous.

When we consider which nuclides we can expect to detect in such samples, we find the range to be wide. Depending upon the source of the material, we can expect:

- Neutron capture nuclides, e.g. ^{54}Mn and ^{60}Co, in constructional steel and concrete that might have been affected by a reactor neutron flux.
- Fission products, e.g. ^{137}Cs, ^{144}Ce and ^{95}Zr, in reprocessing wastes and materials.
- NORM nuclides, and daughters, in soils and building materials.
- Transuranic nuclides and their daughters, e.g. ^{241}Am, ^{237}Np and ^{233}Pa, in recycled uranium and material contaminated by it.
- ^{235}U and ^{238}U in nuclear materials. Materials from a nuclear site containing uranium could be of natural isotopic composition, enriched in ^{235}U or depleted in ^{235}U.

When considering homogenization procedures, such as grinding and crushing, it should be remembered that, apart from any gamma-emitting nuclides, there might be considerable amounts of alpha emitters present. These may pose a considerable airborne hazard during grinding and crushing. An alternative means of homogenization might be complete dissolution of the sample. If it is suspected that the activity of a sample is present as surface contamination, it is tempting merely to leach the sample to remove that activity. Be aware, though, that any sort of chemical treatment could disturb the equilibrium between particular parents and their daughters.

Measurement of neutron capture and fission nuclides is straightforward and, given a normal degree of vigilance with regard to interferences, should present no particular problem. It is worth remembering that nuclear fission creates many parent/daughter beta decay chains, some of which, depending upon the age of the sample, might be relevant. For example, if ^{95}Zr is detected there must also be its daughter, ^{95}Nb.

17.2.1 Measurement of isotopically modified uranium

It is conceivable that uranium within samples that originate from within the nuclear industry could be depleted or enriched in ^{235}U. In principle, gamma spectrometry can easily measure the ^{235}U/^{238}U ratio. If uranium has been isotopically modified, it must have been chemically separated from its ^{226}Ra daughter and one can be sure that the entire 186 keV peak is due to ^{235}U. Measurement of ^{234}Th will provide an estimate of ^{238}U. It may even be possible to measure ^{238}U directly in higher activity samples, albeit with poor precision.

The difficulty arises when one is asked to measure samples that contain both natural uranium and enriched or depleted uranium – contaminated soils, for example – where ^{226}Ra, from the natural soil content, and ^{235}U from the contamination, in addition to the natural content, both contribute to the 186 keV peak. The ^{235}U cannot be determined by mathematical partition of the 186 keV peak, as suggested in Chapter 16, Section 16.3.5, until the effective isotopic ratio is known. One must resort to using the less abundant 143.76 keV peak. However, even that is problematic because, unless the circumstances are exceptional, the activity of contaminated soils is likely to be low and, without extremely long counts, the uncertainty on peak areas will be high. Attempts to make subtle deductions about isotopic composition are likely to founder. Unless the activities are high, it may be preferable to seek other analytical methods for measuring isotopic ratio.

17.2.2 Measurement of transuranic nuclides

During irradiation within the reactor, nuclides heavier than the uranium isotopes are created within nuclear fuel by successive neutron capture reactions followed by beta decay. For example, a small proportion of thermal neutron interactions with ^{235}U generate ^{236}U, rather than induce a fission. This nuclide has a half-life of 2.34×10^7 years and can indulge in further thermal neutron capture to ^{237}U. That decays to ^{237}Np:

$$^{235}\text{U (n, }\gamma\text{) }^{236}\text{U (n, }\gamma\text{) }^{237}\text{U }\beta^- \longrightarrow {}^{237}\text{Np}$$

A significant sequence from ^{238}U is:

$$^{238}\text{U (n, }\gamma\text{) }^{239}\text{U }\beta^- \longrightarrow {}^{239}\text{Np }\beta^-$$
$$\longrightarrow {}^{239}\text{Pu (n, }\gamma\text{) }^{240}\text{Pu (n, }\gamma\text{) }^{241}\text{Pu, etc.}$$

All of the plutonium isotopes, with one exception, decay by alpha particle emission. The exception – ^{241}Pu beta-decays to ^{241}Am. It follows that uranium which has been reprocessed, and materials contaminated by it, may contain all of these transuranic nuclides. Because of their low gamma-ray emission probabilities, low levels of the plutonium isotopes are not easily measured by gamma spectrometry.

Of the transuranic nuclides, only ^{241}Am, using its 59.54 keV gamma-ray, and ^{237}Np, 86.50 keV, are normally determined by gamma spectrometry. ^{237}Np also emits a gamma-ray at 29.37 keV but this is in an inconvenient region of the spectrum and is subject to serious self-absorption in most samples. Even the 86.50 keV gamma-ray is likely to be affected by adjacent X-rays and other interfering gamma-rays. ^{237}Np is usually in radioactive equilibrium with its daughter, 27.0 d ^{233}Pa. If equilibrium is assured, a more reliable measurement of ^{237}Np can be made by using the 311.90 keV gamma-ray of ^{233}Pa. This gamma-ray is interference-free, less subject to self-absorption and provides a lower MDA.

However, dissolving or leaching a sample will almost certainly destroy equilibrium between parent and daughter because ^{233}Pa, which is carrier-free, readily absorbs on any available surface.

17.2.3 Waste drum scanning

Low level radioactive waste is transferred to disposal sites in drums – typically the 220 L steel drum. Before the drum can be removed from the site of origin, it must be monitored so that some idea of the various nuclides present can be established (so-called 'waste characterization') and so that a certificate of radioactive content can be prepared to accompany it. As I noted above when discussing decommissioning wastes, waste material can be very variable – metals, swarf, concrete, sludge, floor sweepings, for example – and possibly mixed within the same drum. The notion of a homogenous sample is clearly not relevant. To make matters more difficult, the radioactivity could be distributed throughout the drum or, equally likely, concentrated in a few highly radioactive fragments. To circumvent these difficulties and achieve a reasonably reliable inventory, the 'segmented drum scanner' has been developed (Figure 17.2).

The drum, containing perhaps 400 kg of compacted waste, is passed along a conveyor to a measuring station comprising a number of detectors, perhaps four, mounted one above the other and collimated by lead shielding so that each only 'sees' a horizontal slice of the drum. In order to cope with both the heterogeneity of the material and the fact that the radioactivity might be confined to a small region, the drum is rotated during counting. If the waste characterization task is simple, for example, to determine whether the material in the drum is active or inactive or, if the pattern of nuclides expected is known

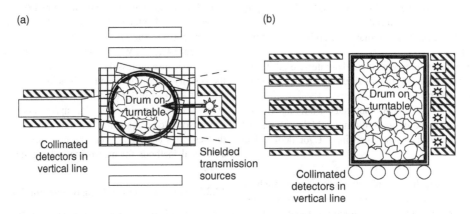

(a)   (b)

Collimated detectors in vertical line   Drum on turntable   Shielded transmission sources   Drum on turntable   Collimated detectors in vertical line

Figure 17.2 One design for a segmented drum (barrel) scanner: (a) view from above; (b) view from the side

and constant, simple measurement of gross activity, the detectors could be NaI(Tl) scintillators. In general, when nuclides are to be positively identified and quantified, HPGe detectors would be preferred. If the latter, there would not be room for liquid-nitrogen cryostats and the detectors would have to be cooled by electrical means (Chapter 3, Section 3.7.5). At some point, usually at the measuring station, the drum would be automatically weighed. The drum might be measured for 3 min or so, giving a throughput of 20 drums h^{-1}, and then the estimated activities in each segment combined to give an overall inventory. An alternative to segmenting the drum for measurement is to move the drum vertically at stages during the measurement period. Such a system would only need one detector, but unless the detector were much larger, would need a longer overall measurement period, and would involve a mechanically more-complex measurement station.

The large size of a drum sample and the density of the material in it mean that correction for self-absorption of the gamma-rays is essential. This is accomplished by mounting transmission sources at the side of the drum opposite to each detector. A source emitting a number of gamma-rays at low and high energy would be used, for example ^{152}Eu, and be of sufficient activity to be easily measured through the drum contents. An automatic shutter would allow the transmission measurement to be made separately from the assay measurement. Measurement of the intensity of the transmission source gamma-rays at different energies, with and without drum present, allows a self-absorption function to be derived which can be applied to the gamma-rays from the sample itself. The measuring station would also have dose rate measuring instrumentation to allow health physics transport documentation to be completed.

In pursuit of high throughput, drum scanner systems have acquired remarkable complexity. Bar coded drums can be automatically read as the sequence begins. The measured activities can be compared to a 'go/no-go' limit and, if passing satisfactorily, be automatically sprayed with a band of coloured paint to indicate that. In one particular system, drums with acceptable activity will pass automatically to a drum crusher to reduce its physical size, significantly reducing the volume to be sent for disposal.

Much waste from nuclear sites will contain significant amounts of transuranic nuclides. While ^{241}Am can readily be measured by gamma spectrometry, the plutonium isotopes cannot. However, some of the even-mass plutonium isotopes undergo spontaneous fission (at sub-critical levels!) and emit neutrons at measurable rates as they do so. Drum scanning systems can be combined with neutron-detection systems that can be used to estimate those plutonium isotopes and, if the normal isotope pattern for the waste stream is known, calculate a '^{240}Pu equivalent' estimate to be included with the gamma-emitting inventory.

17.3 SAFEGUARDS

In this section, I will look at safeguarding special nuclear materials (SNMs) in a general sense. In order to ensure that no material is being diverted to illicit uses, or indeed is simply being lost within the system, it is necessary to have the facilities to check the amounts of SNMs at all stages from manufacture through storage and disposal of waste to use. A brief list of possible applications for instruments for the examination of SNMs might be:

- Portal monitoring, for personnel and vehicles, at exits from nuclear facilities to prevent removal, accidental or deliberate, of SNMs.
- Process control of reprocessing and fabrication.
- Routine monitoring inside plants for inventory purposes.
- Quality control checking of product.
- Monitoring of plant to confirm that no material is being unexpectedly held up within it.
- Monitoring of waste material before disposal.
- From an off-site perspective, monitoring of packages and premises suspected of holding SNMs.
- Monitoring at international borders to control international trafficking of SNMs.

Instruments suitable for all of these purposes are available. I shall confine myself to the general principles on which they operate. Some instruments will aim to measure only uranium, others uranium plus plutonium, possibly in the presence of decay daughters and fission products. Some of these instruments are likely to be used by persons who have no particular reason to be well versed in the arts of gamma spectrometry. In these cases, it is important that the whole system, hardware and especially the software, can provide reliable results without technical attention.

SNMs are defined as all plutonium, uranium enriched to more than 20 % (referred to as Highly Enriched Uranium - HEU) and ^{233}U. These are materials that could be used to make nuclear weapons. Their common feature is that, by virtue of their isotopic composition and/or enrichment, they have high cross-sections for induced fission.

Unfortunately, their gamma-ray emission properties are far from convenient. All of the nuclides in question are alpha emitters and because of that have low, sometimes very low, gamma-ray emission probabilities. For example, ^{239}Pu emits a 5.157 MeV alpha particle in 70.79 % decays but the most intense gamma-ray, at 51.62 keV, has an emission probability of only 0.0269 %. For ^{240}Pu, the figures are 72.74 % for its 5.168 MeV alpha and 0.0450 % for its most intense gamma at 45.24 keV. There is very little gamma-ray emission at high energy and the low-energy gamma-rays must be measured in the presence of many other small peaks, including many overlapping X-ray peaks. Take into account the fact that the material being measured has a high density, meaning serious self-absorption, will be in a container, and sometimes shielded, and the measurement problem seems insurmountable.

As it happens, from a safeguards point of view, the low emission probability is not a particular problem because the amounts of material being measured are usually high. Counting times are often of only a few minutes duration. In some circumstances, an acceptable limit of detection would be in the range of grams. While SNMs are not well endowed with gamma emissions, they do have a significant spontaneous fission rate, which means that they emit neutrons. Figure 17.3 shows the neutron emission rate from spontaneous fission for a number of relevant nuclides. It shows quite clearly that if a significant neutron emission is detected then one or more of the even numbered plutonium isotopes are present. In safeguards applications, neutron detector systems will often

be combined with a gamma spectrometer. The spontaneous emission rate of ^{235}U is not large enough to allow its measurement by passive neutron measurements, but active systems, in which neutrons from an isotope source or neutron generator are use to stimulate fission in the sample, are available.

17.3.1 Enrichment meters

Isotopic measurements of uranium alone are relatively easy. ^{235}U emits three gamma-rays with reasonable probability – 185.72, 143.76 and 163.33 keV – in decreasing order of probability. ^{238}U emits a gamma-ray of 49.55 keV with very low probability that is not easily usable, but its daughters, ^{234}Th and ^{234m}Pa, provide gamma-rays with much greater emission probability. (It should be remembered that the long half-life of ^{234}U blocks the growth of any nuclides beyond it.) At 63.28 keV, and a doublet at 92.37 + 92.79 keV, the ^{234}Th is not easily measured, except by high resolution spectrometry, and in the absence of Pu isotopes. ^{234m}Pa has two gamma-rays at 766.37 and 1001.03 keV, which have low emission probabilities but are measurable. Instruments called 'Enrichment Monitors', which can incorporate low or high resolution detectors, make use of two regions-of-interest in the spectrum. One is centred on the 185.72 keV peak, in which counts are mainly due to ^{235}U, and one at some distance above to measure a portion of the Compton continuum created by the 1001.03 keV gamma-ray of ^{234m}Pa, which is indicative of ^{238}U. This forms the basis of a simple system easily understood that can use a low cost, low resolution detector. Recently, when the introduction of portable cooling systems was introduced, hand-held enrichment meters became feasible.

Limitations of simple enrichment monitors are that they are only relevant to uranium which is in equilibrium with its daughters and samples must be of 'infinite thickness'. By that, is meant that the sample is so massive that all of the gamma-rays from the parts of the sample distant from the detector will be absorbed within the sample. Making the sample larger would have no effect on the spectrum. This avoids inaccuracies due to unknown self-absorption factors.

17.3.2 Plutonium spectra

When both uranium and plutonium (or even plutonium alone) are present, the spectrometry becomes more difficult. By and large, low resolution detectors are not adequate in these circumstances. As pointed out above, gamma emissions from plutonium isotopes are of very low

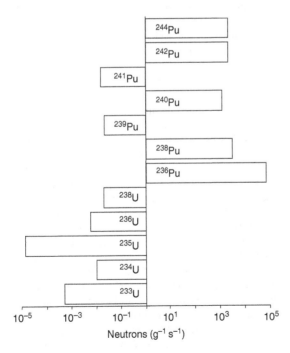

Figure 17.3 Emission of spontaneous fission neutrons by uranium and plutonium isotopes

probability. The most intense peaks in the spectrum will be at low energy and will be accompanied by L X-rays and K X-rays of uranium and neptunium, together with an abundance of germanium escape peaks and X-ray summed peaks. In high-concentration samples, there may also be enhanced Pu X-rays by self-fluorescence induced by the emitted gamma-rays and alpha particles. Not an enticing analytical prospect! So that the analysis software has the best chance of deconvoluting such spectra, a better-than-average resolution detector designed for low energy spectrometry is called for – 550–600 keV FWHM at 122.1 keV is probably the maximum desirable peak width. Figure 17.4 shows the spectrum of a plutonium reference material, containing all of the typical isotopes, and the decay daughters of ^{241}Pu, of which more below.

A NUREG document (Reilly *et al.* (1991)) discusses all of the various methods of passive assay of nuclear materials – neutron and gamma. Of particular relevance to us is a detailed assessment of the decay characteristics of the plutonium isotopes and the various regions of their spectra and their relevance to plutonium measurement (Sampson (1991)). The region around 100 keV is

particularly 'busy' but it contains all of the information necessary to calculate a complete set of isotopic abundances. The region contains about 15 peaks of various sizes, needing sophisticated spectrum analysis. This is a region of the spectrum where the intrinsic shape of X-rays begins to be significantly different from the shape of γ-rays, and the fact that X-rays are Voight distributions rather than Gaussian must be taken into account (Chapter 1, Section 1.7.4). Such spectra are analysed by programs, or codes, designed specifically for plutonium isotopics. One such is MGA – Multi Group Analysis code – written by Gunnink (1960) and developed at the Lawrence Livermore National Laboratory. This code has been licensed to all of the major equipment manufacturers who have extended it in various ways. The code takes account of the nuclear data for the various isotopes and performs fitting operations to the spectrum data, out of which come relative isotopic abundances. Systems which use MGA need no efficiency calibration because the results derived are only relative. That is not necessarily a problem because in the context in which they are used the overall amount of material is almost certainly already known.

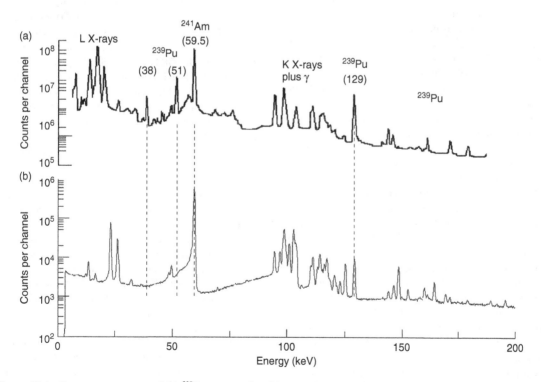

Figure 17.4 Gamma-ray spectrum of (a) ^{239}Pu, compared to (b) an aged sample of a Pu reference material (A-460, mixed Pu isotopes)

Various modifications of MGA are in use for particular purposes – MGAU for isotopic analysis of uranium alone, for example, and MGAHI for analysis using high-energy spectra from a coaxial rather than a low energy detector. One particular version (MGA++, ORTEC) takes into account a low energy spectrum *and* a high-energy spectrum.

17.3.3 Fresh and aged samples

The abbreviated decay schemes of some of the relevant isotopes of Pu are shown in Table 17.1. Of particular note is the fact that some have relatively short half-lives compared to their siblings. This is of importance with computer programs that make use of the relative amounts of different nuclides in analysing the spectra of U + Pu mixtures. ^{238}Pu has a half-life of 87.74 years and decays into ^{234}U, which might also accompany ^{238}U. ^{241}Pu has only a 14.33 year half-life and, very soon after separation, its beta-decay mode will generate sufficient ^{241}Am to be noticeable. Because of the low emission probabilities, after only 30 days decay the 59.54 keV peak of ^{241}Am will be at least 25 times greater than any of those of the ^{241}Pu gamma-rays. In addition to that, the low probability (0.0025 %) alpha decay branch of ^{241}Pu will produce ^{247}U. Both ^{241}Am and ^{247}U decay to ^{247}Np (α and β, respectively) and will emit some gamma-rays in common. In particular, both will emit the 59.54 keV gamma-ray. In fact, because of its short half-life and consequent rapid growth, ^{47}U will be even more of a problem than the ^{241}Am in the short term, although it will have saturated, effectively, after 70 days. On the other hand, the in-growth of ^{237}U does provide a number of extra easily measurable gamma-rays that can be used to determine ^{241}Pu once equilibrium is established.

Figure 17.4 compares the spectra of a relatively fresh ^{241}Pu source (a) and that of a reference source of mixed, aged, Pu isotopes (b). The dominance of the 59.54 keV peak in the spectrum shown in Figure 17.4(b) is obvious. From the spectrometry point of view, these large (in comparison) grown-in activities dominate the count rate and obscure potentially useful peaks of the plutonium nuclides at around 40 keV. The solution to high count rate is to interpose a heavy-metal filter between source and detector to absorb all low energy radiation; 0.8 mm of tin will absorb 80 % or all gamma-rays below 80 keV and 98 % at 59.54 keV.

Clearly, software that assumes decay equilibrium between the various decay products will not be able to cope with non-equilibrium mixtures.

17.3.4 Absorption of gamma-rays

Considering the fact that MGA uses low energy gamma-rays, it is obvious that there will be serious absorption of the gamma-rays before reaching the detector. There will be self-absorption with the sample itself – uranium and plutonium and their compounds have a high density. The sample will be in a container and there may be shielding as well. In general, MGA is able to deduce the absorption factors by taking account of the peak area ratios for ^{239}Pu. This allows correction for absorption of all other gamma-ray count rates. However, if the shielding is too thick, none of the 100 keV region gamma-rays will be measured at all. In that case, reliance has to be placed on the high-energy mode of MGA

17.3.5 Hand-held monitors

For roving monitoring and plant tracing, and suchlike, hand held monitors are essential. NaI(Tl)-based systems are one solution but tend only to provide detection rather than identification. For that, higher resolution systems are needed. Portable HPGe detectors are readily available and with their attached laptop or notebook computer, give a portable, rather than hand-held, instrument. The limitation is, of course, the need to provide the HPGe detector with liquid nitrogen coolant. The recent introduction of a miniature Stirling-cycle electrical cooling system for HPGes, as in the ORTEC trans-SPEC, gives a truly (if large) hand-held high resolution spectrometer. A disadvantage of a portable detector is the lack of external shielding in situations where more than one source of radiation is present. Some designs now incorporate graded shielding *within the detector housing*. In such a location, a lower volume of absorber is needed, keeping the total mass of the system within the portable description. It is said that a well-designed arrangement grading can eliminate the backscatter peak from the spectrum – valuable for systems concentrating on the low energy end of the spectrum.

An alternative to HPGe detectors is the use of room-temperature semiconductor CdZnTe (CZT) detectors. These have resolution much better than scintillation detectors, but worse than HPGe (Figure 17.5). Existing software packages (MGA and PC/FRAM, for example) have already been configured to accommodate CZT. A note illustrating their use in safeguards has been published by Arlt *et al.* (2000). Improvements in the quality of the CdZnTe detectors being produced (driven by space research and medical imaging requirements) and in detector design have improved resolution, increased peak-to-Compton ratio and have limited the tailing

Table 17.1 Decay schemes of some of the plutonium isotopes[a]

^{238}Pu	^{238}Pu	^{240}Pu	^{241}Pu	^{242}Pu	^{244}Pu
^{238}Pu 87.74 years ↓α	^{239}Pu 2.41 × 10^4 years ↓α	^{240}Pu 6561 years ↓α	^{241}Pu 14.33 years ↓β 99.9975% ↓α 0.0025%	^{242}Pu 3.73 × 10^5 years ↓α	^{244}Pu 8.0 × 10^7 years ↓α
^{234}U 2.454 × 10^5 years (then follow ^{238}U decay scheme from step 4, Figure 16.1)	^{235}U 7.037 × 10^8 years (then follow ^{235}U decay scheme, Figure 16.2)	^{236}U 2.342 × 10^7 years ↓α	^{241}Am 432.6 years ↓α	^{238}U 4.468 × 10^9 years (then follow ^{232}Th decay, Figure 16.3)	^{240}U 14.1 h ↓α
		^{232}Th 1.405 × 10^{10} years (then follow ^{232}Th decay, Figure 16.3)	^{237}U 6.75 d ↓β		^{240}Pu 6561 years (then follow ^{240}Pu scheme in 3rd column)
			^{237}Np 2.14 × 10^6 years ↓α		
			^{233}Pa 26.98 d		
			(followed by 10 other decays to ^{209}Bi)		

[a]Small (<0.2%) spontaneous fission branches in the 242 and 244 decay schemes have been omitted for clarity.

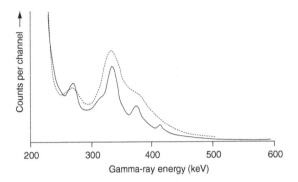

Figure 17.5 Portions of spectra of a mixed Pu isotope sample measured with a CZT detector (continuous line) and a NaI(Tl) detector (dotted line)

on the spectrum peaks, making the detectors much more capable in this field.

CdZnTe detectors are necessarily small because of the charge collection limitations. However, for some purposes a small size can be an advantage. One such is plant monitoring, where the ability to squeeze a detector into small spaces between pipes and valves is a useful asset.

17.4 PINS – PORTABLE ISOTOPIC NEUTRON SPECTROMETRY

My final example of 'gamma spectrometry in pursuit of a safer world' is an example of prompt gamma-ray spectrometry. PINS (Portable Isotopic Neutron Spectrometry) was developed for the on-site analysis of non-nuclear munitions to identify their contents, looking for chemical warfare agents and high explosives, to help suitable destruction processes to be chosen.

A neutron source provides neutrons that can penetrate the shell casing and irradiate the contents, stimulating (n, γ) and $(n, n'\gamma)$ reactions. The prompt gamma-ray is emitted as the nucleus de-excites. The range of energies of prompt gamma-rays is much larger than for conventional decay gammas up to 7 MeV would be typical.

In addition to their carbon, hydrogen and oxygen components, chemical warfare agents can contain nitrogen, fluorine, phosphorus, sulfur, chlorine and arsenic – all of which have a spectral signature. Table 17.2 list the components of a range of agents. The principle is simple: irradiate the item, measure a spectrum for a period of time and analyse the spectrum to identify the components. Systems have been devised by using both ^{252}Cf neutron sources and pulsed-neutron generators, and both large HPGe and NaI(Tl) detectors.

Table 17.2 Elemental compositions of chemical warfare agents and explosives

Agent	Matrix elements				Key elements				
	H	C	N	O	F	P	S	Cl	As
Clark II	*	*	*	—	—	—	—	*	*
Lewisite	*	*	—	—	—	—	—	*	*
Phosgene	—	*	—	*	—	—	—	*	—
Tabun	*	*	*	*	—	*	—	—	—
Sarin	*	*	—	*	*	*	—	—	—
VX	*	*	*	*	—	*	*	—	—
TNT	*	*	*	*	—	—	—	—	—

Some reported alternative practical uses have involved identification of the contents of drums and cylinders that, through age or corrosion, had lost all identification.

FURTHER READING

- Information on the CTBTO can be obtained from their website at *http://www.ctbto.org*. Particular documents are:

De Geer, L.-E. (1999). *CTBT Relevant Nuclides*, CTBTO Technical Report PTS/IDC-1999/02, Comprehensive Nuclear Test Ban Treaty Organization, Vienna, Austria.

CTBTO (2003). *Report of the Executive Secretary to the 21st Session of the Preparatory Commission for the Comprehensive Nuclear-Test-Ban Treaty Organization*, CTBT/PC-21/1/Annex III, Comprehensive Nuclear Test Ban Treaty Organization, Vienna, Austria.

- On a new view of the Currie criteria:

De Geer, L.-E. (2005). *A Decent Currie at the PTS*, CTBT/PTS/TP/2005-1, Comprehensive Nuclear Test Ban Treaty Organization, Vienna, Austria.

- Information on the UK involvement in the IMA (and source of the data used to prepare Figure 17.1):

Comley, C. (2001). Radionuclide monitoring at AWE, *Discovery*, **3**, 10–13.

Comley, C., Price, O. (2003). *Verification Yearbook*, VERTIC, London (Ed. Findlay, T.) pp. 141–150.

- On an improved method to Xe isotope measurements:

Stocki, T.T., Bean, M., Ungar, K.R., Toivonen, H., Zhang, W., Whyte, J. and Meyerhof, D. (2004). Low level noble-gas measurements in the field and laboratory in support of the Comprehensive Nuclear-Test-Ban Treaty, *Appl. Radiat. Isotopes*, **61**, 231–235.

- On measurements on nuclear materials in general:

Reilly, D., Ensslin, N. and Smith, H. (eds) (1991). *Passive Nondestructive Assay of Nuclear Materials*, NUREG/CR-5550 and LA-UR-90-732, NUREG, Washington, DC.

Sampson, T.E. (1991). Plutonium isotopic composition by gamma-ray spectroscopy, in *Passive Nondestructive Assay of Nuclear Materials*, NUREG/CR-5550 and LA-UR-90-732, D. Reilly, N. Ensslin and H. Smith (Eds), NUREG, Washington, DC, pp. 221–272.

• On equipment for safeguards measurements:
Canberra (1996). *IMCA: A Dedicated Instrument for Uranium Enrichment Measurements in Safeguards Inspections*, Canberra Application note (available at *http://www.canberra.com/literature/1048.asp*). This note is followed by a number of useful references to programs for achieving isotopic analysis.
Canberra (1997). *U-Pu InSpector: A Dedicated Instrument for Assessing the Isotopic Composition of Uranium and Plutonium*, Canberra Application note (available at *http://www.canberra.com/literature/1049.asp*).

• Software for isotopic analysis:
Gunnink, R. (1990). MGA: *A Gamma-ray Spectrum Analysis Code for Determining Plutonium Istopic Abundances*, LLNL Report: UCRL-LR-103220, USDOE, Washington, DC.
Koskelo, M.J., Wilkins, C.G. and Fleissner, J.G. (2001). Comparison of the performance of different plutonium isotopic codes using a range of detector types, presented at the *23rd ESARDA Annual Symposium on Safeguards and Nuclear Material Management*, Brugge, Belgium, May 7–11 (available at *http://www.canberra.com/literature/1180.asp*).

• On CZT detectors:
Arlt, R., Ivanov, V. and Parnham, K. (2000). *Advantages and Use of CdZnTe Detectors in Safeguards Measurements* (available at *http://www.evproducts.com/white_papers_news.html*).

Appendix A: Sources of Information

A.1 INTRODUCTION

In this appendix, I list sources of information of value to gamma spectrometrists. While it is very convenient to have nuclear data available in book form on one's desk, it has to be admitted that the high cost of printed material increases the attraction of free information via the Internet. As in the earlier parts of the book, I quote below many internet sources of information. I believe these to be of good quality at this time, but it is up to the user to make sure that the information has not fallen out-of-date. The nuclear-data sources I have selected should remain updated but I am in no position to guarantee that, or to comment on the frequency of updates.

I remind the reader that all the links within this appendix are provided on the Gamma Spectrometry website (*http://www.gammaspectrometry.co.uk*). I also draw attention to the caveat given at the beginning of this book, that as well as holding the up-to-date information, the Internet is also a vast repository of ancient, irrelevant, inaccurate and out-of date information. It is up to the user to check the pedigree of all downloaded material.

A.2 NUCLEAR DATA

An experimental result can only be as good as the data used in its production. Sources of nuclear data are many and it is not always clear which are reliable. With gamma-ray measurements, there are four pieces of nuclear information required for each nuclide to be measured:

- Its gamma-ray energies, which identify the radionuclide – published tables of energies are unlikely to be so inaccurate that misidentifications are caused. However, poor quality energy data is likely to cause errors in the deconvolution of multiplet peaks. Of course, energies used for calibration should be very reliable; errors here will feed into all subsequent measurements.

- Its half-life, to allow us to determine the activity at some specific point in time; in broad terms, published values for half-lives are less reliable than gamma energies, although for most nuclides routinely encountered they are very well known. The half-life is important, even for comparative measurements; in principle, decay corrections should be applied to all measurements. As pointed out in Chapter 15, the published uncertainty in the half-life should be used to calculate its contribution to the overall uncertainty of the result. Thus, we need half-life data that include uncertainties.

- The emission probability for each gamma-ray used – needed to convert count rate to disintegration rate. This information would be irrelevant if comparative measurements are made, but most people will derive their results from efficiency calibrations and good quality data is then essential. As with half-lives, the emission probability must be accompanied by an uncertainty, which is carried through the calculation of activity to make its proper contribution to the overall uncertainty,

- Less used, but sometimes of crucial importance are the decay scheme data, which allow us to see the gamma-ray cascades and determine which would be likely to sum. The reader of Chapter 8 will appreciate the significance of this. Most data sources do not provide decay scheme data. I will list below some useful sources that do.

Other nuclear parameters, such as decay mode probability (e.g. x % β, y % EC), internal conversion coefficients, fluorescence yields and transition multipolarities, are of secondary importance in routine radionuclide measurements, but form a supporting layer of information below the essentials.

A.2.1 Recent developments in the distribution of nuclear data

In the first edition of this book, I recommended that the IAEA-TECDOC-619 data should be taken in preference to all other data sources because it was the most up-to-date, carefully evaluated set of data available at that time. Since then, things have moved on. In 1995, an international collaboration was formed, the Decay Data Evaluation Project (DDEP), which includes members of the BNM-CEA/LNHB (France), PTB (Germany), INEEL (USA), KRI (Russia), LBNL (USA), NPL (United Kingdom) and CIEMAT (Spain), with the objective of providing carefully evaluated recommended data. The DDEP is an ongoing project to evaluate and update nuclear data and make them available on its database, maintained by the Laboratoire National Henri Becquerel (LHNB) and accessible on the internet. Re-evaluation of data for the nuclides in TECDOC-619 was initiated, together with another 27 nuclides of importance: 63 nuclides in all. The results of that exercise were published in the late spring of 2007 (although the cut-off date for evaluated data was much earlier than that). That data, known in short as XGAMMA, is available on the internet and is incorporated into DDEP.

BIPM, the international weights and measures bureau, has recommended that all radiometric laboratories take their data from DDEP. All of the national bodies responsible for maintaining standards within their respective countries have subscribed to that advice. The DDEP is very detailed. It covers alpha, beta and gamma-ray emissions, X-rays and electron emissions and includes the references to the data taken into account in the evaluation. It also carries decay scheme diagrams. The data are available on line as downloadable .pdf files – either individually for each nuclide or as a series of volumes summarizing the updates at particular points in time.

The whole database, nuclide-by-nuclide, is accessible via a 'Recommended Data' page (URLs are given below). This links to the most up-to-date information and is updated whenever a new evaluation is completed. At the time of writing, the most recent update was 28th August 2007. In my opinion, there is a significant inconvenience to the DDEP database as it stands; the .pdf files have been prepared in such a way that the user cannot, using his/her computer keyboard, copy the data items and paste them into their own libraries. That has to be done manually with the accompanying risk of transcription error. It seems an unnecessary restriction. A more convenient access to the DDEP data is offered by the online 'Nucleide Gamma and Alpha Library', also maintained by LNHB. That particular database, called LARA, does allow copy-and-paste of its data.

At the present time, DDEP only holds data for 129 nuclides. Persons requiring data for other nuclides, including such oft-measured nuclides as ^{210}Pb, have to look elsewhere. In the UK, NPL recommend that the ENSDF database should be consulted for data not in DDEP or XGAMMA. Conveniently, the LARA database takes data from ENSDF for nuclides not covered by DDEP. Although LARA is a secondary source, since it and the primary DDEP are maintained by the same organization, we can have reasonable confidence that LARA will be up-to-date. (However, I have found a couple of instances where some XGAMMA data incorporated into LARA have been given the wrong half-life units. *Caveat utilitor!*) In the USA, there is likely to be a different opinion on this issue. In an email copied to the RADCH email list, the American NuDat database administrator, is quoted as saying:

I think that the data in NuDat 2.3 is the most comprehensive resource world wide, people contributing to it have PhDs in nuclear physics with many years of experience. ... Most DDEP contributors also contribute to NuDat, so both databases are not really independent, and DDEP has a far more limited scope. JEFF 3.1 [another commonly used database] is not independent, it is mostly based on ENSDF and DDEP. Also, JEFF is frozen in time, being updated every 5–10 years or so. ENSDF and NuDat are updated monthly

It is not immediately clear, however, whether all of the NuData data has been subject to the same careful, defined, evaluation as the DDEP data. I would support the NPL recommendation of DDEP followed by ENSDF, at least for data for routine gamma spectrometry purposes. For nuclear-level data, as opposed to gamma-ray energies, the online NuData resource is excellent. I have no hesitation in recommending it.

On-line sources of data from routinely updated databases are excellent in the sense that one always has the most up-to-data information – much more reassuring than the book and CD-ROMS we have been used to. There is, however, a down-side to that. Quality-assurance processes may require the data source for a particular measurement to be specified within the written method. How would the QA manager feel about taking data from a continuously changing source that does not have an audit trail? We have a duty to use the most reliable source of data, but we could end up in a situation where, at the time of a query, perhaps twelve months after a measurement, the DDEP value for a particular half-life, say, is different to the value used at the measurement time. This became a particular issue when finalizing the text for this edition; when checking consistency of gamma-ray energies throughout the book, I became aware that the data for ^{214}Bi and ^{214}Pb, taken from the recent XGAMMA data, released early in 2007, had already been updated in DDEP with a more recent evaluation. In practice, data changes are likely to be small in magnitude and would probably not have a significant impact on the result of the measurement, but the consequences need consideration when defining our quality assurance procedures.

A.2.2 On line internet sources of gamma-ray emission data

These items are presented in a suggested order of priority.

• DDEP

Information about the Decay Data Evaluation Project is at *http://www.nucleide.org/DDEP.htm*. Background information from the BIPM about the project can be found at *http://www.bipm.org/en/publications/monographie-ri-5.html*. This page also holds links to composite files containing the complete state of the database at different times.

The list of links to the evaluated data is at *http://www.nucleide.org/DDEP_WG/DDEPdata.htm*. The linked .pdf format files, one for each nuclide, are very comprehensive. At the time of writing, 129 nuclides are covered. Links are also given to the ENSDF data for the same nuclides. When using the .pdf files, for normal gamma spectrometry purposes, users should take care to take emission data from the 'Emission' tables, not from the 'Gamma Transitions' tables.

DDEP also holds a list of recommended half-lives independently of the main body of data. This can be found at: *http://www.nucleide.org/DDEP_WG/Periodes_2006.pdf*.

• LARA

This database is more extensive in terms of number of nuclides, but only includes gamma-ray and X-ray emission data. Energy levels and other minutiae of nuclear decay are not included. The database is in synchronism with the DDEP data; data for other nuclides seem to be taken from ENSDF. Unlike the DDEP data itself, data can be copied and pasted from this database. There is also a useful facility that lists all gamma-rays within an energy range – a useful tool for identifying unexpected interferences. The database itself is at *http://laraweb.free.fr/*.

Other information about LNHB data sources can be found at *http://www.nucleide.org/*.

• XGAMMA

The follow-up to the IAEA TECDOC-619 (1991) data-evaluation exercise, covering all of the original 36 plus additional nuclides, to a total of 63. Information about the data is at *http://www-nds.iaea.org/xgamma_standards/* and includes links to individual html files containing half-lives, gamma-ray emission data and X-ray emission data. There is also a limited amount of evaluated data on emission rates of high-energy gamma-rays beyond the normal gamma spectrometry range. A link to an Excel spreadsheet file containing the whole set of data, *http://www-nds.iaea.org/xgamma_standards/data.xls*, is the easiest way of importing the data into your own libraries.

• ENSDF

The Evaluated Nuclear Structure Data File (ENSDF) is a database maintained by the National Nuclear Data Center (NNDC), based at Brookhaven National Laboratory in the USA. The easiest access to this database, via a Periodic Table of the Elements display, is via the 'Table of Isotopes' website: *http://ie.lbl.gov/education/isotopes.htm*. The user selects an element from the table and then, from a list, the isotope required. A most useful feature of this data source is access to decay scheme plots. To do this, it is necessary to have all of the features of Java installed on the user's computer. The Java display allows gamma-rays in coincidence to be highlighted.

Extracts from ENSDF also form the Nuclear Data Sheets. Nationals of the OECD, and perhaps others, have access to a similar file, the Joint Evaluated File (JEF, see below), kept by the Nuclear Energy Agency in France. There is close collaboration between these organizations, and between them and national bodies, such as NIST, NPL, PTB and other groups like the ICRM (International Committee on Radionuclide Metrology). The DDEP recommended data page referred to above has links to the raw ENSDF data.

• NuDat 2.4 (*http://www.nndc.bnl.gov/nudat2/*)

NuDat is maintained by the National Nuclear Data Center at Berkeley National Laboratory in the USA. This is very comprehensive source of nuclear data. The initial user interface is a Segré Chart from which nuclides can be selected for particular attention, although it is easier to use the search links above the display. The chart can be zoomed and can be colour coded in different ways; one of them coded by decay mode. Perversely, the red is used for neutron rich nuclides and blue for those deficient in neutrons – the opposite of the very well-known Karlsruhe chart. Links are provided to tables of energy levels and decay emissions, and to a decay scheme plot. There is, however, no way of picking out coincident gamma-rays.

A.2.3 Off-line sources of gamma-ray emission data

These are computer resources for accessing nuclear data, which might be downloaded from the internet, but which are Installed and run locally.

● IAEA (1991). *X-Ray and Gamma-Ray standards for Detector Calibration*, IAEA-TECDOC-619, International Atomic Energy Authority, Vienna, Austria.
This is available on PC disk from IAEA Nuclear Data Section, PO Box 100, A-1400, Vienna, Austria, or within the UK, from NPL, Teddington, UK, if the user supplies blank disks (Tel: 0181-943 6424). It can also be downloaded from *http://www.gammaspectrometry.co.uk/iaea*. At the time of publication, this was the best available data, but only for some 36 radionuclides. The data has been superseded by DDEP.

● Isotope Explorer
This is also available as a download via links at *http://ie.lbl.gov/toi.htm*. It uses the ENSDF and TORI (Table Of Radioactive Isotopes) databases. It is a Java-based program that can be set up to access either the on line or locally stored database.

● JANIS 3.0
The Nuclear Energy Agency (NEA) has recently improved and updated its JANIS (JAva-based Nuclear Information Software) to Version 3.0. The program provides access to a number of databases, including JEFF-3.1 and allows plots of gamma-rays by emission probability, nuclear reaction cross-sections, fission yields and much other data not relevant to gamma spectrometry. You can download the software at *http://www.nea.fr/janis/download.html*. Access to an on-line version is at *http://www.nea.fr/janis*. A DVD containing JANIS and a number of databases is available from NEA free of charge. It can be ordered at *http://www.nea.fr/janis/orderform.htm*.

● JEF-2.2 – Joint-Evaluated File
This is a collaborative project between the countries participating in the NEA Data Bank. The JEF-2.2 library comprises sets of evaluated nuclear data, mainly for fission reactor applications; it contains a number of different data types, including neutron interaction data, radioactive decay data, fission yield data, thermal neutron scattering raw data and photoatomic interaction data. JEF-PC is a personal computer package containing selected data from the JEF 2.2 library.

The user interface is via a 'Chart of the Nuclides' format. Available, at a charge, from OECD publications, Electronic Editions, 2 Rue Andre-Pascal, 75775 Paris Cedex 16, France (Tel: +33 (1) 4910 4265, fax: +33 (1) 4910 4299). At the time of writing, information on the internet about JEF is scanty. Try *http://www-nds.iaea.org/indg_nsdd.html*.

● JEFF-3.1 – Joint-Evaluated Fission and Fusion Library
The complete suite of data was released in May 2005, and contains general purpose nuclear-data evaluations compiled at the NEA Data Bank in co-operation with several laboratories in NEA Data Bank member countries (JEFF-3.2 is at beta test stage). JEFF-3.1 also contains radioactive decay data, activation data and fission yield data. The library contains neutron reaction data, incident proton data and thermal neutron scattering raw data in the ENDF-6 format.

The library can be downloaded in its entirety as a single file or one file per isotope from the JEFF-3.1 project page. A CD-ROM of the complete library can also be requested free of charge by sending an e-mail request. See *http://www.nea.fr/html/dbdata/projects/nds_jef.htm*.

● TORI
Browne and Firestone's 'Table Of Radioactive Isotopes' is based on the ENSDF data, described above. Data from this database can be accessed by Isotope Explorer, again described above.

A.2.4 Nuclear data in print

For this section, I have omitted the older publications mentioned in the first edition. In view of the recent developments in the publication of nuclear data, there seems little point in referring to sources of data that are now out-of-date. At the end of this section are two relatively recent sources that I recommended in the first edition. I include them here only to point out that they can no longer be relied upon.

● Browne, E., Firestone, R.B., Shirley, V.S. (Ed.) and Baglin, C.M. (Ed.) (1999). *Table of Isotopes*, John Wiley & Sons, Inc., New York, NY, USA.
Earlier editions of this book have graced the desks of many gamma spectrometrists for a long time, and rightly so. If you want printed data, it doesn't get better than this. It is comprehensive, authoritative and easy to use. Limitations are a lack of general ordering by energy, only very rudimentary display of isobaric decay relationships and there is no cascade information. This book is now available in paperback and is accompanied by a CD containing the TORI nuclear data tables referred to above. See *http://www.wiley.com/products/subject/physics/toi/*.

• Woods, M.J., Collins, S.M. and Woods, S.A. (2004). *Evaluation of Half-Life Data,* NPL Report CAIR 8, National Physical Laboratory, Teddington, UK. Data intended to update the TECDOC-619 data and extend to 63 nuclides. These data are now incorporated into DDEP.

• Smith, D. and Woods, S.A (1995). *Recommended Nuclear Decay Data*, NPL Report RSA(EXT)53, National Physical Laboratory, Taddiugton, UK. Data for 53 nuclides taken from JEF 2, UKPADD and ENSDF databases. Does not include the TECDOC-619 nuclides. Superseded by DDEP.

• Longworth, G. (Ed.), (1998). *The Radiochemical Manual*, AEA Technology plc, HMSO, London, UK. Incorporates the NPL RSA(EXT)53 data and other data taken from the UKHEDD database. Includes the TECDOC-619 nuclides but does not incorporate the TECDOC-619 data. Superseded by DDEP and no longer recommended as a source of nuclear data. That does not diminish the value of this manual from a general informative point of view.

A.3 INTERNET SOURCES OF OTHER NUCLEAR DATA

The International Atomic Energy Authority (IAEA) maintains a very detailed list of nuclear databases of all kinds at *http://www-nds.iaea.org/*, including those of value to persons engaged in prompt gamma-ray measurements at *http://www-nds.iaea.org/pgaa/*.

• **X-ray Energies**
X-ray energies and emission rates of particular nuclides will be found with the gamma-ray data in the sources listed in Section A.2.2. A source of information about X-ray energies and emission rates independent of nuclear decay can be found at *http://ie.lbl.gov/tomic/x2.pdf*.

• **Isobaric-Chain Diagrams**
http://ie.lbl.gov/toia.html.
This URL gives access to diagrams showing simplified decay schemes arranged in isobaric chains, as described in Chapter 1. Diagrams can be downloaded in either Postscript or .pdf format.

• **Nuclear-Energy Levels**
http://www.nndc.bnl.gov/nudat2.
This URL was mentioned above as a source of gamma-ray data. It is particularly useful for nuclear-energy levels.

• **Mass-Attenuation Coefficients**
This site is immensely useful for when absorption corrections must be made. Data are provided element-by-element and for a number of compounds and mixtures. A diagram of the attenuation curve is provided, together with a table of coefficients at a large number of energies in ASCII and html format. The introduction to the site is at *http://physics.nist.gov/PhysRefData/XrayMassCoef/cover.html*.

The element data is at *http://physics.nist.gov/PhysRefData/XrayMassCoef/tab3.html*.

The mixtures/compounds list is at *http://physics.nist.gov/PhysRefData/XrayMassCoef/tab4.html*.

• **Physical Constants**
http://ie.lbl.gov/elem/phys.pdf.
This URL downloads a .pdf file containing values for all of the primary physical constants and a number of conversion factors for day-to-day units.

• **Gamma-Ray Spectra**
In the beginning, there was sodium iodide. In 1964, R.L. Heath published *Scintillation Spectrometry – Gamma-ray Spectrum Catalog* – this contained printed spectra of a large number of nuclides. Invaluable at the time, this is still of interest today and is available on the internet, together with the Ge(Li)–Si(Li) spectrum catalogue that followed. The introductory page is *http://www.inl.gov/gammaray/catalogs/catalogs.shtml*.

The high resolution spectra (89 Mb .pdf) are at *http://www.inl.gov/gammaray/catalogs/pdf/gecat.pdf*.

Low resolution spectra (23 MB .pdf) are at *http://www.inl.gov/gammaray/catalogs/pdf/naicat.pdf*.

A.4 CHEMICAL INFORMATION

• **Periodic Tables**
All of these have links to information about individual elements via the table itself:

• Chemicool Periodic Table (*http://www-tech.mit.edu/Chemicool/*).
• Los Alamos National Laboratory table (*http://periodic.lanl.gov/default.htm*).
• Royal Society of Chemistry-less information, but more Flashy (*http://www.chemsoc.org/viselements/pages/pertable_fla.htm*).
• Wikipedia Periodic Table (*http://en.wikipedia.org/wiki/Periodic_table*).
• Webelements table (*http://www.webelements.com/*).

• **Isotopic Abundances**
The official IUPAC source of isotopic abundances: *http://www.iupac.org/reports/1998/7001rosman/iso.pdf*.

- **Atomic Masses**

  A number of sources – choose the one that you like best. They all refer to the same basic data:

  - IUPAC list (*http://www.chem.qmul.ac.uk/iupac/AtWt/*).
  - 2003 evaluation including full list (*http://ie.lbl.gov/mass/2003AWMass_3.pdf*).
  - This site allows you to search by nuclide or nuclides. If you leave the form blank, you will get the complete list (*http://ie.lbl.gov/toi2003/ MassSearch.asp*).
  - For completeness, the Atomic Mass Data Center site (*http://www.nndc.bnl.gov/amdc/*).
    Access to the data is not as immediate as the other sources. The direct link to a plain text list is at *http://www.nndc.bnl.gov/amdc/nubase/nubtab03.asc*.

- **Elemental Abundances**

  These URLs provide the abundances of the chemical elements in the solar system, earth's crust and the earth's sea:

  - Elemental Abundances: (*http://ie.lbl.gov/elem/chem2.pdf*).
  - Properties of the Elements - Density, MP, BP, CP, Ionization Potential and Specific Heat, followed by list of abundances: (*http://ie.lbl.gov/elem/chem.pdf*).

- **Chemical Properties**

  The Wikipedia Periodic Table (above) links to files for each element. To select a particular element, replace the 'Periodic_Table' in the URL with the name of the element. For example, for copper the URL is *http://en.wikipedia.org/wiki/Copper*.

- **Chemistry of the Elements**

  Many years ago, the US National Academy of Sciences, within the National Research Council Nuclear Science Series, created a useful series of monographs on *Radiochemistry and Radiochemical Techniques*. At a time when much of activation analysis involved chemical separation, these were invaluable. They are now on the Internet at *http://lib-www.lanl.gov/radiochemistry/elements.htm*.

A.5 MISCELLANEOUS INFORMATION

I list here a number of links that do not fit into any previous category but may be of interest to some readers:

- Glossary of Nuclear Science (LBL)(*http://ie.lbl.gov/education/glossary/glossaryf.htm*).
- The Nuclear Spectrometry Users' Forum - an active UK – based group of alpha and gamma spectrometrists (*http://www.npl.co.uk/nsuf/*).

- The NIST Reference on Constants, Units, and Uncertainty (http://physics.nist.gov/cuu/).
- N-BASE Nuclear Information Service (http://www.n-base.org.uk).
- RADCH-L – the radiochemical mailing list. The list is administered by Gilles Montavon from the Laboratory SUBATECH in Nantes. It carries messages of information and requests for assistance from radiochemists around the world – very worthwhile. To subscribe to the list, e-mail the simple message 'SUB RADCH-L your-name' to *listserv@in2p3.fr*. More details at *http://www.gammaspectrometry.co.uk/radch.htm*.

An immensely useful resource that I have used extensively in recent years is *Science Direct* (*http://www.sciencedirect.com*). This provides access to all of the journals published by Elsevier. There is a good search facility to allow relevant articles to be found. It is also possible to put oneself on an e-mail list to be sent the list of contents of selected journals as they are published – that alone has saved me many a wasted journey to the library! Non-registered users only have access to the article abstracts, but full copies can be bought on-line by using a credit card. Persons registered, personally, through their place of work or university, have full download access to .pdf files of the full articles. The facility is pricey, but it does display author's contact e-mail addresses to allow you to make a request for a copy in the time-honoured fashion.

A.6 OTHER PUBLICATIONS IN PRINT

Suggestions for additional reading matter were given in the body of the book at the end of each chapter. Here I list general texts I have found most useful.

- Knoll, G.F. (2000). *Radiation Detection and Measurement*, 3rd Edn, John Wiley & Sons, Inc. New York, NY USA.

An excellent and indispensable text on radiation measurement in general – particularly good on explaining electronics and detection mechanisms. A paperback version is also available.

- Debertin, K. and Helmer, R.G. (1988). *Gamma and X-Ray Spectrometry with Semiconductor Detectors*, North Holland, Amsterdam, The Netherlands.

Packed with interesting information, detailed on analysis, although somewhat sketchy on hardware.

- Tsoulfanidis, N. (1995). *Measurement and Detection of Radiation*, 2nd Edn, Taylor & Francis, Inc., London.

Covers more-or-less the same ground but a little less detailed than Knoll (1989).

- Birks, J.B. (1964). *The Theory and Practice of Scintillation Counting*, Pergamon, Oxford, UK.

A classic text.

- Longworth, G. (Ed.) (1998). *The Radiochemical Manual*, AEA Technology plc, HMSO, London, UK.

This is a useful explanation of a range of radiochemical matters. It comes with a CD containing a large catalogue of nuclear data. That should be regarded as being superseded by DDEP (see Section A.2.4). That does not diminish the value of this manual from a general informative point of view.

- Quittner, P. (1972). *Gamma-Ray Spectrometry with Particular Reference to Detector and Computer Evaluation Techniques*, Hilger, London, UK.

An interesting slim volume.

- Jenkins, R., Gould, R.W. and Gedcke, D. (1981). *Quantitative X-Ray Spectrometry*, Marcel Dekker, New York, NY, USA.

Chapter 4, on dead time losses, is of particular interest.

- Evans, R. D. (1982 and later editions). *The Atomic Nucleus*, Krieger Publishing Company, Melbourne, Florida, USA.

Still authoritative on basics, despite its age. A 2003 edition is listed but appears to be out-of-print.

- Ehmann, W.D. and Vance, D.E. (1991). *Radiochemical and Nuclear Methods of Analysis*, John Wiley & Sons, Inc., New York, NY, USA.

A readable and up-to-date text covering much of the material in Chapter 1 of this book.

Appendix B: Gamma- and X-Ray Standards for Detector Calibration

This appendix contains evaluated and recommended data on a selected set of radionuclides suitable for use in the energy and efficiency calibration of detectors. The data in the first edition of this book were taken from *X-Ray and Gamma-Ray Standards for Detector Calibration*, published by the International Atomic Energy Agency as IAEA TECDOC-619. Since that time, the data have been re-evaluated and extended within the international Decay Date Evaluation Project (DDEP). After much delay, the updated data were published by the IAEA as XGAMMA (*X-ray and Gamma-ray Decay Data Standards for Detector Calibration and Other Applications*) in late spring, 2007.

Rather than reproduce the whole of that report, which is accessible on the internet at *http://www-nds.iaea.org/xgamma_standards/*, I have restricted myself to updating the data for nuclides in the original table. For convenience, the gamma-ray and X-ray standards, listed separately in XGAMMA, are here combined below into one table.

Notes

- The 'decay mode' shown is the major mode but is frequently not the only one.
- All half-lives are in days. This may appear cumbersome for the longer half-lives, but the year is not a unit approved for use with the SI (a calendar year is of variable length; in the long term, 1 year (International symbol, a) = 365.242 198 78 d, and this is sometimes used).
- The uncertainties shown are estimated standard uncertainties, and refer to the uncertainty of the last one or two digits – thus, 950.8 (9) means 950.8 ± 0.9 and 0.999 35 (15) means $0.999\,35 \pm 0.00015$.
- In general, X-rays below 10 keV have been omitted.
- Where appropriate, emissions have been identified as particular X-rays or gamma-rays emitted by a particular daughter nuclide. Unidentified emissions are gamma-rays from the nuclide for which the data are quoted.

Table B.1 Gamma-ray and X-ray standards

Nuclide	Decay Mode	Half-life (d)	Emission, ID	Energy (keV)	Emission probability, P_γ
^{22}Na	EC	950.57 (23)		511.00 1274.537 (3)	1.798 (2) 0.9994 (14)
^{24}Na	β^-	0.623 29 (6)		1368.626 (5) 2754.007 (11)	0.999 935 (5) 0.998 72 (8)

Practical Gamma-ray Spectrometry – 2nd Edition   Gordon R. Gilmore
© 2008 John Wiley & Sons, Ltd

Table B.1　(Continued)

Nuclide	Decay Mode	Half-life (d)	Emission, ID	Energy (keV)	Emission probability, P_γ
^{46}Sc	β^-	83.79 (4)	Ti K	4.51	0.000 047 (2)
				889.271 (2)	0.999 833 (5)
				1120.537 (3)	0.999 86 (−4, +36)
^{51}Cr	EC	27.7009 (20)	V Kα	4.94–4.95	0.202 (3)
			V Kβ	5.43–5.46	0.0269 (7)
				320.0835 (4)	0.0987 (5)
^{54}Mn	EC	312.29 (26)	Cr Kα	5.405–5.415	0.002 27 (3)
			Cr Kβ	5.947	0.000 305 (7)
				834.838 (5)	0.999 746 (11)
^{55}Fe	EC	1002.7 (23)	Mn L	0.556–0.721	0.0066 (10)
			Mn Kα2	5.8877	0.0845 (14)
			Mn Kα1	5.8988	0.1656 (27)
			Mn Kβ′1	6.49–6.54	0.034 (7)
^{56}Co	EC	77.236 (26)	Fe Kα2	6.390 91(5)	0.0753 (10)
			Fe Kα1	6.403 91 (3)	0.1475 (17)
			Fe Kβ′1	7.058–7.108	0.0305 (5)
				846.7638 (19)	0.999 399 (23)
				977.363 (4)	0.014 22 (7)
				1037.8333 (24)	0.1403 (5)
				1175.0878 (22)	0.022 49 (9)
				1238.2736 (22)	0.6641 (16)
				1360.196 (4)	0.0428 (13)
				1771.327 (3)	0.1545 (4)
				2015.176 (5)	0.030 17 (14)
				2034.752 (5)	0.077 41 (13)
				2598.438 (4)	0.1696 (4)
				3009.559 (4)	0.010 38 (19)
				3201.93 (11)	0.032 03 (13)
				3253.402 (5)	0.0787 (3)
				3272.978 (6)	0.018 55 (9)
				3451.119 (4)	0.009 42 (6)
^{57}Co	EC	271.8 (5)	Fe Kα2	6.390 84	0.168 (3)
			Fe Kα1	6.403 84	0.332 (5)
			Fe Kβ′1	7.058–7.108	0.071 (2)
				14.412 95 (31)	0.0915 (17)
				122.060 65 (12)	0.8551 (6)
				136.473 56 (29)	0.1071 (15)
^{58}Co	EC	70.86 (6)	Fe Kα	6.4	0.235 (3)
			Fe Kβ	7.06	0.032 (10)
			Ni Kα	7.46–7.48	0.000 098 (3)
			Ni Kβ	8.26–8.33	0.000 013 6 (5)
				511	0.3 (4)
				810.759 (2)	0.9945 (1)
^{60}Co	β^-	1925.23 (27)	Ni Kα	7.46–7.48	0.000 098 (3)
			Ni Kβ	8.26–8.33	0.000 013 6 (5)
				1173.228 (3)	0.9985 (3)
				1332.492 (4)	0.999 826 (6)

^{65}Zn	EC	243.86 (20)	Cu Kα	8.03–8.05	0.347 (3)
			Cu Kβ	8.90–8.98	0.0482 (7)
				511.00	0.0284 (4)
				1115.539 (2)	0.506 (22)
^{75}Se	EC	119.778 (29)	As L	1.28	0.0206 (7)
			As Kα2	10.508	0.1659 (23)
			As Kα1	10.5437	0.322 (4)
			As Kβ	11.72–11.86	0.0764 (12)
				66.0518 (8)	0.011 12 (12)
				96.734 (9)	0.0342 (3)
				121.1155 (11)	0.172 (3)
				136.0001 (6)	0.582 (7)
				198.606 (12)	0.0148 (4)
				264.6576 (9)	0.589 (3)
				279.5422 (10)	0.2499 (13)
				303.9236 (10)	0.01316 (8)
				400.6572 (8)	0.1147 (9)
^{85}Sr	EC	64.851 (5)	Rb Kα2	13.3359(2)	0.1716 (17)
			Rb Kα1	13.3955 (1)	0.3304 (29)
			Rb Kβ′1	14.95–15.09	0.0804 (10)
			Rb Kβ′2	15.19–15.21	0.0093 (4)
				514.0048 (22)	0.985 (4)
^{88}Y	EC	106.625 (24)	Sr Kα2	14.098(1)	0.173 (22)
			Sr Kα1	14.1652 (2)	0.332 (4)
			Sr Kβ′1	15.8359 (4)	0.0821 (12)
			Sr Kβ′2	16.0847 (6)	0.0107 (4)
				898.036 (4)	0.939 (23)
				1836.052 (13)	0.9938 (3)
^{93m}Nb	IT	5.73 (22) × 10^3	Nb Kα2	16.5213	0.0316 (7)
			Nb Kα1	16.6152	0.0604 (12)
			Nb Kβ′1	18.618	0.0156 (5)
			Nb Kβ′2	18.953	0.0023 (1)
				30.77 (2)	0.000 559 (16)
^{94}Nb	β$^-$	7.3 (9) × 10^6	—	702.639 (4)	0.998 15 (6)
				871.114 (3)	0.998 92 (3)
^{95}Nb	β$^-$	34.991 (6)	Mo Kα2	17.374	0.000 286 (9)
			Mo Kα1	17.479	0.000 546 (17)
			Mo Kβ′1	19.59–19.77	0.000 143 (5)
			Mo Kβ′2	19.96–20.00	0.000 022 (11)
				765.803 (6)	0.998 08 (7)
^{109}Cd	EC	461.4 (12)	Ag Kα2	21.9906(2)	0.2899 (25)
			Ag Kα1	22.1632 (1)	0.547 (4)
			Ag Kβ′1	24.912–25.146	0.1514 (18)
			Ag Kβ′2	25.457–25.512	0.0263 (10)
				88.0336 (11)	0.036 26 (20)
^{111}In	EC	2.8049 (6)	Cd Kα2	22.9843	0.236 (2)
			Cd Kα1	23.1738	0.444 (3)
			Cd Kβ′1	26.061–26.304	0.124 (4)
			Cd Kβ′2	26.64–26.70	0.023 (1)
				171.28 (3)	0.9066 (25)
				245.35 (4)	0.9409 (6)
^{113}Sn	EC	115.09 (4)	In Kα2	24.002	0.2785 (22)
			In Kα1	24.2097	0.522 (4)

Table B.1   (Continued)

Nuclide	Decay Mode	Half-life (d)	Emission, ID	Energy (keV)	Emission probability, P_γ
			In Kβ'1	27.238–27.499	0.146 (12)
			In Kβ'2	27.861–27.940	0.0284 (2)
				255.134 (10)	0.0211 (8)
				391.698 (3)	0.6494 (17)
^{125}Sb	β⁻	1007.48 (21)	Te Kα2	27.202 (2)	0.191 (7)
			Te Kα1	27.4726 (2)	0.357 (12)
			Te Kβ'1	30.945–31.236	0.102 (4)
			Te Kβ'2	31.701–31.774	0.0221 (10)
				176.314 (2)	0.0682 (7)
				380.452 (8)	0.0152 (15)
				427.874 (4)	0.2955 (24)
				463.365 (4)	0.1048 (9)
				600.597 (2)	0.1776 (18)
				606.713 (3)	0.0502 (5)
				635.95 (3)	0.1132 (10)
				671.441 (6)	0.017 83 (16)
^{125}I	EC	59.402 (14)	Te Kα2	27.202 (2)	0.397 (6)
			Te Kα1	27.4726 (2)	0.74 (11)
			Te Kβ'1	30.945–31.241	0.212 (4)
			Te Kβ'2	31.701–31.812	0.0459 (14)
				35.4919 (5)	0.0667 (17)
^{133}Ba	EC	3848.7 (12)	Cs Kα2	30.625	0.34 (4)
			Cs Kα1	30.973	0.628 (7)
			Cs Kβ'1	34.92–35.26	0.182 (2)
			Cs Kβ'2	35.82–35.97	0.046 (1)
				53.1622 (6)	0.0214 (3)
				79.6142 (12)	0.0265 (5)
				80.9979 (11)	0.329 (3)
				276.3989 (12)	0.0716 (5)
				302.8508 (5)	0.1834 (13)
				356.0129 (7)	0.6205 (19)
				383.8485 (12)	0.0894 (6)
^{134}Cs	β⁻	753.5 (10)	—	563.243 (3)	0.0837 (3)
				569.327 (3)	0.1538 (4)
				604.72 (3)	0.9765 (18)
				795.83 (3)	0.855 (3)
				801.945 (4)	0.087 (3)
				1365.186 (4)	0.030 17 (12)
^{137}Cs	β⁻	1.099 (4) × 10⁴	Ba L	3.954–5.973	0.009 (5)
			Ba Kα2	31.8174	0.0195 (4)
			Ba Kα1	32.1939	0.0359 (7)
			Ba Kβ'1	36.31–36.67	0.010 55 (22)
			Ba Kβ'2	37.26–37.43	0.002 66 (8)
				661.657 (3)	0.8499 (20)
^{139}Ce	EC	137.642 (20)	La Kα2	33.0344 (2)	0.225 (3)
			La Kα1	33.4421 (1)	0.412 (4)
			La Kβ'1	37.721–38.095	0.123 (18)
			La Kβ'2	38.730–38.910	0.0311 (6)
				165.8575 (11)	0.799 (4)

^{152}Eu	EC β$^-$	4941 (7)	Sm Kα2	39.5229	0.208 (3)
			Sm Kα1	40.1186	0.377 (5)
			Sm Kβ'1	45.289–45.731	0.1178 (19)
			Sm Kβ'2	46.575–46.813	0.0304 (8)
				121.7817 (3)	0.2841 (13)
				244.6974 (8)	0.0755 (4)
			β$^-$	344.2785 (12)	0.2658 (12)
			β$^-$	411.1165 (12)	0.022 37 (10)
				443.965 (3)	0.031 25 (14)
			β$^-$	778.9045 (24)	0.1296 (6)
				867.38 (3)	0.042 41 (23)
				964.072 (18)	0.1462 (6)
				1085.837 (10)	0.1013 (6)
			β$^-$	1089.737 (5)	0.017 31 (10)
				1112.076 (3)	0.134 (6)
				1212.948 (11)	0.014 15 (9)
			β$^-$	1299.142 (8)	0.016 32 (9)
				1408.013 (3)	0.2085 (9)
^{154}Eu	β$^-$	3138.1 (14)	Gd Kα2	42.3093	0.072 (2)
			Gd Kα1	42.9967	0.13 (3)
			Gd Kβ'1	48.556–49.053	0.041 (1)
			Gd Kβ'2	49.961–50.219	0.0108 (3)
				123.0706 (9)	0.404 (5)
				247.9288 (7)	0.0689 (7)
				591.755 (3)	0.0495 (5)
				692.4205 (18)	0.0179 (3)
				723.3014 (22)	0.2005 (21)
				756.802 (23)	0.0453 (5)
				873.1834 (23)	0.1217 (12)
				996.262 (6)	0.105 (10)
				1004.725 (7)	0.1785 (17)
				1246.121 (4)	0.008 62 (8)
				1274.429 (4)	0.349 (3)
				1596.4804 (28)	0.017 83 (17)
^{155}Eu	β$^-$	1736 (6)	—	26.531 (21)	0.003 16 (22)
			Gd Kα2	42.3093	0.067 (13)
			Gd Kα1	42.9967	0.1205 (23)
				45.299 (10)	0.0131 (5)
			Gd Kβ'1	48.556–49.053	0.0384 (11)
			Gd Kβ'2	49.961–50.219	0.0098 (3)
				60.0086 (10)	0.0122 (5)
				86.0591 (10)	0.001 54 (17)
				86.5479 (10)	0.307 (3)
				105.3083 (10)	0.211 (6)
^{198}Au	β$^-$	2.695 (7)	Hg Kα2	68.8952 (12)	0.008 09 (8)
			Hg Kα1	70.8196 (12)	0.013 72 (12)
			Hg Kβ'1	79.82–80.76	0.004 66 (8)
			Hg Kβ'2	82.43–83.03	0.001 36 (4)
				411.802 05 (17)	0.9554 (7)
				675.8836 (7)	0.008 06 (7)
				1087.6842 (7)	0.001 59 (3)
^{203}Hg	β$^-$	46.594 (12)	Tl L	8.953–14.738	0.0543 (9)
			Tl Kα2	70.8325 (8)	0.0375 (4)

Table B.1   (Continued)

Nuclide	Decay Mode	Half-life (d)	Emission, ID	Energy (keV)	Emission probability, P_γ
			Tl Kα1	72.8725 (8)	0.0633 (6)
			Tl Kβ'1	82.118–83.115	0.0215 (4)
			Tl Kβ'2	84.838–85.530	0.0064 (2)
				279.1952 (10)	0.8148 (8)
^{207}Bi	EC	1.18 (3) $\times 10^4$	Pb L	9.18–15.84	0.332 (14)
			Pb Kα2	72.805	0.2169 (24)
			Pb Kα1	74.97	0.365 (4)
			Pb Kβ'1	84.451–85.470	0.1246 (23)
			Pb Kβ'2	87.238–88.003	0.0376 (10)
				569.698 (2)	0.9776 (3)
				1063.656 (3)	0.7458 (49)
				1770.228 (9)	0.0687 (3)
^{228}Th	α	698.6 (23)	Tl Ll	8.953	0.001 69 (9)
(With its daughters in equilibrium.			Pb L	9.184–15.216	0.0104 (2)
^{208}Tl emission probabilities have			Tl Lα	10.172–10.268	0.0326 (17)
been adjusted for ^{212}Bi branching)			Tl Lβη	10.994–12.643	0.0272 (15)
			Ra Lα	12.196–12.339	0.0286 (15)
			Ra Lβη	13.662–15.447	0.047 (3)
			Tl Lγ	14.291–14.738	0.005 (2)
			Ra Lγ	17.848–18.412	0.0102 (6)
			Pb Kα2	72.8049 (8)	0.0077 (2)
			Bi Kα2	74.8157 (9)	0.107 (3)
			Pb Kα1	74.97 (9)	0.013 (3)
			Bi Kα1	77.1088 (10)	0.179 (5)
				84.373 (3)	0.0117 (5)
			Pb Kβ'1	84.451–85.470	0.0044 (2)
			Bi Kβ'1	86.835–87.862	0.0612 (20)
			Pb Kβ'2	87.238–88.003	0.001 34 (5)
			Bi Kβ'2	89.732–90.522	0.0187 (7)
			^{212}Pb	115.183 (5)	0.006 23 (22)
				131.612 (4)	0.001 24 (6)
				215.985 (4)	0.002 26 (20)
			^{212}Pb	238.632 (2)	0.436 (3)
			^{208}Tl	277.37 (3)	0.0237 (11)
			^{212}Pb	300.09 (1)	0.0318 (13)
			^{208}Tl	583.187 (2)	0.3055 (17)
			^{212}Bi	727.33 (1)	0.0674 (12)
			^{212}Bi	785.37 (9)	0.0111 (1)
			^{208}Tl	860.56 (3)	0.0448 (4)
			^{212}Bi	1620.74 (1)	0.0151 (3)
			^{208}Tl	2614.511 (10)	0.3585 (7)
^{239}Np	β$^-$	2.35 (4)	Pu Kα1	99.525	0.135 (4)
(Data from DDEP)			Pu Kα2	103.734	0.214 (6)
				106.125 (2)	0.259 (3)
				228.183 (1)	0.1132 (22)
				277.599 (1)	0.144 (1)
^{241}Am	α	1.5785 (23) $\times 10^5$	Np Ll	11.89 (2)	0.008 48 (10)
			Np Lα	13.9 (2)	0.1303 (10)
			Np Lβη	17.81 (2)	0.1886 (15)
			Np Lγ	20.82 (2)	0.0481 (4)

				26.3446 (2)	0.024 (3)
				33.1963 (3)	0.001 21 (3)
				59.5409 (1)	0.3578 (9)
^{243}Am	α	2.692 (8) × 10^6	Np Ll	11.871	0.004 45 (14)
			Np Lα	13.761–13.946	0.0705 (20)
			Np Lη	15.861	0.001 26 (4)
			Np Lβ	16.109–17.992	0.0818 (16)
			Np Lγ	20.784–21.491	0.0197 (4)
				43.53 (2)	0.0589 (10)
				74.66 (2)	0.672 (12)

Appendix C: X-Rays Routinely Found in Gamma Spectra

The table below (Table C.1) is intended to assist gamma spectrometrists in identifying X-rays in their spectra. This includes all those associated with nuclides mentioned in the text, those sometimes visible in the background and one or two extra items that have caught the author unawares in the past. The data are taken from an on-line source provided by the US Lawrence Berkeley Laboratory. All energies have been rounded to two decimal places. You should be aware that other lower intensity X-rays will accompany those listed in the table. In particular, the $K\beta1$ and $K\beta2$ X-rays are accompanied by $K\beta3$ and $K\beta4$, respectively, at almost the same energy.

Table C.1 X-ray energies associated with nuclides discussed within the book[a,b]

	Z	X-ray designation						Possible origin[c]
		$L\alpha1$	$L\beta1$	$K\alpha2$	$K\alpha1$	$K\beta1$	$K\beta2$	
V	23	—	—	4.94	4.95	5.43		[51]Cr
Cr	24	—	—	5.41	5.42	5.95		[54]Mn
Mn	25	—	—	5.89	5.90	6.49		[55]Fe
Fe	26	—	—	6.39	6.40	7.06		[56,57,58]Co
Ni	28	—	—	7.46	7.48	8.27		[60]Co
Cu	29	—	—	8.03	8.05	8.91		Fluorescence, [65]Zn
As	33	—	—	10.51	10.54	11.73	11.86	[75]Se
Rb	37	—	—	13.34	13.40	14.96	15.19	[85]Sr
Sr	38	—	—	14.10	14.17	15.84	16.09	[88]Y
Nb	41	—	—	16.52	16.62	18.62	18.95	[95]Zr
Ag	47	—	—	21.99	22.16	24.94	25.46	[109]Cd
Cd	48	—	—	22.98	23.17	26.10	26.64	Fluorescence
In	49	—	—	24.00	24.21	27.28	27.86	Fluorescence, [113]Sn
Sn	50	—	—	25.04	25.27	28.49	29.11	Fluorescence
Ba	56	—	—	31.82	32.19	36.38	37.26	[137]Cs
La	57	—	—	33.03	33.44	37.80	38.73	[139]Ce
Sm	62	—	—	39.52	40.12	45.41	46.58	[152]Eu
W	74	—	—	57.98	59.32	67.24	69.07	'Self-fluorescence' in 'slags'
Tl	81	—	—	70.83	72.87	82.57	84.87	[203]Hg, [211,212]Bi
Pb	82	—	—	72.80	74.97	84.94	87.30	Fluorescence, [208]Tl, [214]Po
Bi	83	—	—	74.82	77.11	87.35	89.78	[211,212,214]Pb
Po	84	—	—	76.86	79.29	89.81	92.32	[212,214]Bi, [219]Rn

Practical Gamma-ray Spectrometry – 2nd Edition   Gordon R. Gilmore
© 2008 John Wiley & Sons, Ltd

Table C.1   (Continued)

	Z	X-ray designation						Possible origin[c]
		Lα1	Lβ1	Kα2	Kα1	Kβ1	Kβ2	
Rn	86			81.07	83.79	94.87	97.53	^{223,224}Ra
Ra	88	12.34	15.24	85.43	88.47	100.13	102.95	^{227}Th
Th	90	12.97	16.20	89.96	93.35	105.60	108.58	^{235}U
Pa	91	13.29	16.71	92.28	95.86	108.42	111.49	^{231,234}Th
U	92	13.62	17.22	94.65	98.43	111.30	114.45	^{233,234m}Pa, Pu, Np isotopes
Np	93	13.95	17.75	97.07	101.06	114.23	117.46	Am isotopes, ^{237}Pu
Pu	94	14.28	18.30	99.53	103.73	117.23	120.54	Np isotopes
Am	95	14.62	18.86	102.03	106.47	120.28	123.68	^{243}Pu

[a] Source of data: http://ie.lbl.gov/atomic/x2.pdf.
[b] Energies are rounded to two decimal places. Kβ1 and Kβ2 are also associated with smaller Kβ3 and Kβ4 X-rays, respectively, at almost the same energy.
[c] 'Fluorescence' in this context, refers to gamma rays from the source, acting on the materials of the detector shielding.

Appendix D: Gamma-Ray Energies in the Detector Background and the Environment

Changes have been made to this list from the first edition (Table D.1 below). Gone are the 'Chernobyl' nuclides, because, under normal circumstances, these are no longer detected in background spectra. The number of gamma-rays emitted by the uranium and thorium decay series has been increased and the excitation products discussed in Chapter 13 included. Most of those are unlikely to be observed unless the count time is long or the detector very large.

Another feature removed from the list is the half-lives of the nuclides. The half-life is little help in identifying the source of a particular gamma-ray in the background. Either the emitting nuclide is of considerable half-life or, in many cases, is being supported by the decay of a longer-lived parent. In the case of excitations of the detector and its surroundings, the nuclide activity is maintained in a state of equilibrium by the flux of particles bombarding them.

The list now represents what is likely to be observed in a 200 000 s background spectrum measured by a 50 % detector housed in a typical commercial shield in a routine ground-level counting room in an 'unremarkable' geological area of the UK. ^{228}Ac and ^{214}Bi emit hundreds of gamma-rays with low emission probability not included in the list. From time to time, particular with very long counts, some of these may be detected.

- Peaks chosen for inclusion in the list are from:

  (1) The primordial nuclides, ^{40}K, ^{235}U, ^{238}U and ^{232}Th and their daughters.

  (2) A few common reactor activation products that are often present in background.
  (3) A number of nuclides created by neutron reactions with the detector and shielding materials – the source of the neutrons involved might be cosmic or proximity to a nuclear reactor or accelerator.
  (4) The major 'fluorescence' X-rays from likely shielding materials – Pb, Sn, Cd and Cu.

- Data are taken from the following sources in order of priority (see Appendix A for details):

  - DDEP data via the LARA database.
  - IAEA XGAMMA data.
  - For nuclides not listed in those sources, the on-line table of isotopes at *http://ie.lbl.gov/education/isotopes.htm*.
  - For excited-state energies – US National Nuclear Data Center, 'Levels and Gamma Search' at *http://www.nndc.bnl.gov/nudat2/*.
  - For X-ray energies and intensities – the LBNL data at *http://ie.lbl.gov/atomic/x2.pdf*.

- With the exception of a few cases where the quoted precision will not allow it, energies are rounded to two decimal places. Emission probabilities are quoted to the precision given in the source. For X-rays, the emission probabilities quoted are 'intensity per 100 K-shell vacancies'.

- The most prominent background peaks seen in a shielded detector are in bold type.

Table D.1   Gamma-ray energies in the background and the environment

Energy	Nuclide[a]	$P\gamma$ (%)	Related peaks	Source of radiation
8.04	CuKα	29.3	8.91	Fluorescence from shielding
8.91	CuKβ	4.7	8.04	Fluorescence from shielding
22.98	CdKα_2	24.5	23.17	Fluorescence from shielding
23.17	CdKα_1	46.1	22.98	Fluorescence from shielding
25.04	SnKα_2	24.7	25.27	Fluorescence from shielding
25.27	SnKα_1	45.7	25.04	Fluorescence from shielding
26.10	CdKβ_1	7.69	22.98	Fluorescence from shielding
26.64	CdKβ_2	1.98	22.98	Fluorescence from shielding
28.49	SnKβ_1	7.99	25.27	Fluorescence from shielding
29.11	SnKβ_2	2.19	25.27	Fluorescence from shielding
46.54	**^{210}Pb**	**4.25**	**none**	**^{238}U (^{226}Ra) series**
53.23	^{214}Pb	1.060	295.22 (18.50), 351.93 (35.60)	^{238}U (^{226}Ra) series
53.44	^{73m}Ge	10.34	—	^{72}Ge(n, γ), ^{74}Ge(n, 2n)
63.28	**^{234}Th**	**4.8**	**92.58 (5.58)**	**^{238}U series**
68.75	^{73*}Ge	—	—	^{73}Ge(n, n$'$) broad asymmetric peak
72.81	PbKα_2	27.7	74.97 (46.2)	Fluorescence and ^{208}Tl decay
74.82	**BiKα_2**	**27.7**	**77.11 (46.2)**	**^{212,214}Pb decay**
74.97	PbKα_1	46.2	72.81 (27.7)	Fluorescence and ^{208}Tl decay
77.11	**BiKα_1**	**46.2**	**74.82 (27.7)**	**^{212,214}Pb decay**
79.29	PoKα_1	46.1	—	Fluorescence and ^{212,214}Bi decay
81.23	^{231}Th	0.90	—	^{235}U series
84.94	PbKβ_1	10.7	74.97 (46.2)	Fluorescence and ^{208}Tl decay
87.30	PbKβ_2	3.91	74.97 (46.2)	Fluorescence and ^{208}Tl decay
87.35	BiKβ_1	10.7	74.82 (27.7)	^{212,214}Pb decay
89.78	BiKβ_2	3.93	74.82 (27.7)	^{212,214}Pb decay
89.96	ThKα_2	28.1	93.35 (45.4)	^{235}U and ^{228}Ac decay
92.58	**^{234}Th**	**5.58**	**63.28 (4.8)**	**^{238}U series – doublet**
93.35	ThKα_1	45.4	89.96 (28.1)	^{235}U and ^{228}Ac decay
105.60	ThKβ_1	10.7	93.35 (45.4)	^{235}U and ^{228}Ac decay
109.16	^{235}U	1.54	185.72 (57.2)	Primordial
112.81	^{234}Th	0.28	63.28 (4.8), 92.58 (5.58)	^{238}U series
122.32	^{223}Ra	1.192	269.49 (13.7)	^{235}U series
129.06	^{228}Ac	2.42	911.20 (25.8), 968.97 (15.8)	^{232}Th series
139.68	^{75m}Ge	39	—	^{74}Ge(n, γ), ^{76}Ge(n, 2n)
143.76	^{235}U	10.96	—	Primordial
159.7	^{77m}Ge	10.33	—	^{76}Ge(n, γ)
163.33	^{235}U	5.08	185.72 (57.2)	Primordial
174.95	^{71m}Ge	Very small	198.39 ($\approx$ 100)	^{70}Ge(n, γ) activation (summed-out)
185.72	**^{235}U**	**57.2**	**143.76 (10.96)**	**Primordial**
186.21	**^{226}Ra**	**3.555**	**none**	**^{238}U series**
198.39	^{71m}Ge	Sum	—	^{70}Ge(n, γ)

205.31	^{235}U	5.01	185.72 (57.2)	Primordial
209.26	^{228}Ac	3.89	911.20 (25.8), 968.97 (15.8)	^{232}Th series
238.63	**^{212}Pb**	**43.6**	**300.09 (3.18)**	**^{232}Th series**
240.89	^{224}Ra	4.12	—	^{232}Th series
242.00	^{214}Pb	7.268	295.22 (18.50), 351.93 (35.6)	^{238}U (^{226}Ra) series
269.49	^{223}Ra	13.7	122.32 (1.192)	^{235}U series
270.24	^{228}Ac	3.46	911.20 (25.8), 968.97 (15.8)	^{232}Th series
277.37	^{208}Tl	2.37	583.19 (30.6), 2614.51 (35.85)	^{232}Th series^b
278.26	^{64*}Cu	—	—	^{63}Cu(n, γ), ^{65}Cu(n, 2n) prompt γ
295.22	**^{214}Pb**	**18.50**	**351.93 (35.60)**	**^{238}U (^{226}Ra) series**
299.98	^{227}Th	2.16	—	^{235}U series
300.07	^{231}Pa	2.47	—	^{235}U series
300.09	^{212}Pb	3.18	238.63 (43.6)	^{232}Th series
328.00	^{228}Ac	2.95	911.20 (25.8), 968.97 (15.8)	^{232}Th series
336.24	^{115m}Cd/^{115m}In	45.9	527.90 (27.5)	Activation of Cd (daughter of ^{115}Cd)
338.28	^{223}Ra	2.79	—	^{235}U series
338.32	^{228}Ac	11.27	911.20 (25.8), 968.97 (15.8)	^{232}Th series
351.06	^{211}Bi	12.91	—	^{235}U series
351.93	**^{214}Pb**	**35.60**	**295.22 (18.50)**	**^{238}U (^{226}Ra) series**
409.46	^{228}Ac	1.92	911.20 (25.8), 968.97 (15.8)	^{232}Th series
416.86	^{116m}In	27.7	—	^{115}In(n, γ) activation of In metal seal
462.00	^{214}Pb	0.213	295.22 (18.50), 351.93 (35.6)	^{238}U (^{226}Ra) series
463.00	^{228}Ac	4.40	911.20 (25.8), 968.97 (15.8)	^{232}Th series
477.60	^{7}Be	10.44	None	Cosmic
510.7	^{208}Tl	6.29	583.19 (30.6), 2614.51 (35.85)	^{232}Th series^b
511.00	**Annihilation**	—	—	**Annihilation radiation (β^+)**
527.90	^{115}Cd	27.5	336.2(45.9)	^{114}Cd(n, γ) activation
558.46	^{114*}Cd	—	—	^{113}Cd(n, γ) prompt γ
569.70	^{207m}Pb	97.87	—	^{207}Pb(n, n')
570.82	^{228}Ac	0.182	911.20 (25.8), 968.97 (15.8)	^{232}Th series
579.2	^{207*}Pb	—	—	^{207}Pb(n, n') prompt γ
583.19	**^{208}Tl**	**30.6**	**2614.51 (35.85)**	**^{232}Th series^b**
595.85	^{74*}Ge	—	—	^{74}Ge(n, n') broad asymmetric peak
609.31	**^{214}Bi**	**45.49**	**1120.29 (14.907), 1764.49 (15.28)**	**^{238}U (^{226}Ra) series**
661.66	^{137}Cs	84.99	None	Fission
669.62	^{63*}Cu	—	—	^{63}Cu(n, n') prompt γ
689.6	^{72*}Ge	—	—	^{72}Ge(n, n') broad asymmetric peak
726.86	^{228}Ac	0.62	911.20 (25.8), 968.97 (15.8)	^{232}Th series
727.33	^{212}Bi	6.74	1620.74 (1.51)	^{232}Th series
755.31	^{228}Ac	1.00	911.20 (25.8), 968.97 (15.8)	^{232}Th series
768.36	^{214}Bi	4.891	609.31 (45.49), 1764.49 (15.28)	^{238}U (^{226}Ra) series
794.95	^{228}Ac	4.25	911.20 (25.8), 968.97 (15.8)	^{232}Th series

Table D.1   (Continued)

Energy	Nuclide[a]	$P\gamma$ (%)	Related peaks	Source of radiation
803.06	^{206*}Pb	—	—	^{206}Pb(n, n') prompt γ
806.17	^{214}Bi	1.262	609.31 (45.49), 1764.49 (15.28)	^{238}U (^{226}Ra) series
832.01	^{211}Pb	3.52	—	^{235}U series
835.71	^{228}Ac	1.61	911.20 (25.8), 968.97 (15.8)	^{232}Th series
839.04	^{214}Pb	0.587	295.22 (18.50), 351.93 (35.6)	^{238}U (^{226}Ra) series
843.76	^{27}Mg	71.8	—	^{26}Mg(n, γ) or ^{27}Al(n, p) of encapsulation
846.77	^{56*}Fe	—	—	^{56}Fe(n, n')
860.56	^{208}Tl	4.48	583.19 (30.6), 2614.51 (35.85)	^{232}Th series[b]
911.20	**^{228}Ac**	**25.8**	**968.97 (15.8)**	**^{232}Th series**
934.06	^{214}Bi	3.096	609.31 (45.49), 1764.49 (15.28)	^{238}U (^{226}Ra) series
962.06	^{63*}Cu	—	—	^{63}Cu(n, n') prompt γ
964.77	^{228}Ac	4.99	911.20 (25.8), 968.97 (15.8)	^{232}Th series
968.97	**^{228}Ac**	**15.8**	**911.20 (25.8)**	**^{232}Th series**
1001.03	^{234m}Pa	1.021	—	^{238}U series (GRG empirical P_γ)
1014.44	^{27}Mg	28.0	—	^{26}Mg(n, γ) or ^{27}Al(n, p) of encapsulation
1063.66	^{207m}Pb	88.5	—	^{207}Pb(n, n')
1097.3	^{116}In	56.2	1293.54 (84.4)	^{115}In(n, γ) activation of In metal seal
1115.56	^{65*}Cu	—	—	^{65}Cu(n, n')
1120.29	^{214}Bi	14.907	609.31 (45.49), 1764.49 (15.28)	^{238}U (^{226}Ra) series
1155.19	^{214}Bi	1.635	609.31 (45.49), 1764.49 (15.28)	^{238}U (^{226}Ra) series
1173.23	^{60}Co	99.85	1332.49 (99.98)	Activation
1238.11	^{214}Bi	5.827	609.31 (45.49), 1764.49 (15.28)	^{238}U (^{226}Ra) series
1293.54	^{116}In	84.4	1097.3 (56.2)	^{115}In(n, γ) activation of In metal seal
1332.49	^{60}Co	99.98	1173.23 (99.85)	Activation
1377.67	^{214}Bi	3.967	609.31 (45.49), 1764.49 (15.28)	^{238}U (^{226}Ra) series
1407.98	^{214}Bi	2.389	609.31 (45.49), 1764.49 (15.28)	^{238}U (^{226}Ra) series
1459.14	^{228}Ac	0.83	911.20 (25.8), 968.97 (15.8)	^{232}Th series
1460.82	**^{40}K**	**10.66**	**None**	**Primordial**
1588.20	^{228}Ac	3.22	911.20 (25.8), 968.97 (15.8)	^{232}Th series
1620.74	^{212}Bi	1.51	727.33 (6.74)	^{232}Th series
1630.63	^{228}Ac	1.51	911.20 (25.8), 968.97 (15.8)	^{232}Th series
1729.60	^{214}Bi	2.843	609.31 (45.49), 1764.49 (15.28)	^{238}U (^{226}Ra) series
1764.49	**^{214}Bi**	**15.28**	**609.31 (45.49), 1764.49 (15.28)**	**^{238}U(^{226}Ra) series**
1847.42	^{214}Bi	2.023	609.31 (45.49), 1764.49 (15.28)	^{238}U (^{226}Ra) series
2204.21	^{214}Bi	4.913	609.31 (45.49), 1764.49 (15.28)	^{238}U (^{226}Ra) series
2224.57	^{2*}H	—	—	^{1}H(n, γ)
2614.51	**^{208}Tl**	**35.85**	**583.19 (30.6)**	**^{232}Th series; ^{208}Pb(n, p)[b]**

[a] Kβ_1 X-ray peaks are always a composite with the Kβ_3, at lower energy, and Kβ_5, at higher. Kβ_2 X-rays are accompanied by Kβ_4 and KO$_{2,3}$, both at higher energy.
[b] Emission probabilities for ^{208}Tl are quoted relative to the ^{228}Th parent and its other daughters.
[c] GRG, Gordon R. Gilmore.

Appendix E: Chemical Names, Symbols and Relative Atomic Masses of the Elements

- Information is largely taken from the IUPAC Technical Report: *Atomic Weights of the Elements 2001*, as published in *Pure Appl. Chem.*, **75**, 1107—1122 (2003).
- In the following tables (Table E.1 and E.2), the relative atomic masses (*RAM*) values have been rounded to two decimal places.
- An integer within brackets thus [227] is strictly not an atomic mass but is the mass number of one particular isotope of the element. The mass number is, however, quite close to the atomic mass of that isotope. It is given in this list when all isotopes of that element are radioactive and indicates the isotope with the longest known half-life. This is not necessarily the most common isotope. Other isotopic half-lives are listed in the IUPAC report referred to above.
  For the heaviest elements, say $Z > 100$, this longest half-life isotope will probably change with time, as experiments may well produce data from more stable parts of the Chart of the Nuclides.
- At the time of writing, darmstadtium is the IUPAC approved name for $Z = 110$, and roentgenium is likely to be the approved name for $Z = 111$.
- Atomic masses depend on the isotopic composition of the element. This can be quite variable. Uranium, for example, is well known in this respect and its provenance is important. Commercial lithium has an atomic mass varying between 6.939 and 6.996 – this is marked '#' in the tables. Non-terrestrial material is particularly susceptible to isotopic fractionation. See the IUPAC report for further information of great interest.

Table E.1 Ordered by name

Name	Symbol	Z	*RAM*
Actinium	Ac	89	[227]
Aluminum	Al	13	26.98
Americium	Am	95	[243]
Antimony	Sb	51	121.76
Argon	Ar	18	39.95
Arsenic	As	33	74.92
Astatine	At	85	[210]
Barium	Ba	56	137.33
Berkelium	Bk	97	[247]
Beryllium	Be	4	9.01
Bismuth	Bi	83	208.98
Bohrium	Bh	107	[264]
Boron	B	5	10.81
Bromine	Br	35	79.90
Cadmium	Cd	48	112.41
Calcium	Ca	20	40.08
Californium	Cf	98	[251]
Carbon	C	6	12.01
Cerium	Ce	58	140.12
Caesium	Cs	55	132.91
Chlorine	Cl	17	35.45
Chromium	Cr	24	52.00
Cobalt	Co	27	58.93
Copper	Cu	29	63.55
Curium	Cm	96	[247]
Darsmstadtium	Ds	110	[281]
Dubnium	Db	105	[262]
Dysprosium	Dy	66	162.50
Einsteinium	Es	99	[252]

Table E.1 (Continued)

Name	Symbol	Z	RAM
Erbium	Er	68	167.26
Europium	Eu	63	151.96
Fermium	Fm	100	[257]
Fluorine	F	9	19.00
Francium	Fr	87	[223]
Gadolinium	Gd	64	157.25
Gallium	Ga	31	69.72
Germanium	Ge	32	72.64
Gold	Au	79	196.97
Hafnium	Hf	72	178.49
Hassium	Hs	108	[277]
Helium	He	2	4.00
Holmium	Ho	67	164.93
Hydrogen	H	1	1.01
Indium	In	49	114.82
Iodine	I	53	126.90
Iridium	Ir	77	192.22
Iron	Fe	26	55.85
Krypton	Kr	36	83.80
Lanthanum	La	57	138.91
Lawrencium	Lr	103	[262]
Lead	Pb	82	207.20
Lithium #	Li	3	6.94
Lutetium	Lu	71	174.97
Magnesium	Mg	12	24.31
Manganese	Mn	25	54.94
Meitnerium	Mt	109	[268]
Mendelevium	Md	101	[258]
Mercury	Hg	80	200.59
Molybdenum	Mo	42	95.94
Neodymium	Nd	60	144.24
Neon	Ne	10	20.18
Neptunium	Np	93	[237]
Nickel	Ni	28	58.69
Niobium	Nb	41	92.91
Nitrogen	N	7	14.01
Nobelium	No	102	[259]
Osmium	Os	76	190.23
Oxygen	O	8	16.00
Palladium	Pd	46	106.42
Phosphorus	P	15	30.97
Platinum	Pt	78	195.08
Plutonium	Pu	94	[244]
Polonium	Po	84	[209]
Potassium	K	19	39.10
Praseodymium	Pr	59	140.91
Promethium	Pm	61	[145]
Protactinium	Pa	91	231.04
Radium	Ra	88	[226]
Radon	Rn	86	[222]
Rhenium	Re	75	186.21
Rhodium	Rh	45	102.91
Roentgenium	Rg	111	[280]

Name	Symbol	Z	RAM
Rubidium	Rb	37	85.47
Ruthenium	Ru	44	101.07
Rutherfordium	Rf	104	[261]
Samarium	Sm	62	150.36
Scandium	Sc	21	44.96
Seaborgium	Sg	106	[266]
Selenium	Se	34	78.96
Silicon	Si	14	28.09
Silver	Ag	47	107.87
Sodium	Na	11	22.99
Strontium	Sr	38	87.62
Sulfur	S	16	32.07
Tantalum	Ta	73	180.95
Technetiuum	Tc	43	[98]
Tellurium	Te	52	127.60
Terbium	Tb	65	158.93
Thallium	Tl	81	204.38
Thorium	Th	90	232.04
Thulium	Tm	69	168.93
Tin	Sn	50	118.71
Titanium	Ti	22	47.87
Tungsten	W	74	183.84
Uranium	U	92	238.03
Vanadium	V	23	50.94
Xenon	Xe	54	131.29
Ytterbium	Yb	70	173.04
Yttrium	Y	39	88.91
Zinc	Zn	30	65.41
Zirconium	Zr	40	91.22

Table E.2 Ordered by chemical symbol

Symbol	Name	Z	RAM
Ac	Actinium	89	[227]
Ag	Silver	47	107.87
Al	Aluminum	13	26.98
Am	Americium	95	[243]
Ar	Argon	18	39.95
As	Arsenic	33	74.92
At	Astatine	85	[210]
Au	Gold	79	196.97
B	Boron	5	10.81
Ba	Barium	56	137.33
Be	Beryllium	4	9.01
Bh	Bohrium	107	[264]
Bi	Bismuth	83	208.98
Bk	Berkelium	97	[247]
Br	Bromine	35	79.90
C	Carbon	6	12.01
Ca	Calcium	20	40.08
Cd	Cadmium	48	112.41
Ce	Cerium	58	140.12

Cf	Californium	98	[251]	No	Nobelium	102	[259]
Cl	Chlorine	17	35.45	Np	Neptunium	93	[237]
Cm	Curium	96	[247]	O	Oxygen	8	16.00
Co	Cobalt	27	58.93	Os	Osmium	76	190.23
Cr	Chromium	24	52.00	P	Phosphorus	15	30.97
Cs	Caesium	55	132.91	Pa	Protactinium	91	231.04
Cu	Copper	29	63.55	Pb	Lead	82	207.20
Ds	Darmstadtium	110	[281]	Pd	Palladium	46	106.42
Db	Dubnium	105	[262]	Pm	Promethium	61	[145]
Dy	Dysprosium	66	162.50	Po	Polonium	84	[209]
Er	Erbium	68	167.26	Pr	Praseodymium	59	140.91
Es	Einsteinium	99	[252]	Pt	Platinum	78	195.08
Eu	Europium	63	151.96	Pu	Plutonium	94	[244]
F	Fluorine	9	19.00	Ra	Radium	88	[226]
Fe	Iron	26	55.85	Rb	Rubidium	37	85.47
Fm	Fermium	100	[257]	Re	Rhenium	75	186.21
Fr	Francium	87	[223]	Rf	Rutherfordium	104	[261]
Ga	Gallium	31	69.72	Rg	Roentgenium	111	[280]
Gd	Gadolinium	64	157.25	Rh	Rhodium	45	102.91
Ge	Germanium	32	72.64	Rn	Radon	86	[222]
H	Hydrogen	1	1.01	Ru	Ruthenium	44	101.07
He	Helium	2	4.00	S	Sulfur	16	32.07
Hf	Hafnium	72	178.49	Sb	Antimony	51	121.76
Hg	Mercury	80	200.59	Sc	Scandium	21	44.96
Ho	Holmium	67	164.93	Se	Selenium	34	78.96
Hs	Hassium	108	[277]	Sg	Seaborgium	106	[266]
I	Iodine	53	126.90	Si	Silicon	14	28.09
In	Indium	49	114.82	Sm	Samarium	62	150.36
Ir	Iridium	77	192.22	Sn	Tin	50	118.71
K	Potassium	19	39.10	Sr	Strontium	38	87.62
Kr	Krypton	36	83.80	Ta	Tantalum	73	180.95
La	Lanthanum	57	138.91	Tb	Terbium	65	158.93
Li	Lithium #	3	6.94	Tc	Technetiuum	43	[98]
Lr	Lawrencium	103	[262]	Te	Tellurium	52	127.60
Lu	Lutetium	71	174.97	Th	Thorium	90	232.04
Md	Mendelevium	101	[258]	Ti	Titanium	22	47.87
Mg	Magnesium	12	24.31	Tl	Thallium	81	204.38
Mn	Manganese	25	54.94	Tm	Thulium	69	168.93
Mo	Molybdenum	42	95.94	U	Uranium	92	238.03
Mt	Meitnerium	109	[268]	V	Vanadium	23	50.94
N	Nitrogen	7	14.01	W	Tungsten	74	183.84
Na	Sodium	11	22.99	Xe	Xenon	54	131.29
Nb	Niobium	41	92.91	Y	Yttrium	39	88.91
Nd	Neodymium	60	144.24	Yb	Ytterbium	70	173.04
Ne	Neon	10	20.18	Zn	Zinc	30	65.41
Ni	Nickel	28	58.69	Zr	Zirconium	40	91.22

Glossary

A glossary of terms used in gamma spectrometry, including some acronyms and abbreviations.

Words in italics are defined elsewhere in the glossary

A

ABSOLUTE EFFICIENCY The efficiency of a detector expressed as number of entities detected compared to number emitted by the source. Can be *full-energy peak* efficiency or *total efficiency*.

ABSORPTION COEFFICIENT The proportion of gamma-ray energy absorbed within an absorbing medium.

ABSORPTION EDGE The sharp changes in *photoelectric absorption* coefficient as a function of energy as the energy of the gamma-rays decreases, related to the energy of electrons within their shells.

ABUNDANCE When discussing the isotopic composition of a material this might mean *isotopic abundance*. In the context of gamma-ray emission might mean gamma-ray *emission probability*. The use of the term in the latter context is deprecated because its meaning is unclear.

AC COUPLING Signal coupling from one part of a circuit to another by means of a capacitor transformer that removes any dc level present in the earlier circuit.

ACCURACY The discrepancy between the true value, or the accepted reference value, and the result obtained by measurement.

ACTIVE REGION The parts of a detector in which *charge carriers* are produced and collected to form the detector signal.

ACTIVITY The rate of *decay* of an assembly of radioactive atoms; unit the *Becquerel*.

ADC See: *Analogue to Digital Converter*.

ADC RESOLUTION The number of *channels* an ADC has available for use.

ADC ZERO The control on an ADC that allows the *energy calibration* intercept of a gamma spectrometer to be set to pass through zero energy.

ALGORITHM A defined set of algebraic rules for solving a problem.

ALPHA PARTICLE A ^{4}He nucleus, consisting of two protons and two neutrons, emitted in radioactive decay.

amu Atomic Mass Unit, one-twelfth of the mass of an atom of ^{12}C; approximately 1.66×10^{-27} kg. Represented by the symbol u.

AMPLIFIER Electronic device for increasing the height of an electronic signal. In gamma spectrometry, the amplifier performs many other pulse processing functions. (See: Chapter 4.)

ANALOGUE [US: analog] A type of signal or pulse processing that does not use digital technology. The signal has a continuous rather than digitised distribution.

ANALOGUE PROCESSING A method of pulse shaping and measurement using analogue, rather than digital, electronic circuits. Traditional amplifier systems are analogue in nature.

ANALOGUE TO DIGITAL CONVERTER A device that generates a digital number that is proportional to the amplitude of an analogue linear signal.

ANNIHILATION In the context of gamma spectrometry, the meeting of a positron and an electron resulting in their disappearance, their mass being converted to photon energy.

ANNIHILATION RADIATION Photons of 511.00 keV energy resulting from the mutual annihilation of positrons and electrons.

ANTHROPOGENIC RADIONUCLIDES *Radionuclides* introduced into nature by man.

ANTICOINCIDENCE CIRCUIT A circuit with two inputs. An output signal is blocked and does NOT appear if both inputs receive a pulse within a certain small time interval of each other. Often used in a *gating circuit* where a logic pulse triggers rejection of a linear pulse.

ANSI The American National Standards Institute.

AQCS Analytic Quality Control Services. An *IAEA* analytical programme of reference and intercomparison materials.

AREA In the context of gamma spectrometry, the counts in a *peak* within a *spectrum*. GROSS AREA is the total number of counts; NET AREA is the sum of counts above the background continuum, often referred to as Peak Area.

ASCII American Standard Code for Information Interchange A method of encoding alphabetical and numerical characters for digital transmission.

ATTENUATION COEFFICIENT The proportion of gamma-ray intensity lost from a beam by passing through an absorbing medium. This takes into account scattering out of the beam and is not the same as *absorption coefficient*.

ATTENUATION CORRECTION See: *Self-absorption Correction*.

ATOMIC NUMBER of an element; the integer, Z, equal to the number of protons in the nucleus.

ATOMIC WEIGHT, ATOMIC MASS The average mass of the atoms of an element at its natural isotopic abundance relative to that of other atoms taking ^{12}C as the basis. The atomic mass of ^{12}C is exactly 12.

AUGER EFFECT A process by which excitation of the electron shells is used to expel *Auger Electrons*. It is an alternative to X-ray emission.

AUGER ELECTRONS Those electrons emitted from an atom due to the filling of a vacancy in an inner electron shell.

B

BACKGROUND The number of counts recorded which are due to *Background Radiation*.

BACKGROUND RADIATION Radiation due to 1) radioactive materials in the local environment, 2) radioactivity in the detector materials, and 3) cosmic ray interactions with the detector and surroundings. (See: Chapter 13.)

BACK SCATTER The process of scattering of radiation through a large angle so that it passes into the sensitive volume of a radiation detector.

BAND GAP The energy difference between the top of the *valence band* and the bottom of the *conduction band*.

BASELINE A constant (ideally) reference voltage level on which a pulse sits. Usually zero volts.

BASELINE RESTORATION (BLR) A circuit at the amplifier output that maintains the baseline at its reference value.

BECQUEREL (Bq) The SI unit of radioactivity, defined as one disintegration per second.

BALLISTIC DEFICIT Peak broadening when the *charge collection* time is long compared to the amplifier *shaping time*.

BETA DECAY The decay processes, β^-, β^+ and *Electron Capture*, in which the *Mass Number* of the daughter is the same as the parent. (See: Chapter 1.)

BETA PARTICLE A charged particle with unit charge emitted from the nucleus of an atom during *beta decay*. The charge may be positive (a positron, β^+) or negative (a negatron, β^-); in general the latter is assumed.

BGO Bismuth Germanate, $Bi(GeO_4)_3$. A material used in a *scintillation detector* whose advantages are high density and high efficiency.

BIAS A more or less persistent tendency for a group of measurements to be too large or too small; it implies a constant error.

BIAS VOLTAGE A voltage applied to a detector, usually semiconductor, to facilitate the collection of the *charge carriers*. Sometimes referred to simply as 'bias'.

BINOMIAL DISTRIBUTION The distribution relating frequency of events to probability of occurrence when considering systems with two possible states. This is the basic distribution underlying radioactive decay.

BIPM Bureau International des Poids et Mesures; the international body responsible for the definition and maintenance of standard quantities to which all national 'weights and measures' bodies are responsible.

BIPOLAR PULSE A pulse with two poles that firstly rises positively from the baseline, then goes negatively below the baseline.

BLR See: *Baseline restoration*.

BNC A type of coaxial cable connector. The standard signal connector in the *NIM* system.

BRANCHING RATIO This term is confusingly used to describe three related aspects of decay processes. 1) The ratio of different modes of decay of a nuclide, 2) The proportions of particles of different energy emitted during the decay process itself, and 3) in the context of gamma-ray emission it might mean gamma-ray *emission probability*. The use of the term in the latter context is deprecated because its meaning is unclear. Use of the term should be restricted to the first meaning.

BREMSSTRAHLUNG A continuum of *electromagnetic radiation* in the X-ray / low energy gamma-ray region of a spectrum usually resulting from the absorption, or sudden deceleration, of high energy *beta particles*.

BUILD-UP The increase in the gamma-ray flux through a shield arising from scattering of the radiation in broad-beam geometry. This makes the shield less effective than predicted from the simple exponential law.

BULLETIZATION The 'rounding-off' of the face edge of a detector to remove weak-field regions that would degrade detector performance.

C

CASCADE SUMMING See: *True Coincidence Summing*.

CdTe Cadmium Telluride A semiconductor detector for X rays and low energy gamma-rays. Good resolution but low efficiency due to small size.

CdZnTe Cadmium Zinc Telluride. As CdTe above, with some superior characteristics.

CERENKOV RADIATION Coherent radiation produced when a charged particle traverses a medium at a velocity greater than the phase velocity of light in that medium.

CENTROID The geometric centre of a peak. It is unlikely to be a whole number of channels.

CEM Channel Electron Multiplier; detects electrons, ions, soft X rays directly. Multiplication of the signal is a similar process to that of a *photomultiplier tube.*

CHANNEL The smallest energy or time slot used in an *MCA.*

CHARGE CARRIERS Collectively electrons and holes created by the interaction of the gamma-ray with the detector that comprise the detector signal.

CHARGE COLLECTION The act of collecting the electrons and holes within the detector. The time to do that is the **Charge Collection Time.**

CHARGE SENSITIVE PREAMPLIFIER A *preamplifier* for which the output signal is proportional to the charge collected at the input.

CHART OF THE NUCLIDES A chart plotting all the known nuclides by *atomic number* on the y-axis and *neutron number* on the x-axis, each nuclide being labelled with a limited amount of nuclear data. Conventionally, this is split into four sections to allow it to be printed on a sheet of manageable size. Also known as a Segré chart.

CHI-SQUARED TEST, χ^2 A procedure for determining the probability that two different distributions are actually samples of the same population. It can be used to check whether a radiation counting system has an appropriate degree of randomness.

CLOCK TIME (CT) The same as *real time* and true time.

COAXIAL detector A cylindrical detector with one contact within a hole drilled into its base. (See: Chapter 3, Section 3.4.4)

COHERENT A scattering process where the phase of the scattered wave has a definite phase relationship to that of the incident wave.

COINCIDENCE The occurrence of one or more events in one or more detectors within a predetermined time interval.

COINCIDENCE CIRCUIT A circuit with two inputs. An output signal only occurs if pulses appear at both inputs within a predetermined time interval.

COMPENSATED SEMICONDUCTOR A semiconductor detector in which the numbers of electronically active p- and n-type impurities are equal, creating, in effect, an *Intrinsic Semiconductor.*

COMPTON CONTINUUM In gamma spectrometry, that part of the spectrum due to incompletely absorbed gamma-rays, and mostly devoid of useful information.

COMPTON EDGE The upper limit of the *Compton continuum* indicated by a marked fall in the continuum level.

COMPTON SCATTERING Inelastic interactions between photons and electrons.

CONDUCTION BAND The electronic band within a solid through which electrons can readily move to create an electric current.

CONFIDENCE LIMITS The limits between which one can have confidence, at a particular degree of confidence, that the true value lies.

CONSTANT FRACTION TIMING A method of timing discrimination that uses a constant fraction of the peak amplitude for each input pulse.

CONTINUUM That part of a pulse height spectrum that has no definite peaks. The continuum adjacent to a peak is used to estimate the [background] continuum under the peak, and hence the net peak *area.*

CONVERSION GAIN of an *ADC* is the number of channels used in a particular spectrum acquisition. Usually a binary multiple of two.

CONVERSION TIME The time required to change an input signal from one form to another. Examples are analogue to digital, or time difference to pulse amplitude.

COSMIC RAYS Radiation, both particulate and electromagnetic, that originates outside the earth's atmosphere.

COSMOGENIC NUCLIDE A nuclide formed by the action of cosmic rays on a target material.

COUNT 1) A decay event registered by a detector, or 2) the number of such events registered within a period of time.

COUNT RATE STABILITY The degree to which the amplitude of a pulse is distorted by variations in the pulse rate.

COVARIANCE A statistical measure of the interrelationship, or correlation, between two variables - analogous to *variance* for a single variable.

COVERAGE The number of *standard uncertainties* corresponding to a particular degree of confidence.

CRITICAL LIMIT (L_C) The level below which a net signal cannot reliably be detected. In radiometrics, the number of counts below which, using the conventional confidence interval, we are 95% certain the count is part of the background distribution.

CRM Count Rate Meter.

CROSSOVER POINT The time at which a bipolar pulse passes through the baseline.

CROSSOVER TIMING A method of timing detection that uses the crossover point of a bipolar pulse.

CROSSOVER TRANSITION A gamma-ray resulting from a transition between two non-adjacent nuclear levels.

CRRC An electronic *shaping* circuit comprising a differentiator followed by an integrator; a common component of a traditional linear amplifier.

CRT Cathode Ray Tube visual display unit.

CRYOSTAT The vacuum enclosure, including the end cap, preamplifier and cold finger, by which the detector is maintained at its operating temperature.

CTBTO Comprehensive Test Ban Treaty Organisation A group, based at the IAEA Vienna, which monitors nuclear explosions worldwide.

CURIE (Ci) An obsolete unit of radioactivity. $1\ Ci = 3.7 \times 10^{10}$ Bq. See: *Becquerel.* [Originally the radioactivity of one gram of ^{226}Ra]

CURSOR A vertical marker in the *MCA* display. Used for setting a *region of interest*, or reading off the number of counts in a channel.

CZT **C**admium **Z**inc **T**elluride. See: *CdZnTe*

D

DAC **D**igital to **A**nalogue **C**onverter The reverse of an *ADC*.

DAUGHTER The product nucleus of the radioactive decay of a *parent* nuclide.

DC COUPLING Coupling from one circuit to another without an intervening capacitor.

DDEP **D**ecay **D**ata **E**valuation **P**roject; a database recommended by *BIPM* holding, amongst many other data, half-life and gamma emission probability data. (See: Appendix A.)

DEAD LAYER The outermost, highly doped contact regions of the detector, which are inactive as far as *charge carrier* production is concerned.

DEAD TIME (DT) The time that a signal processing circuit is busy processing a pulse, and during which is consequently unable to accept another pulse.

DECAY CONSTANT The probability of decay of a radionuclide per unit time.

DECAY, RADIOACTIVE The disintegration (transformation) of the nucleus of an unstable atom by the spontaneous emission of particles and/or electromagnetic radiation.

DECAY SCHEME A diagram showing the nuclear energy levels and the various de-excitation paths available in order to emit gamma-rays.

DECOMMISSIONING All procedures undertaken once an installation has ceased operating – usually applied to nuclear sites. See: also *remediation*.

DEFAULT The value of a parameter used by a program in the absence of a user-supplied value.

DELAY LINE A circuit component used to delay the propagation of a signal from one point to another.

DELAYED NEUTRON A neutron emitted by a fission product in an isobaric chain; the delay caused by the time taken for beta decay in the chain.

DEPLETED URANIUM Uranium containing a smaller proportion of ^{235}U than the proportion in most natural ores.

DEPLETION REGION The region of the detector, created by the application of the detector *bias*, in which the mobile *charge carriers* are created by the gamma-ray interactions.

DETECTION LIMIT (L_D) or Limit of Detection. The minimum amount of material that can be detected with reasonable certainty. In radiometrics, 'What is the minimum number of counts I can be confident of detecting?'.

DETECTOR GEOMETRY A description of the geometric arrangement of detector and sample (distance, sizes of detector and sample), which is sometimes expressed as a solid angle. It influences the efficiency of counting.

DEVIATION The difference between a measurement and some value, such as the average or expected value, calculated from the data. See: *Standard Deviation*.

DIFFERENTIAL NONLINEARITY A measure of quality of an ADC. In essence, it is a measure of the constancy of channel width.

DIFFERENTIATOR A pulse-*shaping* network using a capacitor in series and a resistor to ground, which allows high frequencies to pass undisturbed yet differentiates low frequencies. Also referred to as a high pass filter.

DIGITAL SIGNAL PROCESSING A method of pulse shaping and measurement using digital technology that can result in improved processing speed and stability. An alternative to *analogue processing*.

DIGITAL STABILIZATION Stabilization of the gain and intercept of a spectrometer by measuring control peak positions digitally, rather than in an analogue fashion.

DISINTEGRATION RATE The rate of decay of a nuclide in a radioactive source. See: *Activity*.

DISCRIMINATOR A device that generates a logic output signal when the input signal exceeds a preset threshold level.

DOUBLE ESCAPE PEAK An extra peak in a spectrum due to the loss of two photons of 511 key energy each during the absorption of a high energy gamma-ray. See: *escape peaks*.

DOPING In the context of detector manufacture, the insertion of impurity atoms into a semiconductor material lattice in order to control its properties.

DPM **D**isintegrations **P**er **M**inute.

DSP and DSA See: *Digital Signal Processing*.

DT See: *dead time*.

DWELL TIME The time interval used to collect counts in any one channel of an MCA that is being used in *multiscaling* mode.

E

EFFICIENCY, ABSOLUTE See: *Absolute Efficiency*.

EFFICIENCY, INSTRINSIC See: *Intinsic Efficiency*.

EFFICIENCY, RELATIVE See: *Relative Efficiency*.

EFFICIENCY CALIBRATION The relationship between peak count rate, at a specified gamma-ray energy, and the *disintegration rate* of the source.

EFFECTIVE HALF-LIFE For a nuclide that decays by more than one process, the effective half-life is the time for the activity to decay by a factor of two taking into account all processes.

EGS **E**lectron **G**amma **S**hower – one of the computer programs available for the mathematical generation of efficiency calibrations.

ELASTIC SCATTERING Scattering processes, such as Rayleigh scattering, which do not transfer energy to the scattering medium.

ELECTROMAGNETIC RADIATION Interacting electric and magnetic wave that propagates through vacuum at the speed of light. Examples are visible light, radio waves and *gamma-rays*.

ELECTRON A sub-atomic particle of charge -1 and $1/1836^{th}$ the mass of a proton.

ELECTRON CAPTURE A nuclear transformation whereby a nucleus captures one of its extra-nuclear electrons.

ELECTRON-HOLE PAIR See: *Charge carriers.*

ELECTRONVOLT (eV) Conventional unit of radiation energy in the X-ray and gamma-ray region. Energy gained by an electron as it passes through a potential difference of one volt. Also keV (1000 eV), and MeV (10^6 eV). 1 eV = $1.602\ 177 \times 10^{-19}$ J.

EMI ElectroMagnetic Interference. A potential source of electronic noise pickup.

EMISSION PROBABILITY, Pγ, (sometimes wrongly: '*yield*', or '*abundance*', or '*branching ratio*') The probability that a radioactive decay will be followed by the emission of the specified radiation, e.g. a *gamma-ray*. When using this, take care to note whether published data refer to percentages or not.

ENERGY CALIBRATION The relationship between spectrum peak position, in channels, and gamma-ray energy.

ENSDF Evaluated Nuclear Data Structure File – a recommended source of nuclear data. (See: Appendix A.)

ERROR In metrology, it denotes deviation from a mean. In this book, it is taken to mean 'mistake' or 'blunder' or unavoidable *bias*.

ESCAPE PEAKS Extra peaks often Seen in a gamma spectrum that contains full-energy peaks greater than 1022 key. They are caused by the escape from the detector of one or two annihilation photons of 511 keV each. The single escape peak is at (full-energy - 511) keV ; the double escape peak is at (full-energy - 1022) keV.

EXCITED STATE The state of a molecule, atom or nucleus when it possesses more than its *ground state* energy.

EXCITATION The addition of energy to a system, usually transforming it from its ground state to an excited state.

EXCITON A bound electron-hole pair formed as a result of energy absorption in a crystalline material. It can migrate through the crystal with a definite half-life.

EXTERNAL VARIANCE A weighted measure of variance that takes into account actual variability of data. (Compare *Internal variance*)

F

FANO FACTOR A factor to make theory match observation with regard to the number of charge carriers produced by absorption of a gamma-ray. A 'fudge factor'.

FALL TIME Two definitions are seen. (1) The time required for a pulse to fall from its maximum amplitude to 37 % (1/e) of that amplitude (also called the clipping time constant), or more commonly, (2) the time required for the pulse to go from 90% to 10% of its maximum amplitude.

FEEDBACK The transfer of energy from one part of a system to another in a direction that is opposite to the main flow of energy.

FIRMWARE An unalterable program for a *CPU* stored in *ROM*.

FISSION (NUCLEAR) The decay process that involves the splitting of an atom into two parts.

FISSION BARRIER The energy barrier that must be surmounted before fission can take place.

FISSION FRAGMENT One of the many nuclides created at the moment of fission.

FISSION PRODUCT One of the many nuclides created by nuclear fission, including nuclides formed by decay of other fission products.

FISSION YIELD The percentage of fissions that give rise to a particular nuclide.

FIXED DEAD TIME ADC (Also fixed conversion time ADC) See: *Successive Approximation ADC.*

FLASH ADC A very fast type of ADC using multiple parallel voltage comparisons. (Used in *DSP*.)

FLUENCE The total number of particles or photons crossing a sphere of unit cross section surrounding a point source.

FLUENCE RATE The product of the number of particles or photons per unit volume and their average speed. [Formerly called: flux – the latter term is used in this book]

FLUORESCENCE In gamma spectrometry, the emission of characteristic X-rays of a material when irradiated with gamma-rays.

FLUORESCENCE YIELD For a given transition from an excited state of a specified atom, the ratio of the number of excited atoms that emit a photon to the total number of excited atoms.

FORBIDDEN BAND The energy region in the band structure of a solid between the *Valence Band* and the *Conduction Band*, within which electrons cannot reside.

FRENKEL DEFECT An atom in a lattice not in its normal position – an interstitial atom.

FULL-ENERGY PEAK A peak in the gamma-ray spectrum resulting from complete absorption of the energy of the X-ray or gamma-ray.

FULL WIDTH... See: *FWFM, FWHM* and *FWTM.*

FWFM or FW0.02M Full Width at Fiftieth Maximum; as FWHM, but being the peak width at one fiftieth of the maximum channel content.

FWHM Full Width at Half Maximum; a measure of the resolution of a spectrum, being the width of a peak, in energy (or time) units, at half the maximum peak height.

FWTM or FW0.1M Full Width at Tenth Maximum; as FWHM, but being the peak width at one tenth of the maximum channel content.

G

GAIN The ratio of the amplitude of the output signal of an *amplifier* to the amplitude of its input signal.

GAMMA-RAY Electromagnetic radiation emitted during de-excitation of the atomic nucleus.

GATE An electronic component that allows pulses to pass when 'open', and blocks them completely when 'closed'. Controlled by a *gating pulse*.

GATED INTEGRATOR (GI) A *pulse processor* that, in effect, creates an output pulse of height proportional to the area of the input pulse, rather than its height, in order to overcome *ballistic deficit* at short *shaping* time. Used for high count rate measurements.

GATING PULSE A variable width logic pulse used to enable or prevent other electronic events for a variable time period.

GAUSSIAN DISTRIBUTION Describes the frequency of events when those events are governed by statistical factors alone. It is a satisfactory approximation to the *Poisson Distribution* when the number of events is 100 or more. Also called the *Normal Distribution*.

GAUSSIAN SHAPING A method of processing a signal in an amplifier so that the shape of the output pulse can be approximately described by a *Gaussian distribution*.

Ge(Li) DETECTOR (Pronounced 'jelly') A germanium detector created by *lithium drifting*.

GEDCKE-HALE One of the methods of *live-time correction*.

GESPECOR GErmanium SPEctrum CORrection – Monte Carlo based software developed for the computation of efficiency, of matrix effects and of coincidence summing effects in gamma-ray spectrometry.

GERMANIUM ESCAPE PEAK See: X-ray escape peak.

GROUND LOOP The situation where different parts of an electronic system are grounded (earthed) at different points giving rise to a continuous path for stray electric currents.

GROUND STATE The lowest energy state of a nucleus.

H

HALF LIFE The characteristic time taken for the activity of a particular radioactive substance to decay to half its original value. See: also *effective half-life*.

HARDWARE The physical electronics equipment.

HISTOGRAM Representation of a variable by means of vertical bars; here, the actual shape of a peak in a spectrum due to it being composed of discrete channels.

HLW High Level radioactive Waste; waste generating heat at > 2 kW/m^3 (UK usage).

HOLE An entity created by the removal of an electron, usually by elevation to the conduction band. It is conventionally thought of as being positively charged.

HPGe High Purity Germanium. Material used for a present day high resolution gamma-ray detectors, and the detector itself.

HRGS High Resolution Gamma Spectrometry Usually used to describe gamma-ray measurement techniques involving germanium detectors as opposed to detectors made of other materials.

HVPS High Voltage Power Supply Also referred to as HV.

HYPERPURE GERMANIUM See: HPGe.

I

IAEA The International Atomic Energy Authority. A United Nations organisation based in Vienna.

IC *Internal Conversion*.

ICR Input Count Rate.

ICRM International Committee for Radionuclide Metrology.

ICP-MS, Inductively Coupled Plasma - Mass Spectrometry A technique for multi-element and multi-isotopic analysis that uses a plasma source as input to a mass spectrometer. Can be used for the measurement of nuclides with very long half-lives.

IEC International Electrochemical Committee.

ILW Intermediate Level radioactive Waste; waste whose activity concentration is greater than the upper level of LLW and which generates less heat than HLW (UK usage).

ION An electrically charged atom or molecule.

ION IMPLANTATION The process of embedding atoms into the surface of a solid by irradiation with a beam of energetic ions; used to create the p+ contacts for *semiconductor detectors*.

IODINE ESCAPE PEAK An *X-ray escape* peak that can, in principle, be seen in sodium iodide detector spectra. It will be just below the full-energy peak, differing from it by the iodine $K\alpha$ X-ray of about 28.5 key.

IONIZATION The process by which an electrically neutral entity acquires a positive or negative charge.

IONIZATION CHAMBER A gas filled radiation detector that measures radiation by means of the direct collection of the ions produced. There is no gas amplification.

IMPEDANCE A measure of opposition to time-varying electric current in an electric circuit; depends on the resistance, capacitance and inductance of the circuit.

IMPEDANCE MATCHING The necessity of terminating a signal with the characteristic impedance of the signal line to allow optimum transfer of undistorted signals from one device to another.

INELASTIC SCATTERING A scattering process in which the energy of the scattered entity is less than its initial energy, some of that energy being retained by the scattering medium.

INES International Nuclear Event Scale Defines the safety significance of an event occurring in a nuclear facility. Scale 0 (least serious) to 7 (most serious). Chernobyl would have been a 7.

INTEGRAL NONLINEARITY The performance of an amplifier or ADC expressed as a percentage deviation from linearity. It is related to the non-linearity of the *energy calibration*.

INTEGRATION In gamma spectrometry, the sum of the counts within a *region of interest* – hence PEAK INTEGRATION.

INTEGRATION TIME In an amplifier, the time taken to collect the *electron-hole pairs* and determined by the pulse *shaping time*.

INTEGRATOR A pulse shaping network using a resistor in series and a capacitor to ground, which allows low frequencies through undisturbed yet integrates high frequencies. Also referred to as a low pass filter.

INTERFERENCE (PEAK) The overlap of a peak in a spectrum with one or more other peaks causing difficulties in measurement of the target peak area.

INTERNAL CONVERSION (IC) A process in which an excited nucleus de-excites with the ejection of an extra-nuclear electron from an inner shell. The electrons emitted are conversion electrons.

INTERNAL CONVERSION COEFFICIENT (α) The proportion of nuclear de-excitations that take place by *internal conversion* rather than gamma-ray emission.

INTERNAL VARIANCE A weighted measure of variance that takes into account only the uncertainties of the individual data items. (Compare *External variance*)

INTRINSIC EFFICIENCY The efficiency of a detector in terms of number of entities detected compared to the number entering the detector. It is independent of source-detector distance.

INTRINSIC REGION A region within a semiconductor where there are equal numbers of free holes and electrons.

INTRINSIC SEMICONDUCTOR A semiconductor in which there are no net active impurity sites.

ISOBARS Nuclides of different elements having the same mass numbers but different atomic numbers.

ISOCS In-Situ Object Counting System – commercial computer program for estimating activities without source-based efficiency calibration.

ISOMERIC STATE Energy levels of a nucleus having different energies and half-lives. Often used specifically to refer to energy levels within an atom that have uncharacteristically long half-lives. (See: *Metastable State*)

ISOMERIC TRANSITION (IT) A transition between two *isomeric states* of a nucleus. In particular, used to refer to transitions with uncharacteristically long half-lives.

ISOTOPES Nuclides having the same atomic number but different mass numbers.

J

JANIS JAva-based Nuclear Information Service – nuclear data database. (See: Appendix A.)

JEF Joint Evaluated File – nuclear data database. (See: Appendix A.)

JEFF Joint Evaluated Fission and Fusion Library – nuclear data database. (See: Appendix A.)

JUNCTION Transitional region between regions of different types of semiconductor or between semiconductor and metal.

K

keV 1000 *electronvolts*.

L

LABSOCS LABoratory SOurceless Calibration Software – commercial computer program for estimating activities in a laboratory environment without source-based efficiency calibration.

LARA One of the recommended online nuclear data databases maintained by Laboratoire National Henri Becquerel. (See: Appendix A.)

LCD Liquid Crystal Display visual display unit.

LEAKAGE CURRENT The total current flowing through or across the surface of a detector crystal with operating *bias* applied and no external ionizing radiation.

LEAST SQUARES A mathematical procedure for estimating a parameter from a set of data by making the sum of the squares of the deviations a minimum.

LED Light Emitting Diode.

LFC See: Loss Free Counting.

LIFETIME The mean time that any unstable atom or excited state exists before decay. Numerically, it is the reciprocal of the *decay constant*.

LIMIT OF DETECTION See: *Detection Limit*.

LINEAR AMPLIFIER An amplifier in which the output pulse height is directly proportional to the input pulse height.

LINEAR GATE An electronic *gate* that, when open, allows *linear pulses* to pass unchanged.

LINEAR PULSE A pulse that is carrying information in its maximum amplitude or height.

LITHIUM DRIFTING The process of counter-doping a semiconductor material with lithium atoms to create an *intrinsic region*. Used to prepare silicon and, at one time, germanium detectors.

LIVE TIME (LT) The time when the system is not processing a pulse and is thus available to accept another. It is the effective counting period of the system.

LIVE TIME CLOCK The clock within the *MCA* that allow correction to be made for *dead-time*.

LIVE TIME CORRECTION (LTC) The process of stopping the live time clock in the MCA whenever the pulse processing circuits are busy. A common method of allowing for dead time.

LLD **L**ower **L**evel **D**iscriminator A threshold that is used to eliminate noise and small pulses.

LLW **L**ow **L**evel radioactive **W**aste; waste that contains less than 4×10^9 Bq/tonne of alpha activity, and less than 12×10^9 Bq/tonne of beta/gamma activity (UK usage).

LOGIC PULSE A standard square pulse, used for triggering, switching and counting. In the NIM specification it is +2.5 to +5 V high, and a few μs wide.

LORENTZIAN (Distribution) The energy distribution of excited states within nuclei.

LOSS FREE COUNTING (LFC) A technique for improved dead time correction at high and varying count rates.

LSB **L**east **S**ignificant **B**it The digital bit in a binary number with the least significance.

LT See: *live time*.

LTC See: *Live Time Correction* and *Live Time Clock*.

M

MAGIC NUMBERS The number of protons and neutrons that appear to produce particularly stable nuclides. These are: 2, 8, 20, 28, 50, and 82.

MARINELLI BEAKER A sample container that fits over a detector's endcap, so that the sample material is close to both the top and the sides of the detector crystal.

MASS ABSORPTION COEFFICIENT The *absorption coefficient* divided by the density of the material.

MASS ATTENUATION COEFFICIENT The *attenuation coefficient* divided by the density of the material.

MASS DEFECT The difference between the mass of a nucleus and the sum of the masses of its constituent nucleons.

MASS NUMBER The number of protons plus neutrons in the nucleus of an atom (symbol A).

MCA See: *MultiChannel Analyser*.

MCNP **M**onte **C**arlo **N**-**P**article code – a computer program for simulating the process of gamma-ray detection, used in gamma spectrometry to generate mathematically efficiency curves.

MDA **M**inimum **D**etectable **A**ctivity. The lowest activity in a sample that can be detected with a particular degree of confidence. It is the activity equivalent of the Limit of Detection. It is variously defined and is NOT the minimum activity measurable.

MEAN The arithmetic average of a set of results.

MEAN LIFETIME See: *Lifetime*.

METASTABLE STATE An *excited state* with a *lifetime* long enough for it to be regarded as a nuclide in its own right.

MeV One million *electron volts*.

MGA A gamma-ray spectrum analysis code for determining plutonium isotopic abundances.

MHV An outdated type of high voltage connector.

MICROPHONICS A type of electronic noise caused by mechanical and audio vibration that can degrade a gamma-ray spectrum.

MIXER/ROUTER A device that will accept several inputs and will route each into a separate section of the *MCA* memory.

MOBILITY of electrons and holes is their velocity divided by the field strength.

MOX **M**ixed **Ox**ide fuel. Reactor fuel containing both plutonium and uranium oxides.

MULTICHANNEL ANALYSER A device for separating, measuring and counting pulses as a function of their height. Essential to a traditional gamma spectrometer.

MULTICHANNEL SCALING (MCS) The acquisition of time-correlated data in an *MCA*. Each channel is defined as a time window (See: dwell time); all pulses are stored in one channel, then stepped sequentially to the next.

MULTIPLET In gamma spectrometry, a spectrum feature comprising overlapping peaks. (See: *Interference*, Contrast: *singlet*).

N

NAA **N**eutron **A**ctivation **A**nalysis An excellent method of determining elemental concentrations. Materials are made radioactive by neutron absorption, and the radioactivity is then measured, usually by *HRGS*.

NaI(Tl) See: *sodium iodide*.

NDA **N**on **D**estructive **A**ssay. A number of techniques used for measuring radioactive waste, most of them being based on detecting gamma-rays and/or neutrons.

NEGATRON See: *beta particle* – the corollary of *positron*.

NEUTRON Sub-atomic particle with a mass of 1 u and no charge.

NEUTRON NUMBER The number of neutrons within the *nucleus* of a *nuclide*. (Symbol N)

NIM **N**uclear **I**nstrumentation **M**odule. An American instrumentation standard, now used widely.

NIM BIN Part of the NIM system. A 19 inch rack into which NIM modules are put that supplies mechanical support and electrical power.

NIST **N**ational **I**nstitute of **S**tandards and **T**echnology. US organisation, similar to the **NPL** in the UK, formerly NBS, National Bureau of Standards.

NOISE Unwanted electronic disturbances transmitted with or added to the wanted signal, which may add uncertainty to the information content of the signal.

NORM **N**aturally **O**ccurring **R**adioactive **M**aterial. NORM includes some of the *primordial nuclides* and their daughters and certain nuclides continuously created in the environment. Contrast: *anthropogenic radionuclides*.

NORMAL DISTRIBUTION See: *Gaussian Distribution*.

NORMAL LAW OF ERROR A mathematical equation that in many cases describes the scatter of a collection of measurements around the average for the collection.

NPL National Physical Laboratory. The body in the United Kingdom responsible for the maintenance of standards and their concordance with *BIPM* standards.

n-TYPE SEMICONDUCTOR A semiconductor material in which there is an excess of n-type, 5-valent, impurities. The material is an electron donor.

NUCLEON A *proton* or a *neutron* (as a constituent of the nucleus).

NUCLEUS The central core of the atom, composed of *nucleons* and surrounded by extra-nuclear *electrons*.

NUCLIDE An individual species of atom characterised by its mass number and atomic number. Can be stable or radioactive.

O

OFFSET, ADC A digital shift in the zero channel of a spectrum, performed after the ADC has digitised the data.

OPERATIONAL AMPLIFIER (OP AMP) A high gain amplifier with negative feedback, so that its input circuit forms a low impedance summing junction.

OVERLAP An MCA function that allows one section of memory, or one spectrum, to be displayed adjacent to or on top of another.

P

PAIR PRODUCTION The process by which a photon is converted into an electron-positron pair in the field of an electron or nucleus. The photon energy must be greater than 1022 keV. (An illustration of the conversion of energy into mass.)

PARAMETER A variable that is given a constant value for a specific application.

PARENT A radionuclide that transforms to another by radioactive decay. See: *daughter*.

PARITY Either (1) a nuclear quantum number, or (2) a self checking binary code in which the total number of ones or zeros is always even or odd.

PEAK A narrow region in a gamma spectrum which can be identified with specific gamma-rays emitted by the source.

PEAK AREA See: *Area*.

PEAK CHANNEL The channel number closest to the centroid of a peak in a spectrum.

PEAK BACKGROUND The continuum underlying peaks in a spectrum that must be subtracted when calculating peak *net area*.

PEAKED BACKGROUND CORRECTION (PBC) The subtraction of counts from a peak to correct for the presence of a peak in a background spectrum.

PEAK TO COMPTON RATIO In germanium detectors, the ratio of the maximum count in the 1332.5 key peak from ^{60}Co to the average count per channel in the range 1040 to 1096 key.

PEAK TO TOTAL RATIO Not to be confused with the above. This is the ratio of the number of counts in a full-energy peak to the total number of counts in the whole spectrum.

PEAKING TIME The time taken by a pulse to rise from a (small) threshold to the pulse maximum. Contrast: *rise time*.

PET Pulse Evolution Time. In this context, a mechanism used in the *virtual pulse generator* method of loss free counting.

PHA See: Pulse Height Analyser.

PHOTOELECTRIC ABSORPTION The process in which a photon (i.e. gamma-ray) ejects bound electrons from an atom. See: also *Auger electrons*.

PHOTOMULTIPLIER TUBE (PMT) A device in which light photons are detected and the consequent electron current is amplified by a cascade process based on the secondary emission of electrons.

PHOTON A quantum of *electromagnetic radiation*.

PILE-UP REJECTION (PUR) A circuit for identifying piled-up pulses and preventing them from being recorded in the spectrum.

POISSON DISTRIBUTION The distribution relating frequency of events to probability of occurrence when the total number of events is large and unknown. It is a satisfactory approximation of the *binomial distribution* when the number of events is more than 30 or so.

POLE ZERO (PZ or P/Z) A method for fine-tuning unipolar pulses at the output of an amplifier to compensate for overshoot and undershoot on the trailing edge of the pulse. It matches the amplifier filter circuits to a *resistive feedback preamplifier*.

POOLED VARIANCE See: *internal variance*.

POSITRON A sub-atomic particle of charge +1 and $1/1836^{th}$ the mass of a proton – the anti-particle to the *electron*, with which it *annihilates*.

PREAMPLIFIER A device which **precedes** the **amplifier**. It collects the *charge carriers* from the detector and passes a pulse to the amplifier. It may be sensitive to charge, current or voltage.

PRECISION Refers to the agreement among repeated measurements around the average for the collection.

PRIMORDIAL RADIONUCLIDE A long-lived radionuclide present in the Earth since its formation. See also: *NORM*.

PROMPT Of neutrons or gamma-rays which are emitted immediately during a nuclear reaction such as fission and neutron capture.

PROMPT GAMMA SPECTROMETRY Spectrometry of gamma-rays emitted during nuclear reactions.

PROTON Sub-atomic particle with a mass of 1 u and charge +1.

p-TYPE SEMICONDUCTOR A semiconductor material in which there is an excess of p-type, 3-valent, impurities. The material is an electron acceptor.

PULSE HEIGHT ANALYSER (PHA) An electronic device for measuring the heights of pulses. See: *SCA* and *MCA*.

PULSE PILE-UP A condition where two pulses are generated so close in time that effectively a single composite pulse is produced. The pulse height of the piled up pulse will be greater than either of the components and in a spectrum will most noticeable in the high-energy region. Also referred to as *random summing*.

PULSE PROCESSOR In gamma spectrometry systems, a term interchangeable with '*amplifier*'. In view of the many other functions an amplifier performs this would be a more descriptive term.

PULSE SHAPE DISCRIMINATION (PSD) Electronic methods for separating pulses of differing shape, thus enabling pulses from one type of radiation to be separated from those of another. For example in neutron detectors, neutrons may be separated from gammas by PSD. Same as pulse shape analysis, PSA.

PUR See: *pile up rejection*.

Q

QA Quality Assurance. An administrative or software system ensures that an analysis meets agreed standards of performance.

QUENCHING Various processes that reduce the efficiency of counting; important in scintillation systems and gas counters.

Q-VALUE The energy released in a nuclear reaction or radioactive decay process.

QXAS Quantitative X-ray Analysis Software.

R

RAD An obsolete unit of absorbed dose. 1 rad = 0.01 J kg-1. 1 Gray = 100 rad.

RADIOACTIVE DECAY See: *Decay*.

RADIOACTIVE EQUILIBRIUM The steady state condition in which the rate of decay of a daughter radionuclide becomes equal to its rate of production from the parent radionuclide.

RADIOCHEMICAL PURITY Of a radioactive material, the proportion of the total activity that is present in the stated chemical form.

RADIOISOTOPES *Isotopes* that are radioactive.

RADIONUCLIDE A *nuclide* that is radioactive.

RADIOLYSIS The decomposition of material by ionising radiation; for example, water into hydrogen and oxygen.

RANDOM A procedure for selecting items from a population which gives every member of the population equal opportunity to be chosen.

RANDOM SUMMING See: *pulse pile up*.

RAYLEIGH SCATTERING Coherent elastic scattering of photons by bound electrons. Energy is not lost by the photon.

RC Either 1) RC shaping. A simple circuit containing a resistor (R) and a capacitor (C) that changes the shape of an analogue pulse. Used in conventional amplifiers. 2) RC. As 1) but used as a product of values for resistance [in ohms] and capacitance [in farads] to produce a time constant [in seconds].

REAL TIME (RT) Normal physical time as recorded on a (stationary) clock or watch. Also known as clock time and true time.

REGION OF INTEREST (ROI) A user determined part of a spectrum, which contains information of interest; often set up around a peak in a spectrum.

RELATIVE EFFICIENCY The efficiency of a detector relative to a standard NaI(Tl) detector under standard conditions. (See: Chapter 11, Section 11.4.3)

REMEDIATION The process of restoring land or property that has been contaminated by industrial activity to normal use. In some cases this will involve removal of radioactive waste. On a nuclear site this would be referred to as *decommissioning*.

RESISTIVE FEEDBACK PREAMPLIFER. The conventional preamplifier in which the input voltage step is reset by a feedback resistor. For high count rate systems the *Transistor Reset preamplifier* is preferable.

RESOLUTION A measure of the narrowness of the width of a peak in a spectrum. See: *Full-Width . . .*

RFI Radio-Frequency Interference. A potential source of electronic noise pickup.

RISE TIME The time taken by a pulse to rise from 10 % to 90 % of its maximum amplitude. Contrast *peaking time*.

RG59, RG62 Types of coaxial cable. RG59 is used for high voltages; RG62 for signals.

ROI See: *Region of Interest*.

ROM Read Only Memory.

RSD Relative Standard Deviation.

RT See: *Real Time*.

S

SAL Safeguards Analytical Laboratory. A particular radiochemical laboratory at IAEA Vienna.

SCA See: *Single Channel Analyser*.

SCINTILLATION DETECTOR A range of devices based on the emission of light produced by the absorption of radiation. In general, there is a scintillating material or phosphor coupled to a photomultiplier tube.

SECULAR EQUILIBRIUM The condition that the activity of a daughter radionuclide is equal to that of its parent. A necessary precondition is that the daughter half-life is much smaller than that of the parent.

SEGRĒ CHART See: *Chart Of The Nuclides*.

SELF ABSORPTION CORRECTION The correction to a count rate that needs to be made to take into account the loss of gamma-rays in passing through the sample and other material before reaching the detector.

SEMICONDUCTOR A material with an electronic band structure such that the *band gap* confers limited conductivity.

SEMICONDUCTOR DETECTOR A detector with a p-type/*intrinsic region*/n-type *semiconductor* structure in which the detector signal is collected as electrons and holes.

SEMI GAUSSIAN A unipolar pulse, so called because its shape approximates to a Gaussian distribution.

SGS **S**egmented **G**amma **S**canner. A device for measuring gamma-emitting radionuclides in waste drums.

SHAPING The process of altering the shape of a linear pulse; usually to make it more easily processed by the *ADC*.

SHAPING TIME The effective time constant of the circuits used to perform pulse *shaping*.

SHV **S**afe **H**igh **V**oltage connector. The recommended connector for high voltages in the *NIM* system.

SIEVERT (Sv) The SI unit of radiation dose equivalent.

Si(Li) DETECTOR (Pronounced 'silly') A silicon detector created by *lithium drifting*.

SINGLE CHANNEL ANALYSER (SCA) A device that produces a logic output pulse when an input pulse has a maximum that falls within a user-defined window.

SINGLE ESCAPE PEAK An extra peak in a spectrum due to the loss of a 511 keV photon when a high energy gamma-ray is absorbed by *pair production*. See: *escape peaks*.

SINGLET In this context, a well separated single peak in a spectrum. Contrast: *multiplet*.

SMOOTH An *MCA* function where the count in any one channel is averaged with data in adjacent channels to decrease random fluctuations in the spectrum.

SNM **S**pecial **N**uclear **M**aterial. Plutonium and uranium (usually enriched to greater than 20% in ^{235}U and hence called 'highly-enriched uranium'), other *transuranics*. Often materials suitable for nuclear weapons.

SODIUM IODIDE (NaI) The most common material used in a scintillation detector for measuring gamma-rays. Often shown with its usual added trace of thallium as NaI (Tl).

SOFTWARE An alterable program for a *CPU*.

SPECIFIC RADIOACTIVITY The activity per unit mass of an element or compound containing a radioactive nuclide.

SPECTROMETER A device used to count an emission of radiation of a specific energy or range of energies to the exclusion of all other energies.

SPONTANEOUS FISSION See: *Fission (Nuclear)*.

STABILITY The measure of instrument performance through variations in temperature, line voltage, count rate, time, or other variables.

STANDARD DEVIATION A measure of the variability of a series of measurements about the mean.

STANDARD ERROR Sometimes used for the standard deviation of an average. It is equal to the standard deviation divided by the square root of the number of measurements used to obtain the average.

STANDARD UNCERTAINTY This is numerically equal to the *Standard Deviation*, but, being more descriptive of its origin, is becoming the preferred term.

STOCHASTIC EFFECT Where the probability of occurrence is assumed to be proportional to dose without a threshold.

SUCCESSIVE APPROXIMATION ADC A type of ADC suitable for high throughput rates. The processing time is independent of pulse height; it is also known as the 'fixed conversion time ADC' or 'fixed dead time' ADC.

T

TCS See: *True Coincidence Summing*.

TECDOC-619 A collection of evaluated nuclear data for 35 of the most commonly measured nuclides. (Now Superceded)

TERMINATOR A resistive load at the terminal end of a signal line.

THROUGHPUT The capability of processing data in unit time, e.g. in pulses per second. In the context of pulse processors/amplifiers it is the output pulse rate related to the input pulse rate.

TIME CONSTANT Commonly refers to the shaping time set on an amplifier, which determines the width of the output pulse and the efficiency of charge collection.

TIME GATED SCALER (TGS) A scaler that is started and stopped according to a preset timing schedule.

TIME TO AMPLITUDE CONVERTER (TAC) A device that measures the time interval between two signal pulses, and represents that quantity as an analogue voltage pulse with amplitude proportional to the input time difference.

TIME WALK Variation of precision timing of a discriminator or SCA output pulse caused by the variation in input pulse amplitude.

TIMING SCA An SCA designed to be immune to time walk.

TOTAL EFFICIENCY The efficiency of a gamma spectrometer in terms of the number of gamma-rays, at a particular energy, detected anywhere in the spectrum compared to the number emitted by the source.

TRACER An easily detected material used in small amounts to label a larger quantity so that its subsequent movement can be studied.

TRANSIENT EQUILIBRIUM The condition that the ratio of daughter to parent activities in a given radioactive decay is constant. A necessary prerequisite is that the daughter half-life is less than the parent half-life.

TRANSISTOR RESET PREAMPLIFIER A type of *preamplifier* particularly suited to high count rates. The alternative to the resistive feedback preamplifier. See: Chapter 4.

TRANSMUTATION The conversion of one element to another. This happens all the time in nuclear decay. In the context of nuclear waste treatment it is the process by which radionuclides are irradiated by neutrons or other particles to convert them into nuclides that are either stable or have shorter half-lives.

TRP See: *Transistor Reset Preamplifier.*

TRU TransUranium Often used for waste with atomic number greater than 92.

TRUE COINCIDENCE SUMMING (TCS) The simultaneous detection of two or more photons originating from a single nuclear disintegration that results in only one observed (summed) peak. This results in loss of counts from peaks leading to efficiency calibration errors. (See: Chapter 8.)

U

u The symbol for *atomic mass unit.* Previously *amu.*

UPPER LEVEL DISCRIMINATOR (ULD) The upper discriminator in an SCA. Pulses with heights above this will not be registered.

UPS Uninterruptible Power Supply.

UNIPOLAR PULSE A pulse that goes in only one direction from the baseline (positive or negative). Normally the recommended analogue pulse for spectrometry.

USB Universal Serial Bus.

V

VALENCE BAND Within the band structure of a solid, this is the band where electrons normally reside.

VARIANCE A measure of the uncertainty of a measurement numerically equal to the *standard uncertainty* squared.

VITRIFICATION In the nuclear context, the process by which radioactive waste is immobilised in borosilicate glass.

VLLW Very Low Level radioactive Waste; waste that can be disposed of at landfill sites without special treatment. Allowed radioactivity is $< 4 \times 10^5$ B in 0.1 m^3 (UK usage).

VOIGHT FUNCTION A convolution of *Lorentzian* and *Gaussian* functions – the shape of X-ray peaks in a spectrum.

VPG Virtual Pulse Generator A concept used in a version of *loss free counting.*

W

WELL DETECTOR A detector in which there is a well extending into its body to accommodate the sample. Such detectors have high detection efficiency.

WIDTH CALIBRATION The relationship between peak width and gamma-ray energy.

WILKINSON ADC A type of ADC traditionally noted for its linearity. The processing time varies with pulse height.

WINDOW The region bounded by the LLD and the ULD in an SCA; this is the 'window' through which pulse of acceptable heights are passed.

X

X-RAY Electromagnetic radiation similar to low energy gamma-rays but produced by a different mechanism. Energy is characteristic of the element involved, being equal to the difference between the electronic states of orbital electrons.

X-RAY ESCAPE PEAK This effect can occur near the surface of a detector, where it is possible for induced fluorescence X-radiation of the detector material to escape. This is most likely to be the Kα X-ray. Thus with germanium there will be an additional peak, the germanium escape peak, near the full-energy peak but 9.9 keV lower.

X-RAY FLUORESCENCE (XRF) Characteristic X-ray emission produced by irradiating a material with photons of energy greater than the K-shell binding energy of the element irradiated. Used for chemical analysis.

Y

YIELD In dealing with fission, would refer to *fission yield*. In the context of gamma-ray emission might mean gamma-ray *emission probability*. The use of the term in the latter context is deprecated because its meaning is unclear.

Index

References to images and figures are given in italic type. References to tables are given in bold type.